W9-BCS-452

LINEAR ALGEBRA

EAGLE MATHEMATICS SERIES

A series of textbooks for an undergraduate program in mathematical analysis

PLANNED AND EDITED BY

Ralph Abraham UNIVERSITY OF CALIFORNIA AT SANTA CRUZ

Ernest Fickas HARVEY MUDD COLLEGE

Jerrold Marsden UNIVERSITY OF CALIFORNIA AT BERKELEY

Kenneth McAloon UNIVERSITÉ DE PARIS

Michael O'Nan RUTGERS UNIVERSITY

Anthony Tromba UNIVERSITY OF CALIFORNIA AT SANTA CRUZ

VOLUMES

1AB *Calculus of Elementary Functions*

1BCD *Calculus*

1BC *Calculus of One Variable*

2A *Linear Algebra*
2B *Vector Calculus*

3 *Advanced Calculus*

Other titles in preparation

Volume 2A

LINEAR ALGEBRA

Michael O'Nan

RUTGERS UNIVERSITY

With the assistance of Charles Setzer,
and in collaboration with the Series Editors

HARCOURT BRACE JOVANOVICH, INC.

New York *Chicago* *San Francisco* *Atlanta*

© 1971 by Harcourt Brace Jovanovich, Inc.

All rights reserved. No part of this publication may be reproduced
or transmitted in any form or by any means, electronic or mechanical, including
photocopy, recording, or any information storage and retrieval system,
without permission in writing from the publisher.

ISBN: 0-15-518558-6

Library of Congress Catalog Card Number: 70-130345

Printed in the United States of America

Drawings by Felix Cooper

FOREWORD

The Eagle Mathematics Series was planned at Princeton in the Spring of 1968. At a summer writing conference in Berkeley, the first four manuscripts were drafted. Within two years, seven manuscripts were completed and revised for continuity and notation.

The series was planned to meet the recommendations of CUPM, the curriculum committee of the Mathematics Association of America, and CEEB, the College Entrance Examination Board. The first five volumes comprise two different basic calculus sequences, including the material of CEEB Advanced Placement Tests AB and BC (initiated in 1969) and CUPM courses 1, 2, 3, and 4 (*A General Curriculum in Mathematics for Colleges,* CUPM, 1965). Special attention has been given to the basic CUPM recommendations: semester-length courses, gradually increasing rigor, adequate mathematical and physical motivation, early linear algebra in one of the two equivalent sequences for the freshman and sophomore years, and the inclusion of elementary ordinary differential equations.

The first book, Volume 1AB (*Calculus of Elementary Functions*), conforms exactly to the CEEB syllabus for the year course preparing for the AB test. The next, Volume 1BC (*Calculus of One Variable*) follows CEEB syllabus BC, which is almost identical to the first-year calculus course recommended by CUPM (conventional version). The remaining volumes are for one-semester courses and accommodate the option for early linear algebra: in one sequence linear algebra precedes differentials, multiple integrals, and linear differential equations, which are developed with reference to it; in the other sequence, partial derivatives and multiple integrals precede the linear algebra.

Several volumes will cover more advanced courses recommended by CUPM in its January, 1967, *Recommendations on the Undergraduate Mathematics Program for Engineers and Physicists.*

Ralph Abraham

PREFACE

This book is intended for a one-semester course in linear algebra at the sophomore level. The necessary preparation is a certain amount of mathematical maturity, which need not, though probably will, be obtained in a first-year calculus course.

In order to offset the natural abstractness of the subject, the spirit of the book is geometric. All concepts and examples are concretely related to their appropriate embodiments in two- and three-dimensional space. On the other hand, all proofs are written out in detail, and no steps are skipped. For these reasons, the discussion should be accessible to all students who need a course in linear algebra, regardless of their field of specialization.

The general pattern of the book is to proceed from the concrete to the abstract. The material splits naturally into two parts. In the first part, we treat linear equations, column vectors, matrices, and determinants; in the second part we consider vector spaces, linear transformations, and inner products. Because of this ordering, it is possible to progress gradually from computational methods to more sophisticated concepts, thus sparing the student any traumatic plunge into the depths (or heights, according to point of view) of mathematical abstraction. This arrangement of topics has the additional advantage of providing examples that are relevant when vector spaces, linear transformations, and other abstract objects are introduced. Whenever possible, concepts are given their natural geometric interpretation—for example, vectors as directed line segments, determinants as areas and volumes, linear transformations as rotations, reflections, projections, and so on.

Chapter 1 treats simultaneous equations; in particular, their solution by Gaussian elimination is emphasized.

The second chapter begins with a study of vectors in real three-dimensional space. After defining addition and scalar multiplication geometrically, we proceed to deduce the usual operations on components, which correspond to the operations on vectors. Lines and other geometric objects are studied in this connection. Higher-dimensional spaces are introduced as spaces of column vectors. Next, matrices and the related operations of addition and multiplication are defined. Particular emphasis is given to the notion of an inverse of a matrix and its usefulness in solving systems of linear equations.

Chapter 3 deals with determinants and begins with the 2×2 case. This is done primarily to acquaint students with row and column operations, under circumstances in which the determinant is probably already familiar. For the higher-dimensional case, an inductive definition using cofactor expansions is given. More emphasis is placed on the properties of the determinant (row and column operations) than on the formal definition. The multiplicative property of determinants is proved, and their usefulness in finding inverses of matrices and solving simultaneous systems of linear equations is pointed out. In addition, the student is shown the more practical technique of inverting matrices by synthetic elimination.

Abstract vector spaces are introduced in Chapter 4. Because this may be the student's first contact with abstract mathematical objects, the approach is deliberately slow. Separate sections are devoted to the concepts of the span of a set of vectors, linear independence, subspaces, and basis. In each case, the concept is interpreted in two- and three-dimensional space, and then the principal theorems on dimension and basis are obtained.

In the fifth chapter, linear transformations are defined and explicated. Just as in the preceding chapter, the pace is unhurried; a full section is given to the range space of a linear transformation, while another treats the nullspace. We then consider the technique of calculating the rank of a matrix using row and column operations. (Since this section is not necessary for the subsequent development, it may be omitted.) Following this, the notions of inverse and isomorphism are studied. The matrix representation of a linear transformation, effected by a choice of basis, is also discussed in this chapter. The usual theorems concerning the relation of the rank and nullity of linear transformations are obtained, as are those concerning isomorphism and dimension.

The discussion of general inner products in Chapter 6 is motivated by first studying the dot and cross products in three-dimensional space. Both the geometric and algebraic definitions of these quantities are given. Later sections contain the definition of inner products in $\boldsymbol{R}^n$ *and* $\boldsymbol{C}^n$, and deal with orthogonality, orthogonal complements, orthonormal basis, and the like. Results such as the Cauchy–Schwarz inequality, Bessel's inequality, and the Gram–Schmidt orthogonalization procedure are obtained.

In the final chapter, eigenvalues are defined and a few of their properties are covered. Also, the diagonizibility of symmetric matrices is demonstrated, and a brief introduction of bilinear forms is included.

Each section contains numerous exercises; some allow the student to practice routine computational skills, while others test his understanding of the theory and challenge him to establish deeper results. Answers for selected exercises are found at the end of the book. The symbol ∎ denotes the end of a proof.

Michael O'Nan

CONTENTS

1
SYSTEMS OF LINEAR EQUATIONS, 1

2
VECTORS AND MATRICES, 21

3
DETERMINANTS, 67

4
VECTOR SPACES, 125

5
LINEAR TRANSFORMATIONS, 193

6

PRODUCTS, 297

7

EIGENVALUES AND CANONICAL FORMS, 347

1

SYSTEMS OF LINEAR EQUATIONS

1 INTRODUCTION

There are many problems in mathematics and in the physical, biological, and social sciences whose solution requires the use of a system of simultaneous linear equations. Historically, linear algebra arose in the attempt to formulate criteria for determining the solutions of such systems. It is, therefore, appropriate that we should begin this book with a study of systems of linear equations.

Consider, as an example, the problem of finding all real numbers x and y satisfying the equations

$$\begin{aligned} 2x + 3y &= 7 \\ x + y &= 1 \end{aligned}$$

In this case the problem has a geometric formulation. Each equation may be interpreted as defining a line in the Cartesian plane. Since the problem as stated requires finding a point satisfying both equations, we must locate the point where the lines intersect. In this case the lines are neither parallel nor identical, so there is a unique point of intersection. See Figure 1-1. In order to determine the coordinates of the point of intersection, we resort to the familiar procedure of elimination.

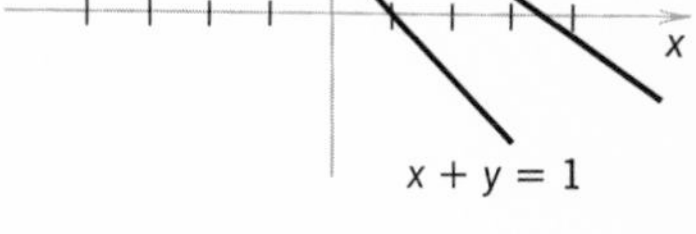

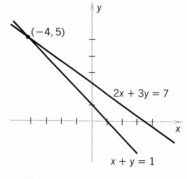

FIGURE 1-1

We start with our original system of equations.

$$2x + 3y = 7$$
$$x + y = 1$$

Multiply the second equation by -2.

$$\downarrow$$

$$2x + 3y = 7$$
$$-2x - 2y = -2$$

Add the first equation to the second equation.

$$\downarrow$$

$$2x + 3y = 7$$
$$y = 5$$

Multiply the second equation by -3 and add the result to the first equation.

$$\downarrow$$

$$2x = -8$$
$$y = 5$$

Thus, we see that the point $(-4, 5)$ is the only solution to the new system of equations. On the other hand, since each step in our sequence of operations is reversible, we see that $(-4, 5)$ is indeed a solution to the given system of equations.

It would be well to note at this point that there are systems of equations which have no solutions and systems of equations which have infinitely many solutions.

As an example of the former we have the system

$$2x + 4y = 3$$
$$3x + 6y = 1$$

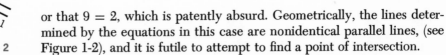

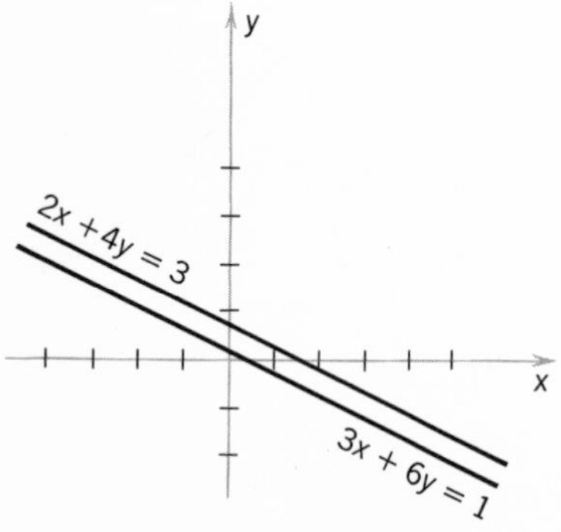

FIGURE 1-2

Indeed, if we assume that a solution to this system exists and multiply the first equation by 3 and the second by 2, we obtain

$$6x + 12y = 9$$
$$6x + 12y = 2$$

or that $9 = 2$, which is patently absurd. Geometrically, the lines determined by the equations in this case are nonidentical parallel lines, (see Figure 1-2), and it is futile to attempt to find a point of intersection.

As an example of a system having infinitely many solutions, we have the equations

$$x + 2y = 3$$
$$3x + 6y = 9$$

In this case the equations, though formally different, represent the same line. It is clear that the entire line is the solution set to the system.

As another example, let us solve the system

$$\begin{aligned} -x + y + z &= 4 \\ 4x - 5y + 2z &= 0 \\ 2x + 3y + z &= 11 \end{aligned}$$

Multiply the first equation by 4 and add it to the second equation; Multiply the first equation by 2 and add it to the third equation.

$$\downarrow$$

$$\begin{aligned} -x + y + z &= 4 \\ - y + 6z &= 16 \\ 5y + 3z &= 19 \end{aligned}$$

Multiply the second equation by 5 and add it to the third equation.

$$\downarrow$$

$$\begin{aligned} -x + y + z &= 4 \\ - y + 6z &= 16 \\ 33z &= 99 \end{aligned}$$

Multiply the third equation by $\frac{1}{33}$.

$$\downarrow$$

$$\begin{aligned} -x + y + z &= 4 \\ - y + 6z &= 16 \\ z &= 3 \end{aligned}$$

Multiply the third equation by -1 and add it to the first equation.

Multiply the third equation by -6 and add it to the second equation.

$$\downarrow$$

$$\begin{aligned} -x + y &= 1 \\ -y &= -2 \\ z &= 3 \end{aligned}$$

Add the second equation to the first equation.

$$\begin{aligned} -x &= -1 \\ -y &= -2 \\ z &= 3 \end{aligned}$$

or

$$x = 1, \qquad y = 2, \qquad z = 3$$

EXERCISES

1. Consider the following systems of linear equations in two unknowns. For each pair of equations, determine whether a solution exists, what the solution is, if it exists, and interpret your results geometrically.

   (a) $\begin{aligned} 3x + y &= -3 \\ x - 10y &= -1 \end{aligned}$ (b) $\begin{aligned} x + y &= 2 \\ 10x + y &= 11 \end{aligned}$

   (c) $\begin{aligned} x - 3y &= 7 \\ 2x - 6y &= 7 \end{aligned}$ (d) $\begin{aligned} 2x - 3y &= 0 \\ 3x - 4y &= 1 \end{aligned}$

   (e) $\begin{aligned} 2x - 3y &= 0 \\ 3x - 4y &= 0 \end{aligned}$ (f) $\begin{aligned} x + 2y &= 1 \\ 2x + 4y &= 2 \end{aligned}$

   (g) $\begin{aligned} 2x + y &= 0 \\ x + 3y &= 0 \end{aligned}$

2. Consider the system of equations

$$\begin{aligned} ax + by &= 0 \\ cx + dy &= 0 \end{aligned}$$

   where a, b, c, and d are fixed real numbers.

   (a) Show that the system always has at least one solution.

   (b) If $ad - bc \neq 0$, show that the system has only one solution. Interpret this condition geometrically.

3. Solve

$$\begin{aligned} x + y + z &= 1 \\ 3x + 3y - 2z &= 3 \\ 7x + 3y - 5z &= 2 \end{aligned}$$

4. Solve

$$\begin{aligned} 3x_1 - 5x_2 + 6x_3 &= 1 \\ 2x_1 - 3x_2 + 4x_3 &= 2 \\ 3x_1 - 4x_2 + 17x_3 &= 3 \end{aligned}$$

5. Solve

$$\begin{aligned} x_1 + 2x_2 + 3x_3 + 3x_4 &= 0 \\ -5x_1 + 2x_2 + x_3 + x_4 &= -8 \\ 3x_1 + 4x_2 + 7x_3 + 2x_4 &= -8 \\ x_1 + 3x_2 + 5x_3 + 4x_4 &= -2 \end{aligned}$$

2 EQUIVALENCE OF LINEAR SYSTEMS

So far we have seen several examples of systems of linear equations and have used various techniques to solve them. Let us now abstract from these examples general principles and techniques for dealing with such systems.

By a **linear equation** we mean an equation of the form $a_1x_1 + a_2x_2 + \cdots + a_nx_n = b$, where $a_1, a_2, \ldots, a_n$ and b are constants and $x_1, x_2, \ldots, x_n$ are the quantities to be determined. The x_i's are called variables. The number a_i is called the coefficient of x_i in the equation. By a **solution** to the equation $a_1x_1 + a_2x_2 + \cdots + a_nx_n = b$, we mean an n-tuple of real numbers $(c_1, c_2, \ldots, c_n)$ such that $a_1c_1 + a_2c_2 + \cdots + a_nc_n = b$. For example, $2x_1 - x_2 + 3x_3 + 10x_4 = 8$ is a linear equation. x_1, x_2, x_3, and x_4 are its variables. The coefficient of x_1 is 2. The coefficient of x_2 is -1, and so on. The student can verify that one solution to the equation is (4, 0, 0, 0). Another solution is (0, -8, 0, 0).

A system of linear equations is a collection of several linear equations. For example,

$$\begin{aligned} 3x_1 + 2x_2 + 3x_3 &= 0 \\ x_1 + x_2 + x_3 &= 0 \\ -x_1 - x_2 + 3x_3 &= 0 \end{aligned}$$

is a system of linear equations.

By a **solution to a system of linear equations** in n variables we mean an n-tuple of numbers which satisfies each of the equations of the system. For example, (3, -1, 0, 2) is a solution to the system

$$\begin{aligned} x_1 + x_2 - x_3 + 4x_4 &= 10 \\ -x_1 - x_2 + 2x_3 + x_4 &= 0 \\ 10x_1 + 3x_2 \qquad + x_4 &= 29 \\ x_2 \qquad - x_4 &= -3 \end{aligned}$$

Suppose we are given two systems of linear equations in n variables. We say that these systems are **equivalent** if any n-tuple of numbers which satisfies the first system of equations satisfies the second system of equations, and any n-tuple which satisfies the second system, also satisfies the first system. Stated otherwise, the two systems have the same solutions.

Thus, according to our definition the system

$$\begin{aligned} 2x + 3y &= 7 \\ x + y &= 1 \end{aligned}$$

is equivalent to the system

$$\begin{aligned} 2x + 3y &= 7 \\ -2x - 2y &= -2 \end{aligned}$$

since in both cases, the solution is $(-4, 5)$.

Note that changing the order in which the equations are presented has no effect on the solutions and hence provides an equivalent system of equations. In addition there are two other ways for changing a system of linear equations into an equivalent system:

(1) If one system of linear equations differs from another only in that one equation in the second system is a nonzero multiple of a single equation in the first system, all other equations in the two systems being identical, the systems are equivalent.

Proof Let the single equation in the first system which is multiplied by a nonzero constant in the second system be $a_1x_1 + a_2x_2 + \cdots + a_nx_n = b$. In the second system it becomes $ca_1x_1 + \cdots + ca_nx_n = cb$, where $c \neq 0$. If $(g_1, g_2, \ldots, g_n)$ is a solution to the first system, $a_1g_1 + a_2g_2 + \cdots + a_ng_n = b$. Therefore, $(ca_1)g_1 + \cdots + (ca_n)g_n = cb$. Since all remaining equations of the two systems are identical, $(g_1, g_2, \ldots, g_n)$ is a solution to the second system. If on the other hand $(h_1, h_2, \ldots, h_n)$ is a solution to the second system of equations, we must have $ca_1h_1 + ca_2h_2 + \cdots + ca_nh_n = cb$. Multiplying by c^{-1}, we have $a_1h_1 + a_2h_2 + \cdots + a_nh_n = b$. Again all remaining equations in the two systems are the same. Therefore, $(h_1, h_2, \ldots, h_n)$ is a solution to the first system. ∎

As examples of this principle we see that the systems

$$\begin{aligned} x + 3y + z &= 0 \\ 4x + y + 2z &= 2 \end{aligned}$$

and

$$\begin{aligned} x + 3y + z &= 0 \\ 8x + 2y + 4z &= 4 \end{aligned}$$

are equivalent. Also, by repeated use of (1) we can show that

$$\begin{aligned} x + y &= 1 \\ x - y &= 0 \end{aligned}$$

and

$$\begin{aligned} 2x + 2y &= 2 \\ 3x - 3y &= 0 \end{aligned}$$

are equivalent.

Our second principle of equivalence for systems of equations is the following:

(2) If one system of linear equations differs from another only in that one equation in the first system is replaced in the second system by the sum of that equation with another equation from the first system, all other equations in the two systems being the same, the systems are equivalent.

According to the second principle, the system of equations

$$\begin{aligned} x + y - z &= 1 \\ x + 3y - 2z &= 0 \\ x + y + 4z &= 1 \end{aligned}$$

is equivalent to the system

$$\begin{aligned} x + y - z &= 1 \\ 2x + 4y - 3z &= 1 \\ x + y + 4z &= 1 \end{aligned}$$

since the second equation in the second system was obtained by adding the first equation in the first system to the second equation in the first system, while all other equations are the same.

Proof Let $a_1x_1 + a_2x_2 + \cdots + a_nx_n = b$ be an equation of the first system. Let $c_1x_1 + c_2x_2 + \cdots + c_nx_n = d$ be the equation in the first system which is to be added to $a_1x_1 + a_2x_2 + \cdots + a_nx_n = b$. Thus, in the second system the equation $a_1x_1 + a_2x_2 + \cdots + a_nx_n = b$ is replaced by $(a_1 + c_1)x_1 + (a_2 + c_2)x_2 + \cdots + (a_n + c_n)x_n = b + d$, while all other equations in the two systems are the same. Now suppose $(g_1, g_2, \ldots, g_n)$ is a solution to the first system. Then we have

$$a_1g_1 + a_2g_2 + \cdots + a_ng_n = b$$
$$c_1g_1 + c_2g_2 + \cdots + c_ng_n = d$$

Therefore, $(a_1 + c_1)g_1 + (a_2 + c_2)g_2 + \cdots + (a_n + c_n)g_n = b + d$. Since all other equations in the first system are identical with those of the second system, it follows that $(g_1, g_2, \ldots, g_n)$ is a solution to the second system.

Next let $(h_1, h_2, \ldots, h_n)$ be a solution to the second system. Then

$$(a_1 + c_1)h_1 + (a_2 + c_2)h_2 + \cdots + (a_n + c_n)h_n = b + d$$
$$c_1h_1 + \qquad c_2h_2 + \cdots + \qquad c_nh_n = d$$

Therefore, $a_1h_1 + a_2h_2 + \cdots + a_nh_n = b$. Again all remaining equations in the two systems are the same, so $(h_1, h_2, \ldots, h_n)$ is a solution to the first system. ▨

By repeated use of these two principles we can obtain simpler systems of equations.

EXERCISES

1. In each of the following use the definition of equivalence of systems of equations to determine which of the two given systems of equations are equivalent.

(a) $\begin{aligned} x + 2y &= 1 \\ x + y &= 0 \end{aligned}$ $\quad$ $\begin{aligned} x + 2y &= 1 \\ 2x + 3y &= 1 \end{aligned}$

(b) $\begin{aligned} x + y &= 0 \\ x - y &= 0 \end{aligned}$ $\quad$ $\begin{aligned} x + y &= 1 \\ x - y &= 0 \end{aligned}$

(c) $\begin{aligned} 3x + 3y &= 2 \\ 2x + 6y &= 0 \end{aligned}$ $\quad$ $\begin{aligned} x + y &= 1 \\ x - y &= 0 \end{aligned}$

(d) $\begin{aligned} x - y &= 1 \\ x + y &= 0 \end{aligned}$ $\quad$ $\begin{aligned} x - y &= 1 \\ 2x &= 1 \end{aligned}$

(e) $\begin{aligned} x + 3y &= 2 \\ 2x + 6y &= 0 \end{aligned}$ $\quad$ $\begin{aligned} x + 3y &= 2 \\ 2x + 6y &= 1 \end{aligned}$

(f) $\begin{aligned} x + y &= 1 \\ x - y &= 1 \end{aligned}$ $\quad$ $\begin{aligned} x + y &= 1 \\ 2x &= 2 \end{aligned}$

2. Consider the system of equations

$$\begin{aligned} a_{11}x_1 + a_{12}x_2 + a_{13}x_3 &= b_1 \\ a_{21}x_1 + a_{22}x_2 + a_{23}x_3 &= b_2 \\ a_{31}x_1 + a_{32}x_2 + a_{33}x_3 &= b_3 \end{aligned}$$

where $b_1^2 + b_2^2 + b_3^2 \neq 0$ and the system of equations

$$\begin{aligned} c_{11}x_1 + c_{12}x_2 + c_{13}x_3 &= 0 \\ c_{21}x_1 + c_{22}x_2 + c_{23}x_3 &= 0 \\ c_{31}x_1 + c_{32}x_2 + c_{33}x_3 &= 0 \end{aligned}$$

Show they are not equivalent.

3 METHOD OF GAUSSIAN ELIMINATION

If we are given a system of m linear equations in n unknowns,

$$
\begin{aligned}
a_{11}x_1 + a_{12}x_2 + a_{13}x_3 + \cdots + a_{1n}x_n &= b_1 \\
a_{21}x_1 + a_{22}x_2 + a_{23}x_3 + \cdots + a_{2n}x_n &= b_2 \\
a_{31}x_1 + a_{32}x_2 + a_{33}x_3 + \cdots + a_{3n}x_n &= b_3 \\
&\vdots \\
a_{m1}x_1 + a_{m2}x_2 + a_{m3}x_3 + \cdots + a_{mn}x_n &= b_m
\end{aligned}
$$

where m and n are positive integers, a_{ij} and b_i are constants for $i = 1, 2, \ldots, m$ and $j = 1, 2, \ldots, n$, and x_j is a variable for $j = 1, 2, \ldots, n$, we can apply the following reduction procedure.

(1) Choose a variable with nonzero coefficient. Suppose x_s is such a variable and that its coefficient in the tth equation is $a_{ts} \neq 0$. Multiply the tth equation by a_{ts}^{-1}.

(2) For each $i \neq t$ and $i = 1, 2, \ldots, m$ add $-a_{is}$ times equation t to the ith equation in the system resulting from step (1). This step eliminates x_s from all equations but the tth. x_s is now called a used variable and the tth equation is called a used equation.

(3) Next choose a new variable, x_r, with a nonzero coefficient in an unused equation. Use steps (1) and (2) to eliminate x_r from all other equations, including those which have already been used. Again the equation we have worked with in this step is called used. Step (3) is repeated until no unused equations remain, or until only equations of the form $0 = c$ are left, where c is some number.

If there are solutions to the system of equations, they will all be obtained from this procedure. If no solutions exist, a contradiction will be obtained using this method.

It is important to note that by the principles of the preceding section, the final system of equations that we obtain is equivalent to our original system.

We now present several examples illustrating this technique.

Example 1

$$\begin{aligned} x + y &= 1 \\ 6x - 2z &= -8 \\ 3y - z &= -3 \end{aligned}$$

Let y be the first variable to be used. Choosing the third equation as the equation to be used, we multiply it by $\frac{1}{3}$.

$$\downarrow$$

$$\begin{aligned} x + y &= 1 \\ 6x - 2z &= -8 \\ y - \tfrac{1}{3}z &= -1 \end{aligned}$$

Multiply the third equation by (-1) and add it to the first; multiply the third equation by 0 and add it to the second.

$$\downarrow$$

$$\begin{aligned} x + \tfrac{1}{3}z &= 2 \\ 6x - 2z &= -8 \\ y - \tfrac{1}{3}z &= -1 \end{aligned}$$

Choose the first equation as the next equation to be used, and z as next variable to be used. Multiply equation one by 3.

$$\downarrow$$

$$\begin{aligned} 3x + z &= 6 \\ 6x - 2z &= -8 \\ y - \tfrac{1}{3}z &= -1 \end{aligned}$$

Add 2 times the first equation to the second; add $\frac{1}{3}$ times the first equation to the third.

$$\downarrow$$

$$\begin{aligned} 3x + z &= 6 \\ 12x &= 4 \\ x + y &= 1 \end{aligned}$$

Choose the second equation and x as equation and variable to be used; multiply the second equation by $\frac{1}{12}$.

$$\downarrow$$

$$\begin{aligned} 3x \qquad + z &= 6 \\ x \qquad\qquad &= \tfrac{1}{3} \\ x + y \qquad &= 1 \end{aligned}$$

Add (-3) times equation two to equation one; add (-1) times equation two to equation three.

$$\downarrow$$

$$\begin{aligned} z &= 5 \\ x \qquad\qquad &= \tfrac{1}{3} \\ y \qquad &= \tfrac{2}{3} \end{aligned}$$

Thus in this case a unique solution is obtained.

Example 2

$$\begin{aligned} 2x + 4y &= 3 \\ 3x + 6y &= 1 \end{aligned}$$

Choose x as the first variable to be used and equation one as the first equation to be used. Multiply the first equation by $\frac{1}{2}$.

$$\downarrow$$

$$\begin{aligned} x + 2y &= \tfrac{3}{2} \\ 3x + 6y &= 1 \end{aligned}$$

Adding (-3) times equation one to equation two.

$$\downarrow$$

$$\begin{aligned} x + 2y &= \tfrac{3}{2} \\ 0 &= -\tfrac{7}{2} \end{aligned}$$

But now what? Recall again the meaning of the equivalence of systems of linear equations: Two systems are equivalent if they have the same solutions. In this case, there are no values of x and y which satisfy the second equation, and since the second system admits no solution, we can only conclude that the first system likewise has no solutions. In terms of our original definition the second equation is really $0 \cdot x + 0 \cdot y = -\frac{7}{2}$, which can never be satisfied.

Example 3

$$\begin{array}{rrrrr} x_1 & & - x_3 & + 2x_4 & = 0 \\ 2x_1 & - 3x_2 & & - 4x_4 & = 0 \\ & x_2 & + 2x_3 & + 3x_4 & = 0 \\ 3x_1 & - 2x_2 & + x_3 & + x_4 & = 0 \end{array}$$

Use x_1 and the first equation.

$$\downarrow$$

$$\begin{array}{rrrrr} x_1 & & - x_3 & + 2x_4 & = 0 \\ & - 3x_2 & + 2x_3 & - 8x_4 & = 0 \\ & x_2 & + 2x_3 & + 3x_4 & = 0 \\ & - 2x_2 & + 4x_3 & - 5x_4 & = 0 \end{array}$$

Use x_2 and the third equation.

$$\downarrow$$

$$\begin{array}{rrrrr} x_1 & & - x_3 & + 2x_4 & = 0 \\ & & 8x_3 & + x_4 & = 0 \\ & x_2 & + 2x_3 & + 3x_4 & = 0 \\ & & 8x_3 & + x_4 & = 0 \end{array}$$

Use x_4 and the second equation.

$$\downarrow$$

$$\begin{array}{rrrrr} x_1 & & - 17x_3 & & = 0 \\ & & 8x_3 & + x_4 & = 0 \\ & x_2 & - 22x_3 & & = 0 \\ & & & 0 & = 0 \end{array}$$

The method can be used no farther since the only unused variable, x_3, does not occur with nonzero coefficient in the unused equation. From the above system of equations we see that if we choose any value for x_3, say $x_3 = c$, and if we then let

$$\begin{aligned} x_1 &= 17x_3 = 17c \\ x_4 &= -8x_3 = -8c \\ x_2 &= 22x_3 = 22c \end{aligned}$$

we obtain a solution to the above system, and that any solution to the system is of this type for some value of c. In other words, the most general solution is the 4-tuple $(17c, 22c, c, -8c)$. Since c may vary over all the real numbers, there are infinitely many solutions to this equation.

Example 4

$$\begin{aligned} x_1 + x_2 + x_3 + x_4 &= 1 \\ x_1 - x_2 + x_3 - x_4 &= 1 \\ x_1 \qquad + x_3 \qquad &= 1 \end{aligned}$$

Use x_1 and the third equation.

$$\downarrow$$

$$\begin{aligned} x_2 \qquad + x_4 &= 0 \\ -x_2 \qquad - x_4 &= 0 \\ x_1 \qquad + x_3 \qquad &= 1 \end{aligned}$$

Use x_2 and the first equation.

$$\downarrow$$

$$\begin{aligned} x_2 \qquad + x_4 &= 0 \\ 0 &= 0 \\ x_1 \qquad + x_3 \qquad &= 1 \end{aligned}$$

Again we cannot proceed any further since both unused variables, x_3 and x_4, have zero coefficient in the only unused equation. This implies that if x_3 and x_4 are chosen arbitrarily, i.e., $x_3 = c$ and $x_4 = d$, and if we let

$$\begin{aligned} x_2 &= -x_4 = -d \\ x_1 &= 1 - x_3 = 1 - c \end{aligned}$$

then $(1 - c, -d, c, d)$ is a solution to the system. Thus, in this case, we obtain a "two parameter family" of solutions. Later we shall see how to make this statement more precise.

Example 5 By applying our elimination procedure, we attempt to determine for what values of a, b, and c a solution to the following system exists.

$$\begin{aligned} 2x - y + z &= a \\ x + 2y + z &= b \\ 3x + y + 2z &= c \end{aligned}$$

Use x and the second equation.

$$\begin{aligned} -5y - z &= a - 2b \\ x + 2y + z &= b \\ -5y - z &= c - 3b \end{aligned}$$

Use z and the first equation.

$$\begin{aligned} -5y - z &= a - 2b \\ x - 3y \quad &= a - b \\ 0 &= c - a - b \end{aligned}$$

The elimination procedure comes to a halt since in our unused equation (the third) all variables have coefficients 0. It follows from the third equation that for any solutions to exist at all, we must have $c = a + b$.

If we assume this condition is satisfied and choose y arbitrarily and let $x = 3y + (a - b)$, $z = -5y - a + 2b$, then we have a solution to the third system. By the equivalence of the systems, likewise we have a solution to the first system.

Thus, a necessary and sufficient condition that solutions exist for the first system is that $c = a + b$.

EXERCISES

1. Solve the following systems of linear equations, indicating if the system has no solutions and giving the general form if it has many solutions.

(a) $$\begin{aligned} x + y + z &= 2 \\ 2x - y + z &= 0 \\ x + 2y - z &= 4 \end{aligned}$$

(b) $$\begin{aligned} x_1 - x_2 - 3x_3 - x_4 &= 1 \\ 2x_1 + 4x_2 \qquad - 2x_4 &= 2 \\ 3x_1 + 4x_2 - 2x_3 \qquad &= 0 \\ x_1 \qquad + 2x_3 - 3x_4 &= 3 \end{aligned}$$

(c) $$\begin{aligned} x_1 - x_2 + x_3 &= -2 \\ x_1 + 3x_2 + x_3 &= 10 \\ 4x_1 + x_2 - x_3 &= 7 \end{aligned}$$

(d) $$\begin{aligned} x_1 + 2x_2 + 3x_3 + x_4 &= 1 \\ 2x_1 + 3x_2 - x_3 - x_4 &= 1 \\ x_2 + x_3 - 2x_4 &= 0 \\ 3x_1 + 7x_2 + 4x_3 - 4x_4 &= 2 \end{aligned}$$

(e) $$\begin{aligned} x_1 + 2x_2 - x_3 + 5x_4 &= 6 \\ -x_1 - x_2 + 3x_3 + x_4 &= 0 \\ x_1 + x_2 + x_3 - x_4 &= 1 \\ 2x_1 + 5x_2 + 4x_3 + 16x_4 &= 0 \end{aligned}$$

(f) $$\begin{aligned} x_1 - x_2 \qquad\qquad &= 3 \\ 2x_1 + x_2 + x_3 - x_4 &= 7 \\ x_2 - x_3 + 5x_4 &= -5 \\ x_1 + x_2 - 6x_3 + x_4 &= -23 \end{aligned}$$

(g) $$\begin{aligned} x - y - z &= 0 \\ x + 2y + 5z &= 1 \end{aligned}$$

(h) $$\begin{aligned} x_1 - x_2 + x_3 - x_4 &= 1 \\ 2x_1 - 3x_2 + 7x_3 + 5x_4 &= 7 \\ x_1 + x_2 - 8x_3 + x_4 &= 0 \\ 4x_1 - 3x_2 \qquad + 5x_4 &= 0 \end{aligned}$$

(i) $\begin{aligned} 3x_1 - x_2 + x_3 + 3x_4 &= 1 \\ x_1 - x_2 + 4x_3 - 8x_4 &= 3 \\ 5x_1 - 3x_2 + 9x_3 - 13x_4 &= 0 \end{aligned}$ (j) $\begin{aligned} x_1 + x_2 + 5x_3 - x_4 &= 0 \\ x_2 + x_3 - 2x_4 &= 0 \\ x_1 + 2x_2 \qquad - 5x_4 &= 0 \end{aligned}$

2. Solve the following systems, where a, b, and c are arbitrary.

(a) $\begin{aligned} x_1 + x_2 + x_3 &= a \\ x_2 + x_3 &= b \\ x_1 + x_2 \qquad &= c \end{aligned}$ (b) $\begin{aligned} x - y &= a \\ x + 2y &= b \end{aligned}$

3. If the systems of equations

$$\begin{aligned} a_{11}x_1 + a_{12}x_2 + \cdots + a_{1n}x_n &= b_1 \\ a_{21}x_1 + a_{22}x_2 + \cdots + a_{2n}x_n &= b_2 \\ &\vdots \\ a_{m1}x_1 + a_{m2}x_2 + \cdots + a_{mn}x_n &= b_m \end{aligned}$$

and

$$\begin{aligned} a_{11}x_1 + a_{12}x_2 + \cdots + a_{1n}x_n &= b_1' \\ a_{21}x_1 + a_{22}x_2 + \cdots + a_{2n}x_n &= b_2' \\ &\vdots \\ a_{m1}x_1 + a_{m2}x_2 + \cdots + a_{mn}x_n &= b_m' \end{aligned}$$

have solutions, show that the systems

$$\begin{aligned} a_{11}x_1 + a_{12}x_2 + \cdots + a_{1n}x_n &= b_1 + b_1' \\ a_{21}x_1 + a_{22}x_2 + \cdots + a_{2n}x_n &= b_2 + b_2' \\ &\vdots \\ a_{m1}x_1 + a_{m2}x_2 + \cdots + a_{mn}x_n &= b_m + b_m' \end{aligned}$$

and

$$\begin{aligned} a_{11}x_1 + a_{12}x_2 + \cdots + a_{1n}x_n &= cb_1 \\ a_{21}x_1 + a_{22}x_2 + \cdots + a_{2n}x_n &= cb_2 \\ &\vdots \\ a_{m1}x_1 + a_{m2}x_2 + \cdots + a_{mn}x_n &= cb_m \end{aligned}$$

also have solutions.

4. Solve to find x and y as functions of t if

$$\begin{aligned} (1 - t)x + y &= 0 \\ -2x + (1 - t)y &= 1 \end{aligned}$$

5. Determine a and b so that the point $(2, -1)$ lies on the intersection of the lines

$$ax + by = 1$$
$$-bx + ay = 1$$

6. Determine a necessary and sufficient condition on a, b, and c, so that the system of equations admits a solution.

(a) $$\begin{aligned} x - 3y + z &= a \\ x - 5y + z &= b \\ -x + 2y - z &= c \end{aligned}$$

(b) $$\begin{aligned} x + y - 2z &= a \\ -x + 8y - 7z &= b \\ x - 2y + z &= c \end{aligned}$$

4 HOMOGENEOUS SYSTEMS

We call a system of equations of the form

$$\begin{aligned} a_{11}x_1 + a_{12}x_2 + a_{13}x_3 + \cdots + a_{1n}x_n &= 0 \\ a_{21}x_1 + a_{22}x_2 + a_{23}x_3 + \cdots + a_{2n}x_n &= 0 \\ a_{31}x_1 + a_{32}x_2 + a_{33}x_3 + \cdots + a_{3n}x_n &= 0 \\ &\vdots \\ a_{m1}x_1 + a_{m2}x_2 + a_{m3}x_3 + \cdots + a_{mn}x_n &= 0 \end{aligned}$$

a **homogeneous** system. In other words, all quantities on the right-hand side of the system vanish.

Any system of homogeneous equations always has at least one solution, namely $x_1 = 0, x_2 = 0, \ldots, x_n = 0$. This solution is often called the **trivial** solution. Any other solution is called a **nontrivial** solution.

In the above system of equations suppose $m < n$, i.e., there are more unknowns than equations. In this case we have the following important theorem.

Theorem In a homogeneous system of linear equations in which the number of variables is greater than the number of equations, a nontrivial solution exists.

Proof Let the system be

$$\begin{aligned} a_{11}x_1 + a_{12}x_2 + \cdots + a_{1n}x_n &= 0 \\ a_{21}x_1 + a_{22}x_2 + \cdots + a_{2n}x_n &= 0 \\ &\vdots \\ a_{m1}x_1 + a_{m2}x_2 + \cdots + a_{mn}x_n &= 0 \end{aligned}$$

where $m < n$.

If all the coefficients of x_1 in the system are zero, i.e., if $a_{11} = a_{21} = \cdots = a_{m1} = 0$, then $x_1 = 1$, $x_2 = x_3 = \cdots = x_n = 0$ is a nontrivial solution. So suppose some coefficient of x_1 is nonzero. By renumbering the system we may suppose $a_{11} \neq 0$. Multiply equation one by a_{11}^{-1} and eliminate x_1 from the remaining equations to obtain

$$\begin{aligned} x_1 + b_{12}x_2 + b_{13}x_3 + \cdots + b_{1n}x_n &= 0 \\ b_{22}x_2 + b_{23}x_3 + \cdots + b_{2n}x_n &= 0 \\ &\vdots \\ b_{m2}x_2 + b_{m3}x_3 + \cdots + b_{mn}x_n &= 0 \end{aligned}$$

If the coefficients of x_2 in the equations other than the first are all zero, i.e., if $b_{22} = b_{32} = \cdots = b_{n2} = 0$, we may find a nontrivial solution by setting $x_3 = x_4 = \cdots = x_n = 0$ and $x_1 = -b_{12}$, $x_2 = 1$.

Now suppose that after renumbering the equations $b_{22} \neq 0$. Multiply the second equation by b_{22}^{-1} and eliminate x_2 from all other equations to obtain a system of the form

$$\begin{aligned} x_1 \qquad + c_{13}x_3 + \cdots + c_{1n}x_n &= 0 \\ x_2 + c_{23}x_3 + \cdots + c_{2n}x_n &= 0 \\ c_{33}x_3 + \cdots + c_{3n}x_n &= 0 \\ &\vdots \\ c_{m3}x_3 + \cdots + c_{mn}x_n &= 0 \end{aligned}$$

By continuing this process and using the fact that the number of variables is greater than the number of equations, we eventually obtain a system of the form

$$\begin{aligned} x_1 \qquad\qquad + d_{1r}x_r + \cdots + d_{1n}x_n &= 0 \\ x_2 \qquad + d_{2r}x_r + \cdots + d_{2n}x_n &= 0 \\ &\vdots \\ x_{r-1} + d_{r-1,r}x_r + \cdots + d_{r-1,n}x_n &= 0 \\ 0 &= 0 \\ &\vdots \end{aligned}$$

where $r < n$. By choosing $x_r = 1$, $x_{r+1} = \cdots = x_n = 0$, and $x_1 = -d_{1r}$, $x_2 = -d_{2r}, \ldots, x_{r-1} = -d_{r-1,\,r}$, we obtain a nontrivial solution to the system. ∎

2

VECTORS AND MATRICES

1 VECTORS

In our chapter on linear equations there were many instances in which we were forced to consider n-tuples of numbers. We are now going to study this concept in greater detail.

We define a **column n-vector** to be an n-tuple of numbers written vertically as

$$\begin{bmatrix} u_1 \\ u_2 \\ \vdots \\ u_n \end{bmatrix}$$

If the numbers u_i are real we have a real column n-vector; if complex we have a complex column n-vector. For example,

$$\begin{bmatrix} 2 \\ -1 \end{bmatrix} \quad \text{is a real column 2-vector}$$

$$\begin{bmatrix} -7 \\ 0 \\ 8 \end{bmatrix} \quad \text{is a real column 3-vector}$$

$$\begin{bmatrix} -1 \\ 0 \\ 1 \\ 0 \end{bmatrix} \quad \text{is a real column 4-vector}$$

$$\begin{bmatrix} i \\ 0 \\ i-1 \end{bmatrix} \quad \text{is a complex column 3-vector}$$

In a similar vein we define a **row n-vector** to be an n-tuple of numbers written horizontally, as in $[u_1, u_2, \ldots, u_n]$. Thus, $[5, -1, 4, 3]$ is a row 4-vector.

We call the collection of all real column n-vectors $\boldsymbol{R}^n$. With this definition $\boldsymbol{R}^2$ consists of all column vectors of the form $\begin{bmatrix} \alpha \\ \beta \end{bmatrix}$, where α and β are arbitrary real numbers, while $\boldsymbol{R}^3$ consists of all column vectors of the form

$$\begin{bmatrix} \alpha \\ \beta \\ \gamma \end{bmatrix}$$

where α, β, and γ can be any real numbers. We call the collection of all complex column n-vectors $\boldsymbol{C}^n$.

In the column vector

$$\begin{bmatrix} u_1 \\ u_2 \\ \vdots \\ u_n \end{bmatrix}$$

the number u_i in the ith slot is called the ith **component** of the vector. Thus in the vector

$$\begin{bmatrix} -6 \\ 0 \\ -1 \\ 7 \end{bmatrix}$$

-6 is the first component, 0 is the second component, -1 is the third component, while 7 is the fourth component.

If $\boldsymbol{a} = \begin{bmatrix} \alpha_1 \\ \alpha_2 \\ \vdots \\ \alpha_n \end{bmatrix}$ and $\boldsymbol{b} = \begin{bmatrix} \beta_1 \\ \beta_2 \\ \vdots \\ \beta_n \end{bmatrix}$ are two column n-vectors, we say $\boldsymbol{a} = \boldsymbol{b}$ if and only if $\alpha_1 = \beta_1$, $\alpha_2 = \beta_2, \ldots, \alpha_n = \beta_n$.

In other words two vectors are equal when and only when they have the same ith components for all i.

We now wish to give the collection of all real column n-vectors an algebraic structure. We define the **sum** of two vectors

$$\boldsymbol{a} = \begin{bmatrix} \alpha_1 \\ \alpha_2 \\ \vdots \\ \alpha_n \end{bmatrix} \quad \text{and} \quad \boldsymbol{b} = \begin{bmatrix} \beta_1 \\ \beta_2 \\ \vdots \\ \beta_n \end{bmatrix}$$

denoted by $\boldsymbol{a} + \boldsymbol{b}$, to be the vector

$$\boldsymbol{a} + \boldsymbol{b} = \begin{bmatrix} \alpha_1 + \beta_1 \\ \alpha_2 + \beta_2 \\ \vdots \\ \alpha_n + \beta_n \end{bmatrix}$$

For example,

$$\begin{bmatrix} -3 \\ 1 \\ 0 \\ 2 \end{bmatrix} + \begin{bmatrix} 6 \\ -1 \\ 0 \\ 8 \end{bmatrix} = \begin{bmatrix} 3 \\ 0 \\ 0 \\ 10 \end{bmatrix}$$

$$\begin{bmatrix} -6 \\ 1 \\ 0 \end{bmatrix} + \begin{bmatrix} 8 \\ 2 \\ 7 \end{bmatrix} = \begin{bmatrix} 2 \\ 3 \\ 7 \end{bmatrix}$$

Notice that addition is only defined when two vectors are of the same size, i.e., when they have the same number of components.

Generally when we are dealing with vectors we refer to numbers as **scalars**. Real numbers are real scalars. Complex numbers are complex scalars.

Another important operation is scalar multiplication. If $\boldsymbol{c}$ is a column n-vector,

$$\boldsymbol{c} = \begin{bmatrix} c_1 \\ c_2 \\ \vdots \\ c_n \end{bmatrix}$$

and α is a scalar, we define the **scalar multiple** of the vector $\boldsymbol{c}$ by the scalar α, written $\alpha\boldsymbol{c}$, by

$$\alpha\boldsymbol{c} = \begin{bmatrix} \alpha c_1 \\ \alpha c_2 \\ \vdots \\ \alpha c_n \end{bmatrix}$$

For example,

$$2\begin{bmatrix} 1 \\ 7 \end{bmatrix} = \begin{bmatrix} 2 \\ 14 \end{bmatrix}, \qquad 3\begin{bmatrix} 8 \\ 0 \\ -31 \end{bmatrix} = \begin{bmatrix} 24 \\ 0 \\ -93 \end{bmatrix}$$

$$(-1)\begin{bmatrix} 8 \\ 1/2 \\ 1/3 \\ 0 \end{bmatrix} = \begin{bmatrix} -8 \\ -1/2 \\ -1/3 \\ 0 \end{bmatrix}$$

Similar definitions can be made for addition of complex n-vectors and multiplication of complex n-vectors by complex scalars.

In terms of components our definitions may be restated as follows: The ith component of the sum of two vectors is the sum of the ith components of each vector. The ith component of $\alpha\boldsymbol{c}$ is α times the ith component of $\boldsymbol{c}$.

The operations of addition and scalar multiplication satisfy several algebraic laws which are worthwhile to remember. In the following we suppose $\boldsymbol{a}$, $\boldsymbol{b}$, and $\boldsymbol{c}$ are n-vectors, μ, λ are scalars, and

$$\boldsymbol{a} = \begin{bmatrix} \alpha_1 \\ \alpha_2 \\ \vdots \\ \alpha_n \end{bmatrix}, \quad \boldsymbol{b} = \begin{bmatrix} \beta_1 \\ \beta_2 \\ \vdots \\ \beta_n \end{bmatrix}, \quad \text{and} \quad \boldsymbol{c} = \begin{bmatrix} \gamma_1 \\ \gamma_2 \\ \vdots \\ \gamma_n \end{bmatrix}$$

First we verify the **commutative law** for vector addition.

Proposition 1 $\boldsymbol{a} + \boldsymbol{b} = \boldsymbol{b} + \boldsymbol{a}$.

Proof By definition of vector addition,

$$\boldsymbol{a} + \boldsymbol{b} = \begin{bmatrix} \alpha_1 + \beta_1 \\ \alpha_2 + \beta_2 \\ \vdots \\ \alpha_n + \beta_n \end{bmatrix} \quad \text{and} \quad \boldsymbol{b} + \boldsymbol{a} = \begin{bmatrix} \beta_1 + \alpha_1 \\ \beta_2 + \alpha_2 \\ \vdots \\ \beta_n + \alpha_n \end{bmatrix}$$

Since $\alpha_i + \beta_i = \beta_i + \alpha_i$ for all real numbers, $\boldsymbol{a} + \boldsymbol{b} = \boldsymbol{b} + \boldsymbol{a}$, by the definition of the equality of vectors. ∎

Next we demonstrate the **associative law** for vector addition.

Proposition 2 $(\boldsymbol{a} + \boldsymbol{b}) + \boldsymbol{c} = \boldsymbol{a} + (\boldsymbol{b} + \boldsymbol{c})$.

Proof By the definition of vector addition,

$$\boldsymbol{a} + \boldsymbol{b} = \begin{bmatrix} \alpha_1 + \beta_1 \\ \alpha_2 + \beta_2 \\ \vdots \\ \alpha_n + \beta_n \end{bmatrix} \quad \text{and} \quad \boldsymbol{b} + \boldsymbol{c} = \begin{bmatrix} \beta_1 + \gamma_1 \\ \beta_2 + \gamma_2 \\ \vdots \\ \beta_n + \gamma_n \end{bmatrix}$$

$$(\boldsymbol{a} + \boldsymbol{b}) + \boldsymbol{c} = \begin{bmatrix} (\alpha_1 + \beta_1) + \gamma_1 \\ (\alpha_2 + \beta_2) + \gamma_2 \\ \vdots \\ (\alpha_n + \beta_n) + \gamma_n \end{bmatrix}$$

and

$$\boldsymbol{a} + (\boldsymbol{b} + \boldsymbol{c}) = \begin{bmatrix} \alpha_1 + (\beta_1 + \gamma_1) \\ \alpha_2 + (\beta_2 + \gamma_2) \\ \vdots \\ \alpha_n + (\beta_n + \gamma_n) \end{bmatrix}$$

Since $(\alpha_i + \beta_i) + \gamma_i = \alpha_i + (\beta_i + \gamma_i)$ for all real numbers, we have $(\boldsymbol{a} + \boldsymbol{b}) + \boldsymbol{c} = \boldsymbol{a} + (\boldsymbol{b} + \boldsymbol{c})$, by the definition of the equality of vectors. ∎

The zero vector, which we will denote by **0**, is the n-vector

$$\begin{bmatrix} 0 \\ 0 \\ \vdots \\ 0 \end{bmatrix}$$

i.e., the vector all of whose components are 0. It will be clear from the context what the size of the vector denoted by **0** is intended to be.

Proposition 3 $\mathbf{0} + \boldsymbol{a} = \boldsymbol{a} + \mathbf{0} = \boldsymbol{a}$.

Proof By definition of vector addition,

$$\mathbf{0} + \boldsymbol{a} = \begin{bmatrix} 0 + \alpha_1 \\ 0 + \alpha_2 \\ \vdots \\ 0 + \alpha_n \end{bmatrix}$$

However, we know that $0 + \alpha_i = \alpha_i$ for all real numbers α_i. So

$$\mathbf{0} + \boldsymbol{a} = \begin{bmatrix} \alpha_1 \\ \alpha_2 \\ \vdots \\ \alpha_n \end{bmatrix} = \boldsymbol{a}$$

Likewise, $\boldsymbol{a} + \mathbf{0} = \boldsymbol{a}$. ∎

We now wish to define what we mean by the **negative** of a vector. If

$$\boldsymbol{a} = \begin{bmatrix} \alpha_1 \\ \alpha_2 \\ \vdots \\ \alpha_n \end{bmatrix}$$

we denote by $-\boldsymbol{a}$, the vector

$$\begin{bmatrix} -\alpha_1 \\ -\alpha_2 \\ \vdots \\ -\alpha_n \end{bmatrix}$$

Thus,

$$-\begin{bmatrix} 6 \\ 2 \\ -1 \\ 2 \end{bmatrix} = \begin{bmatrix} -6 \\ -2 \\ 1 \\ -2 \end{bmatrix}$$

Proposition 4 $\boldsymbol{a} + (-\boldsymbol{a}) = (-\boldsymbol{a}) + \boldsymbol{a} = \boldsymbol{0}.$

Proof By the definition of vector addition,

$$\boldsymbol{a} + (-\boldsymbol{a}) = \begin{bmatrix} \alpha_1 + (-\alpha_1) \\ \alpha_2 + (-\alpha_2) \\ \vdots \\ \alpha_n + (-\alpha_n) \end{bmatrix}$$

But $\alpha_1 + (-\alpha_1) = 0$, for all real numbers. So

$$\boldsymbol{a} + (-\boldsymbol{a}) = \begin{bmatrix} 0 \\ 0 \\ \vdots \\ 0 \end{bmatrix} = \boldsymbol{0}$$

Likewise, $(-\boldsymbol{a}) + \boldsymbol{a} = \boldsymbol{0}.$ ▨

There are also certain algebraic rules satisfied by scalar multiplication.

Proposition 5 $\lambda(\boldsymbol{a} + \boldsymbol{b}) = \lambda\boldsymbol{a} + \lambda\boldsymbol{b}.$

Proof By definition of vector addition,

$$(\boldsymbol{a} + \boldsymbol{b}) = \begin{bmatrix} (\alpha_1 + \beta_1) \\ (\alpha_2 + \beta_2) \\ \vdots \\ (\alpha_n + \beta_n) \end{bmatrix}$$

Also, by definition of scalar multiplication,

$$\lambda(\boldsymbol{a}+\boldsymbol{b}) = \begin{bmatrix} \lambda(\alpha_1+\beta_1) \\ \lambda(\alpha_2+\beta_2) \\ \vdots \\ \lambda(\alpha_n+\beta_n) \end{bmatrix}$$

Again,

$$\lambda\boldsymbol{a} = \begin{bmatrix} \lambda\alpha_1 \\ \lambda\alpha_2 \\ \vdots \\ \lambda\alpha_n \end{bmatrix} \quad \text{and} \quad \lambda\boldsymbol{b} = \begin{bmatrix} \lambda\beta_1 \\ \lambda\beta_2 \\ \vdots \\ \lambda\beta_n \end{bmatrix}$$

Therefore,

$$\lambda\boldsymbol{a} + \lambda\boldsymbol{b} = \begin{bmatrix} \lambda\alpha_1+\lambda\beta_1 \\ \lambda\alpha_2+\lambda\beta_2 \\ \vdots \\ \lambda\alpha_n+\lambda\beta_n \end{bmatrix}$$

by definition of vector addition. Since $\lambda(\alpha_i + \beta_i) = \lambda\alpha_i + \lambda\beta_i$ for all real numbers, we have, using the definition of vector equality, $\lambda(\boldsymbol{a} + \boldsymbol{b}) = \lambda\boldsymbol{a} + \lambda\boldsymbol{b}$. ∎

Other important laws whose proofs are left to the student, are

$$(\lambda + \mu)\boldsymbol{a} = \lambda\boldsymbol{a} + \mu\boldsymbol{a}$$
$$\lambda(\mu\boldsymbol{a}) = (\lambda\mu)\boldsymbol{a}$$
$$1 \cdot \boldsymbol{a} = \boldsymbol{a}$$

We define **subtraction** of vectors as follows: If $\boldsymbol{a}$ and $\boldsymbol{b}$ are vectors, $\boldsymbol{a} - \boldsymbol{b} = \boldsymbol{a} + (-\boldsymbol{b})$.

In terms of components

$$\boldsymbol{a} - \boldsymbol{b} = \boldsymbol{a} + (-\boldsymbol{b}) = \begin{bmatrix} \alpha_1 \\ \alpha_2 \\ \vdots \\ \alpha_n \end{bmatrix} + \begin{bmatrix} -\beta_1 \\ -\beta_2 \\ \vdots \\ -\beta_n \end{bmatrix}$$
$$= \begin{bmatrix} \alpha_1-\beta_1 \\ \alpha_2-\beta_2 \\ \vdots \\ \alpha_n-\beta_n \end{bmatrix}$$

The laws we have derived are useful in so far as they enable us to perform algebraic manipulations on vectors without continually referring back to their components.

For example, to solve the equation $\boldsymbol{x} + \boldsymbol{a} = \boldsymbol{b}$ for the vector $\boldsymbol{x}$, we add $-\boldsymbol{a}$ to both sides to obtain

$$
\begin{aligned}
(\boldsymbol{x} + \boldsymbol{a}) + (-\boldsymbol{a}) &= \boldsymbol{b} + (-\boldsymbol{a}) \\
\boldsymbol{x} + (\boldsymbol{a} - \boldsymbol{a}) &= \boldsymbol{b} - \boldsymbol{a} \\
\boldsymbol{x} + \boldsymbol{0} &= \boldsymbol{b} - \boldsymbol{a} \\
\boldsymbol{x} &= \boldsymbol{b} - \boldsymbol{a}
\end{aligned}
$$

Generally speaking our laws guarantee we can perform addition and scalar multiplication of vectors just as we do with numbers.

EXERCISES

1. Calculate

(a) $\begin{bmatrix} -1 \\ 0 \\ 2 \end{bmatrix} + \begin{bmatrix} 0 \\ 3 \\ 7 \end{bmatrix}$

(b) $3\begin{bmatrix} -6 \\ 8 \\ 9 \\ 21 \end{bmatrix} + \begin{bmatrix} 0 \\ 3 \\ 2 \\ 1 \end{bmatrix}$

(c) $\begin{bmatrix} 4 \\ 0 \\ -1 \\ 2 \end{bmatrix} - 2\begin{bmatrix} 8 \\ 6 \\ 2 \\ 3 \end{bmatrix}$

(d) $6\begin{bmatrix} 0 \\ 1 \end{bmatrix} + 3\begin{bmatrix} 1 \\ 0 \end{bmatrix} + \begin{bmatrix} 1 \\ 1 \end{bmatrix}$

(e) $\begin{bmatrix} 0 \\ 3 \end{bmatrix} + 0\begin{bmatrix} 0 \\ 4 \end{bmatrix} - \begin{bmatrix} 6 \\ 3 \end{bmatrix}$

(f) $\begin{bmatrix} -1 \\ -1 \\ -1 \end{bmatrix} + 7\begin{bmatrix} 1 \\ 1 \\ 1 \end{bmatrix} + \begin{bmatrix} -1 \\ -1 \\ -1 \end{bmatrix}$

(g) $\begin{bmatrix} 0 \\ 2 \end{bmatrix} + \begin{bmatrix} 3 \\ 7 \end{bmatrix} + \begin{bmatrix} 8 \\ 10 \end{bmatrix} - \begin{bmatrix} 6 \\ 0 \end{bmatrix}$

(h) $\begin{bmatrix} -1 \\ 3 \\ 0 \end{bmatrix} + \begin{bmatrix} 8 \\ 1 \\ 2 \end{bmatrix}$

2. Solve each of the following for $\boldsymbol{x}$:

(a) $\boldsymbol{x} + \begin{bmatrix} 1 \\ 0 \\ 2 \\ 1 \end{bmatrix} = \begin{bmatrix} 0 \\ 3 \\ 7 \\ 8 \end{bmatrix}$

(b) $\boldsymbol{x} + \begin{bmatrix} -1 \\ 2 \\ 0 \end{bmatrix} = \begin{bmatrix} 0 \\ 8 \\ 3 \end{bmatrix}$

(c) $5\boldsymbol{x} + \begin{bmatrix} -1 \\ -1 \\ 2 \end{bmatrix} = \begin{bmatrix} 0 \\ 1 \\ 7 \end{bmatrix}$

3. Prove the following laws:

$$
\begin{aligned}
(\mu + \lambda)\boldsymbol{a} &= \mu\boldsymbol{a} + \lambda\boldsymbol{a} \\
\mu(\lambda\boldsymbol{a}) &= (\mu\lambda)\boldsymbol{a} \\
1 \cdot \boldsymbol{a} &= \boldsymbol{a}
\end{aligned}
$$

4. If $\boldsymbol{a}$ and $\boldsymbol{b}$ are n-vectors and $\boldsymbol{a} + \boldsymbol{b} = \boldsymbol{a}$, show that $\boldsymbol{b} = \mathbf{0}$.

5. If $\alpha\boldsymbol{x} = \mathbf{0}$ and $\alpha \neq 0$ show that $\boldsymbol{x} = \mathbf{0}$. If $\alpha\boldsymbol{x} = \beta\boldsymbol{x}$ and $\alpha \neq \beta$ show that $\boldsymbol{x} = \mathbf{0}$.

6. Exhibit 2-vectors $\boldsymbol{a}$ and $\boldsymbol{b}$ such that the equation $\alpha\boldsymbol{a} = \boldsymbol{b}$ cannot be solved for the scalar α.

7. Solve for $\boldsymbol{x}$ in terms of $\boldsymbol{a}$, $\boldsymbol{b}$, and $\boldsymbol{c}$, where $\boldsymbol{x}$, $\boldsymbol{a}$, $\boldsymbol{b}$, and $\boldsymbol{c}$ are n-vectors, $((\boldsymbol{x} + \boldsymbol{a}) + 2\boldsymbol{b}) + \boldsymbol{c} = \boldsymbol{b} - 2\boldsymbol{c}$.

8. Solve the equations

$$\boldsymbol{x} + \boldsymbol{y} = \boldsymbol{a}$$
$$\boldsymbol{x} - \boldsymbol{y} = \boldsymbol{b}$$

to find the vectors $\boldsymbol{x}$ and $\boldsymbol{y}$ in terms of the vectors $\boldsymbol{a}$ and $\boldsymbol{b}$.

9. Show that all the laws for vector addition and scalar multiplication hold for complex vectors.

10. Formulate definitions for addition and scalar multiplication of row n-vectors and prove the corresponding laws.

11. Show that there are 2^n n-vectors such that in each vector all components are 0 or 1.

12. Given the three 2-vectors

$$\boldsymbol{a} = \begin{bmatrix} -2 \\ 1 \end{bmatrix}, \quad \boldsymbol{b} = \begin{bmatrix} 1 \\ 0 \end{bmatrix}, \quad \boldsymbol{c} = \begin{bmatrix} -1 \\ 2 \end{bmatrix}$$

find scalars λ, μ, ν, not all zero, such that $\lambda\boldsymbol{a} + \mu\boldsymbol{b} + \nu\boldsymbol{c} = 0$. Show that this is possible if $\boldsymbol{a}$, $\boldsymbol{b}$, and $\boldsymbol{c}$ are arbitrary 2-vectors.

13. If

$$\boldsymbol{a} = \begin{bmatrix} -1 \\ 0 \\ 2 \end{bmatrix}, \quad \boldsymbol{b} = \begin{bmatrix} 3 \\ 1 \\ 0 \end{bmatrix}, \quad \boldsymbol{c} = \begin{bmatrix} -4 \\ 0 \\ 3 \end{bmatrix}$$

calculate the components of
(a) $x\boldsymbol{a} + y\boldsymbol{b} + z\boldsymbol{c}$
(b) $x\boldsymbol{a} + (1 - x)\boldsymbol{b} + \boldsymbol{c}$
where x, y, and z are real numbers.

14. If

$$\boldsymbol{e}_1 = \begin{bmatrix} 1 \\ 0 \\ 0 \\ 0 \end{bmatrix}, \quad \boldsymbol{e}_2 = \begin{bmatrix} 0 \\ 1 \\ 0 \\ 0 \end{bmatrix}, \quad \boldsymbol{e}_3 = \begin{bmatrix} 0 \\ 0 \\ 1 \\ 0 \end{bmatrix}, \quad \boldsymbol{e}_4 = \begin{bmatrix} 0 \\ 0 \\ 0 \\ 1 \end{bmatrix}$$

show that every $\boldsymbol{x} \in \boldsymbol{R}^4$, can be expressed uniquely in the form $\boldsymbol{x} = a_1\boldsymbol{e}_1 + \alpha_2\boldsymbol{e}_2 + \alpha_3\boldsymbol{e}_3 + \alpha_4\boldsymbol{e}_4$, where α_1, α_2, α_3, and α_4 are real scalars.

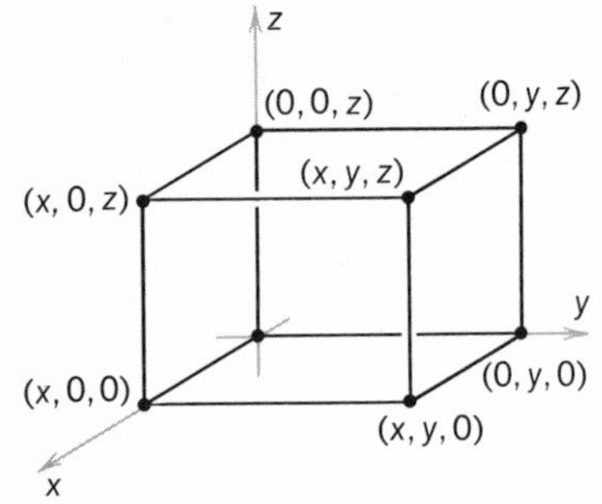

FIGURE 2-1

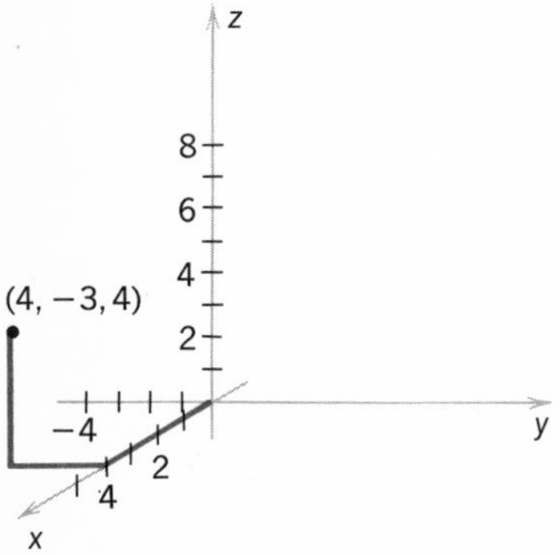

FIGURE 2-2

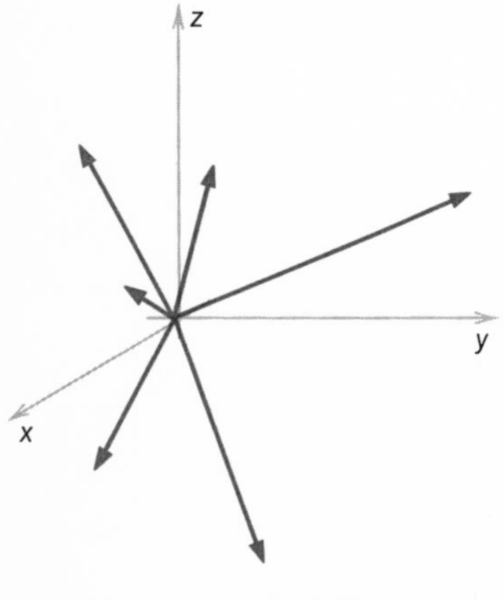

FIGURE 2-3

15. If

$$\begin{aligned} \boldsymbol{x}_1 + \boldsymbol{x}_2 + \boldsymbol{x}_3 + \cdots + \boldsymbol{x}_n &= \boldsymbol{y}_1 \\ \boldsymbol{x}_2 + \boldsymbol{x}_3 + \cdots + \boldsymbol{x}_n &= \boldsymbol{y}_2 \\ \boldsymbol{x}_3 + \cdots + \boldsymbol{x}_n &= \boldsymbol{y}_3 \\ &\vdots \\ \boldsymbol{x}_{n-1} + \boldsymbol{x}_n &= \boldsymbol{y}_{n-1} \\ \boldsymbol{x}_n &= \boldsymbol{y}_n \end{aligned}$$

find $\boldsymbol{x}_1, \ldots, \boldsymbol{x}_n$ in terms of $\boldsymbol{y}_1, \ldots, \boldsymbol{y}_n$, where $\boldsymbol{x}_1, \ldots, \boldsymbol{x}_n$ and $\boldsymbol{y}_1, \ldots, \boldsymbol{y}_n$ are vectors.

2 GEOMETRIC INTERPRETATION OF $\boldsymbol{R}^2$ AND $\boldsymbol{R}^3$

Just as points in the plane may be represented as ordered pairs of real numbers, so too, points in space may be represented as triples of real numbers. To effect this representation, choose three mutually perpendicular lines which meet at a point in space. The lines are called the x axis, y axis, and z axis and the point at which they meet is called the origin. (See Figure 2-1.)

The plane formed by the y axis and z axis is called the y-z plane. (The x-y plane and the x-z plane are determined in a similar manner.) To find the x coordinate of a point P in space construct the plane through P parallel to the y-z plane. The point at which this plane intersects the x axis is called the x coordinate of P. The y coordinate is obtained by determining where the plane through P parallel to the x-z plane intersects the y axis, and so on. Using this procedure we may associate to each point P in space a triple of real numbers (x, y, z). Also, to each triple of real numbers we may associate the point in space which has the given triple as its coordinates. The point $(0, 0, 0)$ where the three axes meet is called the origin of the coordinate system.

For example, to locate the point $(4, -3, 4)$ we go out 4 units on the x axis, -3 unit on the y axis, and up 4 units on the z axis. The situation is illustrated in Figure 2-2.

With this method of representing points in mind we see that the x axis consists of the points of the form $(\alpha, 0, 0)$, where α is any real number. The x-y plane consists of the points of the form $(\alpha, \beta, 0)$. The points of the y axis, z axis, and the remaining planes can be similarly represented.

Geometrically we define a **vector** as a **directed line segment** originating at the origin and proceeding to some point in space. Figure 2-3 shows several vectors. Vectors may be thought of as arrows beginning at the origin.

Using this definition of a vector we associate with the vector $\boldsymbol{v}$, the point in space, say (x, y, z), where $\boldsymbol{v}$ terminates, and write $\boldsymbol{v}(x, y, z)$ to indicate this association.

Geometrically we define vector addition as follows. In the plane formed by the vectors $\boldsymbol{v}_1$ and $\boldsymbol{v}_2$ (see Figure 2-4), form the parallelogram

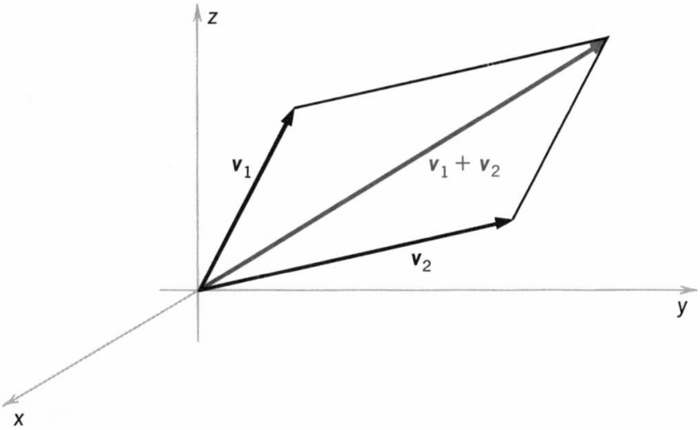

FIGURE 2-4

having $\boldsymbol{v}_1$ as one side and $\boldsymbol{v}_2$ as its adjacent side. We define $\boldsymbol{v}_1 + \boldsymbol{v}_2$ to be the directed line segment along the diagonal. To show that our geometric definition does not contradict our algebraic definition, we must show that $\boldsymbol{v}(x, y, z) + \boldsymbol{v}(x', y', z') = \boldsymbol{v}(x + x', y + y', z + z')$.

We prove this result in the plane, leaving the ambitious reader to formulate the proposition in three dimensional space. Thus we wish to show that $\boldsymbol{v}(x, y) + \boldsymbol{v}(x', y') = \boldsymbol{v}(x + x', y + y')$.

In Figure 2-5 let $\boldsymbol{v}(x, y)$ be the vector ending at the point A, and

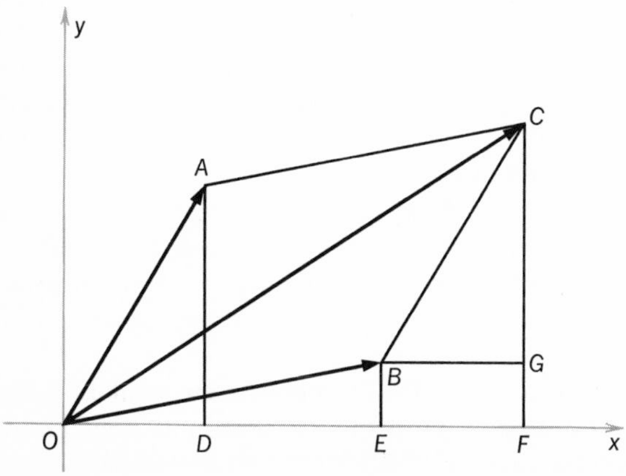

FIGURE 2-5

$\boldsymbol{v}(x', y')$ be the vector ending at the point B. The vector $\boldsymbol{v}(x, y) + \boldsymbol{v}(x', y')$ ends at the vertex C of parallelogram $OBCA$. We wish to show that $\boldsymbol{v}(x, y) + \boldsymbol{v}(x', y') = \boldsymbol{v}(x + x', y + y')$, or in other words, that the coordinates of C are $(x + x', y + y')$.

From our figure the reader may observe that triangle OAD is congruent to triangle CBG. Also observe length $OD = x$, length $OE = x'$. By the congruence relation, length OD = length BG, and since $BGFE$ is a rectangle, we have length OD = length EF.

But length OF = length OE + length EF, and so length $OF = x + x'$. This shows that the x coordinate of C is $x + x'$. The proof for the y coordinate is analogous. From this we see that the geometric definition of vector addition is equivalent to the algebraic definition in which we add components.

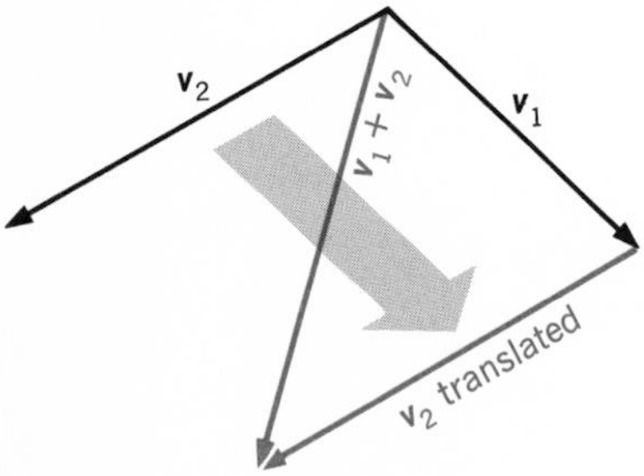

FIGURE 2-6

Figure 2-6 shows that we may also think of vector addition as performed by translating the directed line segment representing the vector $\boldsymbol{v}_2$ so that it begins at the vector $\boldsymbol{v}_1$. The terminal point of the resulting directed segment is the end point of the vector $\boldsymbol{v}_1 + \boldsymbol{v}_2$.

Scalar multiples of vectors have similar geometric interpretations. If α is a scalar and $\boldsymbol{v}$ is a vector, we may define $\alpha\boldsymbol{v}$ to be the vector which is $|\alpha|$ times as long as $\boldsymbol{v}$ and having the same direction as $\boldsymbol{v}$ if $\alpha > 0$ but the opposite direction of $\boldsymbol{v}$ if $\alpha < 0$. Figure 2-7 shows several examples.

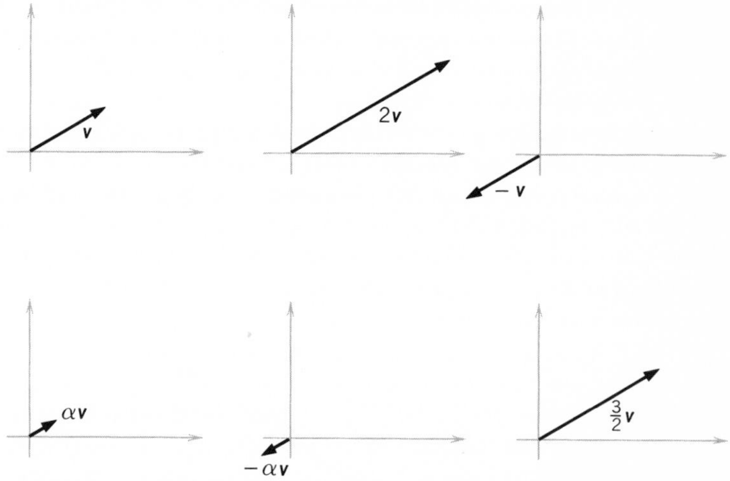

FIGURE 2-7

Using an argument which depends on similar triangles we can prove that $\alpha\boldsymbol{v}(x, y, z) = \boldsymbol{v}(\alpha x, \alpha y, \alpha z)$. Again the geometric definition coincides with the algebraic one.

What is the vector $\boldsymbol{b} - \boldsymbol{a}$? Since $\boldsymbol{a} + (\boldsymbol{b} - \boldsymbol{a}) = \boldsymbol{b}$, it is clear that

$b - a$ is that vector which when added to a gives b. In view of this, $b - a$ is parallel to the directed line segment beginning at the endpoint of a and terminating at the endpoint of b. (See Figure 2-8.)

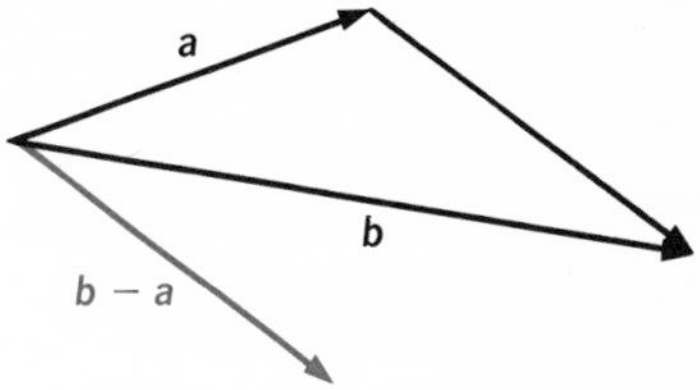

FIGURE 2-8

Let us denote by $\boldsymbol{i}$ the vector which ends at (1, 0, 0), by $\boldsymbol{j}$ that which ends at (0, 1, 0), and by $\boldsymbol{k}$ that which ends at (0, 0, 1). Then,

$$\begin{aligned} v(x, y, z) &= v(x, 0, 0) + v(0, y, 0) + v(0, 0, z) \\ &= xv(1, 0, 0) + yv(0, 1, 0) + zv(0, 0, 1) \\ &= x\boldsymbol{i} + y\boldsymbol{j} + z\boldsymbol{k} \end{aligned}$$

Hence, we can represent any vector in three dimensional space in terms of the unit vectors $\boldsymbol{i}$, $\boldsymbol{j}$, and $\boldsymbol{k}$.

For example the vector which ends at (2, 3, 3) is $2\boldsymbol{i} + 3\boldsymbol{j} + 3\boldsymbol{k}$, that ending at (0, −1, 4) is $-\boldsymbol{j} + 4\boldsymbol{k}$. (See Figure 2-9.) The vectors $\boldsymbol{i}$, $\boldsymbol{j}$, and $\boldsymbol{k}$ are called the standard basis vectors for R^3.

We also have

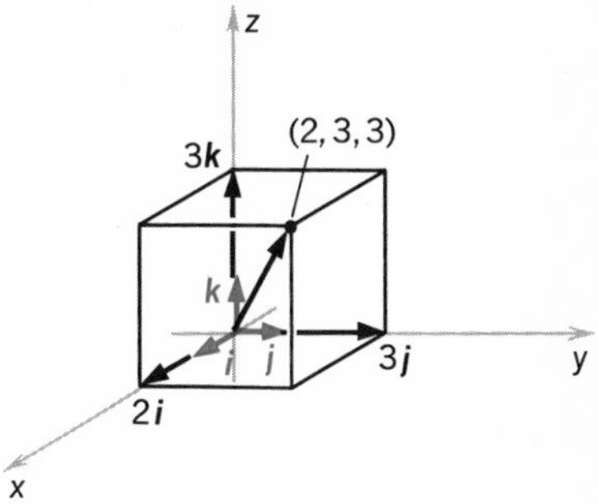

FIGURE 2-9

$$(x\boldsymbol{i} + y\boldsymbol{j} + z\boldsymbol{k}) + (x'\boldsymbol{i} + y'\boldsymbol{j} + z'\boldsymbol{k}) = (x + x')\boldsymbol{i} + (y + y')\boldsymbol{j} + (z + z')\boldsymbol{k}$$

and

$$\alpha(x\boldsymbol{i} + y\boldsymbol{j} + z\boldsymbol{k}) = (\alpha x)\boldsymbol{i} + (\alpha y)\boldsymbol{j} + (\alpha z)\boldsymbol{k}$$

Because of the correspondence between vectors and points we may sometimes refer to a point a, under circumstances in which a has been defined to be a vector. The reader should realize that by this statement we mean the endpoint of the vector a.

As an example let us use vectors to describe the points which lie in the parallelogram whose adjacent sides are the vectors a and b.

If P is any point in the parallelogram and we construct lines l_1 and l_2, parallel to the vectors a and b respectively, we see that l_1 intersects the side of the parallelogram determined by the vector b at some point tb, where $0 \le t \le 1$. Likewise, l_2 intersects the side determined by the vector a at some point sa, where $0 \le s \le 1$. (See Figure 2-10.)

Since P is then the endpoint of the diagonal of a parallelogram having adjacent sides sa and tb, and if v denotes the vector ending at P, we see that $v = sa + tb$. Thus, all the points in the parallelogram are endpoints of vectors of the form $sa + tb$ for $0 \le s \le 1$ and $0 \le t \le 1$. By reversing our steps we easily see that all vectors of this form end within the parallelogram.

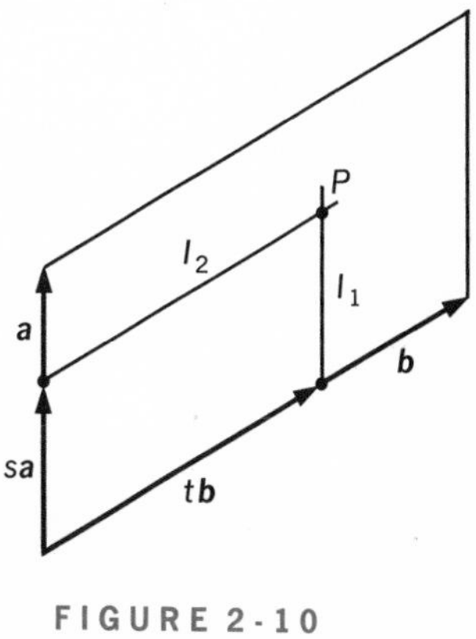

FIGURE 2-10

Using the geometric interpretation of vector addition and scalar multiplication we may find, in parametric form, the equation of a line passing through the endpoint of the vector a in the direction of a vector v. (See Figure 2-11.) As t varies through all real values, the points of the form tv are all scalar multiples of the vector v, and therefore exhaust the points of the line passing through the origin in the direction

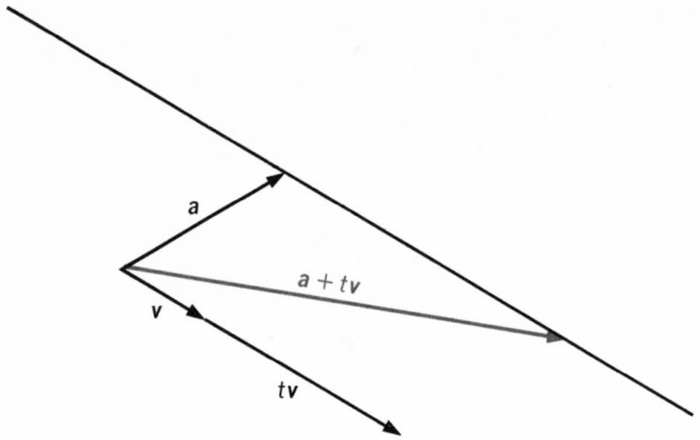

FIGURE 2-11

of $\boldsymbol{v}$. Since every point on the line passing through the endpoint of $\boldsymbol{a}$ in the direction of $\boldsymbol{v}$ is the endpoint of the diagonal of a parallelogram with sides $\boldsymbol{a}$ and $t\boldsymbol{v}$, for some suitable value of t, we see that all the points on the line are of the form $\boldsymbol{a} + t\boldsymbol{v}$. Thus, the line may be expressed parametrically in the form $\boldsymbol{l}(t) = \boldsymbol{a} + t\boldsymbol{v}$. At $t = 0$, $\boldsymbol{l}(0) = \boldsymbol{a}$. As t increases the point $\boldsymbol{l}(t)$ moves away from $\boldsymbol{a}$ in the direction of $\boldsymbol{v}$. As t decreases from $t = 0$ through negative values, $\boldsymbol{l}(t)$ moves away from $\boldsymbol{a}$ in the direction of $-\boldsymbol{v}$.

Of course, there are other parametrizations of the same line. These may be obtained by choosing a different point on the line and forming the parametric equation of the line beginning at that point and in the direction of $\boldsymbol{v}$. For example, the point $\boldsymbol{a} + \boldsymbol{v}$ is on the line $\boldsymbol{a} + t\boldsymbol{v}$, and thus $\boldsymbol{l}'(t) = \boldsymbol{a} + \boldsymbol{v} + t\boldsymbol{v}$ represents the same line.

Other parametrizations may be obtained by observing that if $\alpha \neq 0$, the vector $\alpha\boldsymbol{v}$ has the same or opposite direction as $\boldsymbol{v}$. Thus, $\boldsymbol{l}'(t) = \boldsymbol{a} + \alpha t\boldsymbol{v}$ provides us with another parametrization of $\boldsymbol{l}(t) = \boldsymbol{a} + t\boldsymbol{v}$.

Example 1 Determine the equation of the line passing through (1, 0, 0) in the direction of $\boldsymbol{j}$.

The desired line can be given parametrically as $\boldsymbol{l}(t) = \boldsymbol{i} + t\boldsymbol{j}$. (See Figure 2-12.) In terms of coordinates $x(t) = 1$, $y(t) = t$, $z(t) = 0$. In this case the line is the intersection of the planes $z = 0$ and $x = 1$.

We may also derive the equation of a line passing through the endpoints of two given vectors $\boldsymbol{a}$ and $\boldsymbol{b}$.

In this case since the vector $\boldsymbol{b} - \boldsymbol{a}$ is parallel to the directed line segment from $\boldsymbol{a}$ to $\boldsymbol{b}$, what we really wish to do is calculate the parametric equations of the line passing through $\boldsymbol{a}$ in the direction of $\boldsymbol{b} - \boldsymbol{a}$. (See Figure 2-13.) Thus $\boldsymbol{l}(t) = \boldsymbol{a} + t(\boldsymbol{b} - \boldsymbol{a})$, or $\boldsymbol{l}(t) = (1 - t)\boldsymbol{a} + t\boldsymbol{b}$.

As t increases from 0 to 1, $t(\boldsymbol{b} - \boldsymbol{a})$ starts as the $\boldsymbol{0}$ vector, continues in the direction of $\boldsymbol{b} - \boldsymbol{a}$, increasing in length, until at $t = 1$ it is the

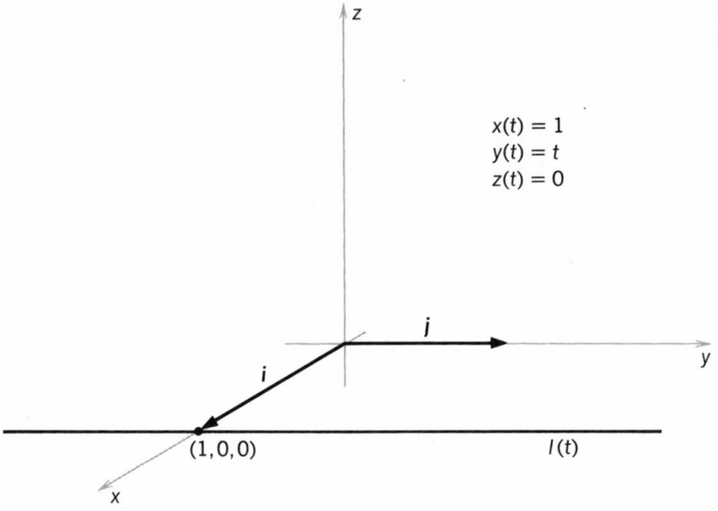

FIGURE 2-12

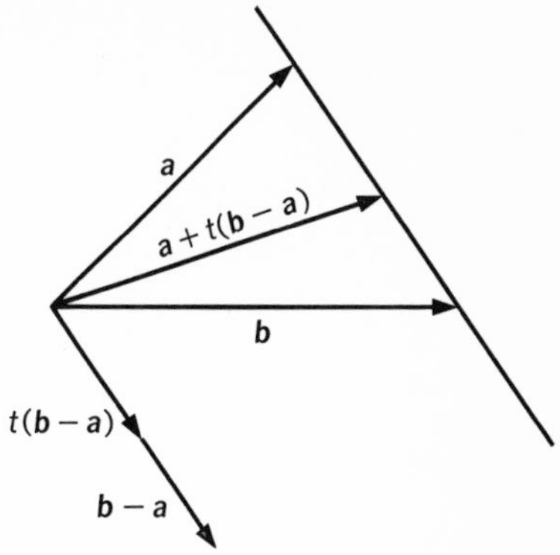

FIGURE 2-13

vector $\boldsymbol{b} - \boldsymbol{a}$. Thus, in $\boldsymbol{l}(t) = \boldsymbol{a} + t(\boldsymbol{b} - \boldsymbol{a})$, as t increases from 0 to 1, $\boldsymbol{l}(t)$ moves from the endpoint of $\boldsymbol{a}$ along the directed line segment from $\boldsymbol{a}$ to $\boldsymbol{b}$ to the endpoint of $\boldsymbol{b}$.

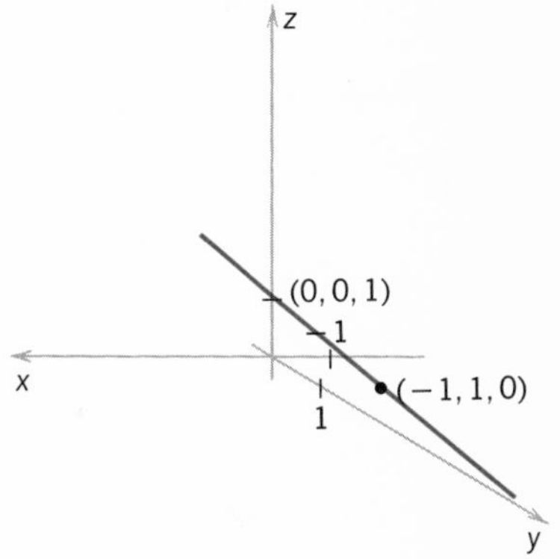

FIGURE 2-14

Example 2 Find the equation of the line passing through $(-1, 1, 0)$ and $(0, 0, 1)$. (See Figure 2-14.)

Letting $\boldsymbol{a} = -\boldsymbol{i} + \boldsymbol{j}$, $\boldsymbol{b} = \boldsymbol{k}$, we have

$$\begin{aligned} \boldsymbol{l}(t) &= (1 - t)(-\boldsymbol{i} + \boldsymbol{j}) + t\boldsymbol{k} \\ &= -(1 - t)\boldsymbol{i} + (1 - t)\boldsymbol{j} + t\boldsymbol{k} \\ x(t) &= -1 + t \\ y(t) &= 1 - t \\ z(t) &= t \end{aligned}$$

Before continuing let us note that any vector $\boldsymbol{c}$ of the form $\boldsymbol{c} = \lambda\boldsymbol{a} + \mu\boldsymbol{b}$, where $\lambda + \mu = 1$, is on the line passing through $\boldsymbol{a}$ and $\boldsymbol{b}$. To see this, observe $\boldsymbol{c} = (1 - \mu)\boldsymbol{a} + \mu\boldsymbol{b} = \boldsymbol{a} + \mu(\boldsymbol{b} - \boldsymbol{a})$, and so $\boldsymbol{c}$ is on the line passing through $\boldsymbol{a}$ and $\boldsymbol{b}$.

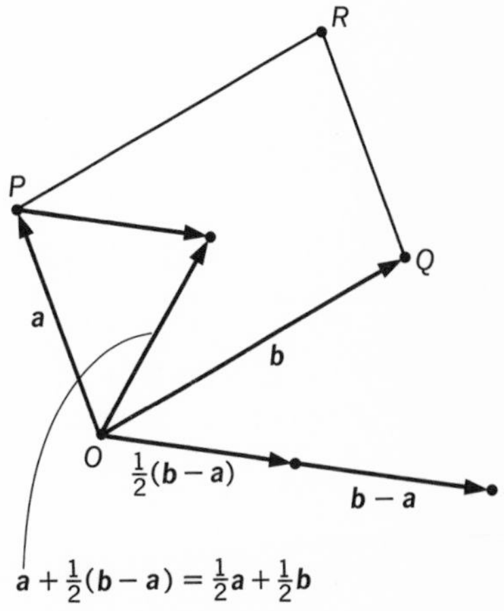

FIGURE 2-15

As another example of vector methods, let us prove that the diagonals of a parallelogram bisect each other.

Let the adjacent sides of the parallelogram be represented by the vectors $\boldsymbol{a}$ and $\boldsymbol{b}$, as in Figure 2-15. We first calculate the vector to the midpoint of PQ. We know $\boldsymbol{b} - \boldsymbol{a}$ is parallel to the directed segment from P to Q, and $\frac{1}{2}(\boldsymbol{b} - \boldsymbol{a})$ is therefore parallel to the directed line segment from P to the midpoint of PQ. Thus, the vector $\boldsymbol{a} + \frac{1}{2}(\boldsymbol{b} - \boldsymbol{a}) = \frac{1}{2}\boldsymbol{a} + \frac{1}{2}\boldsymbol{b}$, ends at the midpoint of PQ.

Next, we calculate the vector to the midpoint of OR. We know $\boldsymbol{a} + \boldsymbol{b}$ ends at R, thus $\frac{1}{2}(\boldsymbol{a} + \boldsymbol{b})$ ends at the midpoint of OR.

Since we know that the vector $\frac{1}{2}\boldsymbol{a} + \frac{1}{2}\boldsymbol{b}$ ends at both the midpoint of OR and the midpoint of PQ, we see that OR and PQ bisect each other.

EXERCISES

1. Sketch the following vectors:
   (a) $-2\boldsymbol{i} + 3\boldsymbol{j} + \boldsymbol{k}$ (b) $\boldsymbol{i} - \boldsymbol{j} + \boldsymbol{k}$
   (c) $-\boldsymbol{i} + 2\boldsymbol{j} - \boldsymbol{k}$ (d) $\boldsymbol{i} + \boldsymbol{j}$
   (e) $\boldsymbol{i} + \boldsymbol{j} + \boldsymbol{k}$
2. Consider the cube bounded by the planes

$$x = 0 \quad \text{and} \quad x = 1$$
$$y = 0 \quad \text{and} \quad y = 1$$
$$z = 0 \quad \text{and} \quad z = 1$$

   Find the vectors which end at its vertices.
3. Find in parametric form the equation of the line passing through the endpoint of $\boldsymbol{j}$ in the direction of $\boldsymbol{i} - \boldsymbol{j} - \boldsymbol{k}$. Sketch the line. At what point does the line intersect the plane $x = 0$; the plane $z = 2$?
4. Find the equation of the line which passes through $(-1, -1, 0)$ and $(1, 1, 1)$. Sketch the line. Where does this line intersect the plane $z = 2$; the plane $x = 0$?
5. Show that

$$x(t) = \alpha + \lambda t$$
$$y(t) = \beta + \mu t$$
$$z(t) = \gamma + \nu t$$

   are the equations of a straight line in space.
6. Using the Pythagorean theorem show that the length of the vector $x\boldsymbol{i} + y\boldsymbol{j} + z\boldsymbol{k}$ is $\sqrt{x^2 + y^2 + z^2}$.
7. What point on the line segment connecting $(-1, -1, 0)$ to $(1, 1, 1)$ is closest to the origin?
8. If $\boldsymbol{x}$, $\boldsymbol{y}$, and $\boldsymbol{z}$ are three noncoplanar vectors, show that $\boldsymbol{x} + (\boldsymbol{y} + \boldsymbol{z})$ is the diagonal of the parallelipiped with sides $\boldsymbol{x}$, $\boldsymbol{y}$, and $\boldsymbol{z}$. Interpret the associative law $\boldsymbol{x} + (\boldsymbol{y} + \boldsymbol{z}) = (\boldsymbol{x} + \boldsymbol{y}) + \boldsymbol{z}$ geometrically.
9. Interpret the laws

$$(\lambda + \mu)\boldsymbol{a} = \lambda\boldsymbol{a} + \mu\boldsymbol{a}$$
$$\lambda(\boldsymbol{a} + \boldsymbol{b}) = \lambda\boldsymbol{a} + \lambda\boldsymbol{b}$$
$$\lambda(\mu\boldsymbol{a}) = (\lambda\mu)\boldsymbol{a}$$

geometrically, where $\boldsymbol{a}$, $\boldsymbol{b}$ are vectors and λ, μ are scalars.

10. Show that the endpoints of the vectors $\boldsymbol{x}$, $\boldsymbol{y}$, and $\boldsymbol{z}$ are collinear if and only if there are scalars λ, μ, and ν, not all zero, such that

$$\lambda \boldsymbol{x} + \mu \boldsymbol{y} + \nu \boldsymbol{z} = 0$$
$$\lambda + \mu + \nu = 0$$

11. If $\boldsymbol{a}$ and $\boldsymbol{b}$ are noncollinear vectors, show that $\boldsymbol{a} + \boldsymbol{b}$, $-\boldsymbol{a} + \boldsymbol{b}$, $\boldsymbol{a} - \boldsymbol{b}$, and $-\boldsymbol{a} - \boldsymbol{b}$ are the vertices of a parallelogram.
12. Show that the figure formed by connecting the midpoints of adjacent sides of a parallelogram is a parallelogram.
13. Find the midpoint of the line segment connecting each of the following pairs of points. Sketch your results.
(a) $(1, 0, 0)$ and $(1, 1, 1)$
(b) $(1, 1, 1)$ and $(1, 1, -1)$
(c) $(1, 1, 2)$ and $(-1, -1, 0)$
14. What point is on the line segment connecting (x_1, y_1, z_1) to (x_2, y_2, z_2) and $1/3$ of the way from (x_1, y_1, z_1) to (x_2, y_2, z_2)?
15. If $\boldsymbol{l}_1(t)$ and $\boldsymbol{l}_2(t)$ are parametric equations of two lines, show that the lines are parallel if and only if there are constants, λ_1 and λ_2, not both zero, such that $\lambda_1 \boldsymbol{l}_1(t) + \lambda_2 \boldsymbol{l}_2(t)$ is a constant vector.
16. Prove that any two medians of a triangle intersect at a point $2/3$ of the way along either median from the vertex which it passes to the opposite side.
17. Let $\boldsymbol{a}$, $\boldsymbol{b}$, and $\boldsymbol{c}$ be three noncoplanar vectors in $\boldsymbol{R}^3$. Show that the points within the parallelepiped determined by $\boldsymbol{a}$, $\boldsymbol{b}$, and $\boldsymbol{c}$ are precisely the endpoints of vectors of the form $\alpha \boldsymbol{a} + \beta \boldsymbol{b} + \gamma \boldsymbol{c}$, where $0 \le \alpha \le 1$, $0 \le \beta \le 1$, $0 \le \gamma \le 1$. Show that a point is on one of the sides of the parallelepiped if and only if one of the numbers α, β, or γ is 0 or 1.

3 MATRICES

We define a **matrix** as a rectangular array of numbers, real or complex:

$$\begin{bmatrix} a_{11} & a_{12} & a_{13} & \cdots & a_{1n} \\ a_{21} & a_{22} & a_{23} & \cdots & a_{2n} \\ a_{31} & a_{32} & a_{33} & \cdots & a_{3n} \\ & & \vdots & & \\ a_{m1} & a_{m2} & a_{m3} & \cdots & a_{mn} \end{bmatrix}$$

The numbers in the array are the **elements** or **entries** of the matrix. The subscripts i and j of the element a_{ij} serve to identify the row and column where a_{ij} is located.

For example,

$$\begin{bmatrix} -3 & 0 & 13 \\ 2 & 1 & 2 \\ 1 & 3 & 0 \\ 0 & -1 & 2 \end{bmatrix}$$

is a matrix. The element a_{23} is located in the second row and third column and in this case is 2. The entry in position (3, 3) is 0.

A matrix which has m rows and n columns is said to be an $m \times n$ matrix, or an m by n matrix. When $m = n$, i.e., when the matrix is square, we simply say that we have a matrix of order n. For example,

$$\begin{bmatrix} -3 & 2 \\ 0 & 1 \\ 4 & 6 \end{bmatrix}$$

is a 3×2 matrix, while

$$\begin{bmatrix} 3 & 7 & -1 \\ 0 & 2 & 3 \end{bmatrix}$$

is a 2×3 matrix.

Often we abbreviate the above notation to $[a_{ij}]_{(mn)}$, i.e., the $m \times n$ matrix whose element in position (i, j) is a_{ij}.

For example, in this notation $[ij]_{(23)}$ means the 2×3 matrix which has the product ij in position (i, j), or

$$\begin{bmatrix} 1 & 2 & 3 \\ 2 & 4 & 6 \end{bmatrix}$$

Matrices arise naturally in linear equations. For example, with the system

$$3x + 2y - z = 0$$
$$-x + 6y + z = 0$$

we may associate the matrix

$$\begin{bmatrix} 3 & 2 & -1 \\ -1 & 6 & 1 \end{bmatrix}$$

We return to this connection shortly.

Two matrices are equal when and only when they are of the same order and have the same entry in each position (i, j); in other words, $[a_{ij}]_{(mn)} = [b_{ij}]_{(pq)}$ if and only if $m = p$, $n = q$, and $a_{ij} = b_{ij}$ for all i and j.

If $A = [a_{ij}]_{(mn)}$ and $B = [b_{ij}]_{(mn)}$ are two $m \times n$ matrices, we define $A + B$ to be the matrix $[a_{ij} + b_{ij}]_{(mn)}$. For example,

$$\begin{bmatrix} 3 & -1 & 2 \\ 0 & 13 & 7 \\ 8 & 10 & 12 \end{bmatrix} + \begin{bmatrix} \frac{1}{2} & -3 & 1 \\ 0 & -\frac{1}{2} & 1 \\ 2 & 0 & 0 \end{bmatrix} = \begin{bmatrix} \frac{7}{2} & -4 & 3 \\ 0 & \frac{25}{2} & 8 \\ 10 & 10 & 12 \end{bmatrix}$$

$$\begin{bmatrix} 1 & 3 \\ 0 & \frac{1}{2} \\ 2 & \frac{1}{3} \end{bmatrix} + \begin{bmatrix} 0 & 0 \\ -\frac{1}{2} & 1 \\ 2 & 7 \end{bmatrix} = \begin{bmatrix} 1 & 3 \\ -\frac{1}{2} & \frac{3}{2} \\ 4 & \frac{22}{3} \end{bmatrix}$$

We define the matrix $-A = [-a_{ij}]_{(mn)}$. For example,

$$-\begin{bmatrix} 1 & 0 & 2 \\ 0 & -1 & 3 \end{bmatrix} = \begin{bmatrix} -1 & 0 & -2 \\ 0 & 1 & -3 \end{bmatrix}$$

In other words to add matrices we add corresponding entries. To take the negative of a matrix we take the negatives of the corresponding entries.

We define the scalar multiple of the matrix A by the scalar α, written αA, by $\alpha A = [\alpha a_{ij}]_{(mn)}$. For example,

$$2\begin{bmatrix} 7 & 0 & 7 \\ -8 & 1 & -3 \end{bmatrix} = \begin{bmatrix} 14 & 0 & 14 \\ -16 & 2 & -6 \end{bmatrix}$$

Addition and scalar multiplication of matrices satisfy the same laws as addition and scalar multiplication of real numbers and vectors.

Proposition 1 $(A + B) + C = A + (B + C)$

Proof Let

$$A = [a_{ij}]_{(mn)}$$
$$B = [b_{ij}]_{(mn)}$$
$$C = [c_{ij}]_{(mn)}$$

$(A + B) + C = [(a_{ij} + b_{ij}) + c_{ij}]_{(mn)}$ while $A + (B + C) = [a_{ij} + (b_{ij} + c_{ij})]_{(mn)}$. But $(a_{ij} + b_{ij}) + c_{ij} = a_{ij} + (b_{ij} + c_{ij})$ for all real numbers. Therefore, $(A + B) + C = A + (B + C)$. ∎

The proofs of all other laws are similar.

Proposition 2 $A + B = B + A$

Proposition 3 $A + (-A) = -A + A = 0$, where 0 denotes the matrix of the same order as A, all of whose entries are 0.

Proposition 4 $A + 0 = 0 + A = A$

Proposition 5

$$(\alpha + \beta)A = \alpha A + \beta A$$
$$\alpha(A + B) = \alpha A + \alpha B$$
$$\alpha(\beta A) = (\alpha\beta)A$$
$$1A = A$$

An $n \times 1$ matrix is, according to the definition given in §2.1, just a column vector. A $1 \times n$ matrix is a row vector.

EXERCISES

1. Write out in the form of an array the indicated matrices.
   (a) $[ij]_{(34)}$ (b) $[i^2 + j^2]_{(33)}$
   (c) $[i + j]_{(22)}$ (d) $[i + j]_{(44)}$
2. Calculate
   (a) $\begin{bmatrix} 1 & -2 & -3 \\ 0 & 1 & 7 \end{bmatrix} + \begin{bmatrix} 8 & 7 & 6 \\ 0 & 2 & 4 \end{bmatrix} - \begin{bmatrix} 6 & 1 & 7 \\ 0 & 2 & 8 \end{bmatrix}$

   (b) $\begin{bmatrix} 3 & 7 \\ 2 & -2 \end{bmatrix} + \begin{bmatrix} 8 & 6 \\ 1 & 0 \end{bmatrix} - \begin{bmatrix} 4 & 6 \\ 0 & 1 \end{bmatrix}$

   (c) $\begin{bmatrix} 1 & 8 & 3 \\ 2 & 4 & 6 \\ 0 & 1 & 7 \end{bmatrix} + \begin{bmatrix} -1 & 0 & 13 \\ -2 & -1 & 4 \\ -3 & 6 & 8 \end{bmatrix}$

   (d) $\begin{bmatrix} 1 & 7 & 6 & -4 \\ 0 & 1 & 2 & 3 \end{bmatrix} + \begin{bmatrix} 8 & 2 & 1 & 7 \\ 0 & 2 & 1 & 8 \end{bmatrix}$
3. Prove the laws of matrix addition and scalar multiplication which were not proven in the text.
4. Solve the equation

$$(X + A) + (B - 2C) = 3B + C$$

for X in terms of A, B, and C, where A, B, and C are $m \times n$ matrices.

5. If A and B are $m \times n$ matrices, solve the equations

$$X + Y = A$$
$$X - Y = B$$

for X and Y.

6. Find 2×2 matrices, X, Y, and Z, not all zero, which satisfy the equations

$$X + Y + Z = 0$$
$$X + 2Y + 3Z = 0$$

7. Given a 2×2 matrix A, show there are numbers α, β, γ, δ, such that

$$A = \alpha\begin{bmatrix}1 & 0\\ 0 & 0\end{bmatrix} + \beta\begin{bmatrix}0 & 1\\ 0 & 0\end{bmatrix} + \gamma\begin{bmatrix}0 & 0\\ 1 & 0\end{bmatrix} + \delta\begin{bmatrix}0 & 0\\ 0 & 1\end{bmatrix}$$

8. How many 3×2 matrices are there all of whose entries are either 0 or 1.

4 MATRIX MULTIPLICATION

In addition to the operations of addition and scalar multiplication defined on matrices there is a third algebraic operation, called matrix multiplication, which is often encountered.

Let $A = [a_{ij}]_{(mn)}$ be an $m \times n$ matrix and $B = [b_{ij}]_{(np)}$ be an $n \times p$ matrix. We define an $m \times p$ matrix AB, called the product of A and B, by

$$AB = \left[\sum_{j=1}^{n} a_{ij}b_{jk}\right]_{(mp)}$$

In the 2×2 case this definition becomes more explicitly

$$\begin{bmatrix}a_{11} & a_{12}\\ a_{21} & a_{22}\end{bmatrix}\begin{bmatrix}b_{11} & b_{12}\\ b_{21} & b_{22}\end{bmatrix} = \begin{bmatrix}a_{11}b_{11} + a_{12}b_{21} & a_{11}b_{12} + a_{12}b_{22}\\ a_{21}b_{11} + a_{22}b_{21} & a_{21}b_{12} + a_{22}b_{22}\end{bmatrix}$$

Notice we go across the ith row of the first matrix and down the kth column of the second matrix to obtain the entry in position (i, k). (See Figure 4-1.)

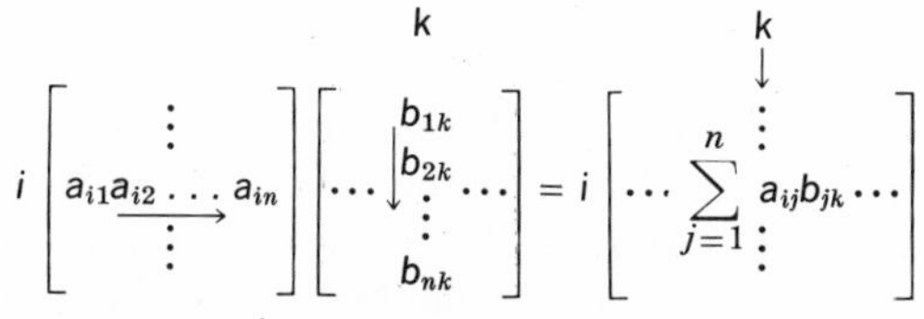

FIGURE 4-1

For example,

$$\begin{bmatrix} 3 & 1 & -2 \\ 6 & 3 & 4 \end{bmatrix} \begin{bmatrix} 4 & 7 \\ 3 & 0 \\ 1 & 2 \end{bmatrix} = \begin{bmatrix} 3\cdot 4 + 1\cdot 3 + (-2)\cdot 1 & 7\cdot 3 + 1\cdot 0 + (-2)(2) \\ 6\cdot 4 + 3\cdot 3 + 4\cdot 1 & 6\cdot 7 + 3\cdot 0 + 4\cdot 2 \end{bmatrix}$$

$$= \begin{bmatrix} 13 & 17 \\ 37 & 50 \end{bmatrix}$$

$$\begin{bmatrix} 3 & -1 \\ 0 & 1 \end{bmatrix} \begin{bmatrix} 4 & 2 \\ 6 & 1 \end{bmatrix} = \begin{bmatrix} 3\cdot 4 + (-1)\cdot 6 & 3\cdot 2 + (-1)(1) \\ 0\cdot 4 + (1)\cdot 6 & 0\cdot 2 + (1)(1) \end{bmatrix}$$

$$= \begin{bmatrix} 6 & 5 \\ 6 & 1 \end{bmatrix}$$

$$\begin{bmatrix} 1 & 0 & 7 \\ 0 & 3 & 1 \\ 2 & 4 & 6 \end{bmatrix} \begin{bmatrix} -1 & 3 & 7 \\ -2 & 1 & 6 \\ 0 & 2 & 8 \end{bmatrix}$$

$$= \begin{bmatrix} 1\cdot(-1) + 0\cdot(-2) + 7\cdot 0 & 1\cdot 3 + 0\cdot 1 + 7\cdot 2 & 1\cdot 7 + 0\cdot 6 + 7\cdot 8 \\ 0\cdot(-1) + 3\cdot(-2) + (1)\cdot 0 & 0\cdot 3 + 3\cdot 1 + 1\cdot 2 & 0\cdot 7 + 3\cdot 6 + 1\cdot 8 \\ 2\cdot(-1) + 4\cdot(-2) + 6\cdot 0 & 2\cdot 3 + 4\cdot 1 + 6\cdot 2 & 2\cdot 7 + 4\cdot 6 + 6\cdot 8 \end{bmatrix}$$

$$= \begin{bmatrix} -1 & 17 & 63 \\ -6 & 5 & 26 \\ -10 & 22 & 86 \end{bmatrix}$$

$$\begin{bmatrix} 4 \\ -1 \\ 6 \end{bmatrix} [2 \quad 3] = \begin{bmatrix} 4\cdot 2 & 4\cdot 3 \\ (-1)\cdot 2 & (-1)\cdot 3 \\ 6\cdot 2 & 6\cdot 3 \end{bmatrix} = \begin{bmatrix} 8 & 12 \\ -2 & -3 \\ 12 & 18 \end{bmatrix}$$

Note that in order to have the product AB defined the number of columns in A must equal the number of rows in B. Note also that the product AB has as many rows as A and as many columns as B.

There are several rules connected with the multiplication of matrices which we now state.

Proposition 1 If A is an $m \times n$ matrix, B is an $n \times p$ matrix, and C is a $p \times q$ matrix, then $(AB)C = A(BC)$. (Associative law of matrix multiplication.)

Proof Let

$$A = [a_{ij}]_{(mn)}$$
$$B = [b_{jk}]_{(np)}$$
$$C = [c_{kl}]_{(pq)}$$

Then,

$$AB = \left[\sum_{j=1}^{n} a_{ij}b_{jk}\right]_{(mp)}$$
$$(AB)(C) = \left[\sum_{k=1}^{p}\left(\sum_{j=1}^{n} a_{ij}b_{jk}\right)c_{kl}\right]_{(mq)}$$
$$= \left[\sum_{k=1}^{p}\sum_{j=1}^{n} a_{ij}b_{jk}c_{kl}\right]_{(mq)}$$
$$BC = \left[\sum_{k=1}^{p} b_{jk}c_{kl}\right]_{(nq)}$$
$$A(BC) = \left[\sum_{j=1}^{n} a_{ij}\left(\sum_{k=1}^{p} b_{jk}c_{kl}\right)\right]$$
$$= \left[\sum_{j=1}^{n}\sum_{k=1}^{p} a_{ij}b_{jk}c_{kl}\right]_{(mq)}$$

Since $\Sigma_{k=1}^{p}\,\Sigma_{j=1}^{n}\,a_{ij}b_{jk}c_{kl} = \Sigma_{j=1}^{n}\,\Sigma_{k=1}^{p}\,a_{ij}b_{jk}c_{kl}$, we have $(AB)C = A(BC)$. ∎

We can verify the distribution laws for matrix multiplication.

Proposition 2

$$A(B + C) = AB + AC$$
$$(A + B)C = AC + BC$$

Proof Let

$$A = [a_{ij}]_{(mn)}$$
$$B = [b_{jk}]_{(np)}$$
$$C = [c_{jk}]_{(np)}$$

Then, $B + C = [b_{jk} + c_{jk}]_{(np)}$.

$$A(B + C) = \left[\sum_{j=1}^{n} a_{ij}(b_{jk} + c_{jk})\right]_{(mp)}$$
$$AB = \left[\sum_{j=1}^{n} a_{ij}b_{jk}\right]_{(mp)}$$
$$AC = \left[\sum_{j=1}^{n} a_{ij}c_{jk}\right]_{(mp)}$$

$$AB + AC = \left[\sum_{j=1}^{n} a_{ij}b_{jk} + \sum_{j=1}^{n} a_{ij}c_{jk}\right]_{(mp)}$$

Since $\Sigma_{j=1}^{n} a_{ij}(b_{jk} + c_{jk}) = \Sigma_{j=1}^{n} a_{ij}b_{jk} + \Sigma_{j=1}^{n} a_{ij}c_{jk}$ we have $A(B + C) = AB + AC$. We leave the proof of the other distributive law as an exercise for the reader. ▨

Although in these respects the multiplication of matrices behaves like the multiplication of scalar quantities, there are many ways in which the two differ.

For instance it is **not** in general true that $AB = BA$. This is easy to see if we suppose that A is an $m \times n$ matrix and B an $n \times p$ matrix. In order to have BA defined we must have $m = p$. Thus, AB may be defined, but BA may not be defined. Moreover, even if both A and B are square of order n, in which case both AB and BA are defined, we still need not have $AB = BA$.

For example,

$$\begin{bmatrix} 1 & 2 \\ 0 & 3 \end{bmatrix} \begin{bmatrix} -1 & 2 \\ 1 & 3 \end{bmatrix} = \begin{bmatrix} 1 & 8 \\ 3 & 9 \end{bmatrix}$$

$$\begin{bmatrix} -1 & 2 \\ 1 & 3 \end{bmatrix} \begin{bmatrix} 1 & 2 \\ 0 & 3 \end{bmatrix} = \begin{bmatrix} -1 & 4 \\ 1 & 11 \end{bmatrix}$$

In mathematical terms we say that the multiplication of matrices is in general noncommutative. If it does happen that both AB and BA are defined and $AB = BA$, we say A and B commute.

There is another striking way in which matrix multiplication differs from scalar multiplication: we may have $AB = AC$ and $A \neq 0$, without having $B = C$. We may even have $AB = 0$ with $A \neq 0$ and $B \neq 0$, as for example,

$$\begin{bmatrix} 1 & 0 \\ 0 & 0 \end{bmatrix} \begin{bmatrix} 0 & 0 \\ 0 & 1 \end{bmatrix} = \begin{bmatrix} 0 & 0 \\ 0 & 0 \end{bmatrix}$$

EXERCISES

1. Calculate the following matrix products:

(a) $\begin{bmatrix} 1 & 3 \\ 2 & 7 \end{bmatrix} \begin{bmatrix} 6 & -1 \\ -1 & 7 \end{bmatrix}$

(b) $\begin{bmatrix} 1 \\ 7 \\ 3 \end{bmatrix} \begin{bmatrix} 6 & 9 & 3 \end{bmatrix}$

(c) $\begin{bmatrix} 6 & 9 & 3 \end{bmatrix}\begin{bmatrix} 1 \\ 7 \\ 3 \end{bmatrix}$

(d) $\begin{bmatrix} 1 & -7 & 3 \\ 6 & -9 & 2 \\ 3 & 4 & 1 \end{bmatrix}\begin{bmatrix} 6 \\ 2 \\ 1 \end{bmatrix}$

(e) $\begin{bmatrix} 3 & 6 & 2 \\ 4 & 3 & 1 \\ -1 & 4 & 0 \end{bmatrix}\begin{bmatrix} 1 & 0 \\ 3 & 2 \\ 7 & 1 \end{bmatrix}$

(f) $\begin{bmatrix} 0 & 2 & 3 \\ -1 & 7 & 6 \\ 0 & 1 & 3 \end{bmatrix}\begin{bmatrix} 0 & 2 & 0 \\ 1 & -1 & 7 \\ 0 & 1 & 2 \end{bmatrix}$

(g) $\begin{bmatrix} 1 & a \\ 0 & 1 \end{bmatrix}\begin{bmatrix} 1 & b \\ 0 & 1 \end{bmatrix}$

(h) $\begin{bmatrix} 1 & b \\ 0 & 1 \end{bmatrix}\begin{bmatrix} 1 & a \\ 0 & 1 \end{bmatrix}$

(i) $\begin{bmatrix} 1 & 3 \\ -4 & 2 \\ 6 & 7 \\ 0 & 1 \end{bmatrix}\begin{bmatrix} 1 & -1 \\ 3 & 0 \end{bmatrix}$

(j) $\begin{bmatrix} -1 & 2 & 0 \\ 1 & 3 & 7 \\ 1 & 6 & 2 \end{bmatrix}\begin{bmatrix} -1 \\ 0 \\ 0 \end{bmatrix}$

2. Let

$$A = \begin{bmatrix} 3 & 4 \\ 5 & 7 \end{bmatrix}, \quad B = \begin{bmatrix} 2 & -1 \\ -2 & 1 \end{bmatrix}, \quad C = \begin{bmatrix} 1 & 1 \\ 0 & 1 \end{bmatrix}$$

$$I = \begin{bmatrix} 1 & 0 \\ 0 & 1 \end{bmatrix}, \quad D = \begin{bmatrix} 1 & 1 \\ 1 & 1 \end{bmatrix}$$

Calculate

(a) AB	(b) $(AB)C$	(c) BC	(d) $A(BC)$
(e) BA	(f) $AB - BA$	(g) IA	(h) AI
(i) IB	(j) BI	(k) BD	(l) DB
(m) CC	(n) $(CC)C$		

3. Find

(a) $\left(\begin{bmatrix} 1 & -1 & 2 \end{bmatrix}\begin{bmatrix} -3 & 0 & 3 \\ 4 & 1 & 7 \\ 0 & 2 & 1 \end{bmatrix}\right)\begin{bmatrix} 1 \\ 0 \\ 2 \end{bmatrix}$

(b) $\left(\begin{bmatrix} 1 & 0 & 3 \\ -3 & 1 & 2 \\ 0 & 5 & 7 \end{bmatrix}\begin{bmatrix} 0 & 3 & 0 \\ -1 & 1 & 1 \\ 0 & 1 & 2 \end{bmatrix}\right)\begin{bmatrix} 1 & 0 & 3 \\ 0 & 1 & 0 \\ 7 & 0 & 0 \end{bmatrix}$

4. If AB and BA are both defined and $AB = BA$, show that A and B are square matrices of the same order.

5. Show $B \cdot 0 = 0 \cdot B = 0$.

6. Let A be a 2×3 matrix, C be a 3×2 matrix, and suppose B is a matrix such that the product $(AB)C$ is defined. What is the order of B?

7. Prove the distributive law $(A + B)C = AC + BC$.

8. Prove that $(\alpha A)B = \alpha(AB) = A(\alpha B)$.

9. Prove that $(\alpha A)(\beta B) = (\alpha\beta)(AB)$.

10. Find a, b, c, and d if

$$\begin{bmatrix} a & b \\ c & d \end{bmatrix}\begin{bmatrix} 1 & 1 \\ 0 & 1 \end{bmatrix} = \begin{bmatrix} 1 & -1 \\ 0 & 1 \end{bmatrix}$$

11. If B is a 2×2 matrix such that

$$\begin{bmatrix} 1 & -1 \\ 3 & -3 \end{bmatrix} B = 0$$

show that

$$B = \begin{bmatrix} a & b \\ a & b \end{bmatrix}$$

for some numbers a and b. If B is any matrix of this form show that

$$\begin{bmatrix} 1 & -1 \\ 3 & -3 \end{bmatrix} B = 0$$

12. Show that a 2×2 matrix commutes with

$$\begin{bmatrix} 1 & 3 \\ 0 & 1 \end{bmatrix}$$

if and only if it is of the form

$$\begin{bmatrix} a & b \\ 0 & a \end{bmatrix}$$

for some a and b.

13. Let

$$A = \begin{bmatrix} 1 & -1 \\ -3 & 3 \end{bmatrix}$$

(a) Find a column vector $\boldsymbol{x}$, $\boldsymbol{x} \neq 0$ such that $A\boldsymbol{x} = \mathbf{0}$.
(b) Find a row vector $\boldsymbol{y}$, $\boldsymbol{y} \neq 0$, such that $\boldsymbol{y}A = \mathbf{0}$.

14. Let

$$\mathcal{C} = \left\{ \begin{bmatrix} a & b \\ -b & a \end{bmatrix} \quad a \text{ and } b \text{ real numbers} \right\}$$

(a) If $A \in \mathcal{C}$ and $B \in \mathcal{C}$, show that $A + B \in \mathcal{C}$.
(b) If $A \in \mathcal{C}$ and $B \in \mathcal{C}$, show that $AB \in \mathcal{C}$.
(c) If $A \in \mathcal{C}$ and $B \in \mathcal{C}$, show that $AB = BA$.

15. Let A, B, and C be $n \times n$ matrices. If A commutes with C and B commutes with C, show that AB commutes with C.

16. Let A and B be 2×2 matrices. If A and B commute with

$$\begin{bmatrix} 0 & 1 \\ -1 & 0 \end{bmatrix}$$

show that A commutes with B.

17. Let A be an $m \times n$ matrix and let

$$e_i = \begin{bmatrix} 0 \\ 0 \\ \vdots \\ 0 \\ 1 \\ 0 \\ \vdots \\ 0 \end{bmatrix} \left.\vphantom{\begin{bmatrix} 0 \\ 0 \\ \vdots \\ 0 \\ 1 \end{bmatrix}}\right\} i$$

denote the n-dimensional vector all of whose components but the ith are zero and whose ith component is 1. Show that Ae_i is the ith column of the matrix A.

18. Let A be an $m \times n$ matrix. If $A\boldsymbol{x} = \boldsymbol{0}$ for all vectors $\boldsymbol{x}$, show that $A = 0$. [Hint: See exercise 17.]

19. Find a 2×3 matrix A, and a 3×2 matrix B, such that

$$AB = \begin{bmatrix} 1 & 0 \\ 0 & 1 \end{bmatrix}$$

20. If A and B are commuting square matrices, show that
(a) $(A + B)^2 = A^2 + 2AB + B^2$
(b) $(A - B)(A + B) = A^2 - B^2$

5 SQUARE MATRICES

The square matrices of a fixed order have properties which make them particularly interesting objects of study. For example, if A and B are square of the same order, then AB and BA are both defined.

We define the **Kronecker delta** δ_{ij} as follows

$$\delta_{ij} = 1 \qquad \text{if } i = j$$
$$\delta_{ij} = 0 \qquad \text{if } i \neq j$$

For example, $\delta_{23} = 0$, $\delta_{11} = 1$, $\delta_{33} = 1$, $\delta_{45} = 0$.

Using this symbol we define the $n \times n$ matrix $I_n = [\delta_{ij}]_{(nn)}$. This matrix has only 1's on its diagonal entries, and 0's off the diagonal. For example,

$$I_2 = \begin{bmatrix} 1 & 0 \\ 0 & 1 \end{bmatrix}, \qquad I_3 = \begin{bmatrix} 1 & 0 & 0 \\ 0 & 1 & 0 \\ 0 & 0 & 1 \end{bmatrix}$$

I_n is called the **identity matrix of order n.**

Theorem 1 If A is an $n \times n$ matrix, $AI_n = I_nA = A$.

Proof Let $A = [a_{ij}]_{(nn)}$.

$$\text{Then, } AI_n = \left[\sum_{j=1}^{n} a_{ij}\delta_{jk}\right]_{(nn)}$$
$$= [a_{ij}]_{(nn)}$$

since in the sum $\Sigma_{j=1}^{n} a_{ij}\delta_{jk}$, $a_{ij}\delta_{jk} = 0$ unless $j = k$, and then $a_{ij}\delta_{kk} = a_{ij}$. Thus, $AI_n = A$. In a similar manner, $I_nA = A$. ∎

If α is a scalar, a matrix of the form αI_n is called a **scalar matrix.** It may be described as a matrix whose diagonal entries are all equal to α, and whose off diagonal entries are 0.

A concept of great importance when dealing with matrices is that of the inverse of a matrix. If A is an $m \times m$ matrix and there is an $m \times m$ matrix B, such that $AB = BA = I_m$, A is said to be **invertible** and B is said to be the **inverse** of A.

Theorem 2 Let A be an $m \times m$ matrix with inverse B. If C is another matrix such that $AC = CA = I_m$, then $C = B$.

Proof

$$\begin{aligned} C = CI_m &= C(AB) \\ &= (CA)B \\ &= I_mB \\ &= B \end{aligned}$$

∎

Thus, if a matrix has an inverse, it has only one inverse. It is customary to denote the inverse of a matrix, if it exists, by A^{-1}, and to say that A is invertible. For example,

$$\begin{bmatrix} 7 & 4 \\ 5 & 3 \end{bmatrix} \begin{bmatrix} 3 & -4 \\ -5 & 7 \end{bmatrix} = \begin{bmatrix} 1 & 0 \\ 0 & 1 \end{bmatrix}$$

$$\begin{bmatrix} 3 & -4 \\ -5 & 7 \end{bmatrix} \begin{bmatrix} 7 & 4 \\ 5 & 3 \end{bmatrix} = \begin{bmatrix} 1 & 0 \\ 0 & 1 \end{bmatrix}$$

so

$$\begin{bmatrix} 7 & 4 \\ 5 & 3 \end{bmatrix}^{-1} = \begin{bmatrix} 3 & -4 \\ -5 & 7 \end{bmatrix}$$

Another example is

$$\begin{bmatrix} 1 & 0 & -1 \\ 0 & 1 & 0 \\ 0 & 0 & 1 \end{bmatrix} \begin{bmatrix} 1 & 0 & 1 \\ 0 & 1 & 0 \\ 0 & 0 & 1 \end{bmatrix} = \begin{bmatrix} 1 & 0 & 1 \\ 0 & 1 & 0 \\ 0 & 0 & 1 \end{bmatrix} \begin{bmatrix} 1 & 0 & -1 \\ 0 & 1 & 0 \\ 0 & 0 & 1 \end{bmatrix} = \begin{bmatrix} 1 & 0 & 0 \\ 0 & 1 & 0 \\ 0 & 0 & 1 \end{bmatrix}$$

Thus,

$$\begin{bmatrix} 1 & 0 & -1 \\ 0 & 1 & 0 \\ 0 & 0 & 1 \end{bmatrix}^{-1} = \begin{bmatrix} 1 & 0 & 1 \\ 0 & 1 & 0 \\ 0 & 0 & 1 \end{bmatrix}$$

In our chapter on determinants we shall determine a necessary and sufficient condition that a matrix have an inverse. At this point we content ourselves with merely pointing out examples of matrices which do not have inverses.

Example 1 Consider the matrix

$$A = \begin{bmatrix} 0 & 1 \\ 0 & 1 \end{bmatrix}$$

and calculate its product with some other matrix, say B, where

$$B = \begin{bmatrix} \beta_{11} & \beta_{12} \\ \beta_{21} & \beta_{22} \end{bmatrix}$$

We have

$$BA = B\begin{bmatrix} 0 & 1 \\ 0 & 1 \end{bmatrix} = \begin{bmatrix} \beta_{11} & \beta_{12} \\ \beta_{21} & \beta_{22} \end{bmatrix}\begin{bmatrix} 0 & 1 \\ 0 & 1 \end{bmatrix}$$

$$= \begin{bmatrix} 0 & \beta_{11} + \beta_{12} \\ 0 & \beta_{21} + \beta_{22} \end{bmatrix}$$

Since the product of A with any other 2×2 matrix has a 0 in entry (1, 1), it is impossible to choose a matrix B, such that

$$\begin{bmatrix} 0 & 1 \\ 0 & 1 \end{bmatrix} B = I_2$$

and so A does not have an inverse.

We can obtain many other examples of matrices without inverses using the following lemma.

Lemma Let A be an $n \times n$ matrix. If there is a nonzero vector $\boldsymbol{x}$, such that $A\boldsymbol{x} = \boldsymbol{0}$, then A does not have an inverse.

Proof We will show that if $A\boldsymbol{x} = \boldsymbol{0}$ and A is invertible, then $\boldsymbol{x} = \boldsymbol{0}$, proving the lemma. In the equality $A\boldsymbol{x} = \boldsymbol{0}$, multiply both sides by A^{-1}. Then, $A^{-1}(A\boldsymbol{x}) = A^{-1} \cdot \boldsymbol{0} = \boldsymbol{0}$. So $(A^{-1}A)\boldsymbol{x} = I\boldsymbol{x} = \boldsymbol{x} = \boldsymbol{0}$. ∎

Example 2 Consider the matrix $A = \begin{bmatrix} 1 & -1 \\ 2 & -2 \end{bmatrix}$ and the vector $\boldsymbol{x} = \begin{bmatrix} 1 \\ 1 \end{bmatrix}$

Since

$$\begin{bmatrix} 1 & -1 \\ 2 & -2 \end{bmatrix}\begin{bmatrix} 1 \\ 1 \end{bmatrix} = \begin{bmatrix} 0 \\ 0 \end{bmatrix}$$

we can conclude, by the lemma, that A does not have an inverse.

The matrix

$$A = \begin{bmatrix} 1 & -3 & 2 \\ 2 & -6 & 4 \\ 0 & 1 & 0 \end{bmatrix}$$

is not invertible, since we can find a nonzero vector $\boldsymbol{x}$ satisfying $A\boldsymbol{x} = \boldsymbol{0}$, i.e.,

and $f(x) = 1 + x + 2x^2$, then

$$f(A) = \begin{bmatrix} 1 & 0 \\ 0 & 1 \end{bmatrix} + \begin{bmatrix} 1 & 1 \\ 0 & 1 \end{bmatrix} + 2\begin{bmatrix} 1 & 2 \\ 0 & 1 \end{bmatrix}$$

$$= \begin{bmatrix} 4 & 5 \\ 0 & 4 \end{bmatrix}$$

If $f(x) = x^2 - 2x + 1$, then

$$f(A) = \begin{bmatrix} 1 & 2 \\ 0 & 1 \end{bmatrix} - 2\begin{bmatrix} 1 & 1 \\ 0 & 1 \end{bmatrix} + \begin{bmatrix} 1 & 0 \\ 0 & 1 \end{bmatrix} = \begin{bmatrix} 0 & 0 \\ 0 & 0 \end{bmatrix}$$

A square matrix is said to be a **diagonal** matrix if all the entries off its diagonal vanish. For example,

$$\begin{bmatrix} -1 & 0 \\ 0 & 1 \end{bmatrix}, \quad \begin{bmatrix} 1 & 0 & 0 \\ 0 & 5 & 0 \\ 0 & 0 & 7 \end{bmatrix}, \quad \begin{bmatrix} -7 & 0 & 0 & 0 \\ 0 & 6 & 0 & 0 \\ 0 & 0 & 1 & 0 \\ 0 & 0 & 0 & 0 \end{bmatrix}$$

are all diagonal matrices.

EXERCISES

1. If A is an $m \times n$ matrix, show that $AI_n = A$ and $I_mA = A$.
2. Verify each of the following:

(a) $\begin{bmatrix} 6 & 7 \\ 5 & 6 \end{bmatrix}^{-1} = \begin{bmatrix} 6 & -7 \\ -5 & 6 \end{bmatrix}$

(b) $\begin{bmatrix} 7 & 4 \\ 3 & 2 \end{bmatrix}^{-1} = \frac{1}{2}\begin{bmatrix} 2 & -4 \\ -3 & 7 \end{bmatrix}$

(c) $\begin{bmatrix} 3 & 6 & 1 \\ 0 & 1 & 8 \\ 2 & 4 & 1 \end{bmatrix}^{-1} = \begin{bmatrix} -31 & -2 & 47 \\ 16 & 1 & -24 \\ -2 & 0 & 3 \end{bmatrix}$

(d) $\begin{bmatrix} 2 & 3 & 4 \\ 4 & 3 & 1 \\ 1 & 2 & 4 \end{bmatrix}^{-1} = \frac{1}{5}\begin{bmatrix} -10 & 4 & 9 \\ 15 & -4 & -14 \\ -5 & 1 & 6 \end{bmatrix}$

$$\begin{bmatrix} 1 & -3 & 2 \\ 2 & -6 & 4 \\ 0 & 1 & 0 \end{bmatrix} \begin{bmatrix} -2 \\ 0 \\ 1 \end{bmatrix} = \begin{bmatrix} 0 \\ \\ \end{bmatrix}$$

Matrices which do not have inverses are sa
or **singular.**

In later sections we shall see that the inverse
exists, is an invaluable tool in solving systems of linear

We define powers of square matrices in the usual

$$A^1 = A$$
$$A^2 = AA$$
$$A^3 = AA^2$$
$$\vdots$$
$$A^n = AA^{n-1}$$

Because of the associative law A^n is just A multiplied together n
regardless of the order in which the multiplications are performed.

Thus, if

$$A = \begin{bmatrix} 1 & 1 \\ 0 & 1 \end{bmatrix}$$

$$A^2 = \begin{bmatrix} 1 & 1 \\ 0 & 1 \end{bmatrix} \begin{bmatrix} 1 & 1 \\ 0 & 1 \end{bmatrix} = \begin{bmatrix} 1 & 2 \\ 0 & 1 \end{bmatrix}$$

$$A^3 = \begin{bmatrix} 1 & 1 \\ 0 & 1 \end{bmatrix} \begin{bmatrix} 1 & 2 \\ 0 & 1 \end{bmatrix} = \begin{bmatrix} 1 & 3 \\ 0 & 1 \end{bmatrix}$$

An induction argument will show that

$$A^n = \begin{bmatrix} 1 & n \\ 0 & 1 \end{bmatrix}$$

We may also consider polynomials of matrices. If $f(x) = \alpha_0 + \alpha_1 x + \cdots + \alpha_n x^n$ is a polynomial, we define $f(A) = \alpha_0 I_n + \alpha_1 A + \cdots + \alpha_n A^n$. Thus for example, if

$$A = \begin{bmatrix} 1 & 1 \\ 0 & 1 \end{bmatrix}$$

(e) $$\begin{bmatrix} 1 & 0 & 0 & 0 \\ 2 & 1 & 0 & 0 \\ 4 & 2 & 1 & 0 \\ -2 & 3 & 1 & 1 \end{bmatrix}^{-1} = \begin{bmatrix} 1 & 0 & 0 & 0 \\ -2 & 1 & 0 & 0 \\ 0 & -2 & 1 & 0 \\ 8 & -1 & -1 & 1 \end{bmatrix}$$

(f) $$\begin{bmatrix} 1 & a & b \\ 0 & 1 & c \\ 0 & 0 & 1 \end{bmatrix}^{-1} = \begin{bmatrix} 1 & -a & ac - b \\ 0 & 1 & -c \\ 0 & 0 & 1 \end{bmatrix}$$

(g) $$\begin{bmatrix} 0 & 0 & 0 & 1 \\ 1 & 0 & 0 & 0 \\ 0 & 1 & 0 & 0 \\ 0 & 0 & 1 & 0 \end{bmatrix}^{-1} = \begin{bmatrix} 0 & 1 & 0 & 0 \\ 0 & 0 & 1 & 0 \\ 0 & 0 & 0 & 1 \\ 1 & 0 & 0 & 0 \end{bmatrix}$$

3. Let

$$A = \begin{bmatrix} 0 & -1 \\ 1 & -1 \end{bmatrix}$$

Calculate A^2 and A^3. Show that A is invertible and $A^{-1} = A^2$.

4. Calculate $f(A)$ if

(a) $A = \begin{bmatrix} 1 & 0 \\ -1 & 1 \end{bmatrix}$ and $f(x) = 1 + 3x + x^2$

(b) $A = \begin{bmatrix} 2 & 1 \\ 1 & 3 \end{bmatrix}$ and $f(x) = x + x^2$

(c) $A = \begin{bmatrix} d_1 & 0 \\ 0 & d_2 \end{bmatrix}$ and $f(x) = \alpha_1 + \alpha_1 x + \alpha_2 x^2$

(d) $A = \begin{bmatrix} 1 & 0 & 0 \\ 0 & 0 & 1 \\ 0 & 1 & 0 \end{bmatrix}$ and $f(x) = x^3 - x^2 - x + 1$

(e) $A = \begin{bmatrix} 0 & 1 & 0 \\ 0 & 0 & 1 \\ 0 & 0 & 0 \end{bmatrix}$ and $f(x) = \alpha_0 + \alpha_1 x + \cdots + \alpha_n x^n$

5. Show by induction or otherwise that

$$\begin{bmatrix} 1 & x & y \\ 0 & 1 & x \\ 0 & 0 & 1 \end{bmatrix}^n = \begin{bmatrix} 1 & nx & ny + \dfrac{n(n-1)}{2}x^2 \\ 0 & 1 & nx \\ 0 & 0 & 1 \end{bmatrix}$$

6. If

$$A = \begin{bmatrix} 0 & 1 & 0 \\ 0 & 0 & 1 \\ 5 & 0 & 0 \end{bmatrix}$$

show that $A^3 = 5I_3$.

7. Show

$$\begin{bmatrix} -1 - 2xy & 2x + 2x^2y \\ -2y & 1 + 2xy \end{bmatrix}^2 = \begin{bmatrix} 1 & 0 \\ 0 & 1 \end{bmatrix}$$

Calculate

$$\begin{bmatrix} -1 - 2xy & 2x + 2x^2y \\ -2y & 1 + 2xy \end{bmatrix}^{-1}$$

8. Find all 3×3 diagonal matrices such that $A^2 = I_3$.

9. Define

$$P_\theta = \begin{bmatrix} \cos\theta & -\sin\theta \\ \sin\theta & \cos\theta \end{bmatrix}$$

(a) Show that $P_{\theta_1}P_{\theta_2} = P_{(\theta_1+\theta_2)}$

(b) Show that $(P_\theta)^n = P_{n\theta}$.

(c) If $\theta = 2\pi/n$, show that $(P_{2\pi/n})^n = I_2$.

10. If A is an $n \times n$ matrix, show that

(a) $A^pA^q = A^{p+q}$

(b) $(A^p)^q = A^{pq}$

11. If $AB = \lambda B$, where A is an $n \times n$ matrix and B is an $n \times p$ matrix, show that $A^mB = \lambda^mB$, for any positive integer m.

12. If

$$D = \begin{bmatrix} d_1 & 0 & 0 & \dots & 0 \\ 0 & d_2 & 0 & \dots & 0 \\ 0 & 0 & d_3 & \dots & 0 \\ & & \vdots & & \\ 0 & 0 & 0 & \dots & d_n \end{bmatrix}$$

is a diagonal matrix, and $f(x)$ is a polynomial, show that

$$f(D) = \begin{bmatrix} f(d_1) & 0 & 0 & \dots & 0 \\ 0 & f(d_2) & 0 & \dots & 0 \\ 0 & 0 & f(d_3) & \dots & 0 \\ & & \vdots & & \\ 0 & 0 & 0 & & f(d_n) \end{bmatrix}$$

13. Is it true in general that $(A + B)^2 = A^2 + 2AB + B^2$?
14. Show that $(A - B)(A + B) = A^2 - B^2$ if and only if A and B commute.
15. Suppose A is an invertible $n \times n$ matrix and $AB = 0$ for some $n \times p$ matrix B. Show that $B = 0$.
16. By finding nonzero vectors $\mathbf{x}$ such that $A\mathbf{x} = \mathbf{0}$, show that the following matrices are not invertible.

(a) $\begin{bmatrix} 1 & -2 \\ 7 & -14 \end{bmatrix}$ (b) $\begin{bmatrix} 1 & 7 & 1 \\ 0 & 2 & 1 \\ 1 & 3 & -1 \end{bmatrix}$ (c) $\begin{bmatrix} 0 & 7 & 6 & 0 \\ 9 & 0 & 4 & 1 \\ -3 & -2 & 0 & 1 \\ 6 & 5 & 10 & 2 \end{bmatrix}$

17. Show that a diagonal matrix is invertible if and only if all its diagonal entries are nonzero. What is its inverse?
18. Show that sums and products of diagonal matrices are again diagonal matrices. Show that any pair of diagonal matrices commutes.
19. If A is invertible and B is invertible, show that AB is invertible and that $(AB)^{-1} = B^{-1}A^{-1}$.
20. Show that if A is invertible, so is A^m, and $(A^m)^{-1} = (A^{-1})^m$.
21. If C is invertible, show that
(a) $C^{-1}(A + B)C = C^{-1}AC + C^{-1}BC$
(b) $C^{-1}(AB)C = (C^{-1}AC)(C^{-1}BC)$
(c) $C^{-1}(A^m)C = (C^{-1}AC)^m$
22. If $P^2 = P$ and P is an invertible $n \times n$ matrix, show that $P = I_n$.
23. Let A be an $n \times n$ matrix and

$$D = \begin{bmatrix} d_1 & 0 & \dots & 0 \\ 0 & d_2 & \dots & 0 \\ & \vdots & & \\ 0 & 0 & \dots & d_n \end{bmatrix}$$

be a diagonal matrix. Show that DA can be obtained from A by multiplying the ith row of A by d_i for $i = 1, 2, \dots, n$. Find a similar characterization for AD.
24. Consider the function $\det A$ defined on 2×2 matrices by

$$\det \begin{bmatrix} a & b \\ c & d \end{bmatrix} = ad - bc$$

(a) Show that $\det I_2 = 1$.
(b) Show that $\det AB = \det A \det B$.
(c) If A is invertible show that $\det A \neq 0$, and $\det A^{-1} = (\det A)^{-1}$.
(d) If $\det A \neq 0$, show that A is invertible and

$$A^{-1} = \frac{1}{ad - bc} \begin{bmatrix} d & -b \\ -c & a \end{bmatrix}$$

(e) Using (c) and (d) determine which of the following matrices are invertible and calculate the inverses of those which are invertible.

(1) $\begin{bmatrix} 1 & 7 \\ 0 & 1 \end{bmatrix}$ (2) $\begin{bmatrix} -1 & 0 \\ 1 & 0 \end{bmatrix}$ (3) $\begin{bmatrix} 7 & 3 \\ 9 & 4 \end{bmatrix}$

(4) $\begin{bmatrix} 2 & 3 \\ -8 & -12 \end{bmatrix}$ (5) $\begin{bmatrix} 7 & 4 \\ 10 & 6 \end{bmatrix}$ (6) $\begin{bmatrix} 7 & 4 \\ 5 & 3 \end{bmatrix}$

25. If a and b are real numbers, and $a \neq 0$ and $b \neq 0$, show that

$$\begin{bmatrix} a & b \\ -b & a \end{bmatrix}$$

is invertible and calculate its inverse.

26. A matrix N is said to be **nilpotent** if for some positive integer k, $N^k = 0$. Show that a nilpotent matrix is not invertible.

27. If N is a nilpotent $n \times n$ matrix, such that $N^k = 0$, show that $I_n - N$ is invertible and $(I_n - N)^{-1} = I_n + N + N^2 + \cdots + N^{k-1}$.

28. Using exercise 27 calculate the inverse of

(a) $\begin{bmatrix} 1 & \alpha \\ 0 & 1 \end{bmatrix}$ (b) $\begin{bmatrix} 1 & 1 & 0 \\ 0 & 1 & 0 \\ 0 & 0 & 1 \end{bmatrix}$

(c) $\begin{bmatrix} 1 & 1 & 0 \\ 0 & 1 & 1 \\ 0 & 0 & 1 \end{bmatrix}$ (d) $\begin{bmatrix} 1 & 0 & 1 & 1 \\ 0 & 1 & 0 & 1 \\ 0 & 0 & 1 & 0 \\ 0 & 0 & 0 & 1 \end{bmatrix}$

29. Let E_{ij} denote the $n \times n$ matrix which has zeroes in all entries except entry (i, j), and 1 in entry (i, j).
(a) Show that $E_{ij}E_{kl} = \delta_{jk}E_{il}$.
(b) Show that $E_{ij}^2 = 0$ if $i \neq j$ and that $E_{ij}^2 = E_{ij}$ if $i = j$.
(c) If $A = [a_{ij}]_{(nn)}$, show that $E_{ij}A$ has all entries zero except in the ith row which is the same as the jth row of A.
(d) Show that AE_{ij} has all entries zero except in the ith column which is identical to the jth column of A. [Hint: If you have trouble, write out things explicitly in the 3×3 case.]

30. Show that an $n \times n$ matrix which commutes with all $n \times n$ matrices must be a scalar multiple of the identity matrix. In addition, show that any scalar multiple of the identity matrix commutes with all $n \times n$ matrices. [Hint: Use exercise 29.]

31. If E_{ij} is as defined in exercise 29 and $i \neq j$, show that $I_n + E_{ij}$ is invertible. [Hint: Use exercise 27.]
32. Let A and B be $n \times n$ matrices and let $C_1 = \alpha_1 A + \beta_1 B$ and $C_2 = \alpha_2 A + \beta_2 B$. Suppose $\alpha_1\beta_2 - \alpha_2\beta_1 \neq 0$. Show that C_1 and C_2 commute if and only if A and B commute.
33. If A and B are square matrices and A is invertible, show that

$$(A + B)A^{-1}(A - B) = (A - B)A^{-1}(A + B).$$

34. If A and B are commuting square matrices, show that A^m and B^n commute, where m and n are positive integers.
35. If A and B are invertible square matrices, show that the following are equivalent.
    (a) A commutes with B.
    (b) A commutes with B^{-1}.
    (c) A^{-1} commutes with B^{-1}.
36. Let A, B, and C be square matrices of order n. Suppose $A = B + C$, $BC = CB$, and $C^2 = 0$. Show that $A^{k+1} = B^k(B + (k + 1)C)$.
37. Find all 3×3 diagonal matrices D which satisfy the equation $f(D) = 0$, with

$$f(x) = x(x - 1)(x - 2) = x^3 - 3x^2 + 2x.$$

38. Show that A and B commute if and only if $A - \lambda I_n$ and $B - \lambda I_n$ commute for some scalar λ.
39. Let A be an $n \times n$ matrix. Let $A_\lambda = (\lambda I_n - A)^{-1}$, wherever $\lambda I_n - A$ is invertible. Show that $(\lambda - \mu)A_\lambda A_\mu = A_\mu - A_\lambda$ and that $A_\lambda A_\mu = A_\mu A_\lambda$.

6 LINEAR EQUATIONS IN MATRIX NOTATION

Suppose we have a system of linear equations

$$\begin{aligned}
a_{11}x_1 + a_{12}x_2 + a_{13}x_3 + \cdots + a_{1n}x_n &= y_1 \\
a_{21}x_1 + a_{22}x_2 + a_{23}x_3 + \cdots + a_{2n}x_n &= y_2 \\
a_{31}x_1 + a_{32}x_2 + a_{33}x_3 + \cdots + a_{3n}x_n &= y_3 \\
&\vdots \\
a_{m1}x_1 + a_{m2}x_2 + a_{m3}x_3 + \cdots + a_{mn}x_n &= y_m
\end{aligned}$$

We may express this system of equations in a more compact form using matrix notation. We define the matrix $A = [a_{ij}]_{(mn)}$ as the matrix

of coefficients, the vector $\boldsymbol{x} = [x_j]_{(n1)}$ as the matrix of variables, and the vector $\boldsymbol{y} = [y_i]_{(m1)}$. In matrix notation the system becomes $A\boldsymbol{x} = \boldsymbol{y}$.

For example, the system

$$\begin{aligned} 3x - 3y + 4z &= 2 \\ x + 2y - z &= 0 \\ x + 3y - z &= 0 \end{aligned}$$

is represented as

$$\begin{bmatrix} 3 & -3 & 4 \\ 1 & 2 & -1 \\ 1 & 3 & -1 \end{bmatrix} \begin{bmatrix} x \\ y \\ z \end{bmatrix} \begin{matrix} = \\ = \\ = \end{matrix} \begin{bmatrix} 2 \\ 0 \\ 0 \end{bmatrix}$$

The system

$$\begin{aligned} 3x_1 - 2x_2 + 4x_3 - x_4 &= 1 \\ x_1 + 3x_2 - x_3 + 2x_4 &= -1 \\ x_1 - x_2 + x_3 - x_4 &= 0 \end{aligned}$$

becomes

$$\begin{bmatrix} 3 & -2 & 4 & -1 \\ 1 & 3 & -1 & 2 \\ 1 & -1 & 1 & -1 \end{bmatrix} \begin{bmatrix} x_1 \\ x_2 \\ x_3 \\ x_4 \end{bmatrix} = \begin{bmatrix} 1 \\ -1 \\ 0 \end{bmatrix}$$

According to our definition of §1.4, systems of the form $A\boldsymbol{x} = \boldsymbol{0}$ are called homogeneous. We have the following theorem:

Theorem If A is an $m \times n$ matrix where $m < n$, there is an n-vector $\boldsymbol{x}$, $\boldsymbol{x} \neq \boldsymbol{0}$, such that $A\boldsymbol{x} = \boldsymbol{0}$.

Proof If we write

$$\begin{aligned} A &= [a_{ij}]_{(mn)} \\ \boldsymbol{x} &= [x_j]_{(n1)} \end{aligned}$$

the associated homogeneous system of linear equations has more variables than equations, and so by the theorem of §1.4, there is a nontrivial solution. ∎

Thus, for example, given the matrix

$$A = \begin{bmatrix} 1 & 0 & 8 & 3 \\ 4 & 2 & -7 & 1 \\ -3 & 7 & 0 & 2 \end{bmatrix}$$

we can find some nonzero 4-vector $\boldsymbol{x}$ such that $A\boldsymbol{x} = \mathbf{0}$.

Suppose we are given a system of n equations in n variables, in which the matrix of coefficients is invertible. In this case we may show that the system of equations is uniquely solvable. In matrix notation the system of equations becomes $A\boldsymbol{x} = \boldsymbol{y}$.

If we let $\boldsymbol{x} = A^{-1}\boldsymbol{y}$, we have

$$A\boldsymbol{x} = A(A^{-1}\boldsymbol{y}) = (AA^{-1})\boldsymbol{y} = I_n\boldsymbol{y} = \boldsymbol{y}$$

Thus, the system is solvable. To see that the solution is unique, suppose $\boldsymbol{x}_1$ and $\boldsymbol{x}_2$ are two solutions to the system, i.e., $A\boldsymbol{x}_1 = \boldsymbol{y}$ and $A\boldsymbol{x}_2 = \boldsymbol{y}$.

Then, $A\boldsymbol{x}_1 = A\boldsymbol{x}_2$ and we can multiply both sides of the equality by A^{-1} to obtain

$$A^{-1}(A\boldsymbol{x}_1) = A^{-1}(A\boldsymbol{x}_2)$$

or

$$(A^{-1}A)\boldsymbol{x}_1 = (A^{-1}A)\boldsymbol{x}_2$$

so that $\boldsymbol{x}_1 = \boldsymbol{x}_2$.

For example,

$$\begin{bmatrix} 1 & 3 & 3 \\ 1 & 3 & 4 \\ 1 & 4 & 3 \end{bmatrix}\begin{bmatrix} 7 & -3 & -3 \\ -1 & 0 & 1 \\ -1 & 1 & 0 \end{bmatrix} = \begin{bmatrix} 1 & 0 & 0 \\ 0 & 1 & 0 \\ 0 & 0 & 1 \end{bmatrix}$$

and

$$\begin{bmatrix} 7 & -3 & -3 \\ -1 & 0 & 1 \\ -1 & 1 & 0 \end{bmatrix}\begin{bmatrix} 1 & 3 & 3 \\ 1 & 3 & 4 \\ 1 & 4 & 3 \end{bmatrix} = \begin{bmatrix} 1 & 0 & 0 \\ 0 & 1 & 0 \\ 0 & 0 & 1 \end{bmatrix}$$

So

$$\begin{bmatrix} 1 & 3 & 3 \\ 1 & 3 & 4 \\ 1 & 4 & 3 \end{bmatrix}^{-1} = \begin{bmatrix} 7 & -3 & -3 \\ -1 & 0 & 1 \\ -1 & 1 & 0 \end{bmatrix}$$

Thus, when confronted with the system of equations,

$$\begin{aligned} x_1 + 3x_2 + 3x_3 &= y_1 \\ x_1 + 3x_2 + 4x_3 &= y_2 \\ x_1 + 4x_2 + 3x_3 &= y_3 \end{aligned}$$

we have

$$\begin{bmatrix} x_1 \\ x_2 \\ x_3 \end{bmatrix} = \begin{bmatrix} 7 & -3 & -3 \\ -1 & 0 & 1 \\ -1 & 1 & 0 \end{bmatrix} \begin{bmatrix} y_1 \\ y_2 \\ y_3 \end{bmatrix}$$

$$= \begin{bmatrix} 7y_1 & -3y_2 & -3y_3 \\ -y_1 & & +y_3 \\ -y_1 & +y_2 & \end{bmatrix}$$

Although in this section we have emphasized the uses of matrices in solving systems of linear equations, there are many other situations in which they arise independently of this connection. They play prominent roles, for example, in quantum mechanics and the analysis of crystal structures, in classical mechanics and numerous branches of engineering.

EXERCISES

1. Convert the following systems of linear equations to matrix form.

(a)
$$\begin{aligned} 3x_1 - 4x_2 + 5x_3 - x_4 + x_5 &= 1 \\ x_1 - x_2 + 10x_3 + x_4 + 2x_5 &= 0 \\ x_1 - x_2 + 3x_3 + x_4 + 3x_5 &= -1 \end{aligned}$$

(b)
$$\begin{aligned} x_1 - x_2 + x_3 &= 1 \\ x_1 - 2x_2 + x_3 &= 0 \\ 3x_1 - x_2 + 4x_3 &= 8 \end{aligned}$$

(c)
$$\begin{aligned} x + 3y &= 0 \\ x - 3y &= 2 \end{aligned}$$

(d)
$$\begin{aligned} 7x_1 + 8x_2 - x_3 &= 1 \\ x_1 - 8x_2 + 6x_3 &= 0 \end{aligned}$$

2. Show that

$$\begin{bmatrix} 1 & 2 & 3 & 1 \\ 1 & 3 & 3 & 2 \\ 2 & 4 & 3 & 3 \\ 1 & 1 & 1 & 1 \end{bmatrix}^{-1} = \begin{bmatrix} 1 & -2 & 1 & 0 \\ 1 & -2 & 2 & -3 \\ 0 & 1 & -1 & 1 \\ -2 & 3 & -2 & 3 \end{bmatrix}$$

and solve the system of equations

$$\begin{aligned} x_1 + 2x_2 + 3x_3 + x_4 &= y_1 \\ x_1 + 3x_2 + 3x_3 + 2x_4 &= y_2 \\ 2x_1 + 4x_2 + 3x_3 + 3x_4 &= y_3 \\ x_1 + x_2 + x_3 + x_4 &= y_4 \end{aligned}$$

3. Show that

$$\begin{bmatrix} 13 & 19 \\ 2 & 3 \end{bmatrix}^{-1} = \begin{bmatrix} 3 & -19 \\ -2 & 13 \end{bmatrix}$$

and use this to solve

$$\begin{aligned} 13x + 19y &= \alpha \\ 2x + 3y &= \beta \end{aligned}$$

4. If $ad - bc = 1$, show that

$$\begin{bmatrix} a & b \\ c & d \end{bmatrix}^{-1} = \begin{bmatrix} d & -b \\ -c & a \end{bmatrix}$$

Obtain the general solution to the equations

$$\begin{aligned} ax_1 + bx_2 &= y_1 \\ cx_1 + dx_2 &= y_2 \end{aligned}$$

5. If A is an $m \times n$ matrix, let $\mathfrak{N} = \{\boldsymbol{x}: \boldsymbol{x} \in \boldsymbol{R}^n \text{ and } A\boldsymbol{x} = \boldsymbol{0}\}$.
   (a) Show that if $\boldsymbol{x}_1 \in \mathfrak{N}$ and $\boldsymbol{x}_2 \in \mathfrak{N}$, then $\boldsymbol{x}_1 + \boldsymbol{x} \in \mathfrak{N}$.
   (b) Show that if $\boldsymbol{x} \in \mathfrak{N}$, then $c\boldsymbol{x} \in \mathfrak{N}$.
   (c) Show that $\boldsymbol{0} \in \mathfrak{N}$.
6. (a) If A is an $m \times n$ matrix and if there is an $n \times m$ matrix B, such that $AB = I_n$, show that the equation $A\boldsymbol{x} = \boldsymbol{y}$ is always solvable.
   (b) Using the fact that

$$\begin{bmatrix} -1 & 2 & 0 \\ 8 & -17 & 1 \end{bmatrix} \begin{bmatrix} -17 & -2 \\ -8 & -1 \\ 0 & 0 \end{bmatrix} = \begin{bmatrix} 1 & 0 \\ 0 & 1 \end{bmatrix}$$

find solutions to the equations

$$\begin{aligned} -x + 2y &= \alpha \\ 8x - 17y + z &= \beta \end{aligned}$$

7. If A is an $m \times n$ matrix and $m < n$, show that there is a nonzero $n \times p$ matrix B such that $AB = 0$.
8. If A is an $m \times n$ matrix, $\boldsymbol{x}$ is an n-vector, and $\boldsymbol{y}_1$, $\boldsymbol{y}_2$ are m-vectors such that $A\boldsymbol{x} = \boldsymbol{y}_1$ and $A\boldsymbol{x} = \boldsymbol{y}_2$ can be solved, show that $A\boldsymbol{x} = \boldsymbol{y}_1 + \boldsymbol{y}_2$ can be solved.

7 THE TRANSPOSE OF A MATRIX

If A is an $m \times n$ matrix, $A = [a_{ij}]_{(mn)}$, we consider a new matrix, called the **transpose** of A, denoted by A^T, defined by $A^T = [b_{ij}]_{(nm)}$, where $b_{ij} = a_{ji}$.

For example,

$$\begin{bmatrix} 1 & 3 \\ 0 & 2 \end{bmatrix}^T = \begin{bmatrix} 1 & 0 \\ 3 & 2 \end{bmatrix}$$

$$\begin{bmatrix} 1 & 0 & 7 \\ -1 & 2 & 8 \end{bmatrix}^T = \begin{bmatrix} 1 & -1 \\ 0 & 2 \\ 7 & 8 \end{bmatrix}$$

$$\begin{bmatrix} 6 & -1 & 8 \\ 0 & 2 & 1 \\ 3 & 7 & 0 \end{bmatrix}^T = \begin{bmatrix} 6 & 0 & 3 \\ -1 & 2 & 7 \\ 8 & 1 & 0 \end{bmatrix}$$

If A is a complex matrix, we may define a matrix which is analogous to the transpose, called the **adjoint**, and denoted by A^*, where $A^* = [b_{ij}]_{(nm)}$ with $b_{ij} = \overline{a_{ji}}$, (the notation $\bar{a}$ denotes the complex conjugate of the number a).

Thus,

$$\begin{bmatrix} i & 0 \\ -i & 1 \\ 0 & 1+i \end{bmatrix}^* = \begin{bmatrix} -i & i & 0 \\ 0 & 1 & 1-i \end{bmatrix}$$

If A is real, then $A^* = A^T$.

The transpose operation satisfies certain laws.

(1) $(A + B)^T = A^T + B^T$

(2) $(\alpha A)^T = \alpha A^T$

(3) $(AB)^T = B^T A^T$

(4) $(A^T)^T = A$

Proof of (1) Let $A = [a_{ij}]_{(mn)}$ and $B = [b_{ij}]_{(mn)}$. Then, $A + B = [a_{ij} + b_{ij}]_{(mn)}$, and $(A + B)^T = [c_{ij}]_{(nm)}$, where $c_{ij} = a_{ji} + b_{ji}$.

Now,

$$\begin{aligned} A^T + B^T &= [\alpha_{ij}]_{(nm)} + [\beta_{ij}]_{(nm)} \\ &= [\alpha_{ij} + \beta_{ij}]_{(nm)} \end{aligned}$$

where $\alpha_{ij} = a_{ji}$ and $\beta_{ij} = b_{ji}$. Since $c_{ij} = \alpha_{ij} + \beta_{ij}$, we have $(A + B)^T = A^T + B^T$. ▨

Proof of (3) Let $A = [a_{ij}]_{(mn)}$ and $B = [b_{ij}]_{(np)}$. So

$$(AB) = \left[\sum_{j=1}^{n} a_{ij}b_{jk}\right]_{(np)}$$

Then, $(AB)^T = [c_{ik}]_{(pn)}$, where $c_{ik} = \Sigma_{j=1}^{n} a_{kj}b_{ji}$. Now, $B^TA^T = [\Sigma_{j=1}^{n} \beta_{ij}\alpha_{jk}]$ where $\beta_{ij} = b_{ji}$ and $\alpha_{jk} = a_{kj}$.

So $B^TA^T = [c'_{ik}]_{(pm)}$, where $c'_{ik} = \Sigma_{j=1}^{n} b_{ji}a_{kj} = \Sigma_{j=1}^{n} a_{kj}b_{ji}$. Since $c'_{ik} = c_{ik}$, we have $(AB)^T = B^TA^T$. ∎

We leave the proof of (2) and (4) as exercises for the reader.

We say a matrix is **symmetric** if $A = A^T$. We say it is **Hermitian** if $A = A^*$. For example,

$$\begin{bmatrix} 1 & 3 \\ 3 & 2 \end{bmatrix}^T = \begin{bmatrix} 1 & 3 \\ 3 & 2 \end{bmatrix}$$

So the matrix

$$A = \begin{bmatrix} 1 & 3 \\ 3 & 2 \end{bmatrix}$$

is symmetric.

Likewise, the matrix

$$\begin{bmatrix} 1 & 7 & 8 \\ 7 & 0 & 4 \\ 8 & 4 & -1 \end{bmatrix}$$

is symmetric, while

$$\begin{bmatrix} -1 & i & 4 \\ -i & 0 & -i \\ 4 & i & 2 \end{bmatrix}$$

is Hermitian.

As an example of the use of the laws of transposition, let us show that AA^T is symmetric.

$$\begin{aligned} \text{Now, } (AA^T)^T &= (A^T)^TA^T && \text{[by (3)]} \\ &= AA^T && \text{[by (4)]} \end{aligned}$$

Since $(AA^T)^T = AA^T$, AA^T is symmetric.

EXERCISES

1. Find A^T if

(a) $A = \begin{bmatrix} 1 & -3 \\ -3 & 0 \end{bmatrix}$ (b) $A = \begin{bmatrix} 1 & 3 \\ -3 & 0 \end{bmatrix}$

(c) $A = \begin{bmatrix} 1 & 4 \\ 0 & 1 \\ 2 & -1 \\ 0 & 3 \end{bmatrix}$ (d) $A = \begin{bmatrix} 1 & -7 & 0 & 6 & 3 \\ 0 & 1 & 0 & 2 & -3 \end{bmatrix}$

(e) $A = \begin{bmatrix} 1 & 7 & 0 \\ 6 & 1 & 2 \\ -1 & 0 & -1 \end{bmatrix}$

2. Prove

$$(\alpha A)^T = \alpha A^T$$
$$(A^T)^T = A$$

3. Prove

$$(A + B)^* = A^* + B^*$$
$$(\alpha A)^* = \bar{\alpha} A^*$$
$$(AB)^* = B^* A^*$$
$$(A^*)^* = A$$

[Hint: Observe $A^* = (\overline{A^T})$; where if $B = [b_{ij}]$, $\bar{B} = [\bar{b}_{ij}]$.]

4. (a) Show that $A^T A$ is symmetric.
   (b) Show that $A + A^T$ is symmetric.
5. Prove that all symmetric matrices are square.
6. Show that A^*A, AA^*, and $A + A^*$ are Hermitian.
7. Prove that all diagonal matrices are symmetric.
8. Show that all the diagonal entries of a Hermitian matrix are real.
9. If A and B are symmetric, show that AB is symmetric if and only if A and B commute.
10. If A is invertible, show that A^T is invertible, and that $(A^T)^{-1} = (A^{-1})^T$.
11. A matrix is said to be **skew-symmetric** if $A^T = -A$. Show that every matrix can be uniquely expressed as the sum of a symmetric and a skew-symmetric matrix.

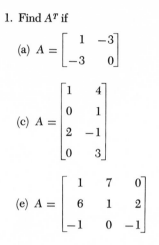

12. If A commutes with B, show that A^T commutes with B^T.
13. Find a 2×2 matrix such that $AA^T \neq A^TA$.
14. If A is an $n \times n$ matrix, and $f(x)$ is a polynomial, show that $f(A^T) = (f(A))^T$.
15. Show that every square matrix A may be expressed in the form $A = H + iK$, where H and K are Hermitian.
    (a) Show that this expression is unique.
    (b) Show that H and K commute if and only if A and A^* commute.

3

DETERMINANTS

1 DETERMINANTS OF 2 × 2 MATRICES

If in the system of equations

$$a_{11}x_1 + a_{12}x_2 = b_1$$
$$a_{21}x_1 + a_{22}x_2 = b_2$$

the quantity $a_{11}a_{22} - a_{12}a_{21}$ is nonzero, it is possible to show that the solutions are

$$x_1 = \frac{b_1a_{22} - b_2a_{12}}{a_{11}a_{22} - a_{12}a_{21}}, \qquad x_2 = \frac{b_2a_{11} - b_1a_{21}}{a_{11}a_{22} - a_{12}a_{21}}$$

Moreover the matrix of coefficients

$$\begin{bmatrix} a_{11} & a_{12} \\ a_{21} & a_{22} \end{bmatrix}$$

is invertible and its inverse is

$$\frac{1}{a_{11}a_{22} - a_{12}a_{21}} \begin{bmatrix} a_{22} & -a_{12} \\ -a_{21} & a_{11} \end{bmatrix}$$

From these facts it is clear that the number $a_{11}a_{22} - a_{12}a_{21}$ plays an important role in the problems of solving systems of linear equations and matrix inversion. For these reasons if

$$A = \begin{bmatrix} a_{11} & a_{12} \\ a_{21} & a_{22} \end{bmatrix}$$

we define det $A = a_{11}a_{22} - a_{12}a_{21}$. The quantity det A is called the **determinant** of A and is often written

$$\begin{vmatrix} a_{11} & a_{12} \\ a_{21} & a_{22} \end{vmatrix}$$

It is important to note that the determinant is a scalar associated with each matrix, while a matrix is an array of numbers.

With this definition of the determinant we may rewrite the solutions to the system

$$\begin{aligned} a_{11}x_1 + a_{12}x_2 &= b_1 \\ a_{21}x_1 + a_{22}x_2 &= b_2 \end{aligned}$$

as

$$x_1 = \frac{\begin{vmatrix} b_1 & a_{12} \\ b_2 & a_{22} \end{vmatrix}}{\begin{vmatrix} a_{11} & a_{12} \\ a_{21} & a_{22} \end{vmatrix}}, \qquad x_2 = \frac{\begin{vmatrix} a_{11} & b_1 \\ a_{21} & b_2 \end{vmatrix}}{\begin{vmatrix} a_{11} & a_{12} \\ a_{21} & a_{22} \end{vmatrix}}$$

In this chapter we define the determinant of an arbitrary $n \times n$ matrix and exhibit methods for calculating it. In §3.5 and §3.6 we derive formulas analogous to the above for computing solutions to systems of linear equations and inverting matrices.

In this chapter we often denote a matrix A by $[A_1, A_2, \ldots, A_n]$, where A_i is the ith column of the matrix A. Thus, in the matrix

$$A = \begin{bmatrix} -1 & 1 & -4 \\ 0 & 2 & 3 \\ 3 & 7 & 2 \end{bmatrix}$$

$$A_1 = \begin{bmatrix} -1 \\ 0 \\ 3 \end{bmatrix}, \qquad A_2 = \begin{bmatrix} 1 \\ 2 \\ 7 \end{bmatrix}, \qquad A_3 = \begin{bmatrix} -4 \\ 3 \\ 2 \end{bmatrix}$$

Before we proceed to the development of the determinant for the general case of $n \times n$ matrices, let us note several important properties of the determinant of 2×2 matrices, important properties which also hold in the general case.

(D1)
$$\det [A_1 + A_1', A_2] = \det [A_1, A_2] + \det [A_1', A_2]$$
$$\det [A_1, A_2 + A_2'] = \det [A_1, A_2] + \det [A_1, A_2']$$

(D2)
$$\det [cA_1, A_2] = c \det [A_1, A_2]$$
$$\det [A_1, cA_2] = c \det [A_1, A_2]$$

(D3)
$$\det [A, A] = 0$$

(D4)
$$\det I_2 = 1$$

Properties (D1) and (D2) say that the determinant depends linearly on one of the columns if the other columns remain fixed. Property (D3) means that a matrix with two equal columns has determinant 0.

To prove (D1), let

$$A_1 = \begin{bmatrix} a_{11} \\ a_{21} \end{bmatrix}, \qquad A_1' = \begin{bmatrix} a_{11}' \\ a_{21}' \end{bmatrix}, \qquad A_2 = \begin{bmatrix} a_{12} \\ a_{22} \end{bmatrix}$$

Then,

$$\begin{aligned} \det [A_1 + A_1', A_2] &= \det \begin{bmatrix} a_{11} + a_{11}' & a_{12} \\ a_{21} + a_{21}' & a_{22} \end{bmatrix} \\ &= (a_{11} + a_{11}')a_{22} - (a_{21} + a_{21}')a_{12} \\ &= (a_{11}a_{22} - a_{21}a_{12}) + (a_{11}'a_{22} - a_{21}'a_{12}) \\ &= \det [A_1, A_2] + \det [A_1', A_2] \end{aligned}$$

The proof of the second half of (D1) is completely analogous.

To obtain (D2),

$$\begin{aligned} \det [cA_1, A_2] &= \det \begin{bmatrix} ca_{11} & a_{12} \\ ca_{21} & a_{22} \end{bmatrix} \\ &= ca_{11}a_{22} - ca_{12}a_{21} \\ &= c(a_{11}a_{22} - a_{12}a_{21}) \\ &= c \det [A_1, A_2] \end{aligned}$$

To obtain (D3),

$$\det [A, A] = \det \begin{bmatrix} a & a \\ b & b \end{bmatrix} = ab - ab = 0$$

To obtain (D4),

$$\det I_2 = \det \begin{bmatrix} 1 & 0 \\ 0 & 1 \end{bmatrix} = 1 \cdot 1 - 0 \cdot 0 = 1$$

Using (D1), (D2), (D3), and (D4), we may formulate additional properties of the determinant.

(D5) $$\det [A_1, A_2 + cA_1] = \det [A_1, A_2]$$
$$\det [A_1 + cA_2, A_2] = \det [A_1, A_2]$$

(D6) $$\det [A_1, A_2] = -\det [A_2, A_1]$$

(D7) $$\det [A, 0] = \det [0, A] = 0$$

(D5) tells us we add a scalar multiple of one column to another without affecting the value of the determinant. (D6) says that interchanging two columns changes the sign of the determinant. (D7) means that if in a determinant one of the columns is zero, the determinant is zero.

To prove (D5), observe that

$$\begin{aligned} \det [A_1, A_2 + cA_1] &= \det [A_1, A_2] + \det [A_1, cA_1] && \text{[by (D1)]} \\ &= \det [A_1, A_2] + c \det [A_1, A_1] && \text{[by (D2)]} \\ &= \det [A_1, A_2] + c \cdot 0 && \text{[by (D3)]} \\ &= \det [A_1, A_2] \end{aligned}$$

To prove (D6), observe that

$$\begin{aligned} \det [A_1, A_2] &= \det [A_1 + A_2, A_2] && \text{[by (D5)]} \\ &= \det [A_1 + A_2, -A_1] && \text{[by (D5)]} \\ &= \det [A_2, -A_1] && \text{[by (D5)]} \\ &= -\det [A_2, A_1] && \text{[by (D2)]} \end{aligned}$$

Using properties (D1)–(D7) we may calculate the determinant without recourse to our original definition, as in

$$\begin{vmatrix} 1 & -7 \\ 2 & 3 \end{vmatrix} = \begin{vmatrix} 1 & -7 + 7 \cdot 1 \\ 2 & 3 + 7 \cdot 2 \end{vmatrix} \qquad \text{[by (D5)]}$$

$$= \begin{vmatrix} 1 & 0 \\ 2 & 17 \end{vmatrix}$$

$$= 17 \begin{vmatrix} 1 & 0 \\ 2 & 1 \end{vmatrix} \qquad \text{[by (D2)]}$$

$$= 17 \begin{vmatrix} 1 + (-2) \cdot 0 & 0 \\ 2 + (-2) \cdot 1 & 1 \end{vmatrix} \qquad \text{[by (D5)]}$$

$$= 17 \begin{vmatrix} 1 & 0 \\ 0 & 1 \end{vmatrix} = 17 \qquad \text{[by (D4)]}$$

In the case of 2 × 2 matrices it is probably simpler to calculate the determinant directly from its definition. For larger determinants, however, repeated use of the analogs of (D1)–(D7) is the more efficient procedure.

EXERCISES

1. Calculate

(a) $\begin{vmatrix} 1 & -3 \\ 7 & 4 \end{vmatrix}$ (b) $\begin{vmatrix} 1 & 3 \\ 5 & 6 \end{vmatrix}$ (c) $\begin{vmatrix} 0 & -1 \\ 2 & 1 \end{vmatrix}$

(d) $\begin{vmatrix} 7 & 5 \\ 4 & 3 \end{vmatrix}$ (e) $\begin{vmatrix} 2 & 1 \\ 3 & 2 \end{vmatrix}$ (f) $\begin{vmatrix} 0 & 1 \\ 1 & 0 \end{vmatrix}$

2. For what values of λ, does

(a) $\begin{vmatrix} \lambda & \lambda \\ 3 & \lambda - 2 \end{vmatrix} = 0$ (b) $\begin{vmatrix} 1 - \lambda & 1 \\ 1 & 1 + \lambda \end{vmatrix} = 0$

(c) $\begin{vmatrix} 1 + \lambda & 1 \\ 2 + 2\lambda & 2 \end{vmatrix} = 0$

3. If $\det \begin{bmatrix} a & b \\ c & d \end{bmatrix} = 0$, show that

$$\begin{bmatrix} d & -b \\ -c & a \end{bmatrix} \begin{bmatrix} a & b \\ c & d \end{bmatrix} = \begin{bmatrix} a & b \\ c & d \end{bmatrix} \begin{bmatrix} d & -b \\ -c & a \end{bmatrix} = 0$$

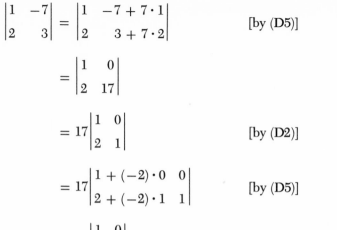

4. Find

(a) $\begin{bmatrix} 1 & 1 \\ 5 & 6 \end{bmatrix}^{-1}$ (b) $\begin{bmatrix} 7 & 4 \\ 3 & 2 \end{bmatrix}^{-1}$ (c) $\begin{bmatrix} 1 & a \\ 0 & 1 \end{bmatrix}^{-1}$

(d) $\begin{bmatrix} 1-a & a \\ -a & 1+a \end{bmatrix}^{-1}$ (e) $\begin{bmatrix} \cos\theta & \sin\theta \\ -\sin\theta & \cos\theta \end{bmatrix}^{-1}$

5. Given a function $D(A)$, defined on 2×2 matrices, and satisfying properties (D1)–(D4), show that $D(A) = \det A$. In other words, (D1)–(D4) completely characterize the determinant.
6. Show that properties (D1)–(D7) hold for rows, as well as for columns.

2 DEFINITION AND PRINCIPAL PROPERTIES OF DETERMINANTS

In this section we give an inductive definition of the determinant of an $n \times n$ matrix. That is, knowing how the determinent is defined for 2×2 matrices, we give a definition for 3×3 matrices, and then using the definition for 3×3 matrices, we give one for 4×4 matrices. In general, using our definition for $(n-1) \times (n-1)$ matrices we give one for $n \times n$ matrices.

If A is an $n \times n$ matrix we use $\hat{A}_{ij}$ to denote the matrix obtained by deleting the ith row and jth column of the matrix A. Thus, if

$$A = \begin{bmatrix} 0 & 0 & -7 \\ 3 & 1 & 2 \\ 4 & -2 & 3 \end{bmatrix}$$

$$\hat{A}_{11} = \begin{bmatrix} 1 & 2 \\ -2 & 3 \end{bmatrix}, \quad \hat{A}_{31} = \begin{bmatrix} 0 & -7 \\ 1 & 2 \end{bmatrix}, \quad \hat{A}_{22} = \begin{bmatrix} 0 & -7 \\ 4 & 3 \end{bmatrix}$$

Assuming we have defined the determinant for $(n-1) \times (n-1)$ matrices we define it for an $n \times n$ matrix, $A = [a_{ij}]$, by

$$\begin{aligned} \det A &= (-1)^{1+1}a_{11} \det \hat{A}_{11} + (-1)^{1+2}a_{12} \det \hat{A}_{12} \\ &\quad + \cdots + (-1)^{1+n}a_{1n} \det \hat{A}_{1n} \\ &= \sum_{j=1}^{n} (-1)^{1+j}a_{1j} \det \hat{A}_{1j} \end{aligned}$$

For 3×3 matrices, this becomes

$$\det \begin{bmatrix} a_{11} & a_{12} & a_{13} \\ a_{21} & a_{22} & a_{23} \\ a_{31} & a_{32} & a_{33} \end{bmatrix} = a_{11} \det \begin{bmatrix} a_{22} & a_{23} \\ a_{32} & a_{33} \end{bmatrix} - a_{12} \det \begin{bmatrix} a_{21} & a_{23} \\ a_{31} & a_{33} \end{bmatrix} + a_{13} \det \begin{bmatrix} a_{21} & a_{22} \\ a_{31} & a_{32} \end{bmatrix}$$

$$= a_{11}a_{22}a_{33} - a_{11}a_{23}a_{32} + a_{12}a_{23}a_{31} - a_{12}a_{21}a_{33} + a_{13}a_{21}a_{32} - a_{13}a_{22}a_{31}$$

We often write

$$\det \begin{bmatrix} a_{11} & a_{12} & \dots & a_{1n} \\ a_{21} & a_{22} & \dots & a_{2n} \\ & & \vdots & \\ a_{n1} & a_{n2} & \dots & a_{nn} \end{bmatrix}$$

as

$$\begin{vmatrix} a_{11} & a_{12} & \dots & a_{1n} \\ a_{21} & a_{22} & \dots & a_{2n} \\ & & \vdots & \\ a_{n1} & a_{n2} & \dots & a_{nn} \end{vmatrix}_n$$

where the n on the lower right-hand corner indicates the order of the determinant.

For 4×4 matrices our definition becomes, in more explicit form,

$$\begin{vmatrix} a_{11} & a_{12} & a_{13} & a_{14} \\ a_{21} & a_{22} & a_{23} & a_{24} \\ a_{31} & a_{32} & a_{33} & a_{34} \\ a_{41} & a_{42} & a_{43} & a_{44} \end{vmatrix} = a_{11} \begin{vmatrix} a_{22} & a_{23} & a_{24} \\ a_{32} & a_{33} & a_{34} \\ a_{42} & a_{43} & a_{44} \end{vmatrix} - a_{12} \begin{vmatrix} a_{21} & a_{23} & a_{24} \\ a_{31} & a_{33} & a_{34} \\ a_{41} & a_{43} & a_{44} \end{vmatrix}$$

$$+ a_{13} \begin{vmatrix} a_{21} & a_{22} & a_{24} \\ a_{31} & a_{32} & a_{34} \\ a_{41} & a_{42} & a_{44} \end{vmatrix} - a_{14} \begin{vmatrix} a_{21} & a_{22} & a_{23} \\ a_{31} & a_{32} & a_{33} \\ a_{41} & a_{42} & a_{43} \end{vmatrix}$$

In calculating determinants, instead of applying the definition directly, it is generally more efficient to utilize certain properties which we now state and prove. These properties are the higher-dimensional analogs of (D1)–(D7) of §3.1.

(D1) $\det [A_1, \dots, A_{i-1}, A_i + A_i', A_{i+1}, \dots, A_n]$

$$= \det [A_1, \dots, A_{i-1}, A_i, A_{i+1}, \dots, A_n] + \det [A_1, \dots, A_{i-1}, A_i', A_{i+1}, \dots, A_n]$$

In words, (D1) says that if each element of the ith column of a matrix A is expressed as the sum of two terms, then det A may be expressed as the sum of two determinants, the elements of whose ith columns are respectively the first and second terms of the corresponding elements of the ith column of A. All other columns are the same.

The proof of (D1) is by induction on n, the order of the matrix. Instead of writing out the general induction step in its necessarily cumbersome notation we confine ourselves to exhibiting the details explicitly in the 3×3 and 4×4 cases. The interested reader can fill in the general induction step.

In the 3×3 case, working on the second column, (D1) becomes

$$\begin{vmatrix} a_{11} & a_{12} + a'_{12} & a_{13} \\ a_{21} & a_{22} + a'_{22} & a_{23} \\ a_{31} & a_{32} + a'_{32} & a_{33} \end{vmatrix} = \begin{vmatrix} a_{11} & a_{12} & a_{13} \\ a_{21} & a_{22} & a_{23} \\ a_{31} & a_{32} & a_{33} \end{vmatrix} + \begin{vmatrix} a_{11} & a'_{12} & a_{13} \\ a_{21} & a'_{22} & a_{23} \\ a_{31} & a'_{32} & a_{33} \end{vmatrix}$$

To prove this observe that, by our definition of the determinant,

$$\Delta = \begin{vmatrix} a_{11} & a_{12} + a'_{12} & a_{13} \\ a_{21} & a_{22} + a'_{22} & a_{23} \\ a_{31} & a_{32} + a'_{32} & a_{33} \end{vmatrix} = a_{11} \begin{vmatrix} a_{22} + a'_{22} & a_{23} \\ a_{32} + a'_{32} & a_{33} \end{vmatrix} - (a_{12} + a'_{12}) \begin{vmatrix} a_{21} & a_{23} \\ a_{31} & a_{33} \end{vmatrix} + a_{13} \begin{vmatrix} a_{21} & a_{22} + a'_{22} \\ a_{31} & a_{32} + a'_{32} \end{vmatrix}$$

Using property (D1) in the 2×2 case,

$$\begin{vmatrix} a_{22} + a'_{22} & a_{23} \\ a_{32} + a'_{32} & a_{33} \end{vmatrix} = \begin{vmatrix} a_{22} & a_{23} \\ a_{32} & a_{33} \end{vmatrix} + \begin{vmatrix} a'_{22} & a_{23} \\ a'_{32} & a_{33} \end{vmatrix}$$

$$\begin{vmatrix} a_{21} & a_{22} + a'_{22} \\ a_{31} & a_{32} + a'_{32} \end{vmatrix} = \begin{vmatrix} a_{21} & a_{22} \\ a_{31} & a_{32} \end{vmatrix} + \begin{vmatrix} a_{21} & a'_{22} \\ a_{31} & a'_{32} \end{vmatrix}$$

Thus,

$$\Delta = a_{11} \begin{vmatrix} a_{22} & a_{23} \\ a_{32} & a_{33} \end{vmatrix} - a_{12} \begin{vmatrix} a_{21} & a_{23} \\ a_{31} & a_{33} \end{vmatrix} + a_{13} \begin{vmatrix} a_{21} & a_{22} \\ a_{31} & a_{32} \end{vmatrix}$$
$$+ a_{11} \begin{vmatrix} a'_{22} & a_{23} \\ a'_{32} & a_{33} \end{vmatrix} - a'_{12} \begin{vmatrix} a_{21} & a_{23} \\ a_{31} & a_{33} \end{vmatrix} + a_{13} \begin{vmatrix} a_{21} & a'_{22} \\ a_{31} & a'_{32} \end{vmatrix}$$

So

$$\Delta = \begin{vmatrix} a_{11} & a_{12} & a_{13} \\ a_{21} & a_{22} & a_{23} \\ a_{31} & a_{32} & a_{33} \end{vmatrix} + \begin{vmatrix} a_{11} & a'_{12} & a_{13} \\ a_{21} & a'_{22} & a_{23} \\ a_{31} & a'_{32} & a_{33} \end{vmatrix}$$

Linearity in the other columns is demonstrated in a similar manner. Again, to go from the 3 × 3 case to the 4 × 4 case, observe that

$$\Delta = \begin{vmatrix} a_{11} & a_{12} & a_{13} + a'_{13} & a_{14} \\ a_{21} & a_{22} & a_{23} + a'_{23} & a_{24} \\ a_{31} & a_{32} & a_{33} + a'_{33} & a_{34} \\ a_{41} & a_{42} & a_{43} + a'_{43} & a_{44} \end{vmatrix}$$

$$= a_{11} \begin{vmatrix} a_{22} & a_{23} + a'_{23} & a_{24} \\ a_{32} & a_{33} + a'_{33} & a_{34} \\ a_{42} & a_{43} + a'_{43} & a_{44} \end{vmatrix} - a_{12} \begin{vmatrix} a_{21} & a_{23} + a'_{23} & a_{24} \\ a_{31} & a_{33} + a'_{33} & a_{34} \\ a_{41} & a_{43} + a'_{43} & a_{44} \end{vmatrix}$$

$$+ (a_{13} + a'_{13}) \begin{vmatrix} a_{21} & a_{22} & a_{24} \\ a_{31} & a_{32} & a_{34} \\ a_{41} & a_{42} & a_{44} \end{vmatrix} - a_{14} \begin{vmatrix} a_{21} & a_{22} & a_{23} + a'_{23} \\ a_{31} & a_{32} & a_{33} + a'_{33} \\ a_{41} & a_{42} & a_{43} + a'_{43} \end{vmatrix}$$

Now using linearity in the 3 × 3 case,

$$\Delta = a_{11} \begin{vmatrix} a_{22} & a_{23} & a_{24} \\ a_{32} & a_{33} & a_{34} \\ a_{42} & a_{43} & a_{44} \end{vmatrix} - a_{12} \begin{vmatrix} a_{21} & a_{23} & a_{24} \\ a_{31} & a_{33} & a_{34} \\ a_{41} & a_{43} & a_{44} \end{vmatrix}$$

$$+ a_{13} \begin{vmatrix} a_{21} & a_{22} & a_{24} \\ a_{31} & a_{32} & a_{34} \\ a_{41} & a_{42} & a_{44} \end{vmatrix} - a_{14} \begin{vmatrix} a_{21} & a_{22} & a_{23} \\ a_{31} & a_{32} & a_{33} \\ a_{41} & a_{42} & a_{43} \end{vmatrix} + a_{11} \begin{vmatrix} a_{22} & a'_{23} & a_{24} \\ a_{32} & a'_{33} & a_{34} \\ a_{42} & a'_{43} & a_{44} \end{vmatrix} - a_{12} \begin{vmatrix} a_{21} & a'_{23} & a_{24} \\ a_{31} & a'_{33} & a_{34} \\ a_{41} & a'_{43} & a_{44} \end{vmatrix}$$

$$+ a'_{13} \begin{vmatrix} a_{21} & a_{22} & a_{24} \\ a_{31} & a_{32} & a_{34} \\ a_{41} & a_{42} & a_{44} \end{vmatrix} - a_{14} \begin{vmatrix} a_{21} & a_{22} & a'_{23} \\ a_{31} & a_{32} & a'_{33} \\ a_{41} & a_{42} & a'_{43} \end{vmatrix} = \begin{vmatrix} a_{11} & a_{12} & a_{13} & a_{14} \\ a_{21} & a_{22} & a_{23} & a_{24} \\ a_{31} & a_{32} & a_{33} & a_{34} \\ a_{41} & a_{42} & a_{43} & a_{44} \end{vmatrix} + \begin{vmatrix} a_{11} & a_{12} & a'_{13} & a_{14} \\ a_{21} & a_{22} & a'_{23} & a_{24} \\ a_{31} & a_{32} & a'_{33} & a_{34} \\ a_{41} & a_{42} & a'_{43} & a_{44} \end{vmatrix}$$

Our next property (D2) can be stated: Let A be an $n \times n$ matrix, and B be another matrix the same as A, except that the elements of one column are c times the corresponding elements of the corresponding column of A. Then $\det B = c \det A$, or

$$\det [A_1, \ldots, A_{i-1}, cA_i, A_{i+1}, \ldots, A_n]$$
$$= c \det [A_1, \ldots, A_{i-1}, A_i, A_{i+1}, \ldots, A_n]$$

The proof is again readily obtained by induction on n, but we confine our attention to the 3×3 case, where we have, by definition,

$$\Delta = \begin{vmatrix} a_{11} & ca_{12} & a_{13} \\ a_{21} & ca_{22} & a_{23} \\ a_{31} & ca_{32} & a_{33} \end{vmatrix} = a_{11} \begin{vmatrix} ca_{22} & a_{23} \\ ca_{32} & a_{33} \end{vmatrix} - ca_{12} \begin{vmatrix} a_{21} & a_{23} \\ a_{31} & a_{33} \end{vmatrix} + a_{13} \begin{vmatrix} a_{21} & ca_{22} \\ a_{31} & ca_{32} \end{vmatrix}$$

Using (D2) in the 2×2 case,

$$\Delta = ca_{11} \begin{vmatrix} a_{22} & a_{23} \\ a_{32} & a_{33} \end{vmatrix} - ca_{12} \begin{vmatrix} a_{21} & a_{23} \\ a_{31} & a_{33} \end{vmatrix} + ca_{13} \begin{vmatrix} a_{21} & a_{22} \\ a_{31} & a_{32} \end{vmatrix} = c \begin{vmatrix} a_{11} & a_{12} & a_{13} \\ a_{21} & a_{22} & a_{23} \\ a_{31} & a_{32} & a_{33} \end{vmatrix}$$

To prove (D3), $\det I_n = 1$, observe that in the identity matrix $a_{ij} = 0$ unless $j = i$, and $a_{ii} = 1$. Thus,

$$\begin{vmatrix} 1 & 0 & \dots & 0 \\ 0 & 1 & \dots & 0 \\ & & \vdots & \\ 0 & 0 & \dots & 1 \end{vmatrix}_n = 1 \cdot \begin{vmatrix} 1 & 0 & \dots & 0 \\ 0 & 1 & \dots & 0 \\ & & \vdots & \\ 0 & 0 & \dots & 1 \end{vmatrix}_{n-1} + 0 \cdot (\text{other terms}) = 1$$

since by the induction assumption $\det I_{n-1} = 1$.

(D4) The determinant of a matrix two of whose columns are equal is 0, or $\det [A_1, \dots, A_{i-1}, B, A_{i+1}, \dots, A_{j-1}, B, A_{j+1}, \dots, A_n] = 0$.

We leave the general case to the reader, restricting ourselves to the 3×3 and 4×4 cases. Thus, consider,

$$\Delta = \begin{vmatrix} a_1 & b_1 & b_1 \\ a_2 & b_2 & b_2 \\ a_3 & b_3 & b_3 \end{vmatrix} = a_1 \begin{vmatrix} b_2 & b_2 \\ b_3 & b_3 \end{vmatrix} - b_1 \begin{vmatrix} a_2 & b_2 \\ a_3 & b_3 \end{vmatrix} + b_1 \begin{vmatrix} a_2 & b_2 \\ a_3 & b_3 \end{vmatrix}$$

By (D4) in the 2×2 case the first term vanishes, while the last two cancel each other.

In a similar manner,

$$\Delta = \begin{vmatrix} b_1 & a_1 & b_1 \\ b_2 & a_2 & b_2 \\ b_3 & a_3 & b_3 \end{vmatrix} = b_1 \begin{vmatrix} a_2 & b_2 \\ a_3 & b_3 \end{vmatrix} - a_1 \begin{vmatrix} b_2 & b_2 \\ b_3 & b_3 \end{vmatrix} + b_1 \begin{vmatrix} b_2 & a_2 \\ b_3 & a_3 \end{vmatrix}$$

By (D6) in the 2×2 case

$$\begin{vmatrix} a_2 & b_2 \\ a_3 & b_3 \end{vmatrix} = - \begin{vmatrix} b_2 & a_2 \\ b_3 & a_3 \end{vmatrix}$$

Thus, the first and third terms cancel each other, while the second is zero by (D4) in the 2×2 case.

In the 4×4 case, the argument becomes

$$\Delta = \begin{vmatrix} a_1 & b_1 & c_1 & b_1 \\ a_2 & b_2 & c_2 & b_2 \\ a_3 & b_3 & c_3 & b_3 \\ a_4 & b_4 & c_4 & b_4 \end{vmatrix} = a_1 \begin{vmatrix} b_2 & c_2 & b_2 \\ b_3 & c_3 & b_3 \\ b_4 & c_4 & b_4 \end{vmatrix} - b_1 \begin{vmatrix} a_2 & c_2 & b_2 \\ a_3 & c_3 & b_3 \\ a_4 & c_4 & b_4 \end{vmatrix} + c_1 \begin{vmatrix} a_2 & b_2 & b_2 \\ a_3 & b_3 & b_3 \\ a_4 & b_4 & b_4 \end{vmatrix} - b_1 \begin{vmatrix} a_2 & b_2 & c_2 \\ a_3 & b_3 & c_3 \\ a_4 & b_4 & c_4 \end{vmatrix}$$

Using (D4) in the 3×3 case, we see that the first and third terms are zero. Observe that since the determinant in the fourth term is obtained from that in the second term by interchanging columns, their sum vanishes, showing $\Delta = 0$.

Another property is (D5). The value of the determinant of a matrix is unchanged if some scalar multiple of one column is added to another.

This may be proved using (D1)–(D4). Thus, we consider $\det [A_1, \ldots, A_{i-1}, A_i + cA_j, A_{i+1}, \ldots, A_n]$ for $i \neq j$, which, by (D1), equals

$$\det [A_1, \ldots, A_{i-1}, A_i, A_{i+1}, \ldots, A_n] + \det [A_1, \ldots, A_{i-1}, cA_j, A_{i+1}, \ldots, A_n]$$

which by (D2) is

$$\det [A_1, \ldots, A_{i-1}, A_i, A_{i+1}, \ldots, A_n] + c \det [A_1, \ldots, A_{i-1}, A_j, A_{i+1}, \ldots, A_n]$$

Since the last determinant has the column A_j occurring twice (in both column i and column j), by (D4) its value is zero, demonstrating the truth of our assertion.

As in the 2×2 case, we have (D6). Interchanging two columns of a matrix changes the sign of its determinant.

To prove (D6)

$$\begin{aligned} &\det [A_1, \ldots, A_i, \ldots, A_j, \ldots, A_n] \\ &\quad = \det [A_1, \ldots, A_i + A_j, \ldots, A_j, \ldots, A_n] && \text{[by (D5)]} \\ &\quad = \det [A_1, \ldots, A_i + A_j, \ldots, -A_i, \ldots, A_n] && \text{[by (D5)]} \end{aligned}$$

$$= \det [A_1, \ldots, A_j, \ldots, -A_i, \ldots, A_n] \qquad \text{[by (D5)]}$$
$$= -\det [A_1, \ldots, A_j, \ldots, A_i, \ldots, A_n] \qquad \text{[by (D2)]}$$

Another important property is (D7). If all entries of some column of a matrix are zero, its determinant is zero.

To see this note,

$$\begin{aligned}
\det [A_1, \ldots, 0, \ldots, A_n] \\
&= \det [A_1, \ldots, 0 \cdot 0, \ldots, A_n] \\
&= 0 \cdot \det [A_1, \ldots, 0, \ldots, A_n] && \text{[by (D2)]} \\
&= 0
\end{aligned}$$

By repeated use of (D1)–(D7) we can evaluate any determinant.

Example 1 Evaluate the following determinant.

$$\Delta = \begin{vmatrix} 3 & 2 & 7 \\ 0 & 1 & -3 \\ 3 & 4 & 1 \end{vmatrix}$$

$$= \begin{vmatrix} 3 + (-1)(2) & 2 & 7 \\ 0 + (-1)(1) & 1 & -3 \\ 3 + (-1)(4) & 4 & 1 \end{vmatrix} \qquad \text{[by (D5)]}$$

$$= \begin{vmatrix} 1 & 2 & 7 \\ -1 & 1 & -3 \\ -1 & 4 & 1 \end{vmatrix}$$

$$= \begin{vmatrix} 1 & 2 + (-2)(1) & 7 + (-7)(1) \\ -1 & 1 + (-2)(-1) & -3 + (-7)(-1) \\ -1 & 4 + (-2)(-1) & 1 + (-7)(-1) \end{vmatrix} \qquad \text{[by (D5)]}$$

$$= \begin{vmatrix} 1 & 0 & 0 \\ -1 & 3 & 4 \\ -1 & 6 & 8 \end{vmatrix}$$

$$= \begin{vmatrix} 3 & 4 \\ 6 & 8 \end{vmatrix} \qquad \text{(by the definition of a determinant)}$$

$$= 24 - 24 = 0$$

Or more cleverly,

$$\begin{vmatrix} 3 & 2 & 7 \\ 0 & 1 & -3 \\ 3 & 4 & 1 \end{vmatrix} = \begin{vmatrix} 3 & 2 & 7+(3\cdot 2) \\ 0 & 1 & -3+(3\cdot 1) \\ 3 & 4 & 1+(3\cdot 4) \end{vmatrix} \qquad \text{[by (D5)]}$$

$$= \begin{vmatrix} 3 & 2 & 13 \\ 0 & 1 & 0 \\ 3 & 4 & 13 \end{vmatrix}$$

$$= 3\cdot 13 \begin{vmatrix} 1 & 2 & 1 \\ 0 & 1 & 0 \\ 1 & 4 & 1 \end{vmatrix} \qquad \text{[by (D2)]}$$

$$= 0 \qquad \text{[by (D4)]}$$

Example 2 Evaluate the following determinant.

$$\Delta = \begin{vmatrix} 1 & 2 & -3 & 1 \\ -3 & 0 & 3 & 0 \\ 4 & 1 & 0 & 0 \\ 1 & 2 & 2 & 1 \end{vmatrix}$$

$$= \begin{vmatrix} 1 & 2+(-2)(1) & -3+(3)(1) & 1+(-1)(1) \\ -3 & 0+(-2)(-3) & 3+(3)(-3) & 0+(-1)(-3) \\ 4 & 1+(-2)(4) & 0+(3)(4) & 0+(-1)(4) \\ 1 & 2+(-2)(1) & 2+(3)(1) & 1+(-1)(1) \end{vmatrix} \qquad \text{[by (D5)]}$$

$$= \begin{vmatrix} 1 & 0 & 0 & 0 \\ -3 & 6 & -6 & 3 \\ 4 & -7 & 12 & -4 \\ 1 & 0 & 5 & 0 \end{vmatrix}$$

$$= \begin{vmatrix} 6 & -6 & 3 \\ -7 & 12 & -4 \\ 0 & 5 & 0 \end{vmatrix} \qquad \text{(using the definition of determinant)}$$

$$= \begin{vmatrix} 6 + (-2)(3) & -6 + (2)(3) & 3 \\ -7 + (-2)(-4) & 12 + (2)(-4) & -4 \\ 0 + (-2)(0) & 5 + (2)(0) & 0 \end{vmatrix} \qquad \text{[by (D5)]}$$

$$= \begin{vmatrix} 0 & 0 & 3 \\ 1 & 4 & -4 \\ 0 & 5 & 0 \end{vmatrix} \qquad \text{(using the definition of determinant)}$$

$$= 3\begin{vmatrix} 1 & 4 \\ 0 & 5 \end{vmatrix} = 15$$

Example 3 Evaluate the following determinant.

$$\Delta = \begin{vmatrix} 8 & 7 & 6 & 9 & 1 \\ 14 & 15 & 4 & 10 & 4 \\ 3 & 1 & 2 & 2 & -1 \\ -2 & 5 & -2 & 3 & 2 \\ -1 & 2 & -1 & 1 & 1 \end{vmatrix} = \begin{vmatrix} 8 + (-1)(6) & 7 + (2)(6) & 6 & 9 & 1 \\ 14 + (-1)(4) & 15 + (2)(4) & 4 & 10 & 4 \\ 3 + (-1)(2) & 1 + (2)(2) & 2 & 2 & -1 \\ -2 + (-1)(-2) & 5 + (2)(-2) & -2 & 3 & 2 \\ -1 + (-1)(-1) & 2 + (2)(-1) & -1 & 1 & 1 \end{vmatrix}$$

$$= \begin{vmatrix} 2 & 19 & 6 & 9 & 1 \\ 10 & 23 & 4 & 10 & 4 \\ 1 & 5 & 2 & 2 & -1 \\ 0 & 1 & -2 & 3 & 2 \\ 0 & 0 & -1 & 1 & 1 \end{vmatrix} \qquad \text{[by (D5)]}$$

$$= \begin{vmatrix} 2 & 19 & 6 + 9 & 9 & 1 \\ 10 & 23 & 4 + 10 & 10 & 4 \\ 1 & 5 & 2 + 2 & 2 & -1 \\ 0 & 1 & -2 + 3 & 3 & 2 \\ 0 & 0 & -1 + 1 & 1 & 1 \end{vmatrix} = \begin{vmatrix} 2 & 19 & 15 & 9 & 1 \\ 10 & 23 & 14 & 10 & 4 \\ 1 & 5 & 4 & 2 & -1 \\ 0 & 1 & 1 & 3 & 2 \\ 0 & 0 & 0 & 1 & 1 \end{vmatrix} \qquad \text{[by (D5)]}$$

$$= \begin{vmatrix} 2 & 19 - 15 & 15 & 9 & 1 \\ 10 & 23 - 14 & 14 & 10 & 4 \\ 1 & 5 - 4 & 4 & 2 & -1 \\ 0 & 1 - 1 & 1 & 3 & 2 \\ 0 & 0 - 0 & 0 & 1 & 1 \end{vmatrix} = \begin{vmatrix} 2 & 4 & 15 & 9 & 1 \\ 10 & 9 & 14 & 10 & 4 \\ 1 & 1 & 4 & 2 & -1 \\ 0 & 0 & 1 & 3 & 2 \\ 0 & 0 & 0 & 1 & 1 \end{vmatrix} \qquad \text{[by (D5)]}$$

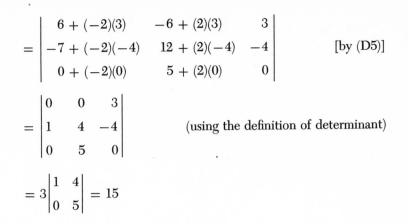

$$= \begin{vmatrix} 2-4 & 4 & 15 & 9-1 & 1 \\ 10-9 & 9 & 14 & 10-4 & 4 \\ 1-1 & 1 & 4 & 2+1 & -1 \\ 0 & 0 & 1 & 3-2 & 2 \\ 0 & 0 & 0 & 1-1 & 1 \end{vmatrix} = \begin{vmatrix} -2 & 4 & 15 & 8 & 1 \\ 1 & 9 & 14 & 6 & 4 \\ 0 & 1 & 4 & 3 & -1 \\ 0 & 0 & 1 & 1 & 2 \\ 0 & 0 & 0 & 0 & 1 \end{vmatrix} \qquad \text{[by (D5)]}$$

$$= \begin{vmatrix} -2 & 4 & 7 & 8 & 1 \\ 1 & 9 & 8 & 6 & 4 \\ 0 & 1 & 1 & 3 & -1 \\ 0 & 0 & 0 & 1 & 2 \\ 0 & 0 & 0 & 0 & 1 \end{vmatrix} = \begin{vmatrix} -2 & -3 & 7 & 8 & 1 \\ 1 & 1 & 8 & 6 & 4 \\ 0 & 0 & 1 & 3 & -1 \\ 0 & 0 & 0 & 1 & 2 \\ 0 & 0 & 0 & 0 & 1 \end{vmatrix} = \begin{vmatrix} 1 & -3 & 7 & 8 & 1 \\ 0 & 1 & 8 & 6 & 4 \\ 0 & 0 & 1 & 3 & -1 \\ 0 & 0 & 0 & 1 & 2 \\ 0 & 0 & 0 & 0 & 1 \end{vmatrix}$$

$$= 1\begin{vmatrix} 1 & 8 & 6 & 4 \\ 0 & 1 & 3 & -1 \\ 0 & 0 & 1 & 2 \\ 0 & 0 & 0 & 1 \end{vmatrix} - (-3)\begin{vmatrix} 0 & 8 & 6 & 4 \\ 0 & 1 & 3 & -1 \\ 0 & 0 & 1 & 2 \\ 0 & 0 & 0 & 1 \end{vmatrix} + 7\begin{vmatrix} 0 & 1 & 6 & 4 \\ 0 & 0 & 3 & -1 \\ 0 & 0 & 1 & 2 \\ 0 & 0 & 0 & 1 \end{vmatrix}$$

$$- 8\begin{vmatrix} 0 & 1 & 8 & 4 \\ 0 & 0 & 1 & -1 \\ 0 & 0 & 0 & 2 \\ 0 & 0 & 0 & 1 \end{vmatrix} + 1\begin{vmatrix} 0 & 1 & 8 & 6 \\ 0 & 0 & 1 & 3 \\ 0 & 0 & 0 & 1 \\ 0 & 0 & 0 & 0 \end{vmatrix} = \begin{vmatrix} 1 & 8 & 6 & 4 \\ 0 & 1 & 3 & -1 \\ 0 & 0 & 1 & 2 \\ 0 & 0 & 0 & 1 \end{vmatrix}$$

By repeating the expansion we eventually obtain the value of the determinant as 1.

Example 4 Show that

$$\begin{vmatrix} a_{11} & a_{12} & a_{13} & \dots & a_{1n} \\ 0 & a_{22} & a_{23} & \dots & a_{2n} \\ 0 & 0 & a_{33} & \dots & a_{3n} \\ 0 & 0 & 0 & \dots & a_{4n} \\ & & \vdots & & \\ 0 & 0 & 0 & \dots & a_{nn} \end{vmatrix} = a_{11}a_{22}a_{33} \dots a_{nn}$$

In words, if a matrix is such that all entries below the diagonal are zero (such a matrix is said to be **upper triangular**), its determinant is the product of its diagonal entries.

If $a_{11} = 0$, both the determinant and $a_{11}a_{22} \ldots a_{nn}$ are zero, and we are finished. If $a_{11} \neq 0$, by adding a suitable multiple of the first column to each of the remaining columns, we find that the determinant equals

$$\begin{vmatrix} a_{11} & 0 & 0 & \ldots & 0 \\ 0 & a_{22} & a_{23} & \ldots & a_{2n} \\ 0 & 0 & a_{33} & \ldots & a_{3n} \\ 0 & 0 & 0 & \ldots & a_{4n} \\ & & \vdots & & \\ 0 & 0 & 0 & \ldots & a_{nn} \end{vmatrix}$$

If $a_{22} = 0$, both the determinant and $a_{11}a_{22} \ldots a_{nn}$ are zero, so equality holds. If $a_{22} \neq 0$ by adding a suitable multiple of column two to each of the remaining columns, we find the determinant equals

$$\begin{vmatrix} a_{11} & 0 & 0 & \ldots & 0 \\ 0 & a_{22} & 0 & \ldots & 0 \\ 0 & 0 & a_{33} & \ldots & a_{3n} \\ 0 & 0 & 0 & \ldots & a_{4n} \\ & & \vdots & & \\ 0 & 0 & 0 & \ldots & a_{nn} \end{vmatrix}$$

By continuing this process we find that the determinant eventually equals

$$\begin{vmatrix} a_{11} & 0 & 0 & \ldots & 0 \\ 0 & a_{22} & 0 & \ldots & 0 \\ 0 & 0 & a_{33} & \ldots & 0 \\ & & \vdots & & \\ 0 & 0 & 0 & \ldots & a_{nn} \end{vmatrix}$$

By repeated use of (D2), this is just

$$a_{11}a_{22} \ldots a_{nn} \begin{vmatrix} 1 & 0 & 0 & \ldots & 0 \\ 0 & 1 & 0 & \ldots & 0 \\ 0 & 0 & 0 & \ldots & 0 \\ & & \vdots & & \\ 0 & 0 & 0 & \ldots & 1 \end{vmatrix} = a_{11}a_{22} \ldots a_{nn}$$

This method highlights a useful technique for evaluating determinants: Pick a nonzero entry a_{ij}. By adding a suitable multiple of the jth column to each of the other columns, we eventually obtain an equal determinant which has 0 in all positions in the ith row, except position (i, j). Then, pick a nonzero entry in another column, say $a_{i'j'}$. By adding a suitable multiple of column j' to each of the remaining columns, we eventually obtain an equal determinant, all of whose entries in row i are zero except in position (i, j) and all of whose entries in row i' are zero except in position (i', j').

By continuing the process we eventually obtain a determinant which has only one nonvanishing entry in each row. By interchaı.ging a suitable number of columns we can obtain a diagonal matrix whose determinant is easily evaluated.

This procedure is very similar to the method of Gaussian elimination discussed in chapter 1.

Example 5 Evaluate the following determinant.

$$\Delta = \begin{vmatrix} -1 & 1 & 0 & -1 & -2 \\ 8 & -10 & 0 & 17 & 20 \\ -2 & 0 & 4 & -2 & -3 \\ 0 & -1 & 1 & 1 & 1 \\ 7 & -7 & -7 & 14 & 14 \end{vmatrix}$$

Adding suitable multiples of column one to each of the remaining columns.

$$= \begin{vmatrix} -1 & 0 & 0 & 0 & 0 \\ 8 & -2 & 0 & 9 & 4 \\ -2 & -2 & 4 & 0 & 1 \\ 0 & -1 & 1 & 1 & 1 \\ 7 & 0 & -7 & 7 & 0 \end{vmatrix}$$

Add suitable multiples of column five to each of the remaining columns.

$$= \begin{vmatrix} -1 & 0 & 0 & 0 & 0 \\ 8 & 2 & -4 & 5 & 4 \\ -2 & -1 & 3 & -1 & 1 \\ 0 & 0 & 0 & 0 & 1 \\ 7 & 0 & -7 & 7 & 0 \end{vmatrix}$$

Add suitable multiples of column four to each of the other columns.

$$= \begin{vmatrix} -1 & 0 & 0 & 0 & 0 \\ 3 & 2 & 1 & 5 & 4 \\ -1 & -1 & 2 & -1 & 1 \\ 0 & 0 & 0 & 0 & 1 \\ 0 & 0 & 0 & 7 & 0 \end{vmatrix}$$

Add multiples of column two to each of the other columns.

$$= \begin{vmatrix} -1 & 0 & 0 & 0 & 0 \\ 1 & 2 & 5 & 3 & 6 \\ 0 & -1 & 0 & 0 & 0 \\ 0 & 0 & 0 & 0 & 1 \\ 0 & 0 & 0 & 7 & 0 \end{vmatrix}$$

Add multiples of column three to the remaining columns.

$$= \begin{vmatrix} -1 & 0 & 0 & 0 & 0 \\ 0 & 0 & 5 & 0 & 0 \\ 0 & -1 & 0 & 0 & 0 \\ 0 & 0 & 0 & 0 & 1 \\ 0 & 0 & 0 & 7 & 0 \end{vmatrix}$$

Interchange column two and three, and column four and five.

$$= \begin{vmatrix} -1 & 0 & 0 & 0 & 0 \\ 0 & 5 & 0 & 0 & 0 \\ 0 & 0 & -1 & 0 & 0 \\ 0 & 0 & 0 & 1 & 0 \\ 0 & 0 & 0 & 0 & 7 \end{vmatrix}$$

$$= 35$$

Example 6 Without evaluating prove that

$$\Delta = \begin{vmatrix} na_1 + b_1 & nb_1 + c_1 & nc_1 + d_1 & nd_1 + e_1 & ne_1 + a_1 \\ na_2 + b_2 & nb_2 + c_2 & nc_2 + d_2 & nd_2 + e_2 & ne_2 + a_2 \\ na_3 + b_3 & nb_3 + c_3 & nc_3 + d_3 & nd_3 + e_3 & ne_3 + a_3 \\ na_4 + b_4 & nb_4 + c_4 & nc_4 + d_4 & nd_4 + e_4 & ne_4 + a_4 \\ na_5 + b_5 & nb_5 + c_5 & nc_5 + d_5 & nd_5 + e_5 & ne_5 + a_5 \end{vmatrix} = (1 + n^5) \begin{vmatrix} a_1 & b_1 & c_1 & d_1 & e_1 \\ a_2 & b_2 & c_2 & d_2 & e_2 \\ a_3 & b_3 & c_3 & d_3 & e_3 \\ a_4 & b_4 & c_4 & d_4 & e_4 \\ a_5 & b_5 & c_5 & d_5 & e_5 \end{vmatrix}$$

Rewrite as

$$\begin{aligned} \Delta &= \det\,[nA + B, nB + C, nC + D, nD + E, nE + A] \\ &= \det\,[nA + B, nB + C - n(nA + B), nC + D, nD + E, nE + A] \\ &= \det\,[nA + B, C - n^2A, nC + D, nD + E, nE + A] \\ &= \det\,[nA + B, C - n^2A, nC + D - n(C - n^2A), nD + E, nE + A] \\ &= \det\,[nA + B, C - n^2A, D + n^3A, nD + E, nE + A] \end{aligned}$$

$$= \det [nA + B, C - n^2A, D + n^3A, nD + E - n(D + n^3A), nE + A]$$
$$= \det [nA + B, C - n^2A, D + n^3A, E - n^4A, nE + A - n(E - n^4A)]$$
$$= \det [nA + B, C - n^2A, D + n^3A, E - n^4A, (1 + n^5)A]$$
$$= (1 + n^5) \det [nA + B, C - n^2A, D + n^3A, E - n^4A, A]$$
$$= (1 + n^5) \det [B, C, D, E, A]$$
$$= (1 + n^5) \det [A, B, C, D, E]$$

where in the last step we interchanged A with E, then D, then C, and lastly B. Since four interchanges occur, there was no sign change.

EXERCISES

1. Calculate

(a) $\begin{vmatrix} 7 & -5 & 3 \\ 12 & 1 & 6 \\ 2 & -1 & 1 \end{vmatrix}$ (b) $\begin{vmatrix} 7 & 15 & 7 & -21 \\ 3 & 6 & 2 & -9 \\ -1 & -2 & -1 & 3 \\ 1 & 4 & 7 & -2 \end{vmatrix}$ (c) $\begin{vmatrix} 2 & 2 & 1 & 0 \\ 6 & 9 & 3 & 3 \\ 2 & 10 & 2 & 8 \\ 2 & 8 & 1 & 6 \end{vmatrix}$

(d) $\begin{vmatrix} 0 & 1 & 0 & 0 & 0 \\ 1 & 0 & 1 & 0 & 0 \\ 0 & 1 & 0 & 1 & 0 \\ 0 & 0 & 1 & 0 & 1 \\ 0 & 0 & 0 & 1 & 0 \end{vmatrix}$ (e) $\begin{vmatrix} 0 & 1 & 0 & 0 \\ 1 & 1 & 1 & 1 \\ 0 & 1 & 1 & 1 \\ 0 & 1 & 2 & 1 \end{vmatrix}$ (f) $\begin{vmatrix} 2 & 2 & 1 \\ -2 & 3 & 2 \\ 2 & 3 & -2 \end{vmatrix}$

(g) $\begin{vmatrix} 4 & 1 & 2 \\ 8 & 5 & 5 \\ 10 & 7 & 7 \end{vmatrix}$ (h) $\begin{vmatrix} 1 & 1 & 1 & 1 & 1 \\ 1 & 0 & 1 & 1 & 1 \\ 1 & 1 & 0 & 1 & 1 \\ 1 & 1 & 1 & 0 & 1 \\ 1 & 1 & 1 & 1 & 0 \end{vmatrix}$ (i) $\begin{vmatrix} 1 & 0 & 0 & 0 & 0 \\ 3 & 2 & 0 & 0 & 0 \\ 6 & 1 & 3 & 1 & 1 \\ 2 & -1 & -1 & 1 & 1 \\ 4 & -4 & 1 & 4 & 3 \end{vmatrix}$

(j) $\begin{vmatrix} 3 & 4 & 5 \\ 1 & 2 & 3 \\ -2 & 5 & -4 \end{vmatrix}$ (k) $\begin{vmatrix} 1 & 1 & 1 & 1 \\ 1 & 2 & 1 & 1 \\ 1 & 1 & 3 & 1 \\ 1 & 1 & 1 & 4 \end{vmatrix}$

2. Show that

$$\begin{vmatrix} a_1 & b_1 & c_1 & d_1 \\ a_2 & b_2 & c_2 & d_2 \\ a_3 & b_3 & c_3 & d_3 \\ a_4 & b_4 & c_4 & d_4 \end{vmatrix} = \begin{vmatrix} -a_1 & -d_1 & -c_1 & b_1 \\ -a_2 & -d_2 & -c_2 & b_2 \\ -a_3 & -d_3 & -c_3 & b_3 \\ -a_4 & -d_4 & -c_4 & b_4 \end{vmatrix}$$

3. Show that

$$\begin{vmatrix} a+b & 1 & b+1 \\ b+c & 1 & c+1 \\ c+d & 1 & d+1 \end{vmatrix} = \begin{vmatrix} a & 1 & b \\ b & 1 & c \\ c & 1 & d \end{vmatrix}$$

4. If one column of a matrix is a multiple of another column, show that the determinant of the matrix is zero.

5. Show that

$$\begin{vmatrix} a_{11} & a_{12} \\ a_{21} & a_{22} \end{vmatrix} = \begin{vmatrix} 1 & \alpha & \beta & \gamma \\ 0 & 1 & \delta & \epsilon \\ 0 & 0 & a_{11} & a_{12} \\ 0 & 0 & a_{21} & a_{22} \end{vmatrix}$$

6. Show that

$$\begin{vmatrix} a_1 & b_1 + \alpha a_1 & c_1 + \beta a_1 + \gamma b_1 & d_1 + \epsilon c_1 \\ a_2 & b_2 + \alpha a_2 & c_2 + \beta a_2 + \gamma b_2 & d_2 + \epsilon c_2 \\ a_3 & b_3 + \alpha a_3 & c_3 + \beta a_3 + \gamma b_3 & d_3 + \epsilon c_3 \\ a_4 & b_4 + \alpha a_4 & c_4 + \beta a_4 + \gamma b_4 & d_4 + \epsilon c_4 \end{vmatrix} = \begin{vmatrix} a_1 & b_1 & c_1 & d_1 \\ a_2 & b_2 & c_2 & d_2 \\ a_3 & b_3 & c_3 & d_3 \\ a_4 & b_4 & c_4 & d_4 \end{vmatrix}$$

7. What is

$$\begin{vmatrix} a_{11} & 0 & 0 & \cdots & 0 \\ a_{21} & a_{22} & 0 & \cdots & 0 \\ a_{31} & a_{32} & a_{33} & \cdots & 0 \\ & & \vdots & & \\ a_{n1} & a_{n2} & a_{n3} & \cdots & a_{nn} \end{vmatrix} ?$$

8. What is

$$\begin{vmatrix} 0 & 0 & 0 & \cdots & 0 & 0 & a_{1n} \\ 0 & 0 & 0 & \cdots & 0 & a_{2,n-1} & a_{2n} \\ 0 & 0 & 0 & \cdots & a_{3,n-2} & a_{3,n-1} & a_{3n} \\ & & & \vdots & & & \\ 0 & a_{n-1,2} & a_{n-1,3} & \cdots & a_{n-1,n-2} & a_{n-1,n-1} & a_{n-1,n} \\ a_{n1} & a_{n2} & a_{n3} & \cdots & a_{n,n-2} & a_{n,n-1} & a_{nn} \end{vmatrix} ?$$

9. Show that $\det \alpha A = \alpha^n \det A$, if n is the order of the matrix A.

10. Show that

$$\det [X_1 + X_2, X_2 + X_3, X_3 + X_4, \ldots, X_{n-1} + X_n, X_n + X_1]$$
$$= (1 + (-1)^{n+1}) \det [X_1, X_2, X_3, \ldots, X_n]$$

11. Show that

$$\det [X_2, X_3, X_4, \ldots, X_n, X_1] = (-1)^{n-1} \det [X_1, X_2, X_3, \ldots, X_n]$$

12. Show that

$$\begin{vmatrix} 1 & \alpha & \beta\gamma \\ 1 & \beta & \gamma\alpha \\ 1 & \gamma & \alpha\beta \end{vmatrix} = \begin{vmatrix} 1 & \alpha & \alpha^2 \\ 1 & \beta & \beta^2 \\ 1 & \gamma & \gamma^2 \end{vmatrix}$$

13. Show that

$$\begin{vmatrix} 1 & a_1 & a_2 & a_3 & a_4 + a_5 \\ 1 & a_2 & a_3 & a_4 & a_1 + a_5 \\ 1 & a_3 & a_4 & a_5 & a_1 + a_2 \\ 1 & a_4 & a_5 & a_1 & a_2 + a_3 \\ 1 & a_5 & a_1 & a_2 & a_3 + a_4 \end{vmatrix} = 0$$

14. Show that

$$\begin{vmatrix} 1 & 1 & 1 \\ a & b & c \\ a^3 & b^3 & c^3 \end{vmatrix} = (b - a)(c - a)(c - b)(a + b + c)$$

15. Show that

$$\begin{vmatrix} x & 0 & 0 & \cdots & 0 & -1 \\ 0 & x & 0 & \cdots & 0 & 0 \\ 0 & 0 & x & \cdots & 0 & 0 \\ & & & \vdots & & \\ 0 & 0 & 0 & \cdots & x & 0 \\ -1 & 0 & 0 & \cdots & 0 & x \end{vmatrix}_n = x^{n-2}(x^2 - 1)$$

16. Let A be a matrix such that each column of A has exactly one nonvanishing entry and that nonvanishing entry is 1. Show that $\det A = +1, -1$, or 0. Exhibit an example of each of the three cases.

17. Show that

$$\begin{aligned} \det [X_1, X_2, X_3] &= -\det [X_2, X_1, X_3] = \det [X_2, X_3, X_1] \\ &= -\det [X_3, X_2, X_1] = \det [X_3, X_1, X_2] \\ &= -\det [X_1, X_3, X_2] \end{aligned}$$

18. Without evaluating, show that

$$\begin{vmatrix} \alpha^2 & (\alpha+1)^2 & (\alpha+2)^2 & (\alpha+3)^2 \\ \beta^2 & (\beta+1)^2 & (\beta+2)^2 & (\beta+3)^2 \\ \gamma^2 & (\gamma+1)^2 & (\gamma+2)^2 & (\gamma+3)^2 \\ \delta^2 & (\delta+1)^2 & (\delta+2)^2 & (\delta+3)^2 \end{vmatrix} = 0$$

19. Find all values of λ, such that

(a) $\begin{vmatrix} \lambda - 1 & 0 & 0 \\ 0 & \lambda - 2 & 0 \\ 0 & 0 & \lambda - 3 \end{vmatrix} = 0$

(b) $\begin{vmatrix} 1 & \lambda & \lambda \\ \lambda & 1 & 0 \\ \lambda & 0 & 1 \end{vmatrix} = 0$

(c) $\begin{vmatrix} 1 & \lambda & \lambda^2 \\ 1 & 1 & 1 \\ 1 & 2 & 4 \end{vmatrix} = 0$

20. Show that

$$\begin{vmatrix} x & y & x+y \\ y & x+y & x \\ x+y & x & y \end{vmatrix} = -2(x^3 + y^3)$$

21. Show that

$$\begin{vmatrix} a & b & d \\ c & 1 & 1 \\ e & 1 & 1 \end{vmatrix} = (e - c)(b - d)$$

22. Show that

$$\begin{vmatrix} 1 & -(x_1 + x_2) & x_1x_2 & 0 \\ 0 & 1 & -(x_1 + x_2) & x_1x_2 \\ 1 & -(y_1 + y_2) & y_1y_2 & 0 \\ 0 & 1 & -(y_1 + y_2) & y_1y_2 \end{vmatrix} = (x_1 - y_1)(x_1 - y_2)(x_2 - y_1)(x_2 - y_2)$$

23. Using exercise 22, show that the polynomials

$$a_0x^2 + a_1x + a_2 = 0, \qquad a_0 \neq 0$$
$$b_0x^2 + b_1x + b_2 = 0, \qquad b_0 \neq 0$$

have a common root if and only if

$$\begin{vmatrix} a_0 & a_1 & a_2 & 0 \\ 0 & a_0 & a_1 & a_2 \\ b_0 & b_1 & b_2 & 0 \\ 0 & b_0 & b_1 & b_2 \end{vmatrix} = 0$$

24. Show that

$$\begin{vmatrix} x & 0 & 0 & 0 & \cdots & 0 & \alpha_0 \\ -1 & x & 0 & 0 & \cdots & 0 & \alpha_1 \\ 0 & -1 & x & 0 & \cdots & 0 & \alpha_2 \\ 0 & 0 & -1 & x & \cdots & 0 & \alpha_3 \\ & & & \vdots & & & \\ 0 & 0 & 0 & 0 & \cdots & -1 & \alpha_{n-1} + x \end{vmatrix} = x^n + \alpha_{n-1}x^{n-1} + \cdots + \alpha_0$$

25. Show that

$$\begin{vmatrix} 1 & 1 & 1 & 1 & 1 & \cdots & 1 & 1 \\ 1 & a_{22} & 1 & 1 & 1 & \cdots & 1 & 1 \\ 1 & a_{32} & a_{33} & 1 & 1 & \cdots & 1 & 1 \\ 1 & a_{42} & a_{43} & a_{44} & 1 & \cdots & 1 & 1 \\ 1 & a_{52} & a_{53} & a_{54} & a_{55} & \cdots & 1 & 1 \\ & & & \vdots & & & & \\ 1 & a_{n2} & a_{n3} & a_{n4} & a_{n5} & \cdots & a_{n,n-1} & a_{nn} \end{vmatrix} = (a_{22} - 1)(a_{33} - 1) \cdots (a_{nn} - 1)$$

26. In $\boldsymbol{R}^3$, let

$$e_1 = \begin{bmatrix} 1 \\ 0 \\ 0 \end{bmatrix}, \quad e_2 = \begin{bmatrix} 0 \\ 1 \\ 0 \end{bmatrix}, \quad e_3 = \begin{bmatrix} 0 \\ 0 \\ 1 \end{bmatrix}$$

(a) If

$$x = \begin{bmatrix} x_1 \\ x_2 \\ x_3 \end{bmatrix}$$

show that $x = x_1 e_1 + x_2 e_2 + x_3 e_3$

(b) If D is a function defined on 3×3 matrices satisfying (D1)–(D4), show that

$$D[A, B, C] = D[\alpha_1 e_1 + \alpha_2 e_2 + a_3 e_3, \beta_1 e_1 + \beta_2 e_2 + \beta_3 e_3, \gamma_1 e_1 + \gamma_2 e_2 + \gamma_3 e_3]$$

$$= \sum_{\substack{1 \leq i \leq 3 \\ 1 \leq j \leq 3 \\ 1 \leq k \leq 3}} \alpha_i \beta_j \gamma_k \, D[e_i, e_j, e_k]$$

(c) With D as in part (b), show that $D[e_i, e_j, e_k] = 0$ if $i = j$, $i = k$, or $j = k$, and

$$\begin{aligned} D[e_1, e_2, e_3] &= -D[e_2, e_1, e_3] \\ = D[e_2, e_3, e_1] &= -D[e_3, e_2, e_1] \\ = D[e_3, e_1, e_2] &= -D[e_1, e_3, e_2] \end{aligned}$$

(d) Using parts (b) and (c), show that $D[A] = \det A$.

27. Find all roots x, to the following equation.

$$\begin{vmatrix} 1 & x_1 & x_2 & x_3 & x_4 \\ 1 & x & x_2 & x_3 & x_4 \\ 1 & x_1 & x & x_3 & x_4 \\ 1 & x_1 & x_2 & x & x_4 \\ 1 & x_1 & x_2 & x_3 & x \end{vmatrix} = 0$$

28. Prove that

$$\begin{vmatrix} x & y & 0 & \cdots & 0 & 0 \\ 0 & x & y & \cdots & 0 & 0 \\ 0 & 0 & x & \cdots & 0 & 0 \\ & & & \vdots & & \\ 0 & 0 & 0 & \cdots & x & y \\ y & 0 & 0 & \cdots & 0 & x \end{vmatrix}_n = x^n - (-y)^n$$

3 MULTIPLICATIVE PROPERTY OF DETERMINANTS

In this section we prove an important multiplicative property of the determinant, namely, if A and B are two $n \times n$ matrices, then $\det AB = \det A \det B$.

Before we embark on a proof of this fact, let us consider some of its consequences. Suppose the $n \times n$ matrix A is invertible, then $AA^{-1} = I_n$ implies that $\det A \det A^{-1} = \det I_n = 1$. Thus, $\det A \neq 0$. Stated otherwise, if the determinant of a matrix is zero, it cannot have an inverse. Later in this chapter we prove that if $\det A \neq 0$, then A does indeed have an inverse.

Theorem $\det A \det B = \det AB$.

Proof Let $A = [A_1, A_2, \ldots, A_n]$ and $B = [b_{ij}]_{(n,n)}$.

Then

$$AB = \left[\sum_{i=1}^{n} b_{i1}A_i, \sum_{i=1}^{n} b_{i2}A_i, \ldots, \sum_{i=1}^{n} b_{in}A_i\right]$$

so

$$\det AB = \det\left[\sum_{i=1}^{n} b_{i1}A_i, \sum_{i=1}^{n} b_{i2}A_i, \ldots, \sum_{i=1}^{n} b_{in}A_i\right]$$

$$= \sum_{\substack{1\le i_1\le n\\ 1\le i_2\le n\\ \vdots\\ 1\le i_n\le n}} b_{i_1 1}b_{i_2 2}, \ldots, b_{i_n n}(\det [A_{i_1}, A_{i_2}, \ldots, A_{i_n}])$$

The last step was obtained by repeated use of (D1) and (D2) of §3.2. The summation is intended to extend over all n-tuples of integers between 1 and n.

If for some term in the above sum $i_k = i_{k'}$, then $\det [A_{i_1}, \ldots, A_{i_n}]$ has two equal columns: the kth and the k'th. Thus, it is zero. So we may assume the sum extends only over the n-tuples $(i_1, i_2, \ldots, i_n)$, such that no two elements of the n-tuple are equal, or, in other words, over all possible rearrangements of the n-tuple $(1, 2, 3, \ldots, n)$.

If the n-tuple $(i_1, \ldots, i_n)$ is a rearrangement of $(1, 2, \ldots, n)$, then by interchanging sufficiently many columns, we have

$$\det [A_{i_1}, \ldots, A_{i_n}] = \pm\det [A_1, \ldots, A_n]$$

Thus,

$$\det AB = \sum_{i_1, i_2, \ldots, n} \pm b_{i_1 1}, b_{i_2 2}, \ldots, b_{i_n n} \det [A_1, \ldots, A_n]$$

$$= \left(\sum_{i_1, i_2, \ldots, i_n} \pm b_{i_1 1}, b_{i_2 2}, \ldots, b_{i_n n} \right)(\det A)$$

This identity holds for all matrices A. In particular if we choose $A = I_n$, we have

$$\det B = \sum_{i_1, i_2, \ldots, i_n} \pm b_{i_1 1}, b_{i_2 2} \ \ldots, b_{i_n n}$$

So $\det AB = \det A \det B$. ∎

The reader who found the above proof too condensed might profit from writing out the details explicitly in the 3×3 case.

The following example demonstrates the usefulness of the above theorem.

Example 1 Show that

$$\Delta = \begin{vmatrix} \cos x & \sin x & \cos x & \sin x \\ \cos 2x & \sin 2x & 2 \cos 2x & 2 \sin 2x \\ \cos 3x & \sin 3x & 3 \cos 3x & 3 \sin 3x \\ \cos 4x & \sin 4x & 4 \cos 4x & 4 \sin 4x \end{vmatrix} = 4 \sin^4 x$$

As a preliminary step, we may subtract the first column from the third column and the second column from the fourth column to obtain

$$\Delta = \begin{vmatrix} \cos x & \sin x & 0 & 0 \\ \cos 2x & \sin 2x & \cos 2x & \sin 2x \\ \cos 3x & \sin 3x & 2 \cos 3x & 2 \sin 3x \\ \cos 4x & \sin 4x & 3 \cos 4x & 3 \sin 4x \end{vmatrix}$$

Observe that

$$\begin{bmatrix} \cos x & \sin x & 0 & 0 \\ \cos 2x & \sin 2x & \cos 2x & \sin 2x \\ \cos 3x & \sin 3x & 2 \cos 3x & 2 \sin 3x \\ \cos 4x & \sin 4x & 3 \cos 4x & 3 \sin 4x \end{bmatrix} \begin{bmatrix} \cos x & -\sin x & 0 & 0 \\ \sin x & \cos x & 0 & 0 \\ 0 & 0 & \cos 2x & -\sin 2x \\ 0 & 0 & \sin 2x & \cos 2x \end{bmatrix} = \begin{bmatrix} 1 & 0 & 0 & 0 \\ \cos x & \sin x & 1 & 0 \\ \cos 2x & \sin 2x & 2 \cos x & 2 \sin x \\ \cos 3x & \sin 3x & 3 \cos 2x & 3 \sin 2x \end{bmatrix}$$

Using the definition of the determinant, we find that the determinant of the last matrix is

$$\sin x(6 \cos x \sin 2x - 6 \sin x \cos 2x) - (3 \sin^2 2x - 2 \sin x \sin 3x)$$

By the use of the addition formulas for sines and cosines this equals $4 \sin^4 x$.

Also, we have

$$\begin{vmatrix} \cos x & -\sin x & 0 & 0 \\ \sin x & \cos x & 0 & 0 \\ 0 & 0 & \cos 2x & -\sin 2x \\ 0 & 0 & \sin 2x & \cos 2x \end{vmatrix} = \cos x \begin{vmatrix} \cos x & 0 & 0 \\ 0 & \cos 2x & -\sin 2x \\ 0 & \cos 2x & \cos 2x \end{vmatrix} + \sin x \begin{vmatrix} \sin x & 0 & 0 \\ 0 & \cos 2x & -\sin 2x \\ 0 & \sin 2x & \cos 2x \end{vmatrix}$$

$$= \cos^2 x(\cos^2 2x + \sin^2 x) + \sin^2 x(\cos^2 x + \sin^2 2x)$$

$$= \cos^2 x + \sin^2 x$$

$$= 1$$

By applying the determinant to the above matrix equation, and using $\det AB = \det A \det B$, we have $\Delta \cdot 1 = 4 \sin^4 x$.

For other examples of this type, the reader may consult exercises 9, 14–17 at the end of this section.

EXERCISES

1. Carry out the argument in the text that $\det AB = \det A \det B$ explicitly in the 3×3 case.
2. Show that $\det A^n = (\det A)^n$ if n is a positive integer.
3. If A is invertible, show that $\det A^{-1} = (\det A)^{-1}$.
4. If A is a real $n \times n$ matrix and $A^k = I_n$ for k odd, show that $\det A = 1$.
5. If $A^2 = I_n$, show that $\det A = \pm 1$.
6. If C is invertible, show that $\det CGC^{-1} = \det G$.
7. By multiplying

$$\begin{bmatrix} a & b \\ -b & a \end{bmatrix} \text{ and } \begin{bmatrix} c & d \\ -d & c \end{bmatrix}$$

show that $(a^2 + b^2)(c^2 + d^2) = (ac - bd)^2 + (ad + bc)^2$.

8. If A is nilpotent (that is, $A^n = 0$ for some positive integer n), show that $\det A = 0$.

9. Let

$$Q = \begin{bmatrix} b & c & 0 \\ a & 0 & c \\ 0 & a & b \end{bmatrix}$$

By calculating det Q, det Q^T, and QQ^T, show that

$$\det \begin{bmatrix} b^2 + c^2 & ab & ac \\ ab & a^2 + c^2 & cb \\ ac & bc & a^2 + b^2 \end{bmatrix} = 4a^2b^2c^2$$

10. Show that the matrix

$$A = \begin{bmatrix} -1 & 0 & 0 \\ 0 & 0 & 1 \\ 0 & -1 & 0 \end{bmatrix}$$

does not have a real square root, i.e., there is no real matrix B, such that $B^2 = A$.

11. Even if $AB \neq BA$, show that det AB = det BA.

12. Let X_1, X_2, and X_3 be column 3-vectors. By showing that

$$[X_1, X_2, X_3] \begin{bmatrix} 0 & a & b \\ a & 0 & c \\ b & c & 0 \end{bmatrix} = [aX_2 + bX_3, aX_1 + cX_3, bX_1 + cX_2]$$

prove that

$$\det [aX_2 + bX_3, aX_1 + cX_3, bX_1 + cX_2] = 2abc \det [X_1, X_2, X_3]$$

13. By multiplying

$$\begin{bmatrix} \alpha & 0 & 0 \\ 0 & \beta & 0 \\ 0 & 0 & \gamma \end{bmatrix} A, \text{ where } A = \begin{bmatrix} a_1 & a_2 & a_3 \\ b_1 & b_2 & b_3 \\ c_1 & c_2 & c_3 \end{bmatrix}$$

show that

$$\det \begin{bmatrix} \alpha a_1 & \alpha a_2 & \alpha a_3 \\ \beta b_1 & \beta b_2 & \beta b_3 \\ \gamma c_1 & \gamma c_2 & \gamma c_3 \end{bmatrix} = \alpha\beta\gamma \det \begin{bmatrix} a_1 & a_2 & a_3 \\ b_1 & b_2 & b_3 \\ c_1 & c_2 & c_3 \end{bmatrix}$$

14. By calculating

$$\begin{bmatrix} 1 & 0 & a \\ 0 & 1 & 0 \\ a & 0 & 1 \end{bmatrix} \begin{bmatrix} 1 & b & 0 \\ b & 1 & b \\ 0 & b & 1 \end{bmatrix}$$

show that

$$\begin{vmatrix} 1 & (a+1)b & a \\ b & 1 & b \\ a & (a+1)b & 1 \end{vmatrix} = (1-a^2)(1-2b^2)$$

15. (a) By calculating

$$AB = \begin{bmatrix} \sin\alpha_1 & \cos\alpha_2 & 0 \\ \sin\alpha_2 & \cos\alpha_2 & 0 \\ \sin\alpha_3 & \cos\alpha_3 & 0 \end{bmatrix} \begin{bmatrix} \cos\alpha_1 & \cos\alpha_2 & \cos\alpha_3 \\ \sin\alpha_1 & \sin\alpha_2 & \sin\alpha_3 \\ 0 & 0 & 0 \end{bmatrix}$$

show that

$$\det \begin{bmatrix} \sin 2\alpha_1 & \sin(\alpha_1+\alpha_2) & \sin(\alpha_1+\alpha_3) \\ \sin(\alpha_1+\alpha_2) & \sin 2\alpha_2 & \sin(\alpha_2+\alpha_3) \\ \sin(\alpha_1+\alpha_3) & \sin(\alpha_2+\alpha_3) & \sin 2\alpha_3 \end{bmatrix} = 0$$

(b) Generalize to the case where AB is an $n \times n$ matrix.

16. (a) Let

$$Q = \begin{bmatrix} 0 & 1 & 1 \\ 1 & 0 & 1 \\ 1 & 1 & 0 \end{bmatrix}$$

Show that

$$Q^{-1} = \tfrac{1}{2}\begin{bmatrix} -1 & 1 & 1 \\ 1 & -1 & 1 \\ 1 & 1 & -1 \end{bmatrix}$$

(b) Let

$$H = \tfrac{1}{2}\begin{bmatrix} b+c & -b+c & b-c \\ -a+c & a+c & a-c \\ -a+b & a-b & a+b \end{bmatrix}$$

By calculating $Q^{-1}HQ$, show that $\det H = abc$.

17. (a) Let

$$Q = \begin{bmatrix} 1 & 1 & 1 & 1 \\ 1 & i & -1 & -i \\ 1 & -1 & 1 & -1 \\ 1 & -i & -1 & i \end{bmatrix}, \text{ with } i^2 = -1$$

Show that

$$Q^{-1} = \tfrac{1}{4}\begin{bmatrix} 1 & 1 & 1 & 1 \\ 1 & -i & -1 & i \\ 1 & -1 & 1 & -1 \\ 1 & i & -1 & -i \end{bmatrix}$$

(b) Let

$$M = \begin{bmatrix} 0 & 0 & 0 & 1 \\ 1 & 0 & 0 & 0 \\ 0 & 1 & 0 & 0 \\ 0 & 0 & 1 & 0 \end{bmatrix}$$

Show that

$$A = \begin{bmatrix} \alpha_0 & \alpha_3 & \alpha_2 & \alpha_1 \\ \alpha_1 & \alpha_0 & \alpha_3 & \alpha_2 \\ \alpha_2 & \alpha_1 & \alpha_0 & \alpha_3 \\ \alpha_3 & \alpha_2 & \alpha_1 & \alpha_0 \end{bmatrix} = \alpha_0 I_4 + \alpha_1 M + \alpha_2 M^2 + \alpha_3 M^3$$

(c) By calculating $Q^{-1}AQ$, show that

$$\begin{vmatrix} \alpha_0 & \alpha_3 & \alpha_2 & \alpha_1 \\ \alpha_1 & \alpha_0 & \alpha_3 & \alpha_2 \\ \alpha_2 & \alpha_1 & \alpha_0 & \alpha_3 \\ \alpha_3 & \alpha_2 & \alpha_1 & \alpha_0 \end{vmatrix} = ((\alpha_0 + \alpha_2)^2 - (\alpha_1 + \alpha_3)^2)((\alpha_0 - \alpha_2)^2 + (\alpha_1 - \alpha_3)^2)$$

18. Two matrices A and B are said to anticommute if $AB = -BA$. If A and B are anticommuting 3×3 matrices, show that one or the other is noninvertible.

19. We define a permutation to be a rearrangement of the integers $(1, 2, \ldots, n)$. Thus, $(1, 2, 3)$, $(1, 3, 2)$, and $(2, 3, 1)$ are all permutations. A transposition is a rearrangement by interchanging two numbers. Thus, $(1, 3, 2)$ is obtained by a transposition of $(1, 2, 3)$.

Let e_i be the vector which has zeroes in all components but the ith and 1 as its ith component.

(a) If $(j_1, j_2, \ldots, j_n)$ is a permutation, we define $\operatorname{sgn}(j_1, j_2, \ldots, j_n) = \det[e_{j_1}, e_{j_2}, \ldots, e_{j_n}]$. Show that $\operatorname{sgn}(j_1, j_2, \ldots, j_n) = \pm 1$.

(b) If $(j_1', j_2', \ldots, j_n')$ is obtained from $(j_1, j_2, \ldots, j_n)$ by a single transposition, show that $\operatorname{sgn}(j_1', j_2', \ldots, j_n') = -\operatorname{sgn}(j_1, j_2, \ldots, j_n)$.

(c) Show that the sign of a permutation is $+1$ if and only if it can be obtained from $(1, 2, \ldots, n)$ by an even number of transpositions, -1 if and only if it can be obtained from $(1, 2, \ldots, n)$ by an odd number of transpositions.

20. By going through the proof that $\det A \det B = \det AB$, show that $\det A = \sum_{(j_1, j_2, \ldots, j_n)} \operatorname{sgn}(j_1, j_2, \ldots, j_n) a_{1j_1} \cdots a_{nj_n}$, where the summation extends over all permutations of $(1, 2, \ldots, n)$.

4 COFACTOR EXPANSIONS AND ROW OPERATIONS

If $A = [a_{ij}]_{(nn)}$ is a matrix, we define its (i, j)th **cofactor**, written A_{ij}, by $A_{ij} = (-1)^{i+j} \det \hat{A}_{ij}$, where $\hat{A}_{ij}$ is the $(n-1) \times (n-1)$ matrix obtained by deleting the ith row and jth column of A.

For example, if

$$A = \begin{bmatrix} 1 & 2 & 1 \\ 0 & 1 & 2 \\ -3 & 4 & 1 \end{bmatrix}$$

then

$$A_{11} = (-1)^{1+1} \det \begin{bmatrix} 1 & 2 \\ 4 & 1 \end{bmatrix} = -7$$

$$A_{13} = (-1)^{1+3} \det \begin{bmatrix} 0 & 1 \\ -3 & 4 \end{bmatrix} = 3$$

$$A_{32} = (-1)^{3+2} \det \begin{bmatrix} 1 & 1 \\ 0 & 2 \end{bmatrix} = -2$$

Using this notation our original definition for the determinant may be rewritten as

$$\det A = \sum_{j=1}^{n} a_{1j} A_{1j}$$

In words the determinant of A is the sum of the entries in the first row times their corresponding cofactors. We intend to extend this formula to

$$\det A = \sum_{j=1}^{n} a_{ij}A_{ij}$$

That is, the determinant of A is the sum of the elements in the ith row times their corresponding cofactors.

This formula is called the **cofactor expansion** along the ith row and it allows us to evaluate a determinant in a variety of ways. For example, in the 3×3 case

$$\begin{vmatrix} a_{11} & a_{12} & a_{13} \\ a_{21} & a_{22} & a_{23} \\ a_{31} & a_{32} & a_{33} \end{vmatrix} = -a_{21}\begin{vmatrix} a_{12} & a_{13} \\ a_{32} & a_{33} \end{vmatrix} + a_{22}\begin{vmatrix} a_{11} & a_{13} \\ a_{31} & a_{33} \end{vmatrix} - a_{23}\begin{vmatrix} a_{11} & a_{12} \\ a_{31} & a_{32} \end{vmatrix}$$

or,

$$\begin{vmatrix} a_{11} & a_{12} & a_{13} \\ a_{21} & a_{22} & a_{23} \\ a_{31} & a_{32} & a_{33} \end{vmatrix} = a_{31}\begin{vmatrix} a_{12} & a_{13} \\ a_{22} & a_{23} \end{vmatrix} - a_{32}\begin{vmatrix} a_{11} & a_{13} \\ a_{21} & a_{23} \end{vmatrix} + a_{33}\begin{vmatrix} a_{11} & a_{12} \\ a_{21} & a_{22} \end{vmatrix}$$

Before we verify the above formula, let us demonstrate the analog of property (D6) for rows. (D′6) states that interchanging two rows changes the sign of the determinant.

To prove (D′6) let $A = [a_{ij}]_{(nn)}$ be a matrix. Let $T_{ii'}$, $i' > i$, be the matrix obtained from the identity matrix by interchanging columns i and i'. Then $T_{ii'}A$ is the same matrix as A except that rows i and i' in A have been interchanged.

For example,

$$T_{24} = \begin{bmatrix} 1 & 0 & 0 & 0 \\ 0 & 0 & 0 & 1 \\ 0 & 0 & 1 & 0 \\ 0 & 1 & 0 & 0 \end{bmatrix}$$

is obtained by interchanging columns 2 and 4 in the identity matrix, and

$$TA = \begin{bmatrix} 1 & 0 & 0 & 0 \\ 0 & 0 & 0 & 1 \\ 0 & 0 & 1 & 0 \\ 0 & 1 & 0 & 0 \end{bmatrix}\begin{bmatrix} a_{11} & a_{12} & a_{13} & a_{14} \\ a_{21} & a_{22} & a_{23} & a_{24} \\ a_{31} & a_{32} & a_{33} & a_{34} \\ a_{41} & a_{42} & a_{43} & a_{44} \end{bmatrix} = \begin{bmatrix} a_{11} & a_{12} & a_{13} & a_{14} \\ a_{41} & a_{42} & a_{43} & a_{44} \\ a_{31} & a_{32} & a_{33} & a_{34} \\ a_{21} & a_{22} & a_{23} & a_{24} \end{bmatrix}$$

By property (D6), det $T_{ii'} = -1$. By the multiplicative property of determinants det $T_{ii'}A$ = det $T_{ii'}$ det A = $-$det A, which was what we desired to prove.

Using this result we can now show the following theorem.

Theorem 1

$$\det A = \sum_{j=1}^{n} a_{ij}A_{ij}$$

Proof If $A = [a_{ij}]$ let B be the matrix obtained from A by interchanging the first and ith rows of A.

Then,

$$\det A = -\det B \qquad \text{[by (D6)]}$$

$$= -\sum_{j=1}^{n}(-1)^{1+j}b_{1j} \det \hat{B}_{1j} \qquad \text{(by definition of the determinant)}$$

Observe that $\hat{B}_{1j}$ is the matrix obtained from $\hat{A}_{ij}$ by interchanging the first row of $\hat{A}_{ij}$ with the second, third, fourth, . . . , $(i - 1)$st rows successively. Thus, det $\hat{B}_{1j} = (-1)^{i-2}$ det $\hat{A}_{ij}$. Also note that $b_{1j} = a_{ij}$. So

$$\det A = \sum_{j=1}^{n}(-1)^{2+j+(i-2)}a_{ij} \det \hat{A}_{ij}$$

$$= \sum_{j=1}^{n}(-1)^{i+j}a_{ij} \det \hat{A}_{ij} = \sum_{j=1}^{n} a_{ij}A_{ij} \blacksquare$$

Using the cofactor expansion along the ith row we can obtain the analog of property (D1) for columns. (D′1) states that if each element of the ith row of a matrix A is expressed as the sum of two terms, then det A may be expressed as the sum of two determinants, the elements of whose ith rows are respectively the first and second terms of the corresponding elements of the ith row of A. All other terms are the same.

To prove this, suppose $a_{ij} = a'_{ij} + a''_{ij}$ is the expression for a_{ij} as the sum of two terms mentioned above. Then, by cofactor expansion along the ith row

$$\det A = \sum_{j=1}^{n} a_{ij}A_{ij} = \sum_{j=1}^{n} a'_{ij}A_{ij} + \sum_{j=1}^{n} a''_{ij}A_{ij}$$

Interpretation of the last two terms gives the desired result.

Of course, (D′2), the analog of (D2), is: If a matrix B differs from a matrix A only in that all of its elements in some row are c times those elements of the corresponding row of A, then det $B = c$ det A.

This can be demonstrated by using a cofactor expansion along the desired row, exactly as in the proof of (D′1).

Taken together, (D′1) and (D′2) yield the 4 × 4 case:

$$\begin{vmatrix} a_1 & a_2 & a_3 & a_4 \\ \alpha b_1 + \beta c_1 & \alpha b_2 + \beta c_2 & \alpha b_3 + \beta c_3 & \alpha b_4 + \beta c_4 \\ d_1 & d_2 & d_3 & d_4 \\ e_1 & e_2 & e_3 & e_4 \end{vmatrix} = \alpha \begin{vmatrix} a_1 & a_2 & a_3 & a_4 \\ b_1 & b_2 & b_3 & b_4 \\ d_1 & d_2 & d_3 & d_4 \\ e_1 & e_2 & e_3 & e_4 \end{vmatrix} + \beta \begin{vmatrix} a_1 & a_2 & a_3 & a_4 \\ c_1 & c_2 & c_3 & c_4 \\ d_1 & d_2 & d_3 & d_4 \\ e_1 & e_2 & e_3 & e_4 \end{vmatrix}$$

(D′4) states that if two rows of a matrix are equal, then the determinant of the matrix is zero.

The proof of (D′4) is as follows. If A is the matrix and if rows i and i', $i < i'$, are equal, then B, the matrix obtained by interchanging rows i and i', equals A. So $\det B = \det A$. On the other hand, by (D′6), interchanging rows changes the sign of the determinant. So $\det B = -\det A$. Thus, $\det A = -\det A$ which implies $\det A = 0$.

Properties (D′5) and (D′7) may be proved from (D′1)–(D′4) exactly as (D5)–(D7) were demonstrated from (D1)–(D4) in §3.2. They may be stated as

(D′5) states that the value of a determinant is unchanged if a scalar multiple of one row is added to another.

(D′7) states that if all entries of some row of a matrix are 0, its determinant is 0.

For example, by (D′5),

$$\begin{vmatrix} a_1 & b_1 & c_1 \\ a_2 & b_2 & c_2 \\ a_3 & b_3 & c_3 \end{vmatrix} = \begin{vmatrix} a_1 & b_1 & c_1 \\ a_2 + \alpha a_1 & b_2 + \alpha b_1 & c_2 + \alpha c_1 \\ a_3 & b_3 & c_3 \end{vmatrix}$$

We have thus demonstrated that all of the operations described for columns in (D1)–(D7) may be performed for rows as well. This provides an additional tool in problems involving the calculation of determinants.

Example 1 Evaluate

$$\Delta = \begin{vmatrix} 1 & -3 & 0 & 7 \\ 2 & 1 & 3 & 5 \\ 4 & 1 & 3 & 5 \\ 6 & 0 & 8 & 1 \end{vmatrix}$$

In this case we see that the second and third rows are nearly identical. So subtracting the second from the third row, we obtain

$$\begin{vmatrix} 1 & -3 & 0 & 7 \\ 2 & 1 & 3 & 5 \\ 2 & 0 & 0 & 0 \\ 6 & 0 & 8 & 1 \end{vmatrix}$$

By using a cofactor expansion along the third row we obtain

$$2\begin{vmatrix} -3 & 0 & 7 \\ 1 & 3 & 5 \\ 0 & 8 & 1 \end{vmatrix} = 2\left((-3)\begin{vmatrix} 3 & 5 \\ 8 & 1 \end{vmatrix} + 7\begin{vmatrix} 1 & 3 \\ 0 & 8 \end{vmatrix}\right) = (-6)(-37) + 7 \cdot 16 = 334$$

There is also a cofactor expansion along columns.

Theorem 2

$$\det A = \sum_{i=1}^{n} a_{ij} A_{ij}$$

For example, in the 4×4 case this might be written as

$$\Delta = \begin{vmatrix} a_{11} & a_{12} & a_{13} & a_{14} \\ a_{21} & a_{22} & a_{23} & a_{24} \\ a_{31} & a_{32} & a_{33} & a_{34} \\ a_{41} & a_{42} & a_{43} & a_{44} \end{vmatrix} = a_{11}\begin{vmatrix} a_{22} & a_{23} & a_{24} \\ a_{32} & a_{33} & a_{34} \\ a_{42} & a_{43} & a_{44} \end{vmatrix} - a_{21}\begin{vmatrix} a_{12} & a_{13} & a_{14} \\ a_{32} & a_{33} & a_{34} \\ a_{42} & a_{43} & a_{44} \end{vmatrix}$$

$$+ a_{31}\begin{vmatrix} a_{12} & a_{13} & a_{14} \\ a_{22} & a_{23} & a_{24} \\ a_{42} & a_{43} & a_{44} \end{vmatrix} - a_{41}\begin{vmatrix} a_{12} & a_{13} & a_{14} \\ a_{22} & a_{23} & a_{24} \\ a_{32} & a_{33} & a_{34} \end{vmatrix}$$

Proof We confine our proof to the 4×4 case leaving the reader to handle the details in the general case. By (D1),

$$\Delta = \begin{vmatrix} a_{11} & a_{12} & a_{13} & a_{14} \\ 0 & a_{22} & a_{23} & a_{24} \\ 0 & a_{32} & a_{33} & a_{34} \\ 0 & a_{42} & a_{43} & a_{44} \end{vmatrix} + \begin{vmatrix} 0 & a_{12} & a_{13} & a_{14} \\ a_{21} & a_{22} & a_{23} & a_{24} \\ 0 & a_{32} & a_{33} & a_{34} \\ 0 & a_{42} & a_{43} & a_{44} \end{vmatrix} + \begin{vmatrix} 0 & a_{12} & a_{13} & a_{14} \\ 0 & a_{22} & a_{23} & a_{24} \\ a_{31} & a_{32} & a_{33} & a_{34} \\ 0 & a_{42} & a_{43} & a_{44} \end{vmatrix} + \begin{vmatrix} 0 & a_{12} & a_{13} & a_{14} \\ 0 & a_{22} & a_{23} & a_{24} \\ 0 & a_{32} & a_{33} & a_{34} \\ a_{41} & a_{42} & a_{43} & a_{44} \end{vmatrix}$$

By using (D2), we may factor a_{11}, a_{21}, a_{31}, and a_{41} from the first columns of each of the above determinants. This leaves the first columns as

$$\begin{bmatrix}1\\0\\0\\0\end{bmatrix}, \quad \begin{bmatrix}0\\1\\0\\0\end{bmatrix}, \quad \begin{bmatrix}0\\0\\1\\0\end{bmatrix}, \quad \begin{bmatrix}0\\0\\0\\1\end{bmatrix}$$

respectively.

Then by adding a suitable multiple of the first column to each of the remaining columns we obtain

$$\Delta = a_{11}\begin{vmatrix}1 & 0 & 0 & 0\\ 0 & a_{22} & a_{23} & a_{24}\\ 0 & a_{32} & a_{33} & a_{34}\\ 0 & a_{42} & a_{43} & a_{44}\end{vmatrix} + a_{21}\begin{vmatrix}0 & a_{12} & a_{13} & a_{14}\\ 1 & 0 & 0 & 0\\ 0 & a_{32} & a_{33} & a_{34}\\ 0 & a_{42} & a_{43} & a_{44}\end{vmatrix} + a_{31}\begin{vmatrix}0 & a_{12} & a_{13} & a_{14}\\ 0 & a_{22} & a_{23} & a_{24}\\ 1 & 0 & 0 & 0\\ 0 & a_{42} & a_{43} & a_{44}\end{vmatrix} + a_{41}\begin{vmatrix}0 & a_{12} & a_{13} & a_{14}\\ 0 & a_{22} & a_{23} & a_{24}\\ 0 & a_{32} & a_{33} & a_{34}\\ 1 & 0 & 0 & 0\end{vmatrix}$$

Interchanging rows we obtain,

$$\begin{aligned}\Delta &= a_{11}\det \hat{A}_{11} + (-1)a_{21}\det \hat{A}_{21} + a_{31}\det \hat{A}_{31} + (-1)a_{41}\det \hat{A}_{41}\\ &= a_{11}A_{11} + a_{21}A_{21} + a_{31}A_{31} + a_{41}A_{41} \quad \blacksquare\end{aligned}$$

Using row and column cofactor expansions, we may prove the following theorem.

Theorem 3 $\det A = \det A^T$.

Proof The argument will be by induction on n. In the 2×2 case we have

$$\det\begin{bmatrix}a & b\\ c & d\end{bmatrix} = \det\begin{bmatrix}a & c\\ b & d\end{bmatrix}$$

If $A = [a_{ij}]_{(nn)}$, let $B = A^T = [b_{ij}]_{(nn)}$, where $b_{ij} = a_{ji}$. Note that $\hat{B}_{ij} = \hat{A}^T_{ji}$. By definition,

$$\begin{aligned}\det B &= \sum_{j=1}^{n} b_{1j}(-1)^{1+j}\det \hat{B}_{1j}\\ &= \sum_{j=1}^{n} a_{j1}(-1)^{1+j}\det \hat{A}^T_{j1}\\ &= \sum_{j=1}^{n} a_{j1}(-1)^{1+j}\det \hat{A}_{j1}\end{aligned}$$

The last step depended on our induction hypothesis for $(n-1) \times (n-1)$ matrices. Thus

$$\det B = \sum_{j=1}^{n} a_{j1}A_{j1} = \det A$$

by the column cofactor expansion of A. ∎

Thus, for example

$$\begin{vmatrix} 1 & a & b \\ d & 1 & c \\ e & f & 1 \end{vmatrix} = \begin{vmatrix} 1 & d & e \\ a & 1 & f \\ b & c & 1 \end{vmatrix}$$

EXERCISES

1. Calculate the following determinants.

(a) $\begin{vmatrix} 1 & -1 & 1 \\ 3 & 4 & 2 \\ 2 & -2 & 2 \end{vmatrix}$ (b) $\begin{vmatrix} 3 & 1 & 7 \\ 2 & 0 & 2 \\ 3 & 1 & 8 \end{vmatrix}$ (c) $\begin{vmatrix} 3 & -1 & 2 \\ 0 & 0 & 3 \\ -1 & 2 & 1 \end{vmatrix}$

(d) $\begin{vmatrix} 1 & 2 & 2 & 2 & 2 \\ 1 & 3 & 3 & 3 & 3 \\ 1 & 1 & 4 & 4 & 4 \\ 1 & 1 & 1 & 5 & 5 \\ 1 & 1 & 1 & 1 & 6 \end{vmatrix}$ (e) $\begin{vmatrix} 1 & 0 & -1 & 0 & 2 \\ 0 & 3 & 0 & 1 & 1 \\ 2 & 0 & -2 & 0 & 3 \\ 0 & 1 & 0 & 0 & 1 \\ 0 & 6 & 0 & 2 & 3 \end{vmatrix}$

(f) $\begin{vmatrix} 2 & 3 & 6 & 3 \\ 0 & 1 & 3 & 1 \\ -1 & -2 & 0 & 4 \\ 1 & 2 & 4 & -1 \end{vmatrix}$ (g) $\begin{vmatrix} 3 & 2 & -1 & 4 \\ 2 & -3 & 4 & 1 \\ -4 & 2 & 0 & 3 \\ 2 & 4 & -1 & 2 \end{vmatrix}$

(h) $\begin{vmatrix} 1 & -1 & -1 & -1 \\ 1 & 1 & -1 & 1 \\ 1 & 1 & 1 & -1 \\ 1 & -1 & 1 & 1 \end{vmatrix}$ (i) $\begin{vmatrix} 0 & -2 & 1 & -1 & 1 \\ -2 & -3 & 2 & -2 & 2 \\ 1 & 2 & 0 & 1 & 1 \\ -1 & -2 & 1 & 0 & -1 \\ 1 & 2 & -1 & 1 & 0 \end{vmatrix}$

(j) $\begin{vmatrix} 5 & 2 & 1 & 2 \\ 6 & 1 & -1 & 2 \\ 3 & 2 & 1 & 2 \\ 4 & 0 & -1 & 1 \end{vmatrix}$ (k) $\begin{vmatrix} -4 & 1 & 4 & 1 \\ -2 & 3 & 6 & 1 \\ 3 & 1 & 1 & 0 \\ 5 & 1 & 4 & 1 \end{vmatrix}$

2. Show that

$$\begin{vmatrix} a_1 & b_1 & c_1 \\ a_2 & b_2 & c_2 \\ a_3 & b_3 & c_3 \end{vmatrix} = \begin{vmatrix} a_1 & a_2 & a_3 \\ a_1 + b_1 & a_2 + b_2 & a_3 + b_3 \\ b_1 + c_1 & b_2 + c_2 & b_3 + c_3 \end{vmatrix}$$

3. If n is odd and A is an $n \times n$ matrix such that $A^T = -A$ (i.e., A is skew-symmetric), show that $\det A = 0$.
4. If Q is a real matrix, show that $\det QQ^T \geq 0$.
5. If Q is a complex matrix, show that $\det QQ^* \geq 0$.
6. If one row of a matrix is a scalar multiple of another row, show that the determinant of the matrix is zero.
7. If A is symmetric, show that $\det [A + B] = \det [A + B^T]$.
8. Show that $\det AB = \det AB^T = \det A^TB$.
9. Let

$$Q = \begin{bmatrix} \alpha_1 & -\alpha_2 & -\alpha_3 & -\alpha_4 \\ \alpha_2 & \alpha_1 & -\alpha_4 & \alpha_3 \\ \alpha_3 & \alpha_4 & \alpha_1 & -\alpha_2 \\ \alpha_4 & -\alpha_3 & \alpha_2 & \alpha_1 \end{bmatrix}$$

By calculating QQ^T, show that $\det Q = (\alpha_1^2 + \alpha_2^2 + \alpha_3^2 + \alpha_4^2)^2$.

10. Let $A = [a_{ij}]_{(nn)}$ be an $n \times n$ matrix. Consider the new matrix B obtained by replacing each entry a_{ij} by $-a_{ij}$ if $i + j$ is odd, and by $+a_{ij}$ if $i + j$ is even. In other words, $B = [(-1)^{i+j}a_{ij}]_{(nn)}$.
(a) Show that $\det B = A$.
(b) If instead in B we replace a_{ij} by $+a_{ij}$ if $i + j$ is even, and a_{ij} by $-a_{ij}$ if $i + j$ is odd, show that $\det B = (-1)^n \det A$.
11. Given that

$$A^TB^T = \begin{bmatrix} 3 & 4 \\ 4 & 6 \end{bmatrix}$$

and $\det B = 2$, what is $\det A$?

12. Without solving, show that the polynomial

$$f(x) = \begin{vmatrix} x & a & b \\ -a & x & c \\ -b & -c & x \end{vmatrix}$$

has $x = 0$ as a root.

13. If A is an $n \times n$ matrix, is invertible, and $A^{-1} = 2A^T$, show that $(\det A)^2 = 1/2^n$.
14. If $QQ^T = I_n$, show that $\det Q = \pm 1$.

15. Show that det $Q^* = \det \overline{Q}$.
16. Using 2×2 matrices, show that

$$(a^2 + b^2)(c^2 + d^2) = (ac + bd)^2 + (ad - bc)^2$$

[Hint:

$$Q = \begin{bmatrix} a & b \\ c & d \end{bmatrix}$$

calculate det Q, det Q^T, and det QQ^T.]

17. Let $A = QQ^T$, where A and Q are real square matrices. Show that
(a) All diagonal entries of A are nonnegative.
(b) $\det A \geq 0$.
18. Show that

$$\begin{bmatrix} 0 & 1 \\ 1 & 0 \end{bmatrix}$$

cannot be written in the form QQ^T, for real matrices Q.

5 INVERSE OF A MATRIX

Associated with a given matrix $A = [a_{ij}]_{(nn)}$, there is a useful matrix called the **adjunct**, denoted by $\mathcal{A}$, and defined by $\mathcal{A} = [A_{ij}]^T$. In other words the adjunct is the transpose of the matrix of cofactors.

For example, if

$$A = \begin{bmatrix} a & b \\ c & d \end{bmatrix}$$

the matrix of cofactors is

$$\begin{bmatrix} d & -c \\ -b & a \end{bmatrix}$$

and the adjunct is

$$\mathcal{A} = \begin{bmatrix} d & -b \\ -c & a \end{bmatrix}$$

If

$$A = \begin{bmatrix} -1 & 0 & 3 \\ 7 & 1 & -1 \\ 2 & 3 & 0 \end{bmatrix}$$

the matrix of cofactors is

$$\begin{bmatrix} 3 & -2 & 19 \\ 9 & -6 & 3 \\ -3 & 20 & -1 \end{bmatrix}$$

So the adjunct is

$$\mathcal{Q} = \begin{bmatrix} 3 & 9 & -3 \\ -2 & -6 & 20 \\ 19 & 3 & -1 \end{bmatrix}$$

In the 2×2 case, note that

$$A\mathcal{Q} = \begin{bmatrix} a & b \\ c & d \end{bmatrix} \begin{bmatrix} d & -b \\ -c & a \end{bmatrix} = \begin{bmatrix} ad - bc & 0 \\ 0 & ad - bc \end{bmatrix}$$

$\mathcal{Q}A = (\det A)I_2$ is demonstrated similarly.

We intend to generalize this result for arbitrary $n \times n$ matrices, but first we need the following lemma.

Lemma The sum of the products of the elements of one line of A by the cofactors of the corresponding elements of a different parallel line of A is always zero.

Proof Observe that in this case we have essentially the determinant of a matrix with two equal parallel lines. By (D4) or (D′4) this determinant is zero. ∎

Theorem 1 Let A be an $n \times n$ matrix. Then

$$A\mathcal{Q} = \mathcal{Q}A = (\det A)I_n$$

Proof Consider

$$A\mathcal{Q} = \begin{bmatrix} a_{11} & a_{12} & \cdots & a_{1n} \\ a_{21} & a_{22} & \cdots & a_{2n} \\ & & \vdots & \\ a_{n1} & a_{n2} & \cdots & a_{nn} \end{bmatrix} \begin{bmatrix} A_{11} & A_{21} & \cdots & A_{n1} \\ A_{12} & A_{22} & \cdots & A_{n2} \\ & & \vdots & \\ A_{1n} & A_{2n} & \cdots & A_{nn} \end{bmatrix}$$

$$= \begin{bmatrix} a_{11}A_{11} + a_{12}A_{12} + \cdots + a_{1n}A_{1n} & a_{11}A_{21} + a_{12}A_{22} + \cdots + a_{1n}A_{2n} & \cdots & a_{11}A_{n1} + a_{12}A_{n2} + \cdots + a_{1n}A_{nn} \\ a_{21}A_{11} + a_{22}A_{12} + \cdots + a_{2n}A_{1n} & a_{21}A_{21} + a_{22}A_{22} + \cdots + a_{2n}A_{2n} & \cdots & a_{21}A_{n1} + a_{22}A_{n2} + \cdots + a_{2n}A_{nn} \\ & \vdots & & \\ a_{n1}A_{11} + a_{n2}A_{12} + \cdots + a_{nn}A_{1n} & a_{n1}A_{21} + a_{n2}A_{22} + \cdots + a_{nn}A_{2n} & \cdots & a_{n1}A_{n1} + a_{n2}A_{n2} + \cdots + a_{nn}A_{nn} \end{bmatrix}$$

By our lemma all off diagonal terms are zero, while each of the diagonal terms is just det A. So $A\mathcal{A} = (\det A)I_n$. A similar proof shows that $\mathcal{A}A = (\det A)I_n$. ∎

Theorem 2 If $\det A \neq 0$, A^{-1} exists, and $A^{-1} = (\det A)^{-1}\mathcal{A}$.

Proof Choosing $B = (\det A)^{-1}\mathcal{A}$, we have $AB = BA = I_n$. Thus, A^{-1} exists and equals B. ∎

Thus the determinant gives an accurate indication of whether or not a given matrix is invertible. It is invertible if and only if its determinant is nonzero. The determinant also yields a means, not necessarily the most efficient, for calculating this inverse.

Example 1 If

$$A = \begin{bmatrix} a & b \\ c & d \end{bmatrix}$$

and $ad - bc \neq 0$,

$$A^{-1} = \frac{1}{ad - bc}\mathcal{A} = \frac{1}{ad - bc}\begin{bmatrix} d & -b \\ -c & a \end{bmatrix}$$

Example 2 If

$$A = \begin{bmatrix} -1 & 0 & 3 \\ 7 & 1 & -1 \\ 2 & 3 & 0 \end{bmatrix}$$

then $\det A = 54$, and

$$\mathscr{Q} = \begin{bmatrix} 3 & 9 & -3 \\ -2 & -6 & 20 \\ 19 & 3 & -1 \end{bmatrix}$$

Therefore

$$A^{-1} = \frac{1}{54}\begin{bmatrix} 3 & 9 & -3 \\ -2 & -6 & 20 \\ 19 & 3 & -1 \end{bmatrix}$$

Example 3

$$A = \begin{bmatrix} 1 & x & 0 \\ -x & 1 & x \\ 0 & -x & 1 \end{bmatrix}$$

In this case

$$\begin{aligned} \det A &= 1 \cdot (1 + x^2) + (-x)(-x) \\ &= 1 + 2x^2 \end{aligned}$$

Thus if x is real, A is invertible.

$$[A_{ij}] = \begin{bmatrix} 1 + x^2 & x & x^2 \\ -x & 1 & x \\ x^2 & -x & 1 + x^2 \end{bmatrix}$$

So

$$A^{-1} = \frac{1}{1 + 2x^2}\begin{bmatrix} 1 + x^2 & -x & x^2 \\ x & 1 & -x \\ x^2 & x & 1 + x^2 \end{bmatrix}$$

EXERCISES

[In the following all matrices are assumed to be square.]

1. Calculate the inverse of each of the following matrices.

(a) $\begin{bmatrix} 1 & 0 & -2 \\ 3 & 1 & 2 \\ 1 & -1 & 0 \end{bmatrix}$ (b) $\begin{bmatrix} 1 & 3 & 3 \\ 1 & 4 & 3 \\ 1 & 3 & 4 \end{bmatrix}$

(c) $\begin{bmatrix} \frac{1}{5} & \frac{1}{5} & \frac{1}{5} \\ \frac{1}{5} & -\frac{1}{5} & -\frac{4}{5} \\ -\frac{2}{5} & \frac{1}{10} & \frac{1}{10} \end{bmatrix}$ (d) $\begin{bmatrix} \frac{1}{2} & -\frac{1}{2} & \frac{1}{2} \\ 0 & 0 & 1 \\ \frac{1}{2} & \frac{1}{2} & -\frac{1}{2} \end{bmatrix}$

(e) $\begin{bmatrix} 2 & -5 \\ -5 & 13 \end{bmatrix}$ (f) $\begin{bmatrix} 1 & -6 \\ -3 & 19 \end{bmatrix}$

(g) $\begin{bmatrix} -1 & 0 & -1 & 3 \\ -1 & -1 & -1 & 1 \\ 1 & 3 & 3 & 3 \\ -4 & 0 & -5 & 14 \end{bmatrix}$ (h) $\begin{bmatrix} -1 & 0 & -2 & -3 \\ -2 & -1 & -4 & -6 \\ 3 & 1 & 7 & 9 \\ 1 & 1 & 4 & 4 \end{bmatrix}$

(i) $\begin{bmatrix} 2 & 3 & 2 & 4 \\ 4 & 6 & 5 & 5 \\ 3 & 5 & 2 & 14 \\ 2 & 2 & -3 & 14 \end{bmatrix}$ (j) $\begin{bmatrix} 1 & 3 & -5 & 7 \\ 0 & 1 & 2 & -3 \\ 0 & 0 & 1 & 2 \\ 0 & 0 & 0 & 1 \end{bmatrix}$

(k) $\begin{bmatrix} 1 & 1 & 1 & 1 \\ 1 & 1 & -1 & -1 \\ 1 & -1 & 1 & -1 \\ 1 & -1 & -1 & 1 \end{bmatrix}$

2. If $A = G_1G_2 \ldots G_n$ is invertible, show that $G_1, G_2, \ldots, G_n$ are invertible.
3. Show that

$$\begin{bmatrix} \cos\theta & -\sin\theta \\ -\sin\theta & -\cos\theta \end{bmatrix}^{-1} = \begin{bmatrix} \cos\theta & -\sin\theta \\ -\sin\theta & -\cos\theta \end{bmatrix}$$

4. If

$$Q = \begin{bmatrix} \alpha_1 & -\alpha_2 & -\alpha_3 & -\alpha_4 \\ \alpha_2 & \alpha_1 & -\alpha_4 & \alpha_3 \\ \alpha_3 & \alpha_4 & \alpha_1 & -\alpha_2 \\ \alpha_4 & -\alpha_3 & \alpha_2 & \alpha_1 \end{bmatrix}$$

show that

$$Q^{-1} = \frac{1}{(\alpha_1^2 + \alpha_2^2 + \alpha_3^2 + \alpha_4^2)} Q^T$$

5. If $\det A \neq 0$, and $AB = AC$, show that $B = C$.
6. If $AB = I_n$, show that A^{-1} exists and equals B.
7. If AB is a nonzero multiple of the identity matrix, show that $AB = BA$.
8. If A and B are nonzero matrices and $AB = 0$, show that $\det A = 0$ and $\det B = 0$.

9. If $P^2 = P$ and $P \neq I_n$, show that $\det P = 0$.
10. (a) Suppose $P^2 = P$. If $\lambda \neq 1$, prove $I_n - \lambda P$ is invertible and

$$(I_n - \lambda P)^{-1} = I_n + (\lambda/(1-\lambda))P$$

(b) Using the result from (a) find

$$\begin{bmatrix} 1-\lambda\cos^2\theta & \lambda\cos\theta\sin\theta \\ \lambda\cos\theta\sin\theta & 1-\lambda\sin^2\theta \end{bmatrix}^{-1}$$

11. Let A be an $n \times n$ matrix with cofactors A_{ij}. Show that

$$\begin{vmatrix} A_{11} & A_{12} & \dots & A_{1n} \\ A_{21} & A_{22} & \dots & A_{2n} \\ & & \vdots & \\ A_{n1} & A_{n2} & \dots & A_{nn} \end{vmatrix} = (\det A)^{n-1}$$

12. Let A be a symmetric matrix with cofactors A_{ij}. Show that $A_{ij} = A_{ji}$.
13. Show that an upper triangular matrix is invertible if and only if all its diagonal entries are nonzero.
14. Determine under what circumstances

$$\begin{bmatrix} x & 1 & 1 \\ 1 & x & 1 \\ 1 & 1 & x \end{bmatrix}$$

is invertible and find its inverse.

15. If $x - x^3 \neq 0$, show that

$$\begin{bmatrix} x & 0 & 0 & \dots & 0 & 1 \\ 0 & x & 0 & \dots & 0 & 0 \\ 0 & 0 & x & \dots & 0 & 0 \\ & & & \vdots & & \\ 0 & 0 & 0 & \dots & x & 0 \\ 1 & 0 & 0 & \dots & 0 & x \end{bmatrix}^{-1} = \frac{1}{x-x^3}\begin{bmatrix} -x^2 & 0 & 0 & \dots & 0 & x \\ 0 & 1-x^2 & 0 & \dots & 0 & 0 \\ 0 & 0 & 1-x^2 & \dots & 0 & 0 \\ & & \vdots & & & \\ 0 & 0 & 0 & \dots & 1-x^2 & 0 \\ x & 0 & 0 & \dots & 0 & -x^2 \end{bmatrix}$$

16. Find 2×2 matrices X and Y, such that

$$\begin{bmatrix} 2 & 1 \\ 3 & 2 \end{bmatrix} X - \begin{bmatrix} 4 & 3 \\ 1 & 1 \end{bmatrix} Y = \begin{bmatrix} 1 & 0 \\ 0 & 1 \end{bmatrix}$$

$$\begin{bmatrix} 2 & 1 \\ 3 & 2 \end{bmatrix} X + \begin{bmatrix} 4 & 3 \\ 1 & 1 \end{bmatrix} Y = \begin{bmatrix} 1 & 0 \\ 0 & 1 \end{bmatrix}$$

17. If a, b, and c are real, show that

$$\begin{bmatrix} 1 & a & b \\ -a & 1 & c \\ -b & -c & 1 \end{bmatrix}$$

is invertible and calculate its inverse.

18. If A and B are invertible and symmetric, and if $AB = BA$, show that AB, $A^{-1}B$, $A^{-1}B^{-1}$, and AB^{-1} are symmetric.

19. If A, B, and C are $n \times n$ matrices and $AB = C$, show that $\mathcal{C} = \mathcal{B}\mathcal{A}$, where $\mathcal{A}$, $\mathcal{B}$, and $\mathcal{C}$ are the adjuncts of A, B, and C, respectively.

20. Show that the adjunct of a symmetric matrix is symmetric.

21. Find a nonzero 3×3 matrix whose adjunct is zero.

22. If an upper triangular matrix has an inverse, show that the inverse is upper triangular.

23. Calculate

$$\begin{bmatrix} 1 + x & -x - \frac{1}{2}x^2 & \frac{1}{2}x^2 \\ x & 1 - \frac{1}{2}x^2 & -x + \frac{1}{2}x^2 \\ x & -\frac{1}{2}x^2 & 1 - x + \frac{1}{2}x^2 \end{bmatrix}^{-1}$$

24. If

$$P_\alpha = \begin{bmatrix} 1 & \alpha & \alpha^2 & \alpha^3 & \alpha^4 \\ 0 & 1 & 2\alpha & 3\alpha^2 & 4\alpha^3 \\ 0 & 0 & 1 & 3\alpha & 6\alpha^2 \\ 0 & 0 & 0 & 1 & 4\alpha \\ 0 & 0 & 0 & 0 & 1 \end{bmatrix}$$

show that $(P_\alpha)^{-1} = P_{-\alpha}$.

25. In each of the following find a 2×2 matrix X satisfying the given equation.

(a) $\begin{bmatrix} 3 & 2 \\ 1 & 1 \end{bmatrix} X \begin{bmatrix} 7 & 3 \\ 9 & 4 \end{bmatrix} = \begin{bmatrix} 0 & 1 \\ 1 & 0 \end{bmatrix}$ (b) $\begin{bmatrix} 5 & 3 \\ 6 & 4 \end{bmatrix} X \begin{bmatrix} 2 & 5 \\ 1 & 3 \end{bmatrix} = \begin{bmatrix} 1 & 1 \\ 1 & 1 \end{bmatrix}$

26. Let U be a matrix with $\det U = 1$. Suppose U_{ij} denotes the cofactor in position (i, j) and u_{ij} the entry in position (i, j). Show that $U_{ij} = u_{ij}$ if and only if $UU^T = I_n$.

27. Let A, B, and C be $n \times n$ matrices, with C invertible. Suppose $A = C^{-1}BC$. Show that A is invertible if and only if B is invertible.

28. If a and b are complex numbers and the matrix

$$\begin{bmatrix} a & b \\ -\bar{b} & \bar{a} \end{bmatrix}$$

is nonzero, show that it is invertible.

6 CRAMER'S RULE

With the concept of the inverse of a matrix to aid us we may attain significant results in the problem of solving a system of n linear equations in n variables.

Let the system be

$$\begin{aligned} a_{11}x_1 + a_{12}x_2 + \cdots + a_{1n}x_n &= b_1 \\ a_{21}x_1 + a_{22}x_2 + \cdots + a_{2n}x_n &= b_2 \\ &\vdots \\ a_{n1}x_1 + a_{n2}x_2 + \cdots + a_{nn}x_n &= b_n \end{aligned}$$

We let $A = [a_{ij}]_{(nn)}$ be the coefficient matrix of the system, $\boldsymbol{b} = [b_i]_{(n1)}$, and $\boldsymbol{x} = [x_i]_{(n1)}$, the vector of variables. Then, in matrix notation the system becomes

$$A\boldsymbol{x} = \boldsymbol{b}$$

If $\det A \neq 0$, we know A^{-1} exists. Letting $\boldsymbol{x} = A^{-1}\boldsymbol{b}$, we see

$$\begin{aligned} A(A^{-1}\boldsymbol{b}) &= (AA^{-1})\boldsymbol{b} \\ &= I_n\boldsymbol{b} \\ &= \boldsymbol{b} \end{aligned}$$

Thus, the system of equations is solvable, with solution $\boldsymbol{x} = A^{-1}\boldsymbol{b}$.

The solution is also unique. For if $\boldsymbol{x}_1$ and $\boldsymbol{x}_2$ are two solutions to $A\boldsymbol{x} = \boldsymbol{b}$, then

$$A\boldsymbol{x}_1 = \boldsymbol{b} = A\boldsymbol{x}_2, \quad \text{so } A\boldsymbol{x}_1 = A\boldsymbol{x}_2$$

and

$$\begin{aligned} A^{-1}(A\boldsymbol{x}_1) &= A^{-1}(A\boldsymbol{x}_2) \\ \boldsymbol{x}_1 &= \boldsymbol{x}_2 \end{aligned}$$

Using the explicit formula for the inverse of a matrix obtained in §3.5, we see

$$\boldsymbol{x} = A^{-1}\boldsymbol{b} = \frac{1}{\det A}\begin{bmatrix} A_{11} & A_{21} & \cdots & A_{n1} \\ A_{12} & A_{22} & \cdots & A_{n2} \\ & & \vdots & \\ A_{1n} & A_{2n} & \cdots & A_{nn} \end{bmatrix}\begin{bmatrix} b_1 \\ b_2 \\ \\ b_n \end{bmatrix}$$

$$\begin{bmatrix} x_1 \\ x_2 \\ \vdots \\ x_n \end{bmatrix} = \frac{1}{\det A} \begin{bmatrix} b_1A_{11} + b_2A_{21} + \cdots + b_nA_{n1} \\ b_1A_{12} + b_2A_{22} + \cdots + b_nA_{n2} \\ \vdots \\ b_1A_{1n} + b_2A_{2n} + \cdots + b_nA_{nn} \end{bmatrix}$$

Writing $A = [A_1, A_2, \ldots, A_n]$, and using cofactor expansions, the solutions may be written as

$$x_1 = \frac{\det [\boldsymbol{b}, A_2, A_3, \ldots, A_n]}{\det [A_1, A_2, A_3, \ldots, A_n]} = \frac{b_1A_{11} + b_2A_{21} + \cdots + b_nA_n}{\det A}$$

$$x_2 = \frac{\det [A_1, \boldsymbol{b}, A_3, \ldots, A_n]}{\det [A_1, A_2, A_3, \ldots, A_n]} = \frac{b_1A_{12} + b_2A_{22} + \cdots + b_nA_{n2}}{\det A}$$

$$\vdots$$

$$x_n = \frac{\det [A_1, A_2, \ldots, \boldsymbol{b}]}{\det [A_1, A_2, \ldots, A_n]} = \frac{b_1A_{1n} + b_2A_{2n} + \cdots + b_nA_{nn}}{\det A}$$

This method of solution for linear equations is called **Cramer's Rule**. In the 3×3 case we have, for example

$$x_1 = \frac{\begin{vmatrix} b_1 & a_{12} & a_{13} \\ b_2 & a_{22} & a_{23} \\ b_3 & a_{32} & a_{33} \end{vmatrix}}{\begin{vmatrix} a_{11} & a_{12} & a_{13} \\ a_{21} & a_{22} & a_{23} \\ a_{31} & a_{32} & a_{33} \end{vmatrix}} \qquad x_2 = \frac{\begin{vmatrix} a_{11} & b_1 & a_{13} \\ a_{21} & b_2 & a_{23} \\ a_{31} & b_3 & a_{33} \end{vmatrix}}{\begin{vmatrix} a_{11} & a_{12} & a_{13} \\ a_{21} & a_{22} & a_{23} \\ a_{31} & a_{32} & a_{33} \end{vmatrix}} \qquad x_3 = \frac{\begin{vmatrix} a_{11} & a_{12} & b_1 \\ a_{21} & a_{22} & b_2 \\ a_{31} & a_{32} & b_3 \end{vmatrix}}{\begin{vmatrix} a_{11} & a_{12} & a_{13} \\ a_{21} & a_{22} & a_{23} \\ a_{31} & a_{32} & a_{33} \end{vmatrix}}$$

In particular if the quantity on the right-hand side of the system of equations vanishes, $\boldsymbol{b} = \boldsymbol{0}$, or in other words, if we have a system of homogeneous equations, we see that if the determinant of the coefficient matrix is nonzero, the only solution to $A\boldsymbol{x} = \boldsymbol{0}$ is $\boldsymbol{x} = \boldsymbol{0}$.

Example 1 Let us solve the system of equations

$$\begin{aligned} x_1 + x_2 + x_3 + x_4 &= y_1 \\ x_1 + x_2 + x_3 - x_4 &= y_2 \\ x_1 + x_2 - x_3 + x_4 &= y_3 \\ x_1 - x_2 + x_3 + x_4 &= y_4 \end{aligned}$$

By inverting the matrix of coefficients

$$A = \begin{bmatrix} 1 & 1 & 1 & 1 \\ 1 & 1 & 1 & -1 \\ 1 & 1 & -1 & 1 \\ 1 & -1 & 1 & 1 \end{bmatrix}$$

Let Δ be the determinant of A. By subtracting the first column from each of the remaining columns we find

$$\Delta = \begin{vmatrix} 1 & 0 & 0 & 0 \\ 1 & 0 & 0 & -2 \\ 1 & 0 & -2 & 0 \\ 1 & -2 & 0 & 0 \end{vmatrix} = \begin{vmatrix} 0 & 0 & -2 \\ 0 & -2 & 0 \\ -2 & 0 & 0 \end{vmatrix} = (-2)\begin{vmatrix} 0 & -2 \\ -2 & 0 \end{vmatrix} = 8$$

Also,

$$A_{11} = (-1)^2\begin{vmatrix} 1 & 1 & -1 \\ 1 & -1 & 1 \\ -1 & 1 & 1 \end{vmatrix} = \begin{vmatrix} 1 & 0 & 0 \\ 1 & -2 & 2 \\ -1 & 2 & 0 \end{vmatrix} = \begin{vmatrix} -2 & 2 \\ 2 & 0 \end{vmatrix} = -4$$

$$A_{12} = (-1)^3\begin{vmatrix} 1 & 1 & -1 \\ 1 & -1 & 1 \\ 1 & 1 & 1 \end{vmatrix} = -\begin{vmatrix} 1 & 0 & -2 \\ 1 & -2 & 0 \\ 1 & 0 & 0 \end{vmatrix} = -\begin{vmatrix} 0 & -2 \\ -2 & 0 \end{vmatrix} = 4$$

$$A_{13} = (-1)^4\begin{vmatrix} 1 & 1 & -1 \\ 1 & 1 & 1 \\ 1 & -1 & 1 \end{vmatrix} = \begin{vmatrix} 1 & 0 & -2 \\ 1 & 0 & 0 \\ 1 & -2 & 0 \end{vmatrix} = (-1)\begin{vmatrix} 0 & -2 \\ -2 & 0 \end{vmatrix} = 4$$

More calculations of the same sort show that the matrix of cofactors is

$$\begin{bmatrix} -4 & 4 & 4 & 4 \\ 4 & 0 & 0 & -4 \\ 4 & 0 & -4 & 0 \\ 4 & -4 & 0 & 0 \end{bmatrix}$$

Thus, the inverse is

$$\frac{1}{2}\begin{bmatrix} -1 & 1 & 1 & 1 \\ 1 & 0 & 0 & -1 \\ 1 & 0 & -1 & 0 \\ 1 & -1 & 0 & 0 \end{bmatrix}$$

and

$$\begin{bmatrix} x_1 \\ x_2 \\ x_3 \\ x_4 \end{bmatrix} = \frac{1}{2}\begin{bmatrix} -1 & 1 & 1 & 1 \\ 1 & 0 & 0 & -1 \\ 1 & 0 & -1 & 0 \\ 1 & -1 & 0 & 0 \end{bmatrix}\begin{bmatrix} y_1 \\ y_2 \\ y_3 \\ y_4 \end{bmatrix} = \frac{1}{2}\begin{bmatrix} -y_1 + y_2 + y_3 + y_4 \\ y_1 - y_4 \\ y_1 - y_3 \\ y_1 - y_2 \end{bmatrix}$$

or

$$\begin{aligned} x_1 &= \tfrac{1}{2}(-y_1 + y_2 + y_3 + y_4) \\ x_2 &= \tfrac{1}{2}(y_1 - y_4) \\ x_3 &= \tfrac{1}{2}(y_1 - y_3) \\ x_4 &= \tfrac{1}{2}(y_1 - y_2) \end{aligned}$$

Example 2 We will derive the law of cosines.

Consider a triangle with sides a, b, and c, and opposite angles α, β, and γ. (See Figure 6-1.) Then, using trigonometric definitions,

$$\begin{aligned} c(\cos\beta) + b(\cos\gamma) &= a \\ c(\cos\alpha) \qquad\qquad + a(\cos\gamma) &= b \\ b(\cos\alpha) + a(\cos\beta) \qquad\qquad &= c \end{aligned}$$

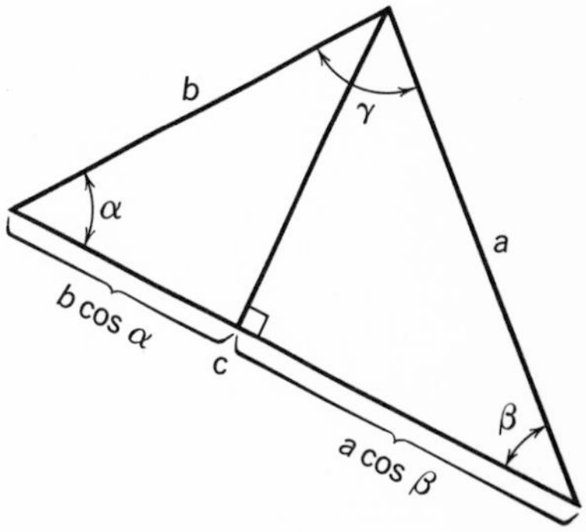

FIGURE 6-1

We wish to solve for $\cos\alpha$, $\cos\beta$, and $\cos\gamma$. That is, we wish to invert the matrix

$$A = \begin{bmatrix} 0 & c & b \\ c & 0 & a \\ b & a & 0 \end{bmatrix}$$

Its determinant is $2abc$. Hence A is invertible if $a \neq 0$, $b \neq 0$, and $c \neq 0$, which certainly hold for a triangle.

In this case

$$A^{-1} = \frac{1}{2abc}\begin{bmatrix} -a^2 & ab & ac \\ ab & -b^2 & bc \\ ac & bc & -c^2 \end{bmatrix}$$

So

$$\cos\alpha = \frac{-a^3 + ab^2 + ac^2}{2abc}$$

$$\cos\alpha = \frac{b^2 + c^2 - a^2}{2bc}$$

$\cos\beta$ and $\cos\gamma$ are given by similar expressions.

EXERCISES

1. Solve the following systems.

(a)
$$\begin{aligned} x + 3y - z &= 0 \\ x + y + z &= 1 \\ -x + 2y - z &= 1 \end{aligned}$$

(b)
$$\begin{aligned} 2x + 3y + z &= 9 \\ x + 2y + 3z &= 6 \\ 3x + y + 2z &= 8 \end{aligned}$$

(c)
$$\begin{aligned} 2x_1 - x_2 + 3x_3 &= 1 \\ 3x_1 + x_2 - x_3 &= 2 \\ x_1 + 2x_2 + 3x_3 &= -6 \end{aligned}$$

(d)
$$\begin{aligned} x_1 - 2x_2 + 2x_3 &= -1 \\ 2x_1 - 3x_2 - 3x_3 &= 1 \\ 3x_1 + x_2 + 2x_3 &= 3 \end{aligned}$$

(e)
$$\begin{aligned} x_1 - x_2 + 2x_3 &= 2 \\ x_1 + 3x_2 - x_3 &= 3 \\ 3x_1 + x_2 + 2x_3 &= 6 \end{aligned}$$

(f)
$$\begin{aligned} 4x + 3y + 2z &= 3 \\ 5x + 2y + 5z &= 8 \\ 8x + 6y + 11z &= 13 \end{aligned}$$

(g)
$$\begin{aligned} 3x_1 + 4x_2 + x_3 + 2x_4 &= 7 \\ 2x_1 \qquad - x_3 + 2x_4 &= 7 \\ 3x_1 - x_2 - x_3 + 3x_4 &= 10 \\ 7x_1 - x_2 - x_3 + 7x_4 &= 22 \end{aligned}$$

(h)
$$\begin{aligned} x_1 + x_2 + x_3 + x_4 &= 3 \\ -x_1 + x_2 + x_3 + x_4 &= 1 \\ x_1 - x_2 + x_3 + x_4 &= 2 \\ x_1 + x_2 - x_3 + x_4 &= 2 \end{aligned}$$

(i)
$$\begin{aligned} 2x_1 + x_2 + 5x_3 + x_4 &= 5 \\ x_1 + x_2 - 3x_3 - 4x_4 &= -1 \\ 3x_1 + 6x_2 - 2x_3 + x_4 &= 8 \\ 2x_1 + 2x_2 + 2x_3 - 3x_4 &= 2 \end{aligned}$$

2. By calculating the inverse of the coefficient matrix, solve the following systems.

(a)
$$\begin{aligned} 3x - 4y + 2z &= u \\ 2x - y &= v \\ 2x - 2y + z &= w \end{aligned}$$

(b)
$$\begin{aligned} 3x + 2y + 2z &= u \\ -4x - y - 2z &= v \\ 2x + z &= w \end{aligned}$$

(c)
$$\begin{aligned} x_1 + x_2 + x_3 + x_4 &= y_1 \\ x_1 + x_3 + x_4 &= y_2 \\ x_1 + x_2 + x_4 &= y_3 \\ x_1 + x_2 + x_3 &= y_4 \end{aligned}$$

(d)
$$\begin{aligned} x_2 + 2x_3 + 2x_4 &= y_1 \\ x_1 + x_2 + 2x_3 + 3x_4 &= y_2 \\ 2x_1 + 2x_2 + 2x_3 + 3x_4 &= y_3 \\ 2x_1 + 3x_2 + 3x_3 + 3x_4 &= y_4 \end{aligned}$$

(e)
$$\begin{aligned} 3x_1 + 4x_2 + 2x_3 + 7x_4 &= y_1 \\ 2x_1 + 3x_2 + 3x_3 + 2x_4 &= y_2 \\ 5x_1 + 7x_2 + 3x_3 + 9x_4 &= y_3 \\ 2x_1 + 3x_2 + 2x_3 + 3x_4 &= y_4 \end{aligned}$$

3. In the system

$$\begin{aligned} a_1x_1 + a_2x_2 + a_3x_3 + a_4x_4 &= a_1 \\ b_1x_1 + b_2x_2 + b_3x_3 + b_4x_4 &= b_1 \\ c_1x_1 + c_2x_2 + c_3x_3 + c_4x_4 &= c_1 \\ d_1x_1 + d_2x_2 + d_3x_3 + d_4x_4 &= d_1 \end{aligned}$$

if the determinant of the coefficient matrix is nonzero, show that the only solution is $x_1 = 1$, $x_2 = x_3 = x_4 = 0$.

4. Given two lines

$$\begin{aligned} a_{11}x_1 + a_{12}x_2 &= b_1 \\ a_{21}x_1 + a_{22}x_2 &= b_2 \end{aligned}$$

and assuming their slopes are unequal, show that

$$\begin{vmatrix} a_{11} & a_{12} \\ a_{21} & a_{22} \end{vmatrix} \neq 0$$

5. Consider the lines

$$\begin{aligned} a_{11}x_1 + a_{12}x_2 &= \beta + \gamma t \\ a_{21}x_1 + a_{22}x_2 &= \mu + \lambda t \end{aligned}$$

where t is a parameter. Assuming

$$\begin{vmatrix} a_{11} & a_{12} \\ a_{21} & a_{22} \end{vmatrix} \neq 0$$

show that the set of all solutions as t varies is a line.

6. For what value of t is the point of intersection of the lines $x + y = t$ and $x - y = t$ closest to the point (1, 1).
7. Show that the system of equations, in which λ is real,

$$\begin{aligned} (1+\lambda)x_1 + 2x_2 + 2x_3 + x_4 &= y_1 \\ -x_1 + (\lambda - 1)x_2 - x_3 - 2x_4 &= y_2 \\ (1+\lambda)x_3 + 2x_4 &= y_3 \\ -x_3 + (\lambda - 1)x_4 &= y_4 \end{aligned}$$

is solvable regardless of the value of λ.

7 SYNTHETIC ELIMINATION

The formula described in §3.5 for calculating the inverse of a matrix is a very inefficient procedure from the computational point of view. It is mainly useful in that it provides us with a proof of the fact that A^{-1} exists if $\det A \neq 0$. A more practical method of matrix inversion is synthetic elimination, which we now describe.

Given the matrix $A = [a_{ij}]_{(nn)}$, consider the system of equations

$$\begin{aligned} a_{11}x_1 + a_{12}x_2 + \cdots + a_{1n}x_n &= y_1 \\ a_{21}x_1 + a_{22}x_2 + \cdots + a_{2n}x_n &= y_2 \\ &\vdots \\ a_{n1}x_1 + a_{n2}x_2 + \cdots + a_{nn}x_n &= y_n \end{aligned}$$

where we regard $x_1, x_2, \ldots, x_n$ and $y_1, y_2, \ldots, y_n$ as variables. If we solve this system by Gaussian elimination, we only perform the operations of multiplying equations by nonzero constants and adding scalar multiples of one equation to another. Thus, if $\det A \neq 0$, the system is solvable and the solutions are of the form

$$\begin{aligned} x_1 &= b_{11}y_1 + b_{12}y_2 + \cdots + b_{1n}y_n \\ x_2 &= b_{21}y_1 + b_{22}y_2 + \cdots + b_{2n}y_n \\ &\vdots \\ x_n &= b_{n1}y_1 + b_{n2}y_2 + \cdots + b_{nn}y_n \end{aligned}$$

We claim that the matrix $B = [b_{ij}]_{(nn)}$ is then the inverse of A.

To see this observe that $Ax = y$ and $By = x$ together imply $(AB)y = y$, or $(AB - I_n)y = 0$, for all n-vectors y. It will follow that $AB = I_n$, after we prove the following lemma.

Lemma If C is an $m \times n$ matrix, and $Cx = 0$ for all n-vectors x, then $C = 0$.

Proof Letting $C = [c_{ij}]_{(nn)}$, then

$$\begin{bmatrix} c_{11} & c_{12} & \cdots & c_{1n} \\ c_{21} & c_{22} & \cdots & c_{2n} \\ & \vdots & & \\ c_{m1} & c_{m2} & \cdots & c_{mn} \end{bmatrix} \begin{bmatrix} 1 \\ 0 \\ \vdots \\ 0 \end{bmatrix} = \begin{bmatrix} c_{11} \\ c_{21} \\ \vdots \\ c_{m1} \end{bmatrix}$$

By hypothesis,

$$C \begin{bmatrix} 1 \\ 0 \\ \vdots \\ 0 \end{bmatrix} = 0$$

Thus, $c_{11} = c_{21} = \cdots = c_{m1} = 0$. In a similar manner, by multiplying C by the n-vectors

$$\begin{bmatrix} 0 \\ 1 \\ 0 \\ 0 \\ \vdots \\ 0 \end{bmatrix}, \quad \begin{bmatrix} 0 \\ 0 \\ 1 \\ 0 \\ \vdots \\ 0 \end{bmatrix}, \ldots, \begin{bmatrix} 0 \\ 0 \\ 0 \\ \vdots \\ 0 \\ 1 \end{bmatrix}$$

successively, we find all other columns are 0. ∎

Thus, we see that $AB = I_n$. A similar argument shows that $BA = I_n$. Thus, $B = A^{-1}$.

Example 1 In Example 1, §3.6, we inverted the matrix

$$A = \begin{bmatrix} 1 & 1 & 1 & 1 \\ 1 & 1 & 1 & -1 \\ 1 & 1 & -1 & 1 \\ 1 & -1 & 1 & 1 \end{bmatrix}$$

using the formula for the inverse of a matrix. Let us solve the same problem by synthetic elimination.

We consider the system

$$\begin{aligned} x_1 + x_2 + x_3 + x_4 &= y_1 \\ x_1 + x_2 + x_3 - x_4 &= y_2 \\ x_1 + x_2 - x_3 + x_4 &= y_3 \\ x_1 - x_2 + x_3 + x_4 &= y_4 \end{aligned}$$

Use the first equation and x_1.

$$\begin{aligned} x_1 + x_2 + x_3 + x_4 &= y_1 \\ -2x_4 &= y_2 - y_1 \\ -2x_3 \qquad &= y_3 - y_1 \\ -2x_2 \qquad\qquad &= y_4 - y_1 \end{aligned}$$

Add multiples of equations two, three, and four to the first equation.

$$\begin{aligned} x_1 \qquad\qquad\qquad &= y_1 + \tfrac{1}{2}(y_2 - y_1) + \tfrac{1}{2}(y_3 - y_1) \\ &\quad + \tfrac{1}{2}(y_4 - y_1) \\ x_4 &= \tfrac{1}{2}y_1 - \tfrac{1}{2}y_2 \\ x_3 \qquad &= \tfrac{1}{2}y_1 - \tfrac{1}{2}y_3 \\ x_2 \qquad\qquad &= \tfrac{1}{2}y_1 - \tfrac{1}{2}y_4 \end{aligned}$$

or

$$\begin{aligned} x_1 &= -\tfrac{1}{2}y_1 + \tfrac{1}{2}y_2 + \tfrac{1}{2}y_3 + \tfrac{1}{2}y_4 \\ x_2 &= \tfrac{1}{2}y_1 \qquad\qquad\qquad\quad - \tfrac{1}{2}y_4 \\ x_3 &= \tfrac{1}{2}y_1 \qquad\quad - \tfrac{1}{2}y_3 \\ x_4 &= \tfrac{1}{2}y_1 - \tfrac{1}{2}y_2 \end{aligned}$$

Thus,

$$A^{-1} = \begin{bmatrix} -1/2 & 1/2 & 1/2 & 1/2 \\ 1/2 & 0 & 0 & -1/2 \\ 1/2 & 0 & -1/2 & 0 \\ 1/2 & -1/2 & 0 & 0 \end{bmatrix}$$

which was achieved with considerably less trouble than in the previous section.

Example 2

$$A = \begin{bmatrix} 0 & 1 & 0 & 0 & 0 & 0 \\ 1 & 0 & 1 & 0 & 0 & 0 \\ 0 & 1 & 0 & 1 & 0 & 0 \\ 0 & 0 & 1 & 0 & 1 & 0 \\ 0 & 0 & 0 & 1 & 0 & 1 \\ 0 & 0 & 0 & 0 & 1 & 0 \end{bmatrix}$$

becomes the system

$$\begin{aligned} x_2 &= y_1 \\ x_1 + x_3 &= y_2 \\ x_2 + x_4 &= y_3 \\ x_3 + x_5 &= y_4 \\ x_4 + x_6 &= y_5 \\ x_5 &= y_6 \end{aligned}$$

$$\downarrow$$

$$\begin{aligned} x_2 &= y_1 \\ x_1 + x_3 &= y_2 \\ x_4 &= y_3 - y_1 \\ x_3 &= y_4 - y_6 \\ x_4 + x_6 &= y_5 \\ x_5 &= y_6 \end{aligned}$$

$$\downarrow$$

$$\begin{aligned} x_2 &= y_1 \\ x_1 &= y_2 - y_4 + y_6 \\ x_4 &= y_3 - y_1 \\ x_3 &= y_4 - y_6 \\ x_6 &= y_5 - y_3 + y_1 \\ x_5 &= y_6 \end{aligned}$$

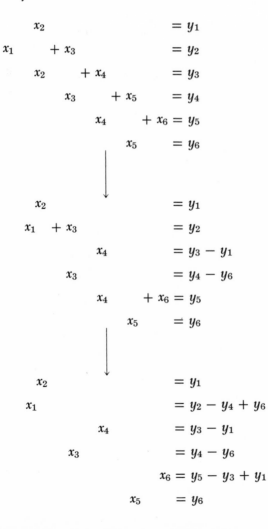

So

$$A^{-1} = \begin{bmatrix} 0 & 1 & 0 & -1 & 0 & 1 \\ 1 & 0 & 0 & 0 & 0 & 0 \\ 0 & 0 & 0 & 1 & 0 & -1 \\ -1 & 0 & 1 & 0 & 0 & 0 \\ 0 & 0 & 0 & 0 & 0 & 1 \\ 1 & 0 & -1 & 0 & 1 & 0 \end{bmatrix}$$

EXERCISES

1. Calculate the inverses of the following matrices.

(a) $\begin{bmatrix} 1 & 3 & -1 \\ 2 & -1 & 2 \\ 0 & 1 & 1 \end{bmatrix}$ (b) $\begin{bmatrix} 0 & 1 & 3 \\ 2 & 1 & 0 \\ 1 & 0 & 2 \end{bmatrix}$ (c) $\begin{bmatrix} 0 & 0 & 1 \\ 1 & 0 & 0 \\ 0 & 1 & 0 \end{bmatrix}$

(d) $\begin{bmatrix} 1 & 1 & 1 \\ 1 & 2 & 1 \\ -1 & 1 & 1 \end{bmatrix}$ (e) $\begin{bmatrix} 1 & 2 & 3 \\ 3 & 4 & 6 \\ 3 & 7 & 9 \end{bmatrix}$ (f) $\begin{bmatrix} 1 & 2 & -1 \\ 1 & -1 & 1 \\ 1 & 0 & 1 \end{bmatrix}$

(g) $\begin{bmatrix} 1 & a & b \\ \mu & 1+\mu a & \mu b \\ \lambda & \lambda a & 1+\lambda b \end{bmatrix}$ (h) $\begin{bmatrix} 8 & -6 & 7 & -7 \\ -3/7 & 1/7 & 0 & 0 \\ -4/7 & -8/7 & 2 & -1 \\ 4/7 & 1/7 & 0 & 0 \end{bmatrix}$

(i) $\begin{bmatrix} 0 & 1 & 0 & 0 \\ 1 & 0 & 1 & 0 \\ 0 & 1 & 0 & 1 \\ 1 & 0 & 0 & 1 \end{bmatrix}$ (j) $\begin{bmatrix} 1 & 2 & 2 & 4 \\ 3 & 2 & 2 & 4 \\ 1 & 3 & 2 & 4 \\ 1 & 2 & 3 & 4 \end{bmatrix}$

(k) $\begin{bmatrix} 1 & -5 & -7/3 & -2/3 \\ 0 & 1 & -1/3 & 1/3 \\ 0 & 0 & 1 & 0 \\ 0 & 0 & 0 & 1 \end{bmatrix}$

2. For what values of λ is $\lambda I_n - A$ invertible? What is its inverse if

(a) $A = \begin{bmatrix} 1 & 0 \\ 0 & -1 \end{bmatrix}$ (b) $A = \begin{bmatrix} 0 & 1 & 0 \\ 1 & 0 & 0 \\ 0 & 0 & 1 \end{bmatrix}$

(c) $A = \begin{bmatrix} 2 & 1 & -1 \\ 2 & 1 & -1 \\ -2 & -1 & 1 \end{bmatrix}$ (d) $A = \begin{bmatrix} 1 & 3 & -1 \\ 0 & 2 & 1 \\ 0 & 0 & -1 \end{bmatrix}$

3. Using synthetic elimination, calculate the inverse of

$$\begin{bmatrix} 1 & a_2 & a_3 & \dots & a_n \\ \lambda_2 & 1+\lambda_2 a_2 & \lambda_2 a_3 & \dots & \lambda_2 a_n \\ \lambda_3 & \lambda_3 a_2 & 1+\lambda_3 a_3 & \dots & \lambda_3 a_n \\ & & \vdots & & \\ \lambda_{n-1} & \lambda_{n-1} a_2 & \lambda_{n-1} a_3 & \dots & \lambda_{n-1} a_n \\ \lambda_n & \lambda_n a_2 & \lambda_n a_3 & \dots & 1+\lambda_n a_n \end{bmatrix}$$

4. Let A be an $n \times n$ matrix having exactly one nonzero entry in each row and column. Show that A is invertible and that its inverse is a matrix of the same form.

5. Find the inverse of the $n \times n$ matrix

$$\begin{bmatrix} 0 & 0 & 0 & \dots & 0 & -1 \\ 1 & 0 & 0 & \dots & 0 & -1 \\ 0 & 1 & 0 & \dots & 0 & -1 \\ 0 & 0 & 1 & \dots & 0 & -1 \\ & & & \vdots & & \\ 0 & 0 & 0 & \dots & 0 & -1 \\ 0 & 0 & 0 & \dots & 1 & -1 \end{bmatrix}$$

6. Find the inverse of the $n \times n$ matrix

$$\begin{bmatrix} 0 & 0 & 0 & \dots & 0 & 1 \\ 1 & 0 & 0 & \dots & 0 & 0 \\ 0 & 1 & 0 & \dots & 0 & 0 \\ 0 & 0 & 1 & \dots & 0 & 0 \\ & & & \vdots & & \\ 0 & 0 & 0 & \dots & 0 & 0 \\ 0 & 0 & 0 & \dots & 1 & 0 \end{bmatrix}$$

7. Find the inverse of the $n \times n$ matrix

$$\begin{bmatrix} 1 & 1 & 1 & 1 & \cdots & 1 & 1 \\ -1 & 1 & 1 & 1 & \cdots & 1 & 1 \\ 1 & -1 & 1 & 1 & \cdots & 1 & 1 \\ 1 & 1 & -1 & 1 & \cdots & 1 & 1 \\ & & & \vdots & & & \\ 1 & 1 & 1 & 1 & \cdots & -1 & 1 \end{bmatrix}$$

8. If $a \neq 0$, find the inverse of

$$\begin{bmatrix} a & b & c & d \\ 0 & a & b & c \\ 0 & 0 & a & b \\ 0 & 0 & 0 & a \end{bmatrix}$$

4

VECTOR SPACES

1 DEFINITION OF A VECTOR SPACE

In an earlier chapter we gave rules for addition and scalar multiplication of vectors and matrices. We saw that these operations obeyed certain laws which enabled us to perform algebraic manipulations without continually referring back to the original definitions. In this chapter we study abstract entities called vector spaces for which the operations of addition and scalar multiplication are also defined: these operations satisfy the same rules of addition and scalar multiplication as the vectors in chapter 2.

We shall study these objects abstractly, using only the basic postulates to develop the theory. By adopting such an approach we simplify and clarify the proofs of many theorems, as well as achieve a considerable extension of the range of applicability of our results.

Throughout this chapter, then, we deal with sets of objects called vectors merely because they fit into our abstract framework. Column vectors, polynomials in a variable x, matrices, and functions on an interval may all be called vectors, since after certain definitions have been given, we see that each is a member of a set which can be designated as a vector space.

A **vector space** is a set V, consisting of objects called vectors, with two operations defined: addition and scalar multiplication. Addition of vectors means that given two vectors $\boldsymbol{x}$ and $\boldsymbol{y}$ in V, there is a rule determining a vector $\boldsymbol{x} + \boldsymbol{y}$ also in V, and this vector is called the **sum** of $\boldsymbol{x}$ and $\boldsymbol{y}$. By scalar multiplication we mean that there is a rule which assigns to each vector $\boldsymbol{x}$ in V and each real scalar α a vector $\alpha\boldsymbol{x}$ also in V. This vector is called the **scalar multiple** of the vector $\boldsymbol{x}$ by the scalar α.

For example, if the set V is the collection of all polynomials with real coefficients, we might consider the sum $f + g$ of two polynomials, f and g, to be the ordinary sum and scalar multiplication to be the ordinary product of a number α with a polynomial f, the result being αf. It is possible to define addition and scalar multiplication differently, but if we want the set of polynomials to constitute a vector space, the operations we define must satisfy all the axioms of a vector space.

Let V be a set for which addition and scalar multiplication are defined and let $\boldsymbol{x}$, $\boldsymbol{y}$, and $\boldsymbol{z}$ belong to V; α and β are real numbers. The axioms of a vector space are

(V1) $\boldsymbol{x} + \boldsymbol{y} = \boldsymbol{y} + \boldsymbol{x}$. (Commutative law of vector addition.)

(V2) $(\boldsymbol{x} + \boldsymbol{y}) + \boldsymbol{z} = \boldsymbol{x} + (\boldsymbol{y} + \boldsymbol{z})$. (Associative law of vector addition.)

(V3) There is an element in V, denoted by $\mathbf{0}$, such that $\mathbf{0} + \boldsymbol{x} = \boldsymbol{x} + \mathbf{0} = \boldsymbol{x}$.

(V4) For each $\boldsymbol{x}$ in V, there is an element $-\boldsymbol{x}$ in V, such that $\boldsymbol{x} + (-\boldsymbol{x}) = (-\boldsymbol{x}) + \boldsymbol{x} = \mathbf{0}$.

(V5) $(\alpha + \beta)\boldsymbol{x} = \alpha\boldsymbol{x} + \beta\boldsymbol{x}$

(V6) $\alpha(\boldsymbol{x} + \boldsymbol{y}) = \alpha\boldsymbol{x} + \alpha\boldsymbol{y}$

(V7) $(\alpha\beta)\boldsymbol{x} = \alpha(\beta\boldsymbol{x})$

(V8) $1 \cdot \boldsymbol{x} = \boldsymbol{x}$

A set V whose operations satisfy the above list of requirements is said to be a **real vector space** or a **vector space over the reals.** The vector $\mathbf{0}$, whose existence is asserted in V, is called the **zero vector.** The vector $-\boldsymbol{x}$ is called the **negative** of the vector $\boldsymbol{x}$.

Complex vector spaces are defined in a similar way. We merely postulate that the scalar multiple of the vector $\boldsymbol{x}$ and the scalar α be defined for each $\boldsymbol{x}$ in V and each complex scalar α.

The importance of the vector space concept stems largely from the extensive list of examples of objects satisfying the vector space axioms.

Example 1 Let $\boldsymbol{R}^n$ be the space of column n-vectors with addition and scalar multiplication as defined in chapter 2.

$$\begin{bmatrix} \alpha_1 \\ \alpha_2 \\ \vdots \\ \alpha_n \end{bmatrix} + \begin{bmatrix} \beta_1 \\ \beta_2 \\ \vdots \\ \beta_n \end{bmatrix} = \begin{bmatrix} \alpha_1 + \beta_1 \\ \alpha_2 + \beta_2 \\ \vdots \\ \alpha_n + \beta_n \end{bmatrix}$$

$$\mu \begin{bmatrix} \alpha_1 \\ \alpha_2 \\ \vdots \\ \alpha_n \end{bmatrix} = \begin{bmatrix} \mu\alpha_1 \\ \mu\alpha_2 \\ \vdots \\ \mu\alpha_n \end{bmatrix}$$

At this point the reader may find it a profitable exercise to repeat the proofs offered in chapter 2 showing that the addition and scalar multiplication defined above satisfy the requisite vector space axioms.

In many ways $\boldsymbol{R}^n$ is the model real vector space. The axioms of a vector space were formulated by singling out the most important properties of addition and scalar multiplication of column vectors. These properties are used to develop in a more general setting theorems analogous to those which hold in $\boldsymbol{R}^n$. Moreover, in a sense which will later be made more precise, real vector spaces which are not "too big" are, in a natural algebraic way, equivalent to $\boldsymbol{R}^n$ for some integer n.

Example 2 The space of complex column n-vectors is a complex vector space, denoted by $\boldsymbol{C}^n$. Addition is defined by

$$\begin{bmatrix} \alpha_1 \\ \alpha_2 \\ \vdots \\ \alpha_n \end{bmatrix} + \begin{bmatrix} \beta_1 \\ \beta_2 \\ \vdots \\ \beta_n \end{bmatrix} = \begin{bmatrix} \alpha_1 + \beta_1 \\ \alpha_2 + \beta_2 \\ \vdots \\ \alpha_n + \beta_n \end{bmatrix}$$

where $\alpha_1, \alpha_2, \ldots, \alpha_n, \beta_1, \beta_2, \ldots, \beta_n$ are complex numbers, and scalar multiplication is defined by

$$\mu \begin{bmatrix} \alpha_1 \\ \alpha_2 \\ \vdots \\ \alpha_n \end{bmatrix} = \begin{bmatrix} \mu\alpha_1 \\ \mu\alpha_2 \\ \vdots \\ \mu\alpha_n \end{bmatrix}$$

with μ a complex scalar.

In the same way that $\boldsymbol{R}^n$ is the model real vector space, $\boldsymbol{C}^n$ is the model complex vector space.

Example 3 Let M_{mn} denote the collection of $m \times n$ matrices with real entries. The addition and scalar multiplication are as defined in §2.3:

$$\begin{bmatrix} a_{11} & a_{12} & \ldots & a_{1n} \\ a_{21} & a_{22} & \ldots & a_{2n} \\ & \vdots & & \\ a_{m1} & a_{m2} & \ldots & a_{mn} \end{bmatrix} + \begin{bmatrix} b_{11} & b_{12} & \ldots & b_{1n} \\ b_{21} & b_{22} & \ldots & b_{2n} \\ & \vdots & & \\ b_{m1} & b_{m2} & \ldots & b_{mn} \end{bmatrix} = \begin{bmatrix} a_{11} + b_{11} & a_{12} + b_{12} & \ldots & a_{1n} + b_{1n} \\ a_{21} + b_{21} & a_{22} + b_{22} & \ldots & a_{2n} + b_{2n} \\ & \vdots & & \\ a_{m1} + b_{m1} & a_{m2} + b_{m2} & \ldots & a_{mn} + b_{mn} \end{bmatrix}$$

$$\mu \begin{bmatrix} a_{11} & a_{12} & \ldots & a_{1n} \\ a_{21} & a_{22} & \ldots & a_{2n} \\ & \vdots & & \\ a_{m1} & a_{m2} & \ldots & a_{mn} \end{bmatrix} = \begin{bmatrix} \mu a_{11} & \mu a_{12} & \ldots & \mu a_{1n} \\ \mu a_{21} & \mu a_{22} & \ldots & \mu a_{2n} \\ & \vdots & & \\ \mu a_{m1} & \mu a_{m2} & \ldots & \mu a_{mn} \end{bmatrix}$$

Since column n-vectors are just $n \times 1$ matrices, Example 1 is a special case of this example.

Example 4 Let P_n denote the collection of all polynomials with real coefficients, of degree less than or equal to n. If f and g belong to P_n, we add them in the ordinary way: let

$$f = a_0 + a_1x + \cdots + a_nx^n = \sum_{k=0}^{n} a_kx^k \quad \text{and} \quad g = b_0 + b_1x + \cdots + b_nx^n = \sum_{k=0}^{n} b_kx^k$$

then

$$\begin{aligned} f + g &= a_0 + b_0 + (a_1 + b_1)x + \cdots + (a_n + b_n)x^n \\ &= \sum_{k=0}^{n} (a_k + b_k)x^k \end{aligned}$$

Scalar multiplication is, likewise, defined in the usual way:

$$\begin{aligned} \alpha f &= \alpha a_0 + (\alpha a_1)x + \cdots + (\alpha a_n)x^n \\ &= \sum_{k=0}^{n} (\alpha a_k)x^k \end{aligned}$$

To explicitly verify the vector space axioms in this case, we let

$$f = \sum_{k=0}^{n} a_kx^k, \qquad g = \sum_{k=0}^{n} b_kx^k, \qquad h = \sum_{k=0}^{n} c_kx^k$$

and α, β be real scalars.

(V1)

$$\begin{aligned} f + g &= \sum_{k=0}^{n} (a_k + b_k)x^k \\ g + f &= \sum_{k=0}^{n} (b_k + a_k)x^k \end{aligned}$$

Since a_k and b_k are real, $a_k + b_k = b_k + a_k$, and so we obtain $f + g = g + f$.

(V2)

$$\begin{aligned} (f + g) + h &= \sum_{k=0}^{n} ((a_k + b_k) + c_k)x^k \\ f + (g + h) &= \sum_{k=0}^{n} (a_k + (b_k + c_k))x^k \end{aligned}$$

a_k, b_k, and c_k being real, we have $(a_k + b_k) + c_k = a_k + (b_k + c_k)$, and so $(f + g) + h = f + (g + h)$.

(V3) Let 0 be the zero polynomial, i.e., the polynomial all of whose coefficients are zeroes. Then $f + 0 = 0 + f = f$.

(V4) If

$$f = \sum_{k=0}^{n} a_k x^k, \quad \text{let } -f = \sum_{k=0}^{n} (-a_k) x^k$$

Then,

$$f + (-f) = \sum_{k=0}^{n} (a_k + (-a_k)) x^k = \sum_{k=0}^{n} 0 \cdot x^k = 0$$

(V5) By definition,

$$(\alpha + \beta) f = \sum_{k=0}^{n} (\alpha + \beta) a_k x^k$$

while

$$\alpha f = \sum_{k=0}^{n} \alpha a_k x^k, \qquad \beta f = \sum_{k=0}^{n} \beta a_k x^k$$

and so

$$\alpha f + \beta f = \sum_{k=0}^{n} (\alpha a_k + \beta a_k) x$$

For real numbers we know $(\alpha + \beta) a_k = \alpha a_k + \beta a_k$, and so $(\alpha + \beta) f = \alpha f + \beta f$.

(V6) We have $f + g = \Sigma_{k=0}^{n} (a_k + b_k) x^k$, and so $\alpha(f + g) = \Sigma_{k=0}^{n} (\alpha(a_k + b_k)) x^k$. Also $\alpha f + \alpha g = \Sigma_{k=0}^{n} (\alpha a_k + \alpha b_k) x^k$. Since $\alpha(a_k + b_k) = \alpha a_k + \alpha b_k$, we have $\alpha(f + g) = \alpha f + \alpha g$.

(V7) $\beta f = \Sigma_{k=0}^{n} (\beta a_k) x^k$, by definition, and so $\alpha(\beta f) = \Sigma_{k=0}^{n} (\alpha(\beta a_k)) x^k$, while $(\alpha\beta) f = \Sigma_{k=0}^{n} ((\alpha\beta) a_k) x^k$.

Now, $\alpha(\beta a_k) = (\alpha\beta) a_k$ holds for real numbers, and so $\alpha(\beta f) = (\alpha\beta) f$.

(V8) $1 \cdot f = \Sigma_{k=0}^{n} (1 \cdot a_k) x^k = \Sigma_{k=0}^{n} a_k x^k = f$

Thus, the polynomials of degree less than or equal to n, having real coefficients, with addition and scalar multiplication as defined above, form a real vector space. Because of this all theorems proved about vector spaces in general hold for the vector space of polynomials.

In an analogous manner, it is possible to show that the polynomials of degree less than or equal to n with complex coefficients form a complex vector space under the usual addition and scalar multiplication.

Although in the examples above the addition and scalar multiplication are in some sense "natural," it is possible to create examples which are not so transparent, as in the following.

Example 5 On the ordered pairs of real numbers (x, y) we define an operation of addition, which we denote by $\oplus$ to distinguish it from ordinary addition, by

$$(x, y) \oplus (x', y') = (x + x' + 1, y + y' + 1)$$

We define scalar multiplication by

$$\alpha * (x, y) = (\alpha x + \alpha - 1, \alpha y + \alpha - 1)$$

We now verify (V1)–(V8).

(V1)
$$(x, y) \oplus (x', y') = (x + x' + 1, y + y' + 1)$$
$$(x', y') \oplus (x, y) = (x' + x + 1, y' + y + 1)$$

Since $x + x' + 1 = x' + x + 1$ and $y + y' + 1 = y' + y + 1$, we see that $(x, y) \oplus (x', y') = (x', y') \oplus (x, y)$.

(V2)
$$\begin{aligned}((x, y) \oplus (x', y')) \oplus (x'', y'')& \\ &= (x + x' + 1, y + y' + 1) \oplus (x'', y'') \\ &= (x + x' + x'' + 2, y + y' + y'' + 2)\end{aligned}$$

$$\begin{aligned}(x, y) \oplus ((x', y') \oplus (x'', y''))& \\ &= (x, y) \oplus (x' + x'' + 1, y' + y'' + 1) \\ &= (x + x' + x'' + 2, y + y' + y'' + 2)\end{aligned}$$

So

$$((x, y) \oplus (x', y')) \oplus (x'', y'') = (x, y) \oplus ((x', y') \oplus (x'', y''))$$

(V3) What is the zero element of our space? Observe that

$$\begin{aligned}(x, y) \oplus (-1, -1) &= (x + (-1) + 1, y + (-1) + 1) \\ &= (x, y)\end{aligned}$$

Hence, we see that $(-1, -1)$ plays the role of the zero element.

(V4) Given (x, y), we let its negative be $(-x - 2, -y - 2)$. Then

$$\begin{aligned}(x, y) \oplus (-x - 2, -y - 2) &= (x + (-x - 2) + 1, y + (-y - 2) + 1) \\ &= (-1, -1)\end{aligned}$$

which is the zero element.

(V5) $(\alpha + \beta) * (x, y)$

$$= ((\alpha + \beta)x + (\alpha + \beta) - 1, (\alpha + \beta)y + (\alpha + \beta) - 1)$$

(by definition of $*$)

Also, by definition,

$$\alpha * (x, y) = (\alpha x + \alpha - 1, \alpha y + \alpha - 1)$$
$$\beta * (x, y) = (\beta x + \beta - 1, \beta y + \beta - 1)$$

By definition of $\oplus$,

$$\begin{aligned}(\alpha * (x, y)) \oplus (\beta * (x, y)) &= ((\alpha x + \alpha - 1) + (\beta x + \beta - 1) + 1,\\ &\qquad (\alpha y + \alpha - 1) + (\beta y + \beta - 1) + 1)\\ &= ((\alpha + \beta)x + (\alpha + \beta) - 1,\\ &\qquad (\alpha + \beta)y + (\alpha + \beta) - 1)\\ &= (\alpha + \beta) * (x, y)\end{aligned}$$

(V6) $\alpha * ((x, y) \oplus (x', y')) = \alpha * (x + x' + 1, y + y' + 1)$

$$= (\alpha(x + x' + 1) + \alpha - 1, \alpha(y + y' + 1) + \alpha - 1)$$

while

$$\begin{aligned}&\alpha * (x, y) \oplus \alpha * (x', y')\\ &= (\alpha x + \alpha - 1, \alpha y + \alpha - 1) \oplus (\alpha x' + \alpha - 1, \alpha y' + \alpha - 1)\\ &= (\alpha x + \alpha - 1 + \alpha x' + \alpha - 1 + 1, \alpha y + \alpha - 1 + \alpha y' + \alpha - 1 + 1)\\ &= \alpha * ((x, y) \oplus (x', y'))\end{aligned}$$

(V7) $$\begin{aligned}\alpha * (\beta * (x, y)) &= \alpha * (\beta x + \beta - 1, \beta y + \beta - 1)\\ &= (\alpha\beta x + \alpha\beta - \alpha + \alpha - 1, \alpha\beta y + \alpha\beta - \alpha + \alpha - 1)\\ &= (\alpha\beta x + \alpha\beta - 1, \alpha\beta y + \alpha\beta - 1)\\ &= (\alpha\beta) * (x, y)\end{aligned}$$

(V8) $1 * (x, y) = (x + 1 - 1, y + 1 - 1) = (x, y)$

Thus, the collection of ordered pairs of real numbers with operation of addition $\oplus$ and scalar multiplication $*$ forms a vector space over the real numbers.

EXERCISES

1. The following is a list of sets with operations of addition and scalar multiplication defined on them. For each set show that the set together with its indicated operations forms a vector space over the reals.
   (a) The real matrices of the form

$$\begin{bmatrix} a & -b \\ b & a \end{bmatrix}$$

with

$$\begin{bmatrix} a & -b \\ b & a \end{bmatrix} + \begin{bmatrix} c & -d \\ d & c \end{bmatrix} = \begin{bmatrix} a+c & -(b+d) \\ b+d & a+c \end{bmatrix}$$

$$\alpha\begin{bmatrix} a & -b \\ b & a \end{bmatrix} = \begin{bmatrix} \alpha a & -\alpha b \\ \alpha b & \alpha a \end{bmatrix}$$

   (b) The real matrices of the form

$$\begin{bmatrix} a & b \\ 0 & c \end{bmatrix}$$

with

$$\begin{bmatrix} a & b \\ 0 & c \end{bmatrix} + \begin{bmatrix} a' & b' \\ 0 & c \end{bmatrix} = \begin{bmatrix} a+a' & b+b' \\ 0 & c+c' \end{bmatrix}$$

$$\alpha\begin{bmatrix} a & b \\ 0 & c \end{bmatrix} = \begin{bmatrix} \alpha a & \alpha b \\ 0 & \alpha c \end{bmatrix}$$

   (c) The set of triples of real numbers (x, y, z), such that $z = x + y$, with $(x, y, z) + (x', y', z') = (x + x', y + y', z + z')$ and $\alpha(x, y, z) = (\alpha x, \alpha y, \alpha z)$.
   (d) The even polynomials of degree less than or equal to n, a positive integer, with usual addition and scalar multiplication for polynomials.
   (e) The odd polynomials of degree less than or equal to n, a positive integer, with usual addition and scalar multiplication for polynomials.
   (f) The polynomials f, of degree less than or equal to n, such that $f(1) = 0$. Addition and scalar multiplication are defined in the usual manner.
   (g) Ordered pairs of real numbers (x, y) with

$$(x, y) \oplus (x', y') = (x + x' + 1, y + y')$$

$$\alpha * (x, y) = (\alpha x + \alpha - 1, \alpha y)$$

(h) Differentiable functions on the interval (0, 1) with

$$(f + g)(x) = f(x) + g(x)$$
$$(\alpha f)(x) = \alpha(f(x))$$

and the usual addition and scalar multiplication for functions.

2. On the triples of real numbers (x, y, z) define addition by $(x, y, z) + (x', y', z') = (x + x', y + y', z + z')$, and scalar multiplication by $\alpha(x, y, z) = (0, 0, 0)$. Show that all axioms for a vector space are satisfied except (V8).
3. On the ordered pairs of real numbers (x, y) define addition by $(x, y) + (x', y') = (x + x', y + y')$ and scalar multiplication by $\alpha(x, y) = (\alpha^2 x, \alpha^2 y)$. Show that all axioms for a vector space are satisfied except (V5).
4. On the ordered pairs of real numbers define addition by $(x, y) + (x', y') = (x + x', y + y')$ and scalar multiplication by $\alpha(x, y) = (3\alpha x, 3\alpha y)$. Show that all axioms for a vector space are satisfied except (V7) and (V8).
5. Let X be a set. Consider the family $\mathfrak{F}$ of all functions from X into the real numbers. Define

$$(f + g)(x) = f(x) + g(x)$$
$$(\alpha f)(x) = \alpha(f(x))$$

Show that with these indicated operations the given family of functions forms a vector space.

6. Let U be a complex vector space. Show that U is a real vector space if we use the same operation of addition while the scalar product is $\alpha \boldsymbol{x}$, for $\boldsymbol{x}$ in U and α a real number. (This definition makes sense, since the real numbers are contained in the complex numbers.)
7. Let V be real vector space, with addition denoted by $+$ and scalar multiplication by $\cdot$. Let $\boldsymbol{t}$ be a fixed vector in V. Define a new addition on V by $\boldsymbol{x} \oplus \boldsymbol{y} = \boldsymbol{x} + \boldsymbol{y} + \boldsymbol{t}$ and scalar multiplication by $\alpha * \boldsymbol{x} = \alpha \boldsymbol{x} + (\alpha - 1)\boldsymbol{t}$. Show that V with $\oplus$ and $*$ as its operations is a vector space.
8. On the ordered pairs of real numbers (x, y) define $\oplus$ addition by $(x, y) \oplus (x', y') = ((x^3 + (x')^3)^{1/3}, (y^3 + (y')^3)^{1/3})$ and scalar multiplication by $\alpha(x, y) = (\alpha x, \alpha y)$. Show that with the indicated operations the ordered pairs of real numbers form a vector space.

2 ADDITIONAL PROPERTIES OF VECTOR SPACES

Given any vector space, we can use the axioms (V1)–(V8) to obtain additional rules for algebraic manipulation of vectors.

Theorem 1 Let V be a vector space and $\boldsymbol{x}$ and $\boldsymbol{y}$ be vectors in V. Then, there is one and only one $\boldsymbol{u}$ in V, such that $\boldsymbol{x} + \boldsymbol{u} = \boldsymbol{y}$.

Proof First, we must show there is one such $\boldsymbol{u}$. To accomplish this, let $\boldsymbol{u} = (-\boldsymbol{x}) + \boldsymbol{y}$. Then,

$$\begin{aligned} \boldsymbol{x} + \boldsymbol{u} &= \boldsymbol{x} + ((-\boldsymbol{x}) + \boldsymbol{y}) \\ &= (\boldsymbol{x} + (-\boldsymbol{x})) + \boldsymbol{y} && \text{[by (V2)]} \\ &= \boldsymbol{0} + \boldsymbol{y} && \text{[by (V4)]} \\ &= \boldsymbol{y} && \text{[by (V3)]} \end{aligned}$$

Next we show that there is only one such $\boldsymbol{u}$. Suppose $\boldsymbol{u}_1$ and $\boldsymbol{u}_2$ are vectors in V, such that

$$\boldsymbol{x} + \boldsymbol{u}_1 = \boldsymbol{y} \quad \text{and} \quad \boldsymbol{x} + \boldsymbol{u}_2 = \boldsymbol{y}$$

Then, $\boldsymbol{x} + \boldsymbol{u}_1 = \boldsymbol{x} + \boldsymbol{u}_2$.

Thus,

$$\begin{aligned} (-\boldsymbol{x}) + (\boldsymbol{x} + \boldsymbol{u}_1) &= (-\boldsymbol{x}) + (\boldsymbol{x} + \boldsymbol{u}_2) \\ ((-\boldsymbol{x}) + \boldsymbol{x}) + \boldsymbol{u}_1 &= ((-\boldsymbol{x}) + \boldsymbol{x}) + \boldsymbol{u}_2 && \text{[by (V2)]} \\ \boldsymbol{0} + \boldsymbol{u}_1 &= \boldsymbol{0} + \boldsymbol{u}_2 && \text{[by (V4)]} \\ \boldsymbol{u}_1 &= \boldsymbol{u}_2 && \text{[by (V3)]} \ \blacksquare \end{aligned}$$

It is customary to denote the vector $\boldsymbol{y} + (-\boldsymbol{x})$ considered above simply by $\boldsymbol{y} - \boldsymbol{x}$. $\boldsymbol{y} - \boldsymbol{x}$ is called the vector obtained from $\boldsymbol{y}$ by subtracting $\boldsymbol{x}$.

As another example of results which can be obtained from the vector space axioms we have the following theorem.

Theorem 2 Let V be a vector space, $\boldsymbol{x}$ be a vector in V, and α be a scalar. Then,

(i) $\alpha \cdot \boldsymbol{0} = \boldsymbol{0}$
(ii) $0 \cdot \boldsymbol{x} = \boldsymbol{0}$
(iii) $\alpha \boldsymbol{x} = \boldsymbol{0}$

implies that either $\alpha = 0$ or $\boldsymbol{x} = \boldsymbol{0}$.

Proof To prove (i), observe that

$$\boldsymbol{0} + \boldsymbol{0} = \boldsymbol{0} \qquad \text{[by (V3)]}$$

Thus,

$$\begin{aligned} \alpha(\boldsymbol{0} + \boldsymbol{0}) &= \alpha \cdot \boldsymbol{0} \\ \alpha \cdot \boldsymbol{0} + \alpha \cdot \boldsymbol{0} &= \alpha \cdot \boldsymbol{0} && \text{[by (V6)]} \end{aligned}$$

We know by Theorem 1 that there is only one vector $\boldsymbol{u}$, such that $\alpha \cdot \mathbf{0} + \boldsymbol{u} = \alpha \cdot \mathbf{0}$.

By (V3), one such vector is $\mathbf{0}$. Thus, $\alpha \cdot \mathbf{0} = \mathbf{0}$. Alternatively, by adding $-\alpha \cdot \mathbf{0}$ to the equation $\alpha \cdot \mathbf{0} + \alpha \cdot \mathbf{0} = \alpha \cdot \mathbf{0}$, we obtain

$$((\alpha \cdot \mathbf{0}) + (\alpha \cdot \mathbf{0})) + (-\alpha \cdot \mathbf{0}) = (-\alpha \cdot \mathbf{0}) + (\alpha \cdot \mathbf{0})$$
$$= \mathbf{0} \qquad \text{[by (V4)]}$$
$$(\alpha \cdot \mathbf{0}) + (\alpha \cdot \mathbf{0} + (-\alpha \cdot \mathbf{0})) = \mathbf{0} \qquad \text{[by (V2)]}$$
$$(\alpha \cdot \mathbf{0}) + \mathbf{0} = \mathbf{0} \qquad \text{[by (V4)]}$$
$$\alpha \cdot \mathbf{0} = \mathbf{0} \qquad \text{[by (V3)]}$$

To prove (ii) observe that $0 \cdot \boldsymbol{x} = (0 + 0)\boldsymbol{x}$, since $0 = 0 + 0$. Thus, $0 \cdot \boldsymbol{x} = 0 \cdot \boldsymbol{x} + 0 \cdot \boldsymbol{x}$ [by (V5)]. As before, by adding $-0 \cdot \boldsymbol{x}$ to both sides of the forgoing equation, we obtain $0 \cdot \boldsymbol{x} = \mathbf{0}$.

To prove (iii) suppose that $\alpha\boldsymbol{x} = \mathbf{0}$. If $\alpha \neq 0$, we may multiply both sides of the above equation by α^{-1} to obtain

$$\alpha^{-1}(\alpha\boldsymbol{x}) = \alpha^{-1} \cdot \mathbf{0} = \mathbf{0} \qquad \text{[by (i)]}$$
$$(\alpha^{-1}\alpha)\boldsymbol{x} = \mathbf{0} \qquad \text{[by (V7)]}$$
$$1 \cdot \boldsymbol{x} = \mathbf{0}$$
$$\boldsymbol{x} = \mathbf{0} \qquad \text{[by (V8)]}$$

Thus, if $\alpha \neq 0$, we have $\boldsymbol{x} = \mathbf{0}$. So we must have either $\alpha = 0$ or $\boldsymbol{x} = \mathbf{0}$. ∎

The following equates the scalar multiple $(-1)\boldsymbol{x}$ of the vector $\boldsymbol{x}$ with its negative $-\boldsymbol{x}$.

Theorem 3 If V is a vector space and $\boldsymbol{x}$ is a vector in V, then

$$(-1)\boldsymbol{x} = -\boldsymbol{x}$$

Proof Since $1 + (-1) = 0$, we have

$$(1 + (-1))\boldsymbol{x} = 0 \cdot \boldsymbol{x} = \mathbf{0} \qquad \text{[by (ii)]}$$

Thus, by (V5),

$$(1)\boldsymbol{x} + (-1)\boldsymbol{x} = \mathbf{0}$$

Or using (V8),

$$\boldsymbol{x} + (-1)\boldsymbol{x} = \mathbf{0}$$

If we add $-\boldsymbol{x}$ to both sides of this equation, we obtain $(-1)\boldsymbol{x} = -\boldsymbol{x}$. ∎

Generally speaking the vector space axioms enable us to perform algebraic operations with abstract vectors in much the same way we do with column vectors. Keeping this in mind will make it unnecessary to refer back to the vector space axioms (V1)–(V8).

EXERCISES

1. Let V be a vector space. Suppose $\boldsymbol{x}$ and $\boldsymbol{e}$ are members of V, and suppose $\boldsymbol{x} + \boldsymbol{e} = \boldsymbol{x}$. Show that $\boldsymbol{e} = \boldsymbol{0}$.
2. Prove by induction that if V is a vector space, $\boldsymbol{x} \in V$, and $\alpha_1, \alpha_2, \ldots, \alpha_n$ are scalars, then $(\alpha_1 + \alpha_2 + \cdots + \alpha_n)\boldsymbol{x} = \alpha_1\boldsymbol{x} + \alpha_2\boldsymbol{x} + \cdots + \alpha_n\boldsymbol{x}$.
3. If V is a vector space, $\boldsymbol{x}_1, \boldsymbol{x}_2, \ldots, \boldsymbol{x}_n$ belong to V, and α is a scalar, show by induction that $\alpha(\boldsymbol{x}_1 + \boldsymbol{x}_2 + \cdots + \boldsymbol{x}_n) = \alpha\boldsymbol{x}_1 + \alpha\boldsymbol{x}_2 + \cdots + \alpha\boldsymbol{x}_n$.
4. If V is a real vector space, $\boldsymbol{x}$ belongs to V, and $\boldsymbol{x} + \boldsymbol{x} = \boldsymbol{0}$, show that $\boldsymbol{x} = \boldsymbol{0}$.
5. If V is a vector space, $\boldsymbol{x}_1, \boldsymbol{x}_2, \boldsymbol{y}_1, \boldsymbol{y}_2$ belong to V,

$$a\boldsymbol{x}_1 + b\boldsymbol{x}_2 = \boldsymbol{y}_1$$
$$c\boldsymbol{x}_1 + d\boldsymbol{x}_2 = \boldsymbol{y}_2$$

   and

$$\begin{vmatrix} a & b \\ c & d \end{vmatrix} \neq 0$$

   find $\boldsymbol{x}_1$ and $\boldsymbol{x}_2$ in terms of $\boldsymbol{y}_1$ and $\boldsymbol{y}_2$.
6. If V is a vector space, $\boldsymbol{x}$ belongs to V, and α and β are scalars, show that if $\alpha\boldsymbol{x} = \beta\boldsymbol{x}$ and $\alpha \neq \beta$, then $\boldsymbol{x} = \boldsymbol{0}$.
7. If V is a vector space, $\boldsymbol{x}_1, \boldsymbol{x}_2, \boldsymbol{x}_3, \boldsymbol{y}_1, \boldsymbol{y}_2, \boldsymbol{y}_3$ are vectors in V, and

$$\begin{aligned} \boldsymbol{x}_1 + \boldsymbol{x}_2 + \boldsymbol{x}_3 &= \boldsymbol{y}_1 \\ -2\boldsymbol{x}_1 + \boldsymbol{x}_2 - 2\boldsymbol{x}_3 &= \boldsymbol{y}_2 \\ -\boldsymbol{x}_1 + 2\boldsymbol{x}_2 - \boldsymbol{x}_3 &= \boldsymbol{y}_3 \end{aligned}$$

   show that $\boldsymbol{y}_1 + \boldsymbol{y}_2 - \boldsymbol{y}_3 = \boldsymbol{0}$.
8. If V is a vector space, $\boldsymbol{x}_1, \boldsymbol{x}_2, \ldots, \boldsymbol{x}_n, \boldsymbol{y}_1, \boldsymbol{y}_2, \boldsymbol{y}_3, \ldots, \boldsymbol{y}_n$ belong to V, and

$$\begin{aligned} \boldsymbol{x}_1 + \boldsymbol{x}_2 + \cdots + \boldsymbol{x}_n &= \boldsymbol{y}_1 \\ \boldsymbol{x}_2 + \cdots + \boldsymbol{x}_n &= \boldsymbol{y}_2 \\ &\vdots \\ \boldsymbol{x}_n &= \boldsymbol{y}_n \end{aligned}$$

   find $\boldsymbol{x}_1, \boldsymbol{x}_2, \ldots, \boldsymbol{x}_n$ in terms of $\boldsymbol{y}_1, \boldsymbol{y}_2, \ldots, \boldsymbol{y}_n$.
9. Find a vector space V which has two vectors $\boldsymbol{x}$ and $\boldsymbol{y}$ neither of which is a scalar multiple of the other.
10. Show that any set V with operations of addition and scalar multiplication which satisfies (V2)–(V8) must also satisfy (V1). [Hint: Calculate $(1 + 1)(\boldsymbol{x} + \boldsymbol{y})$ two ways by using (V5) and (V6).]

3 SUBSPACES

If V is a vector space over the reals (or complexes), there are certain subsets of V, called subspaces, which are again vector spaces under the same algebraic operations. The purpose of this section is to study these objects.

Definition If V is a vector space and H is a subset of V having the properties

(i) Whenever $\boldsymbol{x}$ and $\boldsymbol{y}$ belong to H, then $\boldsymbol{x} + \boldsymbol{y}$ belongs to H.
(ii) If $\boldsymbol{x}$ belongs to H and α is a scalar, then $\alpha\boldsymbol{x}$ belongs to H.

H is said to be a **subspace** of the vector space V.

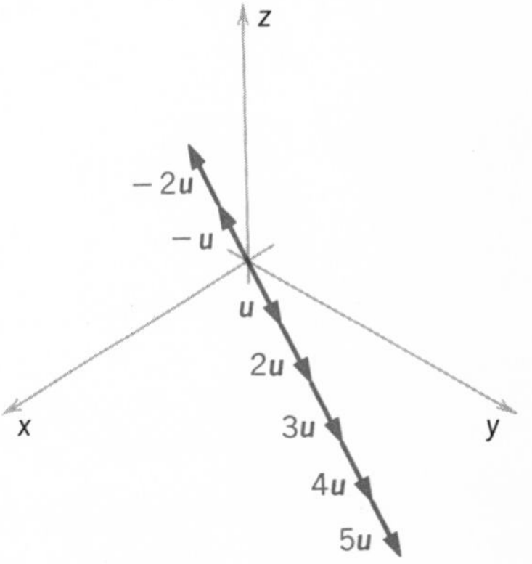

FIGURE 3-1

In other terms, a subspace of V is a subset closed under the algebraic operations of addition and scalar multiplication.

As an example, in $\boldsymbol{R}^3$, let L be the set of vectors lying on some line passing through the origin. (See Figure 3-1.) From our geometric formulation of the process of scalar multiplication, it is clear that all vectors in L are scalar multiples of a single nonzero vector in L, for definiteness, say $\boldsymbol{u}$. If $\boldsymbol{x}$ and $\boldsymbol{y}$ belong to L, we have $\boldsymbol{x} = \alpha\boldsymbol{u}$ and $\boldsymbol{y} = \beta\boldsymbol{u}$ for suitable scalars α and β. Then, $\boldsymbol{x} + \boldsymbol{y} = (\alpha + \beta)\boldsymbol{u}$. Thus, $\boldsymbol{x} + \boldsymbol{y}$ being a scalar multiple of $\boldsymbol{u}$ must necessarily belong to L. Moreover, since $\lambda\boldsymbol{x} = \lambda(\alpha\boldsymbol{u}) = (\lambda\alpha)\boldsymbol{u}$, we also see that if $\boldsymbol{x}$ belongs to L and λ is a scalar, $\lambda\boldsymbol{x}$ belongs to L. Thus, having verified conditions (i) and (ii) in the definition of a subspace above, we see that L is a subspace of $\boldsymbol{R}^3$.

Another example of a subspace of $\boldsymbol{R}^3$ may be obtained by considering the vectors which lie in the xy plane. If

$$\begin{bmatrix} x_1 \\ y_1 \\ 0 \end{bmatrix} \quad \text{and} \quad \begin{bmatrix} x_2 \\ y_2 \\ 0 \end{bmatrix}$$

are two such vectors, their sum,

$$\begin{bmatrix} x_1 + x_2 \\ y_1 + y_2 \\ 0 \end{bmatrix}$$

is again a vector lying in the xy plane. Alternatively, using the geometric formulation of vector addition presented in chapter 2, it is possible to see that the sum of two vectors lying in the xy plane is again a vector in the xy plane. (See Figure 3-2.) We saw it was possible to define the sum of $\boldsymbol{a}$ and $\boldsymbol{b}$, two vectors, to be the vector along the diagonal of the parallelogram, in the plane formed by the vectors $\boldsymbol{a}$ and $\boldsymbol{b}$, having

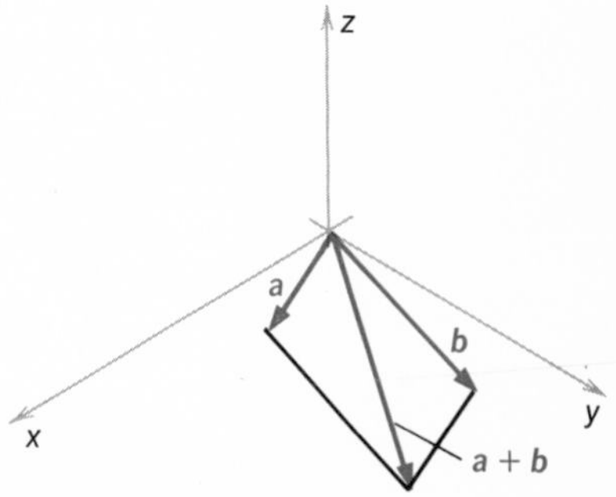

FIGURE 3-2

as adjacent sides the directed line segments representing the vectors $\boldsymbol{a}$ and $\boldsymbol{b}$. Thus, the sum $\boldsymbol{a} + \boldsymbol{b}$ of two vectors $\boldsymbol{a}$ and $\boldsymbol{b}$ in the xy plane is again a vector in the xy plane.

Condition (ii) in the definition of a subspace may be handled in a similar manner. Proceeding algebraically, we see that if

$$\begin{bmatrix} x \\ y \\ 0 \end{bmatrix}$$

is a vector in the xy plane, and α is a scalar, then

$$\alpha \begin{bmatrix} x \\ y \\ 0 \end{bmatrix} = \begin{bmatrix} \alpha x \\ \alpha y \\ 0 \end{bmatrix}$$

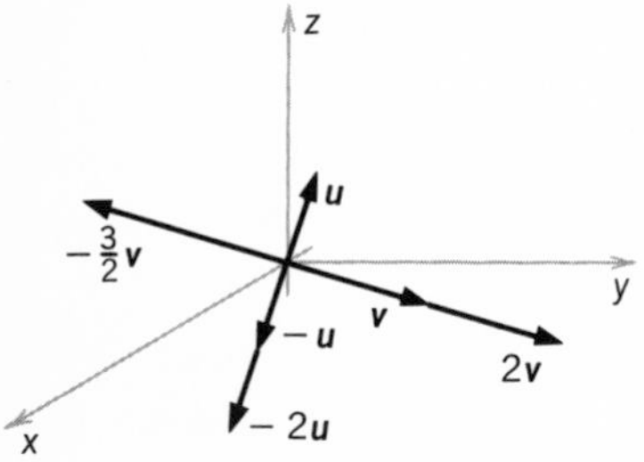

FIGURE 3-3

is still a vector in the xy plane. From the geometric definition of scalar multiplication we see that the scalar multiple $\alpha \boldsymbol{u}$ of some vector $\boldsymbol{u}$ lying in the xy plane must be on the line through the origin determined by the vector $\boldsymbol{u}$. (See Figure 3-3.) Thus, $\alpha \boldsymbol{u}$ must necessarily lie in any plane through the origin in which the vector $\boldsymbol{u}$ lies, or applying this to the case in point, the xy plane.

Having verified in two different ways both conditions of a subspace, we conclude that the set of vectors lying in the xy plane is a subspace of $\boldsymbol{R}^3$.

Any vector space V has at least two subspaces. One subspace is called the **zero subspace**, which consists solely of the zero vector, $\{\boldsymbol{0}\}$. It is clear that this subset is closed under addition and scalar multiplication. Another subspace of V is that which consists of all the vectors in V. A subspace which is neither the zero subspace nor the whole space is said to be a **proper subspace.**

We can prove several interesting results about subspaces.

Proposition 1 If H is a subspace of a vector space V, then $\boldsymbol{0}$ belongs to H.

Proof Let $\boldsymbol{x}$ be any element of H. Then, since scalar multiples of vectors in H belong to H, $0 \cdot \boldsymbol{x} = \boldsymbol{0}$ belongs to H. ∎

Proposition 2 If H is a subspace of a vector space V, and $\boldsymbol{x}$ belongs to H, then $-\boldsymbol{x}$ belongs to H.

Proof If $\boldsymbol{x}$ belongs to H, since H is closed under scalar multiplication, $(-1)\boldsymbol{x} = -\boldsymbol{x}$ also belongs to H. ∎

Theorem If H is a subspace of a vector space V, H is a vector space under the operations of addition and scalar multiplication defined on V.

This theorem tells us that since H is a subspace we may define an addition and scalar multiplication on H using the addition and scalar multiplication on V. Because of conditions (i) and (ii) in the definition of subspace, these operations again yield elements of H. The two preceding propositions guarantee that H has a zero element and that H contains the negative of any element belonging to H. All other vector space axioms are satisfied in H by virtue of the fact that they hold in V.

This theorem enables us to construct many new examples of vector spaces without going through a tedious verification of the vector space axioms (V1)–(V8).

Example 1 Let L be the subset of $\boldsymbol{R}^n$ consisting of those vectors whose first component is 0. Since the sum of two vectors with first component 0 again has first component 0, L is closed under addition. A scalar multiple of a vector with first component 0 likewise has first component 0. Thus, scalar multiples of vectors in L are still in L. Therefore, L is a subspace.

By the theorem above we can conclude that the n-tuples belonging to L form a real vector space.

Example 2 Let A be a fixed $m \times n$ matrix with real entries. Let $N = \{\boldsymbol{x} \mid \boldsymbol{x} \in \boldsymbol{R}^n \text{ and } A\boldsymbol{x} = \boldsymbol{0}\}$. We claim N is a subspace of $\boldsymbol{R}^n$.

First, suppose $\boldsymbol{x}_1 \in N$ and $\boldsymbol{x}_2 \in N$, then

$$\begin{aligned} A(\boldsymbol{x}_1 + \boldsymbol{x}_2) &= A\boldsymbol{x}_1 + A\boldsymbol{x}_2 \\ &= \boldsymbol{0} + \boldsymbol{0} \\ &= \boldsymbol{0} \end{aligned}$$

Thus, $\boldsymbol{x}_1 + \boldsymbol{x}_2 \in N$. Next if $\boldsymbol{x} \in N$ and α is a scalar,

$$\begin{aligned} A(\alpha\boldsymbol{x}) &= \alpha A(\boldsymbol{x}) \\ &= \alpha\boldsymbol{0} \\ &= \boldsymbol{0} \end{aligned}$$

Thus, $\alpha\boldsymbol{x} \in N$.

Having verified conditions (i) and (ii) in the definition of subspace, we see that N is a subspace of $\boldsymbol{R}^n$. Our preceding theorem guarantees that with the proper definitions of addition and scalar multiplication, N is a vector space in its own right.

A more concrete formulation of Example 2 may be found by considering solutions to the system of homogeneous equations

$$\begin{aligned} 3x_1 + 4x_2 - x_3 + 2x_4 &= 0 \\ x_1 - x_2 + 3x_3 + x_4 &= 0 \\ 4x_1 + 3x_2 - x_3 + 2x_4 &= 0 \end{aligned}$$

If

$$\begin{bmatrix} x_1 \\ x_2 \\ x_3 \\ x_4 \end{bmatrix} \quad \text{and} \quad \begin{bmatrix} x_1' \\ x_2' \\ x_3' \\ x_4' \end{bmatrix}$$

are two such solutions, we define their sum by

$$\begin{bmatrix} x_1 \\ x_2 \\ x_3 \\ x_4 \end{bmatrix} + \begin{bmatrix} x_1' \\ x_2' \\ x_3' \\ x_4' \end{bmatrix} = \begin{bmatrix} x_1 + x_1' \\ x_2 + x_2' \\ x_3 + x_3' \\ x_4 + x_4' \end{bmatrix}$$

and a scalar product by

$$\alpha \begin{bmatrix} x_1 \\ x_2 \\ x_3 \\ x_4 \end{bmatrix} = \begin{bmatrix} \alpha x_1 \\ \alpha x_2 \\ \alpha x_3 \\ \alpha x_4 \end{bmatrix}$$

Converting the system of equations to matrix notation and applying Example 2, we see that the set of solutions to the system of equations together with the above addition and scalar multiplication forms a vector space over the reals. Of course, it is possible to reach the same conclusion by explicit verification of the vector space axioms (V1)–(V8).

Example 3 In the space P_n, the set of polynomials of degree less than or equal to n, with real coefficients, the subset

$$H = \left\{ f \mid f \in P_n \text{ and } \int_0^1 f(x)\, dx = 0 \right\}$$

is a subspace. First, suppose $f \in H$ and $g \in H$, then

$$\begin{aligned} \int_0^1 (f(x) + g(x))\, dx &= \int_0^1 f(x)\, dx + \int_0^1 g(x)\, dx \\ &= 0 + 0 \\ &= 0 \end{aligned}$$

So $f + g \in H$. Next suppose $f \in H$ and α is a scalar, then

$$\begin{aligned} \int_0^1 (\alpha f(x))\, dx &= \alpha \int_0^1 f(x)\, dx \\ &= \alpha \cdot 0 \\ &= 0 \end{aligned}$$

So, $\alpha f \in H$.

Having verified conditions (i) and (ii) in the definition of subspace, we conclude that H is a subspace of P_n. Then, from our theorem we conclude that H, with the proper definitions of addition and scalar multiplication, is itself a vector space.

Now that we have seen several examples of subsets of vector spaces which are subspaces, it might be profitable to examine some subsets of $\boldsymbol{R}^2$ which are not subspaces.

Example 4 In $\boldsymbol{R}^2$, let

$$H = \left\{ \begin{bmatrix} \alpha \\ \beta \end{bmatrix} \middle| \begin{bmatrix} \alpha \\ \beta \end{bmatrix} \in \boldsymbol{R}^2 \text{ and } \alpha \geq 0 \right\}$$

Geometrically this set corresponds to the right half of the plane. (See Figure 3-4(a).) If $\boldsymbol{x}_1 \in H$ and $\boldsymbol{x}_2 \in H$, then

$$\boldsymbol{x}_1 = \begin{bmatrix} \alpha_1 \\ \beta_1 \end{bmatrix} \quad \text{and} \quad \boldsymbol{x}_2 = \begin{bmatrix} \alpha_2 \\ \beta_2 \end{bmatrix}$$

where $\alpha_1 \geq 0$ and $\alpha_2 \geq 0$.

Thus,

$$\boldsymbol{x}_1 + \boldsymbol{x}_2 = \begin{bmatrix} \alpha_1 + \alpha_2 \\ \beta_1 + \beta_2 \end{bmatrix}$$

where $\alpha_1 + \alpha_2 \geq 0$, and so $\boldsymbol{x}_1 + \boldsymbol{x}_2 \in H$. Thus H is additively closed.

However, since

$$\begin{bmatrix} 1 \\ 0 \end{bmatrix} \in H$$

and

$$(-1)\begin{bmatrix} 1 \\ 0 \end{bmatrix} = \begin{bmatrix} -1 \\ 0 \end{bmatrix}$$

is not in H, scalar multiples of vectors in H are not in H. (See Figure 3-4(b).)

Thus, H is not a subspace of $\boldsymbol{R}^2$. In this case, it is interesting to note that if $\boldsymbol{x} \in H$ and $\alpha \geq 0$, $\alpha\boldsymbol{x}$ also belongs to H.

Example 5 In $\boldsymbol{R}^2$, let

$$H = \left\{ \begin{bmatrix} \alpha \\ \beta \end{bmatrix} \middle| \begin{bmatrix} \alpha \\ \beta \end{bmatrix} \in \boldsymbol{R}^2 \quad \text{and} \quad \alpha = 0 \text{ or } \beta = 0 \right\}$$

Geometrically this set consists of all points on the two coordinate

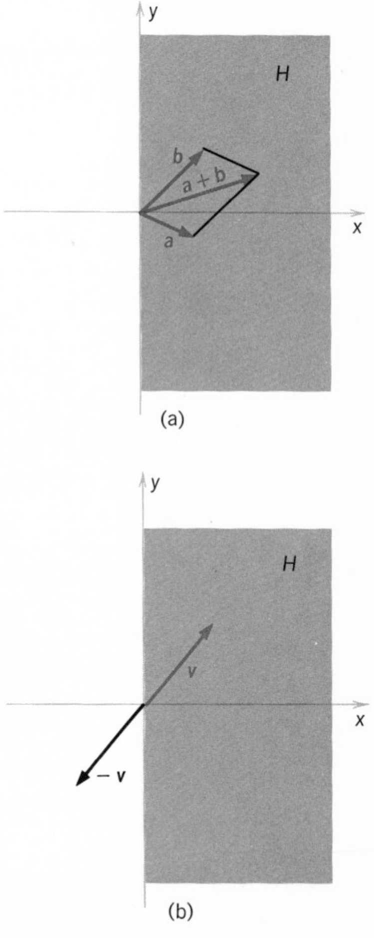

FIGURE 3-4

axes. (See Figure 3-5). Scalar multiples of vectors in H still belong to H, but H is not additively closed. For we have

$$\begin{bmatrix}1\\0\end{bmatrix} \in H \quad \text{and} \quad \begin{bmatrix}0\\1\end{bmatrix} \in H$$

but

$$\begin{bmatrix}1\\0\end{bmatrix} + \begin{bmatrix}0\\1\end{bmatrix} = \begin{bmatrix}1\\1\end{bmatrix}$$

does not belong to H.

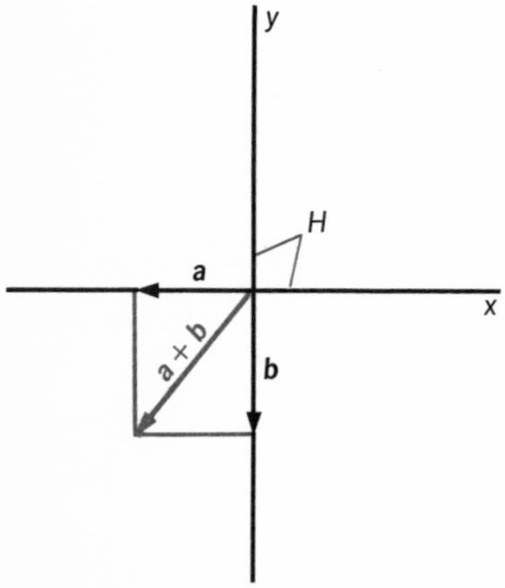

FIGURE 3-5

EXERCISES

1. Which of the following subsets of $\boldsymbol{R}^2$ are subspaces?
   (a) $\{(x, y) \mid x = 3y\}$
   (b) $\{(x, y) \mid x^2 = y^2\}$
   (c) $\{(x, y) \mid x + y = 1\}$
   (d) $\{(x, y) \mid x = y^3\}$
   (e) $\{(x, y) \mid y = 0\}$
   (f) $\{(x, y) \mid x \geq 0, y \geq 0\}$
   (g) $\{(x, y) \mid x = y\}$
2. Which of the following subsets of $\boldsymbol{R}^3$ are subspaces?
   (a) $\{(x, y, z) \mid x + 3y = 0 \text{ and } y + z = 0\}$
   (b) $\{(x, y, z) \mid x = 0 \text{ or } y = 0\}$
   (c) $\{(x, y, z) \mid x + y + z = 0\}$
   (d) $\{(x, y, z) \mid x \geq 0, y \geq 0, z \geq 0\}$
   (e) $\{(x, y, z) \mid x^2 + y^2 \leq 1\}$
   (f) $\{(x, y, z) \mid z = 2x + 2y\}$
   (g) $\{(x, y, z) \mid z = 0 \text{ and } x^2 + y^2 \leq 1\}$
3. Let H_i be the subset of $\boldsymbol{R}^n$ consisting of those vectors whose ith component is 0. Show that H_i is a subspace of $\boldsymbol{R}^n$.
4. Let P_n be the polynomials of degree less than or equal to n. Let H_e be a collection of the even polynomials in P_n and H_o be a collection of the odd polynomials in P_n. Show that H_e and H_o are subspaces of P_n.
5. Show that the following subsets of the space of 2×2 matrices M_{22} are subspaces.
   (a) All matrices of the form

$$\begin{bmatrix} a & b \\ b & a \end{bmatrix}, \qquad a, b \text{ real numbers}$$

   (b) All matrices of the form

$$\begin{bmatrix} a & b \\ -b & a \end{bmatrix}, \qquad a, b \text{ real numbers}$$

   (c) All matrices of the form

$$\begin{bmatrix} a & b \\ 0 & c \end{bmatrix}, \qquad a, b, c \text{ real numbers}$$

(d) All matrices of the form

$$\begin{bmatrix} a & c \\ c & b \end{bmatrix}, \qquad a, b, c \text{ real numbers}$$

(e) All matrices of the form

$$\begin{bmatrix} 0 & a \\ 0 & b \end{bmatrix}, \qquad a, b \text{ real numbers}$$

(f) All matrices of the form

$$\begin{bmatrix} a & c \\ -c & b \end{bmatrix}, \qquad a, b, c \text{ real numbers}$$

(g) All matrices of the form

$$\begin{bmatrix} a & 0 \\ 0 & b \end{bmatrix}, \qquad a, b \text{ real numbers}$$

6. Let P denote the collection of all polynomials with real coefficients.
   (a) Show that P is a vector space over the reals.
   (b) Show that the following subsets of P are subspaces.
      (1) $\{f \mid \text{degree of } f \leq n$, where n is a fixed positive integer$\}$
      (2) $\{f \mid f(\alpha) = 0$, where α is a fixed number$\}$
      (3) $\{f \mid f(\alpha) = f(\beta)$, where α and β are fixed numbers$\}$
      (4) $\{f \mid f'(\alpha) = 0$, where α is a fixed number$\}$
      (5) $\{f \mid f$ is divisible by $(x - 1)\}$
      (6) $\{f \mid f(0) = f'(0) = f''(0) = 0\}$
      (7) $\{f \mid f(0) - \int_0^1 f(x)\, dx = 0\}$

7. In the space of $n \times n$ matrices with real coefficients $A = [a_{ij}]_{(nn)}$, we say that a matrix is upper triangular if $a_{ij} = 0$ for $i > j$. Show that the upper triangular matrices form a subspace of the $n \times n$ matrices.

8. Show that the symmetric matrices form a subspace of the space of $n \times n$ matrices. Do the Hermitian matrices form a subspace of the $n \times n$ matrices with complex entries?

9. Show that the diagonal matrices form a subspace of the space of $n \times n$ matrices.

10. Let V be a vector space over the reals and H be a subset of V. Show that (a), (b), and (c) are equivalent.
   (a) H is a subspace.
   (b) If $\boldsymbol{x}, \boldsymbol{y} \in H$, then $\boldsymbol{x} + \boldsymbol{y} \in H$.
   If $\boldsymbol{x} \in H$, then $-\boldsymbol{x} \in H$.
   If $\boldsymbol{x} \in H$ and $\alpha \geq 0$ is a real scalar then $\alpha\boldsymbol{x} \in H$.
   (c) If $\boldsymbol{x}, \boldsymbol{y} \in H$ and β is a scalar then $\boldsymbol{x} + \beta\boldsymbol{y} \in H$.

11. Let M_{nn} denote the vector space of $n \times n$ matrices with real entries.
   (a) Show that the subset of invertible matrices is not a subspace of M_{nn}.

(b) Show that the subset of noninvertible matrices is closed under scalar multiplication, i.e., a scalar multiple of a noninvertible matrix is noninvertible, but that the subset of noninvertible matrices is not a subspace of M_{nn}.

12. Let B be a fixed matrix in M_{nn}. Show that the following subsets of M_{nn} are subspaces.
(a) $\{A \mid A \in M_{nn} \text{ and } AB = BA\}$ (b) $\{A \mid A \in M_{nn} \text{ and } AB + BA = 0\}$
(c) $\{A \mid A \in M_{nn} \text{ and } AB = 0\}$ (d) $\{A \mid A \in M_{nn} \text{ and } BA = 0\}$
Give examples to show that (c) and (d) do not necessarily represent the same subspace.

13. Let $a_1, a_2, a_3, \ldots, a_n$ be real numbers. Show that the vectors

$$\begin{bmatrix} x_1 \\ x_2 \\ \vdots \\ x_n \end{bmatrix}$$

in $\boldsymbol{R}^n$, such that $a_1x_1 + a_2x_2 + \cdots + a_nx_n = 0$ form a subspace of $\boldsymbol{R}^n$. Why is this a special case of Example 2 in the text?

14. Let V be a vector space and let H and K be subspaces of V. Show that the set $H \cap K = \{\boldsymbol{x} \mid \boldsymbol{x} \in H \text{ and } \boldsymbol{x} \in K\}$ is a subspace of V.

15. Let V be a vector space and let H and K be subspaces of V. Show that the set of $H + K = \{\boldsymbol{x} \mid \boldsymbol{x} = \boldsymbol{h} + \boldsymbol{k}, \text{ where } \boldsymbol{h} \in H \text{ and } \boldsymbol{k} \in K\}$ is a subspace of V.

16. Prove by induction: If H is a subspace of a vector space V, and $\boldsymbol{x}_1, \boldsymbol{x}_2, \ldots, \boldsymbol{x}_n$ belong to H, while $\alpha_1, \alpha_2, \ldots, \alpha_n$ are scalars, then $\alpha_1\boldsymbol{x}_1 + \alpha_2\boldsymbol{x}_2 + \cdots + \alpha_n\boldsymbol{x}_n$ belongs to H.

4 SPAN

Let V be a vector space, $\boldsymbol{x}_1, \boldsymbol{x}_2, \boldsymbol{x}_3, \ldots, \boldsymbol{x}_n$ be vectors in V, and $\alpha_1, \alpha_2, \ldots, \alpha_n$ be scalars. If a vector $\boldsymbol{y}$ in V can be written in the form $\boldsymbol{y} = \alpha_1\boldsymbol{x}_1 + \alpha_2\boldsymbol{x}_2 + \cdots + \alpha_n\boldsymbol{x}_n$, then $\boldsymbol{y}$ is said to be a **linear combination** of $\boldsymbol{x}_1, \boldsymbol{x}_2, \ldots, \boldsymbol{x}_n$. For example, in P_2, the polynomials of degree less than or equal to 2, any element can be expressed as a linear combination of 1, x, and x^2. For we have, $p(x) = \alpha_0 + \alpha_1 x + \alpha_2 x^2$. In $\boldsymbol{R}^3$, if a vector $\boldsymbol{v}$ ends at the point (x, y, z), we have $\boldsymbol{v} = x\boldsymbol{i} + y\boldsymbol{j} + z\boldsymbol{k}$, and so $\boldsymbol{v}$ is a linear combination of $\boldsymbol{i}, \boldsymbol{j}$, and $\boldsymbol{k}$.

Using the idea of linear combination, we may characterize subspaces as follows: A subset H of a vector space V is a subspace if and only if whenever $\boldsymbol{x}_1, \boldsymbol{x}_2, \ldots, \boldsymbol{x}_n$ belongs to H, any linear combination of $\boldsymbol{x}_1, \boldsymbol{x}_2, \ldots, \boldsymbol{x}_n$ also belongs to H.

There is a natural way to generate subspaces of a vector space V using linear combinations. If S is a subset of a vector space V, we define the span of S, written sp(S), to be the set of vectors, each of which can be written as a linear combination of vectors in S.

For example, in $\boldsymbol{R}^3$, the span of the set

$$\left\{\begin{bmatrix}1\\0\\0\end{bmatrix}, \begin{bmatrix}0\\1\\0\end{bmatrix}\right\}$$

consists of all vectors of the form

$$\alpha\begin{bmatrix}1\\0\\0\end{bmatrix} + \beta\begin{bmatrix}0\\1\\0\end{bmatrix} = \begin{bmatrix}\alpha\\\beta\\0\end{bmatrix}$$

In this case the span is precisely that subset of $\boldsymbol{R}^3$ consisting of vectors whose third component is 0. If we consider the span of the set

$$\left\{\begin{bmatrix}1\\0\\0\end{bmatrix}, \begin{bmatrix}0\\1\\0\end{bmatrix}, \begin{bmatrix}1\\1\\0\end{bmatrix}\right\}$$

we find it consists of precisely the same collection of vectors. Indeed, any linear combination of the above three vectors has third component 0. It is clear, on the other hand, that any vector with third component 0 can be expressed as a linear combination of just the first two vectors in the given set and thus necessarily as a linear combination of all three vectors.

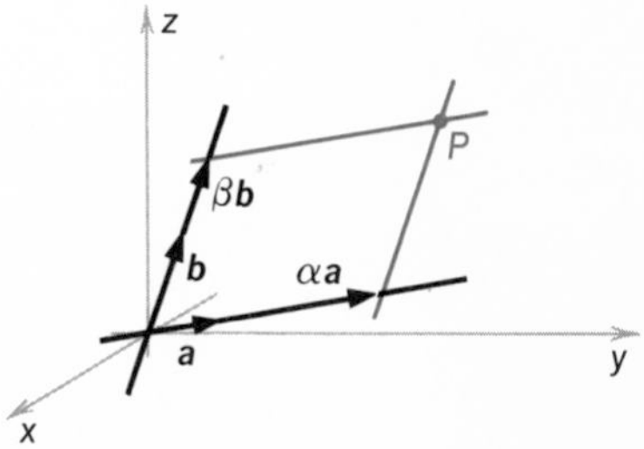

FIGURE 4-1

To elucidate the geometric meaning of the concept of span, we characterize the span of two noncollinear vectors $\boldsymbol{a}$ and $\boldsymbol{b}$ in $\boldsymbol{R}^3$. We will show that the span of $\boldsymbol{a}$ and $\boldsymbol{b}$ consists of those vectors which lie in the plane which passes the origin and is determined by the vectors $\boldsymbol{a}$ and $\boldsymbol{b}$. (See Figure 4-1.)

To begin with we note that, by the geometric meaning of vector addition and scalar multiplication, any linear combination of $\boldsymbol{a}$ and $\boldsymbol{b}$ must lie in the plane determined by $\boldsymbol{a}$ and $\boldsymbol{b}$. For what is the vector $\alpha\boldsymbol{a} + \beta\boldsymbol{b}$? It is clear that $\alpha\boldsymbol{a}$ and $\beta\boldsymbol{b}$, each lying on the line through the origin determined by the vectors $\boldsymbol{a}$ and $\boldsymbol{b}$, respectively, each lie in the plane through the origin determined by the vectors $\boldsymbol{a}$ and $\boldsymbol{b}$. Since $\alpha\boldsymbol{a} + \beta\boldsymbol{b}$ lies in any plane in which $\alpha\boldsymbol{a}$ and $\beta\boldsymbol{b}$ both lie, $\alpha\boldsymbol{a} + \beta\boldsymbol{b}$ lies in the plane determined by the vectors $\boldsymbol{a}$ and $\boldsymbol{b}$.

On the other hand let P be a point in the plane determined by the noncollinear vectors $\boldsymbol{a}$ and $\boldsymbol{b}$. In this plane, construct a line l_a, passing through the point P and parallel to the directed line segment associated with the vector $\boldsymbol{a}$. Since $\boldsymbol{a}$ and $\boldsymbol{b}$ are noncollinear, l_a is not parallel to the line generated by $\boldsymbol{b}$ and therefore intersects it in some point, say the endpoint of the vector $\beta\boldsymbol{b}$. Likewise, if we construct a line l_b through the point P and parallel to the vector $\boldsymbol{b}$, we see that it intersects the line generated by the vector $\boldsymbol{a}$ at the endpoint of some vector $\alpha\boldsymbol{a}$. From this construction P is necessarily the endpoint of the diagonal of a parallelogram having adjacent sides represented by the vectors $\alpha\boldsymbol{a}$ and $\beta\boldsymbol{b}$. Thus, if $\boldsymbol{v}$ is the vector which ends at P, $\boldsymbol{v} = \alpha\boldsymbol{a} + \beta\boldsymbol{b}$. Hence, any vector lying in the plane determined by $\boldsymbol{a}$ and $\boldsymbol{b}$ is a linear combination of $\boldsymbol{a}$ and $\boldsymbol{b}$.

The reader may have, perhaps, noted the similarity between this procedure and that for constructing Cartesian coordinates in the plane. From the foregoing examples we perceive an important property of sp(S).

Theorem If V is a vector space and S is a subset of V, then sp(S) is a subspace of V.

Proof Suppose $\boldsymbol{x}$ and $\boldsymbol{y}$ belong to sp(S). Then,

$$\boldsymbol{x} = \alpha_1\boldsymbol{x}_1 + \alpha_2\boldsymbol{x}_2 + \cdots + \alpha_n\boldsymbol{x}_n$$

$$\boldsymbol{y} = \beta_1\boldsymbol{y}_1 + \beta_2\boldsymbol{y}_2 + \cdots + \beta_m\boldsymbol{y}_m$$

where $\alpha_1, \alpha_2, \ldots, \alpha_n, \beta_1, \beta_2, \ldots, \beta_m$ are scalars and $\boldsymbol{x}_1, \boldsymbol{x}_2, \ldots, \boldsymbol{x}_n, \boldsymbol{y}_1, \boldsymbol{y}_2, \ldots, \boldsymbol{y}_m$ are vectors in S. Then,

$$\boldsymbol{x} + \boldsymbol{y} = \alpha_1\boldsymbol{x}_1 + \cdots + \alpha_n\boldsymbol{x}_n + \beta_1\boldsymbol{y}_1 + \cdots + \beta_m\boldsymbol{y}_m$$

Thus, $\boldsymbol{x} + \boldsymbol{y}$ is a linear combination of vectors in S and so $\boldsymbol{x} + \boldsymbol{y}$ belongs to sp(S). Moreover, if α is a scalar, we have

$$\alpha\boldsymbol{x} = \alpha(\alpha_1\boldsymbol{x}_1 + \cdots + \alpha_n\boldsymbol{x}_n) = (\alpha\alpha_1)\boldsymbol{x}_1 + \cdots + (\alpha\alpha_n)\boldsymbol{x}_n$$

$\alpha\boldsymbol{x}$ thus being a linear combination of vectors in S, necessarily belongs to sp(S). ∎

If we apply this theorem to a previous example, we see that the totality of vectors lying in some plane through the origin constitutes a subspace of $\boldsymbol{R}^3$.

As another example, let P denote the space of all polynomials in the variable x with real coefficients. The span of the polynomials 1, x, $x^2, \ldots, x^n$ in P is precisely P_n, the polynomials of degree less than or equal to n, which is, of course, a subspace of P.

If H is a subspace of a vector space V and $\boldsymbol{x}_1, \boldsymbol{x}_2, \ldots, \boldsymbol{x}_n$ is a collection of vectors in V, such that $\text{sp}(\boldsymbol{x}_1, \boldsymbol{x}_2, \ldots, \boldsymbol{x}_n) = H$, we say $\boldsymbol{x}_1, \boldsymbol{x}_2, \ldots, \boldsymbol{x}_n$ span H. Stated otherwise, the linear combinations of $\boldsymbol{x}_1, \boldsymbol{x}_2, \ldots, \boldsymbol{x}_n$ fill out the subspace H.

It is important to note that a given collection S of vectors which spans a subspace H of a vector space V may contain redundancies in the sense that it may be possible to delete certain members of S and obtain a subset with the same span. Consider, for example, the set of vectors $\{\boldsymbol{i}, \boldsymbol{k}, \boldsymbol{i} + \boldsymbol{k}\}$ in $\boldsymbol{R}^3$. (See Figure 4-2.) These vectors span the xz

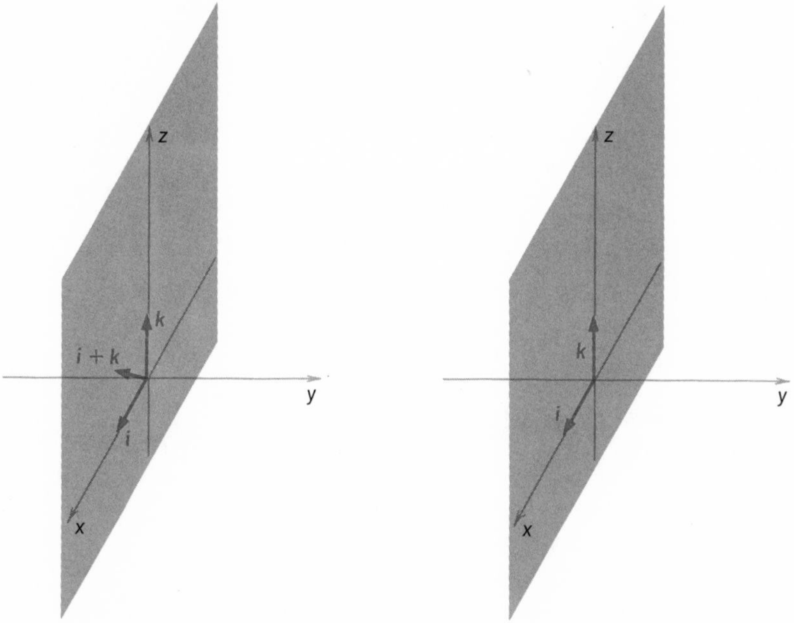

FIGURE 4-2

plane in $\boldsymbol{R}^3$. If, however, we delete any member of the set, we obtain a pair of noncollinear vectors in the same plane. By the preceding example, this pair of vectors again spans the xz plane.

In the next section we investigate a condition on a given set of vectors $\{\boldsymbol{x}_1, \boldsymbol{x}_2, \ldots, \boldsymbol{x}_n\}$ which insures that these repetitions do not occur.

EXERCISES

1. Show that the following sets of vectors span the indicated vector spaces.

(a) $\begin{bmatrix}1\\0\\0\end{bmatrix}, \begin{bmatrix}0\\1\\0\end{bmatrix}, \begin{bmatrix}0\\0\\1\end{bmatrix}; \quad \boldsymbol{R}^3$

(b) $\begin{bmatrix}1\\0\\0\end{bmatrix}, \begin{bmatrix}1\\1\\0\end{bmatrix}, \begin{bmatrix}2\\1\\1\end{bmatrix}; \quad \boldsymbol{R}^3$

(c) $\begin{bmatrix}1\\-1\end{bmatrix}, \begin{bmatrix}1\\1\end{bmatrix}, \begin{bmatrix}2\\1\end{bmatrix}; \quad \boldsymbol{R}^2$

(d) $\begin{bmatrix}1\\1\end{bmatrix}$, $\begin{bmatrix}-1\\2\end{bmatrix}$; $\boldsymbol{R}^2$

(e) $1, x, x^2$; P_2

(f) $1, (1 + x), (1 + x)^2$; P_2

(g) $\begin{bmatrix}1 & 0\\0 & 0\end{bmatrix}$, $\begin{bmatrix}0 & 0\\0 & 1\end{bmatrix}$, $\begin{bmatrix}0 & 1\\1 & 0\end{bmatrix}$

space of 2×2 symmetric matrices.

(h) $\begin{bmatrix}0 & 1\\-1 & 0\end{bmatrix}$

space of 2×2 skew-symmetric matrices.

(i) $1 + x$, $1 + 2x$, $1 + x^2$, $1 + 2x^2$ P_2

(j) $\begin{bmatrix}1 & 0 & 0\\0 & 1 & 0\\0 & 0 & 1\end{bmatrix}$, $\begin{bmatrix}1 & 0 & 0\\0 & -1 & 0\\0 & 0 & 0\end{bmatrix}$, $\begin{bmatrix}1 & 0 & 0\\0 & 1 & 0\\0 & 0 & -1\end{bmatrix}$

space of 3×3 diagonal matrices.

2. Find two vectors which span the indicated planes in $\boldsymbol{R}^3$.
   (a) The plane $x = 0$.
   (b) The plane $x = y$.
   (c) The plane $y = z$.
3. Find n vectors which span $\boldsymbol{R}^n$.
4. Find four elements which span the space of 2×2 matrices.
5. Let $\{\boldsymbol{x}_1, \boldsymbol{x}_2, \ldots, \boldsymbol{x}_n\}$ be a set of vectors in a vector space V. If $\boldsymbol{y}_1, \boldsymbol{y}_2, \ldots, \boldsymbol{y}_n$ belong to $\text{sp}(\boldsymbol{x}_1, \boldsymbol{x}_2, \ldots, \boldsymbol{x}_n)$ and $\boldsymbol{y}_1, \boldsymbol{y}_2, \ldots, \boldsymbol{y}_n$ span V, show that $\{\boldsymbol{x}_1, \ldots, \boldsymbol{x}_n\}$ also span V.
6. Let V be a vector space and $\boldsymbol{x}_1, \boldsymbol{x}_2, \ldots, \boldsymbol{x}_n$ be vectors in V. Show that $\text{sp}(\boldsymbol{x}_1, \boldsymbol{x}_2, \ldots, \boldsymbol{x}_n)$ is the smallest subspace of V containing $\boldsymbol{x}_1, \boldsymbol{x}_2, \ldots, \boldsymbol{x}_n$. In other words, show that if H is a subspace of V containing $\boldsymbol{x}_1, \boldsymbol{x}_2, \ldots, \boldsymbol{x}_n$, then H also contains $\text{sp}(\boldsymbol{x}_1, \boldsymbol{x}_2, \ldots, \boldsymbol{x}_n)$.
7. Let V be a vector space and $\boldsymbol{x}_1, \boldsymbol{x}_2, \boldsymbol{y}_1, \boldsymbol{y}_n$ be vectors in V. If

$$\boldsymbol{x}_1 = \alpha_1\boldsymbol{y}_1 + \alpha_2\boldsymbol{y}_2$$
$$\boldsymbol{x}_2 = \beta_1\boldsymbol{y}_1 + \beta_2\boldsymbol{y}_2$$

where

$$\begin{vmatrix}\alpha_1 & \alpha_2\\ \beta_1 & \beta_2\end{vmatrix} \neq 0$$

show that $\text{sp}(\boldsymbol{x}_1, \boldsymbol{x}_2) = \text{sp}(\boldsymbol{y}_1, \boldsymbol{y}_2)$.

8. If $A \subset B$ are subsets of a vector space V, show that $\text{sp}(A) \subset \text{sp}(B)$.
9. If $\boldsymbol{y}$ belongs to $\text{sp}(\boldsymbol{x}_1, \boldsymbol{x}_2, \ldots, \boldsymbol{x}_n, \boldsymbol{z})$ but $\boldsymbol{y}$ does not belong to $\text{sp}(\boldsymbol{x}_1, \boldsymbol{x}_2, \ldots, \boldsymbol{x}_n)$, show that $\boldsymbol{z}$ belongs to $\text{sp}(\boldsymbol{x}_1, \boldsymbol{x}_2, \ldots, \boldsymbol{x}_n, \boldsymbol{y})$.

10. Show that

$$\text{sp}(\mathbf{x}_1, \mathbf{x}_2, \ldots, \mathbf{x}_n, \mathbf{y}) = \text{sp}(\mathbf{x}_1, \mathbf{x}_2, \ldots, \mathbf{x}_n)$$

if and only if $\mathbf{y}$ is a linear combination of $\mathbf{x}_1, \mathbf{x}_2, \ldots, \mathbf{x}_n$.

11. Show that the invertible 2×2 matrices span the space of 2×2 matrices. Show that the noninvertible 2×2 matrices span the space of 2×2 matrices.

12. Show that the matrices

$$\begin{bmatrix} 1 & 1 \\ 0 & 1 \end{bmatrix}, \quad \begin{bmatrix} -1 & 1 \\ 0 & -1 \end{bmatrix}, \quad \begin{bmatrix} 0 & 1 \\ 0 & 0 \end{bmatrix}$$

do not span the space of 2×2 matrices.

13. Show that the vector space P of all polynomials cannot be spanned by a finite number of elements.

14. Show that the matrices of the form $AB - BA$ do not span the space of $n \times n$ matrices.

15. Is it possible to span the space of $n \times n$ matrices using the powers of a single matrix A, i.e., $I_n, A, A^2, \ldots, A^n, \ldots$?

5 LINEAR INDEPENDENCE

Closely related to the notion of span discussed in the preceding section is the concept of linear independence.

A collection $\{\mathbf{x}_1, \mathbf{x}_2, \ldots, \mathbf{x}_n\}$ of vectors in a vector space is said to be **linearly independent** if whenever

$$\alpha_1\mathbf{x}_1 + \alpha_2\mathbf{x}_2 + \cdots + \alpha_n\mathbf{x}_n = \mathbf{0}$$

we must have $\alpha_1 = \alpha_2 = \cdots = \alpha_n = 0$. In other words, the **only** linear combination of $\mathbf{x}_1, \mathbf{x}_2, \ldots, \mathbf{x}_n$ which vanishes is the obvious one:

$$0 \cdot \mathbf{x}_1 + 0 \cdot \mathbf{x}_2 + \cdots + 0 \cdot \mathbf{x}_n = \mathbf{0}$$

For example, in $\mathbf{R}^2$, the vectors

$$\begin{bmatrix} 1 \\ 0 \end{bmatrix} \text{ and } \begin{bmatrix} 0 \\ 1 \end{bmatrix}$$

are linearly independent. For suppose

$$\alpha\begin{bmatrix} 1 \\ 0 \end{bmatrix} + \beta\begin{bmatrix} 0 \\ 1 \end{bmatrix} = \begin{bmatrix} 0 \\ 0 \end{bmatrix}$$

Then,

$$\begin{bmatrix} \alpha \\ \beta \end{bmatrix} = \begin{bmatrix} 0 \\ 0 \end{bmatrix}$$

and so $\alpha = 0$ and $\beta = 0$. We have shown, therefore, that if

$$\alpha\begin{bmatrix} 1 \\ 0 \end{bmatrix} + \beta\begin{bmatrix} 0 \\ 1 \end{bmatrix} = \begin{bmatrix} 0 \\ 0 \end{bmatrix}$$

then $\alpha = 0$ and $\beta = 0$. Thus, the vectors

$$\begin{bmatrix} 1 \\ 0 \end{bmatrix} \quad \text{and} \quad \begin{bmatrix} 0 \\ 1 \end{bmatrix}$$

are linearly independent.

As another example, in P_2, the polynomials with real coefficients of degree less than or equal to 2, the vectors 1, $1 + x$, $(1 + x)^2$ are independent. For suppose,

$$\alpha \cdot 1 + \beta(1 + x) + \gamma(1 + x)^2 = 0$$

Then,

$$\alpha \cdot 1 + \beta \cdot 1 + \beta x + \gamma \cdot 1 + 2\gamma x + \gamma x^2 = 0$$

or

$$(\alpha + \beta + \gamma) + (\beta + 2\gamma)x + \gamma x^2 = 0$$

Since a polynomial is zero only if all its coefficients are zero, we must have

$$\begin{aligned} \alpha + \beta + \gamma &= 0 \\ \beta + 2\gamma &= 0 \\ \gamma &= 0 \end{aligned}$$

This immediately implies $\alpha = \beta = \gamma = 0$. Having shown that $\alpha \cdot 1 + \beta(1 + x) + \gamma(1 + x)^2 = 0$ implies $\alpha = \beta = \gamma = 0$, we may conclude that $\{1, 1 + x, (1 + x)^2\}$ is a linearly independent set of vectors.

If a set of vectors is not linearly independent, we say it is **linearly dependent**. Thus, if $\{\boldsymbol{x}_1, \boldsymbol{x}_2, \cdots \boldsymbol{x}_n\}$ is a linearly dependent set of vectors, there are scalars $\alpha_1, \alpha_2, \ldots, \alpha_n$, **not all zero,** such that $\alpha_1\boldsymbol{x}_1 + \alpha_2\boldsymbol{x}_2 + \cdots + \alpha_n\boldsymbol{x}_n = \mathbf{0}$.

Thus, in $\boldsymbol{R}^2$, the vectors

$$\begin{bmatrix}1\\1\end{bmatrix}, \quad \begin{bmatrix}1\\0\end{bmatrix}, \quad \begin{bmatrix}0\\-2\end{bmatrix}$$

are linearly dependent, for we have,

$$(1)\begin{bmatrix}1\\1\end{bmatrix} + (-1)\begin{bmatrix}1\\0\end{bmatrix} + \tfrac{1}{2}\begin{bmatrix}0\\-2\end{bmatrix} = \begin{bmatrix}0\\0\end{bmatrix}$$

As another example, in $\boldsymbol{R}^3$, the vectors

$$\begin{bmatrix}1\\-3\\7\end{bmatrix}, \quad \begin{bmatrix}2\\0\\1\end{bmatrix}, \quad \begin{bmatrix}6\\-6\\16\end{bmatrix}$$

are dependent since

$$(-2)\begin{bmatrix}1\\-3\\7\end{bmatrix} + (-2)\begin{bmatrix}2\\0\\1\end{bmatrix} + (+1)\begin{bmatrix}6\\-6\\16\end{bmatrix} = \begin{bmatrix}0\\0\\0\end{bmatrix}$$

The following evinces the relationship between span and linear independence.

Theorem Let $\{\boldsymbol{x}_1, \boldsymbol{x}_2, \ldots, \boldsymbol{x}_n\}$ be a collection of vectors in a vector space V. Then, $\{\boldsymbol{x}_1, \boldsymbol{x}_2, \ldots, \boldsymbol{x}_n\}$ is a linearly dependent set of vectors if and only if one of the vectors is a linear combination of the remaining vectors in the set.

Proof Suppose $\{\boldsymbol{x}_1, \boldsymbol{x}_2, \ldots, \boldsymbol{x}_n\}$ is linearly dependent. Then, $\alpha_1\boldsymbol{x}_1 + \cdots + \alpha_n\boldsymbol{x}_n = \boldsymbol{0}$, where $\alpha_1, \alpha_2, \ldots, \alpha_n$ are not all zero. Suppose $\alpha_i \neq 0$, then

$$\alpha_i\boldsymbol{x}_i = (-\alpha_1)\boldsymbol{x}_1 + \cdots + (-\alpha_{i-1}\boldsymbol{x}_{i-1}) + (-\alpha_{i+1}\boldsymbol{x}_{i+1}) + \cdots + (-\alpha_n\boldsymbol{x}_n)$$

$$\boldsymbol{x}_i = (-\alpha_i^{-1}\alpha_1)\boldsymbol{x}_1 + \cdots + (-\alpha_i^{-1}\alpha_{i-1}\boldsymbol{x}_{i-1}) + (-\alpha_i^{-1}\alpha_{i+1}\boldsymbol{x}_{i+1}) + \cdots + (-\alpha_i^{-1}\alpha_n)\boldsymbol{x}_n$$

Hence, $\boldsymbol{x}_i$ is a linear combination of $\boldsymbol{x}_1, \ldots, \boldsymbol{x}_{i-1}, \boldsymbol{x}_{i+1}, \ldots, \boldsymbol{x}_n$.

On the other hand, suppose one of the vectors, say $\boldsymbol{x}_i$, is a linear combinations of $\{\boldsymbol{x}_1, \ldots, \boldsymbol{x}_{i-1}, \boldsymbol{x}_{i+1}, \ldots, \boldsymbol{x}_n\}$. Then,

$$\boldsymbol{x}_i = \beta_1\boldsymbol{x}_1 + \cdots + \beta_{i-1}\boldsymbol{x}_{i-1} + \beta_{i+1}\boldsymbol{x}_{i+1} + \cdots + \beta_n\boldsymbol{x}_n$$

or

$$\beta_1\boldsymbol{x}_1 + \cdots + \beta_{i-1}\boldsymbol{x}_{i-1} + (-1)\boldsymbol{x}_i + \beta_{i+1}\boldsymbol{x}_{i+1} + \cdots + \beta_n\boldsymbol{x}_n = \boldsymbol{0}$$

Since the coefficient of $\boldsymbol{x}_i$ is $-1 \neq 0$, the vectors $\{\boldsymbol{x}_1, \boldsymbol{x}_2, \ldots, \boldsymbol{x}_n\}$ are linearly dependent. ∎

For example, in P_2, the polynomials 1, $1 + x$, $(1 + x)^2$, x^2 are linearly dependent, since $(1 + x)^2 = 1 \cdot x^2 + 2 \cdot (1 + x) + (-1) \cdot 1$.

Let $\boldsymbol{a}$ and $\boldsymbol{b}$ be two vectors in $\boldsymbol{R}^3$. What does it mean geometrically for $\boldsymbol{a}$ and $\boldsymbol{b}$ to be linearly independent? By the preceding theorem it is evident that the vectors $\boldsymbol{a}$ and $\boldsymbol{b}$ are linearly dependent if and only if one is a scalar multiple of the other. Thus, $\boldsymbol{a}$ and $\boldsymbol{b}$ are linearly dependent if and only if they both lie on some line passing through the origin. From this it is clear that the vectors $\boldsymbol{a}$ and $\boldsymbol{b}$ are linearly independent if and only if they are noncollinear. (See Figure 5-1.)

Thus, for example the pairs of vectors $\{\boldsymbol{i} - \boldsymbol{j} + \boldsymbol{k}, \boldsymbol{i} + \boldsymbol{j}\}$ and $\{\boldsymbol{i} - \boldsymbol{j}, \boldsymbol{j} + \boldsymbol{k}\}$ are linearly independent. (See Figure 5-2.) On the other hand, the pairs of vectors $\{\boldsymbol{i} + \boldsymbol{j}, -\boldsymbol{i} - \boldsymbol{j}\}$ and $\{\boldsymbol{i} + 2\boldsymbol{j} + \boldsymbol{k}, 2\boldsymbol{i} + 4\boldsymbol{j} + 2\boldsymbol{k}\}$ are linearly dependent. (See Figure 5-3.)

Having given a geometric condition on two vectors in $\boldsymbol{R}^3$ which

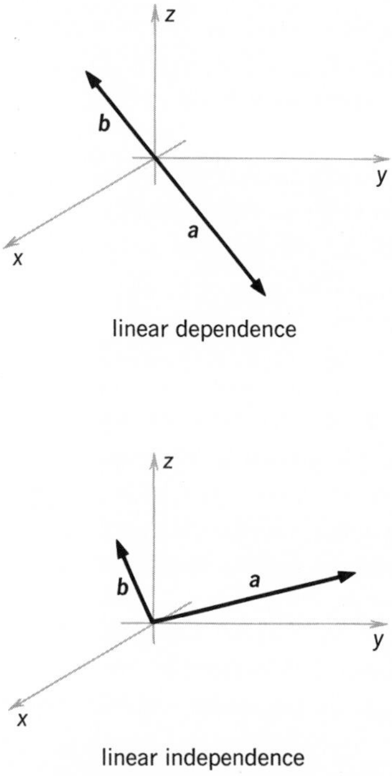

FIGURE 5-1

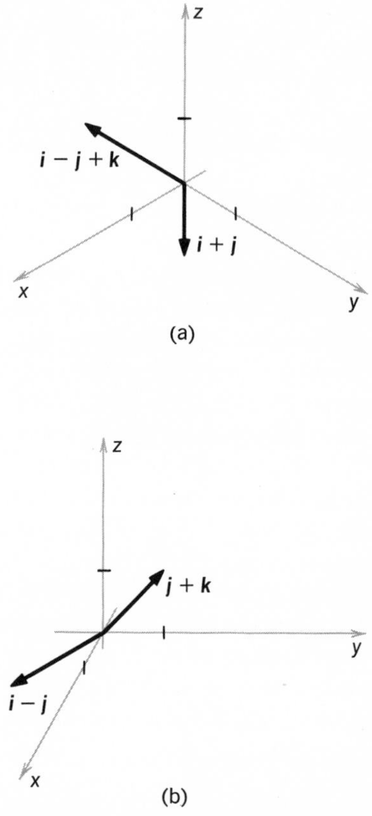

FIGURE 5-2

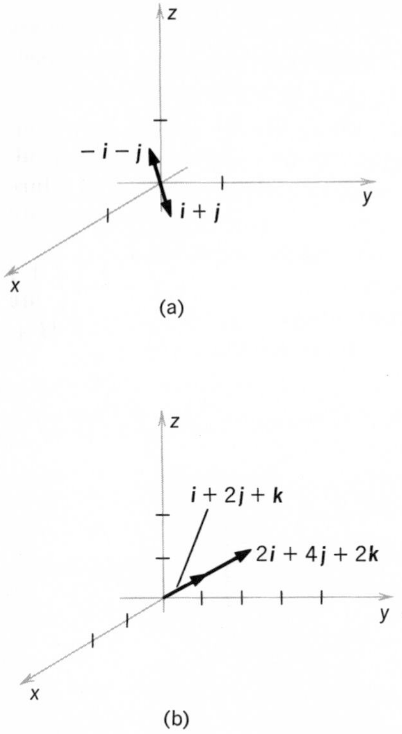

FIGURE 5-3

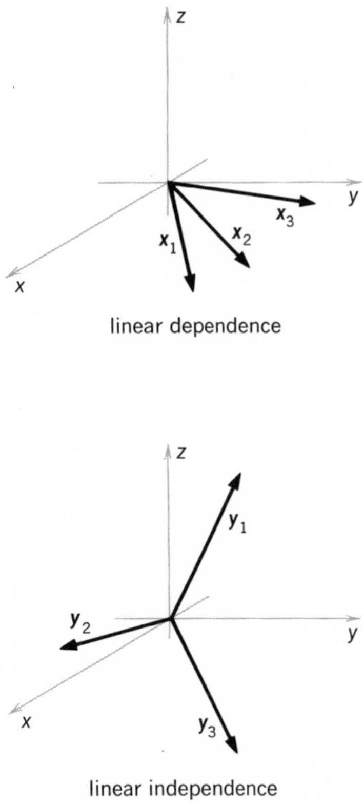

FIGURE 5-4

guarantees their linear independence, it would be interesting to determine under what circumstances a set of three vectors in $\boldsymbol{R}^3$ is linearly independent. We shall see that three vectors in $\boldsymbol{R}^3$ are linearly independent if and only if they are noncoplaner, or equivalently, that three vectors in $\boldsymbol{R}^3$ are linearly dependent when and only when they lie in some plane passing through the origin.

Suppose we are given three linearly dependent vectors $\boldsymbol{x}_1$, $\boldsymbol{x}_2$, and $\boldsymbol{x}_3$ in $\boldsymbol{R}^3$. (See Figure 5-4.) We must show they all lie in some plane. By the theorem of this section, one of the vectors, say $\boldsymbol{x}_1$, is a linear combination of $\boldsymbol{x}_2$ and $\boldsymbol{x}_3$, i.e., $\boldsymbol{x}_1 = \alpha_2\boldsymbol{x}_2 + \alpha_3\boldsymbol{x}_3$. There are two cases. Either $\boldsymbol{x}_2$ and $\boldsymbol{x}_3$ are collinear or they are not collinear. If $\boldsymbol{x}_2$ and $\boldsymbol{x}_3$ are not collinear, we have seen in an example of the previous section that they span a plane through the origin. Since $\boldsymbol{x}_1$ is a linear combination of $\boldsymbol{x}_2$ and $\boldsymbol{x}_3$, it necessarily lies on the plane spanned by $\boldsymbol{x}_2$ and $\boldsymbol{x}_3$. Thus, $\boldsymbol{x}_1$, $\boldsymbol{x}_2$, and $\boldsymbol{x}_3$ all lie in some plane through the origin. If $\boldsymbol{x}_2$ and $\boldsymbol{x}_3$ are collinear, they span a line through the origin. $\boldsymbol{x}_1$, being a linear combination of $\boldsymbol{x}_2$ and $\boldsymbol{x}_3$, necessarily lies on this line. Since $\boldsymbol{x}_1$, $\boldsymbol{x}_2$, and $\boldsymbol{x}_3$ all lie on some line, they necessarily lie in some plane, indeed, any plane containing the line in which they lie, and are necessarily coplaner.

Next, we see that if $\boldsymbol{x}_1$, $\boldsymbol{x}_2$, and $\boldsymbol{x}_3$ all lie in some plane, they are linearly dependent. If two of the vectors, say $\boldsymbol{x}_2$ and $\boldsymbol{x}_3$, are not collinear, by the example of the previous section, they span the plane in which they lie. But by hypothesis $\boldsymbol{x}_1$ belongs to this plane, and is, therefore, a linear combination of $\boldsymbol{x}_2$ and $\boldsymbol{x}_3$. Thus $\boldsymbol{x}_1$, $\boldsymbol{x}_2$, and $\boldsymbol{x}_3$ are linearly dependent. If, on the other hand, each pair of vectors in the set $\{\boldsymbol{x}_1, \boldsymbol{x}_2, \boldsymbol{x}_3\}$ is collinear, each vector must lie on some line passing through the origin. Thus, each is a scalar multiple of a single vector, and so $\boldsymbol{x}_1$, $\boldsymbol{x}_2$, and $\boldsymbol{x}_3$ are linearly dependent.

Using this criterion we see that the sets of vectors $\{\boldsymbol{i}, \boldsymbol{i} + \boldsymbol{j}, \boldsymbol{i} + \boldsymbol{j} + \boldsymbol{k}\}$, $\{\boldsymbol{i} + \boldsymbol{j}, -\boldsymbol{i} + \boldsymbol{j}, -\boldsymbol{j} + \boldsymbol{k}\}$, and $\{\boldsymbol{i} + \boldsymbol{j} - \boldsymbol{k}, \boldsymbol{i} - \boldsymbol{j} - \boldsymbol{k}, \boldsymbol{k}\}$ are independent (see Figure 5-5) while the sets $\{\boldsymbol{i} + \boldsymbol{j}, \boldsymbol{i} - \boldsymbol{j}, \boldsymbol{i}\}$ and $\{\boldsymbol{i} + \boldsymbol{j} + \boldsymbol{k}, -\boldsymbol{i} + \boldsymbol{j} + \boldsymbol{k}, -\boldsymbol{j} - \boldsymbol{k}\}$ are dependent (see Figure 5-6).

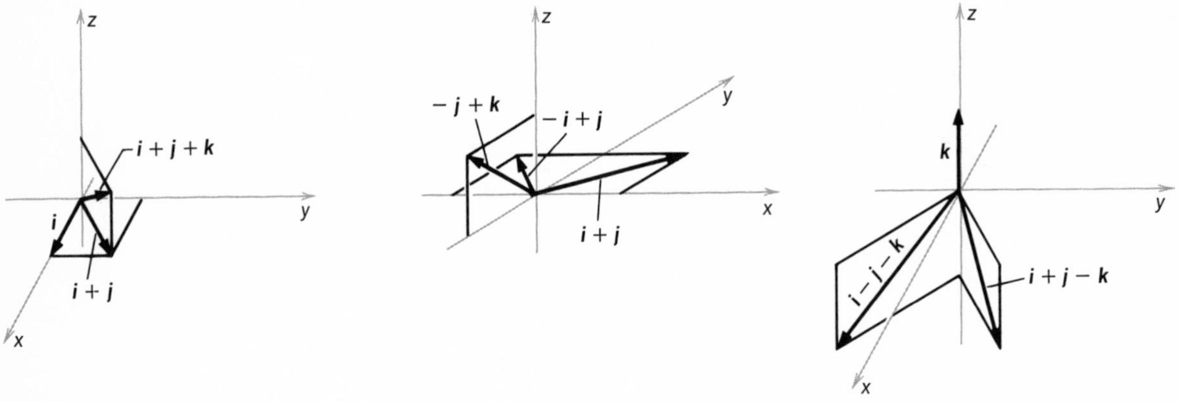

FIGURE 5-5

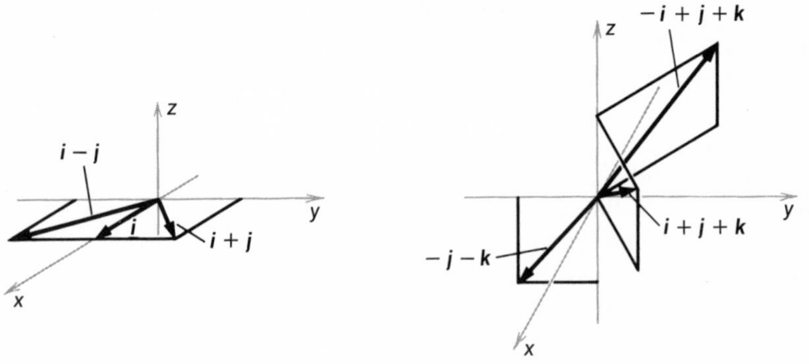

FIGURE 5-6

Our next example is an important illustration of the concepts of span and linear independence. In $\boldsymbol{R}^n$, we consider $\{\boldsymbol{e}_1, \boldsymbol{e}_2, \ldots, \boldsymbol{e}_n\}$, where $\boldsymbol{e}_i$ is the vector which has zeroes in all components except the ith and a 1 in the ith component:

$$\boldsymbol{e}_i = \begin{bmatrix} 0 \\ \vdots \\ 0 \\ 1 \\ 0 \\ \vdots \\ 0 \end{bmatrix} \left.\vphantom{\begin{matrix} 0 \\ \vdots \\ 0 \\ 1 \end{matrix}}\right\} i$$

First, we observe that these vectors are linearly independent. For suppose $\alpha_1\boldsymbol{e}_1 + \alpha_2\boldsymbol{e}_2 + \cdots + \alpha_n\boldsymbol{e}_n = \boldsymbol{0}$. Then,

$$\alpha_1 \begin{bmatrix} 1 \\ 0 \\ \vdots \\ 0 \end{bmatrix} + \alpha_2 \begin{bmatrix} 0 \\ 1 \\ \vdots \\ 0 \end{bmatrix} + \cdots + \alpha_n \begin{bmatrix} 0 \\ 0 \\ \vdots \\ 1 \end{bmatrix} = \begin{bmatrix} 0 \\ 0 \\ \vdots \\ 0 \end{bmatrix}$$

or

$$\begin{bmatrix} \alpha_1 \\ \alpha_2 \\ \vdots \\ \alpha_n \end{bmatrix} = \begin{bmatrix} 0 \\ 0 \\ \vdots \\ 0 \end{bmatrix}$$

So $\alpha_1 = \alpha_2 = \cdots = \alpha_n = 0$.

It is also of interest to note that the vectors $\boldsymbol{e}_1, \boldsymbol{e}_2, \ldots, \boldsymbol{e}_n$ span $\boldsymbol{R}^n$. For if $\boldsymbol{x} \in \boldsymbol{R}^n$,

$$\boldsymbol{x} = \begin{bmatrix} \alpha_1 \\ \alpha_2 \\ \vdots \\ \alpha_n \end{bmatrix} = \alpha_1 \begin{bmatrix} 1 \\ 0 \\ \vdots \\ 0 \end{bmatrix} + \alpha_2 \begin{bmatrix} 0 \\ 1 \\ \vdots \\ 0 \end{bmatrix} + \cdots + \alpha_n \begin{bmatrix} 0 \\ 0 \\ \vdots \\ 1 \end{bmatrix}$$
$$= \alpha_1\boldsymbol{e}_1 + \alpha_2\boldsymbol{e}_2 + \cdots + \alpha_n\boldsymbol{e}_n$$

Since every vector $\boldsymbol{x}$ is a suitable linear combination of $\boldsymbol{e}_1, \boldsymbol{e}_2, \ldots, \boldsymbol{e}_n$, we see that $\boldsymbol{e}_1, \boldsymbol{e}_2, \ldots, \boldsymbol{e}_n$ span $\boldsymbol{R}^n$.

EXERCISES

1. Which of the following sets of vectors in $\boldsymbol{R}^2$ are linearly independent. Illustrate geometrically.

   (a) $\begin{bmatrix} -1 \\ 1 \end{bmatrix}, \begin{bmatrix} 0 \\ 1 \end{bmatrix}$ (b) $\begin{bmatrix} 2 \\ 1 \end{bmatrix}, \begin{bmatrix} 1 \\ 2 \end{bmatrix}$

   (c) $\begin{bmatrix} 3 \\ 3 \end{bmatrix}, \begin{bmatrix} -5 \\ -5 \end{bmatrix}$ (d) $\begin{bmatrix} 1 \\ 0 \end{bmatrix}, \begin{bmatrix} 2 \\ 1 \end{bmatrix}, \begin{bmatrix} 3 \\ 1 \end{bmatrix}$

   (e) $\begin{bmatrix} -7 \\ 2 \end{bmatrix}, \begin{bmatrix} 0 \\ 3 \end{bmatrix}, \begin{bmatrix} 1 \\ 6 \end{bmatrix}$ (f) $\begin{bmatrix} 1 \\ 5 \end{bmatrix}, \begin{bmatrix} 0 \\ 2 \end{bmatrix}, \begin{bmatrix} 7 \\ 1 \end{bmatrix}$

3. We present a list of vector spaces and subsets of them. Which sets are linearly independent?

   (a) 2×2 matrices

$$\begin{bmatrix} 1 & 0 \\ 0 & 0 \end{bmatrix}, \begin{bmatrix} 0 & 1 \\ 0 & 0 \end{bmatrix}, \begin{bmatrix} 0 & 0 \\ 1 & 0 \end{bmatrix}, \begin{bmatrix} 0 & 0 \\ 0 & 1 \end{bmatrix}$$

   (b) P_2

$$1 + t,\ t,\ t^2$$

   (c) P_2

$$1 + t^2,\ t^2,\ 3 + t^2$$

   (d) 2×3 matrices

$$\begin{bmatrix} 1 & 0 & 1 \\ 0 & 0 & 0 \end{bmatrix}, \begin{bmatrix} 1 & 2 & 3 \\ 1 & 0 & 0 \end{bmatrix}, \begin{bmatrix} 2 & 2 & 4 \\ 1 & 0 & 0 \end{bmatrix}$$

   (e) P_n

$$1, t, t^2, \ldots, t^n$$

   (f) P_n

$$1, 1 + t, (1 + t)^2, \ldots, (1 + t)^n$$

4. If V is a vector space and $\boldsymbol{x}$, $\boldsymbol{y}$, and $\boldsymbol{z}$ are linearly independent vectors in V, show that $\boldsymbol{x} + \boldsymbol{y}$, $\boldsymbol{y} + \boldsymbol{z}$, and $\boldsymbol{x} + \boldsymbol{z}$ are linearly independent.
5. Let l be a line in the plane which does not pass through the origin. If $\boldsymbol{v}_1$ and $\boldsymbol{v}_2$ are two different vectors which end on l, show that $\boldsymbol{v}_1$ and $\boldsymbol{v}_2$ are linearly independent.
6. Find four vectors in $\boldsymbol{R}^3$ such that each subset of two vectors is linearly independent, while any subset of three vectors is linearly dependent.

7. Let $A_1, A_2, \ldots, A_l$ be $m \times n$ matrices and B be an $n \times p$ matrix. If $A_1B, A_2B, \ldots, A_lB$ are linearly independent in the space of $m \times p$ matrices, show that $A_1, A_2, \ldots, A_l$ are linearly independent in the space of $m \times n$ matrices.
8. If $\boldsymbol{x}$ and $\boldsymbol{y}$ are linearly independent vectors, show that the vectors

$$\alpha_1\boldsymbol{x} + \alpha_2\boldsymbol{y}$$
$$\beta_1\boldsymbol{x} + \beta_2\boldsymbol{y}$$

are independent if and only if

$$\begin{vmatrix} \alpha_1 & \alpha_2 \\ \beta_1 & \beta_2 \end{vmatrix} \neq 0$$

9. Show that no linearly independent set of vectors can contain the zero vector.
10. Show that any subset of a linearly independent set of vectors is linearly independent.
11. Show that the vectors

$$\boldsymbol{e}_1,\ \boldsymbol{e}_1 + \boldsymbol{e}_2,\ \boldsymbol{e}_1 + \boldsymbol{e}_2 + \boldsymbol{e}_3, \ldots, \boldsymbol{e}_1 + \boldsymbol{e}_2 + \boldsymbol{e}_3 + \cdots + \boldsymbol{e}_n$$

are linearly independent in $\boldsymbol{R}^n$.
12. Let S be a collection of vectors in a vector space with the property that any subset of two elements is linearly dependent. Show that all vectors in S are scalar multiples of a single vector.
13. Let $\{\boldsymbol{x}_1, \boldsymbol{x}_2, \ldots, \boldsymbol{x}_n\}$ be a set of linearly independent vectors in a vector space V. Let

$$\boldsymbol{y} = \alpha_1\boldsymbol{x}_1 + \alpha_2\boldsymbol{x}_2 + \cdots + \alpha_n\boldsymbol{x}_n$$

What condition on the scalars α_i will guarantee that for each i, the vectors $\boldsymbol{x}_1, \boldsymbol{x}_2, \ldots, \boldsymbol{x}_{i-1}, \boldsymbol{y}, \boldsymbol{x}_{i+1}, \ldots, \boldsymbol{x}_n$ are linearly independent.
14. Let $X_1, X_2, \ldots, X_k$ be $m \times n$ matrices which are linearly independent in the space of $m \times n$ matrices. If A is an invertible $m \times m$ matrix and B is an invertible $n \times n$ matrix, show that the matrices $AX_1B, AX_2B, \ldots, AX_kB$ are linearly independent in the space of $m \times n$ matrices.
15. Let V be a vector space and $\boldsymbol{x}_1, \boldsymbol{x}_2, \ldots, \boldsymbol{x}_n$ be vectors in V. If $\boldsymbol{x}_1 \neq \boldsymbol{0}$, $\boldsymbol{x}_2 \notin \text{sp}(\{\boldsymbol{x}_1\})$, $\boldsymbol{x}_3 \notin \text{sp}(\{\boldsymbol{x}_1, \boldsymbol{x}_2\}), \ldots, \boldsymbol{x}_n \notin \text{sp}(\{\boldsymbol{x}_1, \ldots, \boldsymbol{x}_{n-1}\})$, show that the vectors $\boldsymbol{x}_1, \boldsymbol{x}_2, \ldots, \boldsymbol{x}_n$ are linearly independent.
16. Let f_1 and f_2 be two polynomials and suppose there are points x_1 and x_2 such that

$$f_1(x_1) = 1, \qquad f_2(x_1) = 0$$
$$f_1(x_2) = 0, \qquad f_2(x_2) = 1$$

Show that f_1 and f_2 are linearly independent in the space of all polynomials.

17. If $A \neq 0$ is a symmetric matrix and $B \neq 0$ is a skew-symmetric matrix in the space of $n \times n$ matrices, show that A and B are linearly independent.
18. If f and g are polynomials and

$$\begin{vmatrix} f(0) & g(0) \\ f'(0) & g'(0) \end{vmatrix} \neq 0$$

show that f and g are linearly independent in the space of polynomials.
19. Let S be the set of nonzero vectors in $\boldsymbol{R}^n$ which consists of all vectors all of whose components are 0 or 1. Show that any two vectors in S are linearly independent.
20. Let $\boldsymbol{x}_1, \boldsymbol{x}_2, \boldsymbol{y}_1, \boldsymbol{y}_2$ be vectors in $\boldsymbol{R}^2$.
    (a) If $\boldsymbol{x}_1$ and $\boldsymbol{x}_2$ are linearly independent, show that $\boldsymbol{x}_1\boldsymbol{y}_1^T + \boldsymbol{x}_2\boldsymbol{y}_2^T = 0$ implies $\boldsymbol{y}_1 = \boldsymbol{y}_2 = \boldsymbol{0}$.
    (b) If $\boldsymbol{x}_1$ and $\boldsymbol{x}_2$ are linearly independent and $\boldsymbol{y}_1$ and $\boldsymbol{y}_2$ are linearly independent, show that $\boldsymbol{x}_1\boldsymbol{y}_1^T, \boldsymbol{x}_2\boldsymbol{y}_2^T, \boldsymbol{x}_2\boldsymbol{y}_1^T, \boldsymbol{x}_2\boldsymbol{y}_2^T$ are linearly independent in the space of 2×2 matrices.

6 BASIS

In the last example of the previous section we noted that the set of vectors $\{\boldsymbol{e}_1, \boldsymbol{e}_2, \ldots, \boldsymbol{e}_n\}$ is both linearly independent and spans $\boldsymbol{R}^n$. Such sets are very important in the theory of vector spaces and hence we have the following:

Definition Let V be a vector space and $\{\boldsymbol{x}_1, \boldsymbol{x}_2, \ldots, \boldsymbol{x}_n\}$ be a collection of vectors in V. The set $\{\boldsymbol{x}_1, \boldsymbol{x}_2, \ldots, \boldsymbol{x}_n\}$ is said to be a **basis** for V if

(i) $\{\boldsymbol{x}_1, \boldsymbol{x}_2, \ldots, \boldsymbol{x}_n\}$ is a linearly independent set of vectors
(ii) $\{\boldsymbol{x}_1, \boldsymbol{x}_2, \ldots, \boldsymbol{x}_n\}$ spans V

According to this definition the vectors $\boldsymbol{e}_1, \boldsymbol{e}_2, \ldots, \boldsymbol{e}_n$ form a basis for $\boldsymbol{R}^n$. This particular basis is of such frequent occurrence that it is called the **standard basis** for $\boldsymbol{R}^n$. As another example consider the set of vectors $\{\boldsymbol{x}_1, \boldsymbol{x}_2, \boldsymbol{x}_3\}$, where

$$\boldsymbol{x}_1 = \begin{bmatrix} 1 \\ -1 \\ 1 \end{bmatrix}, \qquad \boldsymbol{x}_2 = \begin{bmatrix} 0 \\ 1 \\ 1 \end{bmatrix}, \qquad \boldsymbol{x}_3 = \begin{bmatrix} 2 \\ 3 \\ 0 \end{bmatrix}$$

To show that this set of vectors spans $\boldsymbol{R}^3$, we must show that given any vector

$$\boldsymbol{x} = \begin{bmatrix} x \\ y \\ z \end{bmatrix} \text{ in } \boldsymbol{R}^3$$

there exist scalars α_1, α_2, and α_3, such that $\boldsymbol{x} = \alpha_1\boldsymbol{x}_1 + \alpha_2\boldsymbol{x}_2 + \alpha_3\boldsymbol{x}_3$. Referring back to components, this becomes

$$\begin{bmatrix} x \\ y \\ z \end{bmatrix} = \begin{bmatrix} \alpha_1 \qquad + 2\alpha_3 \\ -\alpha_1 + \alpha_2 + 3\alpha_3 \\ \alpha_1 + \alpha_2 \qquad \end{bmatrix}$$

In other words we must solve the following system of linear equations:

$$\begin{aligned} x &= \alpha_1 \qquad + 2\alpha_3 \\ y &= -\alpha_1 + \alpha_2 + 3\alpha_3 \\ z &= \alpha_1 + \alpha_2 \end{aligned}$$

for α_1, α_2, and α_3.

Since the determinant of the system

$$\begin{vmatrix} 1 & 0 & 2 \\ -1 & 1 & 3 \\ 1 & 1 & 0 \end{vmatrix} = -7$$

we see that the system of equations is solvable, and from this that $\{\boldsymbol{x}_1, \boldsymbol{x}_2, \boldsymbol{x}_3\}$ spans $\boldsymbol{R}^3$.

To prove independence, we suppose there are scalars, α_1, α_2, α_3, such that $\alpha_1\boldsymbol{x}_1 + \alpha_2\boldsymbol{x}_2 + \alpha_3\boldsymbol{x}_3 = \boldsymbol{0}$. Converting this to a system of linear equations, we obtain

$$\begin{aligned} \alpha_1 \qquad + 2\alpha_3 &= 0 \\ -\alpha_1 + \alpha_2 + 3\alpha_3 &= 0 \\ \alpha_1 + \alpha_2 \qquad &= 0 \end{aligned}$$

Since the determinant of the coefficient matrix is nonzero, the only solution to the system is $\alpha_1 = 0$, $\alpha_2 = 0$, $\alpha_3 = 0$. Thus, the vectors $\boldsymbol{x}_1$, $\boldsymbol{x}_2$, $\boldsymbol{x}_3$ are linearly independent, and taken together with the fact that they span $\boldsymbol{R}^3$, we see that $\{\boldsymbol{x}_1, \boldsymbol{x}_2, \boldsymbol{x}_3\}$ is a basis for $\boldsymbol{R}^3$.

An interesting property of bases is indicated in the next result.

Theorem A set $\{\boldsymbol{x}_1, \boldsymbol{x}_2, \ldots, \boldsymbol{x}_n\}$ in a vector space V is a basis for V if and only if for each $\boldsymbol{x}$ in V, there are unique scalars, $\alpha_1, \alpha_2, \ldots, \alpha_n$ such that $\boldsymbol{x} = \alpha_1\boldsymbol{x}_1 + \alpha_2\boldsymbol{x}_2 + \cdots + \alpha_n\boldsymbol{x}_n$.

Proof Let us suppose $\{\boldsymbol{x}_1, \boldsymbol{x}_2, \ldots, \boldsymbol{x}_n\}$ is a basis. Then, $\{\boldsymbol{x}_1, \boldsymbol{x}_2, \ldots, \boldsymbol{x}_n\}$ spans V. Thus, there exist scalars $\alpha_1, \alpha_2, \ldots, \alpha_n$, such that

$$\boldsymbol{x} = \alpha_1\boldsymbol{x}_1 + \alpha_2\boldsymbol{x}_2 + \cdots + \alpha_n\boldsymbol{x}_n$$

To show that these scalars are unique, suppose we have, as well,

$$\boldsymbol{x} = \beta_1\boldsymbol{x}_1 + \beta_2\boldsymbol{x}_2 + \cdots + \beta_n\boldsymbol{x}_n$$

Then,

$$\alpha_1\boldsymbol{x}_1 + \cdots + \alpha_n\boldsymbol{x}_n = \beta_1\boldsymbol{x}_1 + \cdots + \beta_n\boldsymbol{x}_n$$

or

$$(\alpha_1 - \beta_1)\boldsymbol{x}_1 + \cdots + (\alpha_n - \beta_n)\boldsymbol{x}_n = \boldsymbol{0}$$

Since $\{\boldsymbol{x}_1, \boldsymbol{x}_2, \ldots, \boldsymbol{x}_n\}$ is an independent set, we have

$$\alpha_1 - \beta_1 = 0, \ldots, \alpha_n - \beta_n = 0$$

or

$$\alpha_1 = \beta_1, \ldots, \alpha_n = \beta_n$$

thus demonstrating the uniqueness of the scalars $\alpha_1, \ldots, \alpha_n$.

On the other hand, suppose $\{\boldsymbol{x}_1, \boldsymbol{x}_2, \ldots, \boldsymbol{x}_n\}$ is a set such that every $\boldsymbol{x}$ in V can be expressed uniquely in the form $\boldsymbol{x} = \alpha_1\boldsymbol{x}_1 + \cdots + \alpha_n\boldsymbol{x}_n$. It is then clear that $\{\boldsymbol{x}_1, \boldsymbol{x}_2, \ldots, \boldsymbol{x}_n\}$ spans V. To see that $\{\boldsymbol{x}_1, \ldots, \boldsymbol{x}_n\}$ is independent, suppose $\alpha_1\boldsymbol{x}_1 + \cdots + \alpha_n\boldsymbol{x}_n = \boldsymbol{0}$. We know that $0 \cdot \boldsymbol{x}_1 + \cdots + 0 \cdot \boldsymbol{x}_n = \boldsymbol{0}$. Thus, the zero vector is expressed as two linear combinations of $\boldsymbol{x}_1, \ldots, \boldsymbol{x}_n$, and since the coefficients in these linear combinations are, by hypothesis, uniquely determined, we must have $\alpha_1 = 0$, $\alpha_2 = 0, \ldots, \alpha_n = 0$. Thus, we have demonstrated linear independence. ∎

For example, in P_n every element may be expressed as a unique linear combination of $\{1, x, x^2, \ldots, x^n\}$. Thus, we see that $\{1, x, \ldots, x^n\}$ forms a basis for P_n.

Let L be a subspace of $\boldsymbol{R}^3$ consisting of those vectors which lie on some fixed plane passing through the origin.

In §4.4 on span, we saw that any two noncollinear vectors in L span L. In §4.5 on linear independence, we saw that any two noncollinear vectors in $\boldsymbol{R}^3$ are linearly independent. Thus, if $\boldsymbol{a}$ and $\boldsymbol{b}$ are noncollinear vectors in L, we see that $\boldsymbol{a}$ and $\boldsymbol{b}$ span L and are linearly independent. Thus, in order to obtain a basis for L, we need only to choose two noncollinear vectors in L.

For example, in $\boldsymbol{R}^3$, consider the plane consisting of the points (x, y, z), where $y = z$. (See Figure 6-1.)

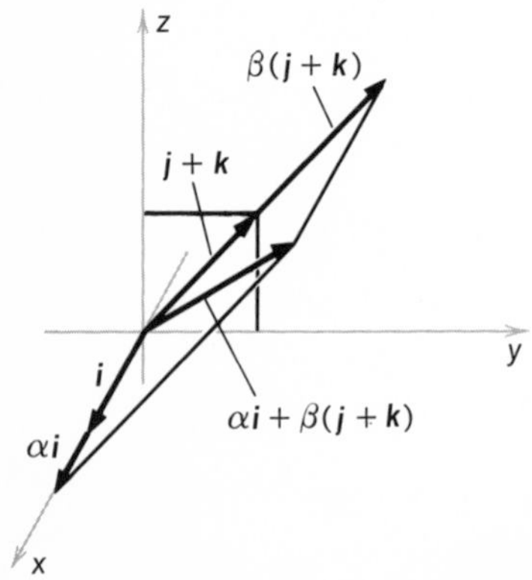

FIGURE 6-1

A pair of noncollinear vectors on this plane is $\{\boldsymbol{i}, \boldsymbol{j} + \boldsymbol{k}\}$. Thus, any vector whatsoever lying on the plane $y = z$ may be expressed uniquely as $\alpha\boldsymbol{i} + \beta(\boldsymbol{j} + \boldsymbol{k})$.

Other bases for the same plane are provided by the pairs of vectors $\{-\boldsymbol{i}, \boldsymbol{i}+\boldsymbol{j}+\boldsymbol{k}\}$ and $\{2\boldsymbol{i}, \boldsymbol{i}+2(\boldsymbol{j}+\boldsymbol{k})\}$. (See Figure 6-2.)

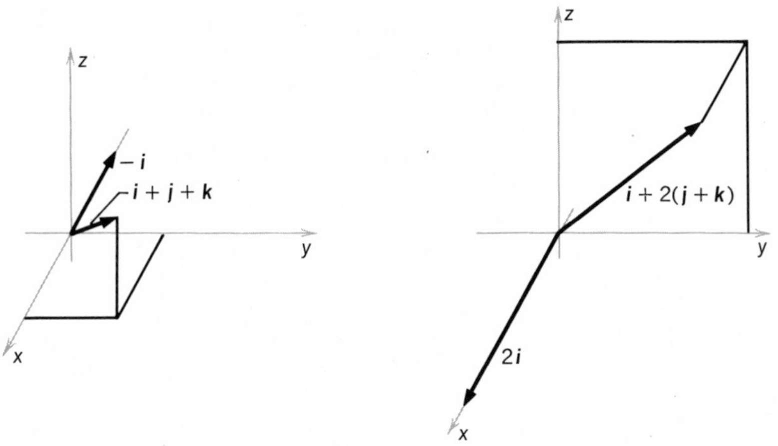

FIGURE 6-2

EXERCISES

1. Which of the following subsets are bases for $\boldsymbol{R}^2$?

(a) $\begin{bmatrix}1\\1\end{bmatrix}, \begin{bmatrix}1\\2\end{bmatrix}$ (b) $\begin{bmatrix}2\\3\end{bmatrix}, \begin{bmatrix}3\\5\end{bmatrix}$

(c) $\begin{bmatrix}1\\0\end{bmatrix}, \begin{bmatrix}0\\1\end{bmatrix}, \begin{bmatrix}1\\3\end{bmatrix}$ (d) $\begin{bmatrix}1\\2\end{bmatrix}, \begin{bmatrix}1\\0\end{bmatrix}$

(e) $\begin{bmatrix}-3\\2\end{bmatrix}, \begin{bmatrix}6\\-4\end{bmatrix}$ (f) $\begin{bmatrix}1\\1\end{bmatrix}, \begin{bmatrix}-3\\-3\end{bmatrix}, \begin{bmatrix}0\\2\end{bmatrix}$

2. Which of the following subsets are bases for $\boldsymbol{R}^3$?

(a) $\begin{bmatrix}1\\3\\0\end{bmatrix}, \begin{bmatrix}0\\1\\0\end{bmatrix}, \begin{bmatrix}0\\-2\\1\end{bmatrix}$ (b) $\begin{bmatrix}1\\1\\0\end{bmatrix}, \begin{bmatrix}0\\-1\\-1\end{bmatrix}, \begin{bmatrix}1\\1\\1\end{bmatrix}$

(c) $\begin{bmatrix}1\\3\\7\end{bmatrix}, \begin{bmatrix}1\\0\\2\end{bmatrix}$ (d) $\begin{bmatrix}1\\0\\1\end{bmatrix}, \begin{bmatrix}0\\1\\0\end{bmatrix}, \begin{bmatrix}1\\1\\1\end{bmatrix}$

(e) $\begin{bmatrix}0\\1\\0\end{bmatrix}, \begin{bmatrix}3\\-4\\0\end{bmatrix}, \begin{bmatrix}2\\1\\0\end{bmatrix}$ (f) $\begin{bmatrix}1\\3\\7\end{bmatrix}, \begin{bmatrix}1\\1\\0\end{bmatrix}, \begin{bmatrix}0\\3\\1\end{bmatrix}$

3. Show that the given subsets are bases for the indicated subspaces of the space of 2×2 matrices.

(a) 2×2 symmetric matrices

$$\begin{bmatrix} 1 & 0 \\ 0 & 0 \end{bmatrix}, \quad \begin{bmatrix} 0 & 0 \\ 0 & 1 \end{bmatrix}, \quad \begin{bmatrix} 0 & 1 \\ 1 & 0 \end{bmatrix}$$

(b) 2×2 skew-symmetric matrices

$$\begin{bmatrix} 0 & 1 \\ -1 & 0 \end{bmatrix}$$

(c) Subspace of matrices of the form

$$\begin{bmatrix} a & b \\ b & a \end{bmatrix}$$

$$\begin{bmatrix} 1 & 0 \\ 0 & 1 \end{bmatrix}, \quad \begin{bmatrix} 1 & 1 \\ 1 & 1 \end{bmatrix}$$

(d) 2×2 symmetric matrices

$$\begin{bmatrix} 1 & 0 \\ 0 & 1 \end{bmatrix}, \quad \begin{bmatrix} 1 & 0 \\ 0 & -1 \end{bmatrix}, \quad \begin{bmatrix} 1 & 1 \\ 1 & 1 \end{bmatrix}$$

(e) 2×2 upper triangular matrices

$$\begin{bmatrix} 1 & 0 \\ 0 & 0 \end{bmatrix}, \quad \begin{bmatrix} 0 & 1 \\ 0 & 0 \end{bmatrix}, \quad \begin{bmatrix} 0 & 0 \\ 0 & 1 \end{bmatrix}$$

(f) All 2×2 matrices

$$\begin{bmatrix} 1 & 0 \\ 0 & 0 \end{bmatrix}, \quad \begin{bmatrix} 0 & 1 \\ 0 & 0 \end{bmatrix}, \quad \begin{bmatrix} 0 & 0 \\ 1 & 0 \end{bmatrix}, \quad \begin{bmatrix} 0 & 0 \\ 0 & 1 \end{bmatrix}$$

(g) 2×2 matrices A, such that

$$A\begin{bmatrix} 1 \\ 0 \end{bmatrix} = \begin{bmatrix} 0 \\ 0 \end{bmatrix}, \quad \begin{bmatrix} 0 & 1 \\ 0 & 0 \end{bmatrix}, \quad \begin{bmatrix} 0 & 0 \\ 0 & 1 \end{bmatrix}$$

(h) 2×2 matrices which commute with

$$B = \begin{bmatrix} 0 & 1 \\ -1 & 0 \end{bmatrix}$$

I_2, B

4. Show that the indicated subsets are bases for P_3, the polynomials in a variable t.

(a) $1, t, t^2, 1 + t + t^2 + t^3$

(b) $1, 1 + t, (1 + t)^2, (1 + t)^3$
(c) $1, 1 + t, 1 + t^2, 1 + t^3$
(d) $1, 1 + t, 1 + t + t^2, 1 + t + t^2 + t^3$
(e) $1 + t, 1 - t, t^2 + t^3, t^2 - t^3$

5. Consider the collection of all polynomials in two variables of degree less than or equal to 2:

$$a_0 + b_1x + b_2y + c_1x^2 + c_2xy + c_3y^2$$

With the usual definitions of addition and scalar multiplication, show that these polynomials form a vector space and determine a basis for this space.

6. Find a basis for each of the following spaces.
   (a) 3×3 symmetric matrices.
   (b) 3×3 skew-symmetric matrices.
   (c) 2×3 matrices.
   (d) 3×2 matrices.
   (e) Even polynomials of degree less than or equal to n.
   (f) Subspace of the space of polynomials of degree less than or equal to 3 which vanish at $x = 1$.
   (g) Subspace of $\boldsymbol{R}^n$ consisting of those vectors whose first two components vanish.
   (h) 4×4 diagonal matrices.
7. In the space of $n \times m$ matrices, let E_{ij} denote that matrix with 0 in all entries except in position (i, j), and 1 in position (i, j).
   (a) Show that the matrices E_{ij} form a basis for the space of $n \times m$ matrices.
   (b) By choosing a suitable subcollection of the E_{ij}'s, find a basis for the upper triangular matrices and diagonal matrices.
8. Find a basis for the space of 2×2 matrices which consists of
   (a) Only matrices such that $A^2 = A$ (a matrix is said to be **idempotent** if $A^2 = A$);
   (b) only invertible matrices.
9. Show that it is not possible to find a basis for the space of $n \times n$ matrices such that each pair of elements in the basis commute.
10. The matrices

$$\sigma_x = \begin{bmatrix} 0 & 1 \\ 1 & 0 \end{bmatrix}, \quad \sigma_y = \begin{bmatrix} 0 & -i \\ i & 0 \end{bmatrix}, \quad \sigma_z = \begin{bmatrix} 1 & 0 \\ 0 & -1 \end{bmatrix}$$

are called the Pauli-spin matrices.
   (a) Show that

$$\sigma_x\sigma_y = -\sigma_y\sigma_x$$
$$\sigma_x\sigma_z = -\sigma_z\sigma_x$$
$$\sigma_y\sigma_z = -\sigma_z\sigma_y$$

   (b) Show that along with the identity matrix these matrices form a basis for the space of 2×2 matrices with complex entries.

11. Is it possible to find a basis for P_n, such that every element of the basis is divisible by the polynomial $f(x) = x$?
12. Let $A_1, A_2, \ldots, A_m$ be a collection of $n \times n$ matrices. If $\boldsymbol{x} \neq 0$ is an n-vector such that $A_1\boldsymbol{x} = A_2\boldsymbol{x} = \cdots = A_m\boldsymbol{x} = 0$, show that the matrices $A_1, A_2, \ldots, A_m$ do not form a basis for the space of $n \times n$ matrices.
13. If $\boldsymbol{x}_1$, $\boldsymbol{x}_2$, and $\boldsymbol{x}_3$ form a basis for some real vector space V, show that $(1 + t)\boldsymbol{x}_1 + (1 + t^2)\boldsymbol{x}_2 + (t^4 + t^2)\boldsymbol{x}_3$ is nonzero for all real t.

7 DIMENSION

For purposes of further study, we single out a collection of vector spaces which are in some sense "small" and particularly suited for applications. Thus, we say a vector space V is **finite dimensional** if it can be spanned by a finite number of vectors. Previous examples show that both the vector spaces $\boldsymbol{R}^n$ and P_n are finite dimensional.

Our first objective is to show that any space spanned by finitely many vectors has a finite basis. Preliminary to this we prove:

Lemma Let V be a vector space. Suppose $\boldsymbol{x}_1, \boldsymbol{x}_2, \ldots, \boldsymbol{x}_n$ span V, and that the vectors $\boldsymbol{x}_1, \boldsymbol{x}_2, \ldots, \boldsymbol{x}_n$ are linearly dependent. Then by deleting a suitable vector of the set $\{\boldsymbol{x}_1, \boldsymbol{x}_2, \ldots, \boldsymbol{x}_n\}$, say $\boldsymbol{x}_i$, we may obtain a set $\{\boldsymbol{x}_1, \boldsymbol{x}_2, \ldots, \boldsymbol{x}_{i-1}, \boldsymbol{x}_{i+1}, \ldots, \boldsymbol{x}_n\}$ which still spans V.

Proof Since the vectors $\boldsymbol{x}_1, \boldsymbol{x}_2, \ldots, \boldsymbol{x}_n$ are linearly dependent, by the theorem in §4.5, one of the vectors, say $\boldsymbol{x}_i$, is a linear combination of the remaining vectors in the set:

$$\boldsymbol{x}_i = \beta_1\boldsymbol{x}_1 + \cdots + \beta_{i-1}\boldsymbol{x}_{i-1} + \beta_{i+1}\boldsymbol{x}_{i+1} + \cdots + \beta_n\boldsymbol{x}_n$$

for some scalars $\beta_1, \ldots, \beta_{i-1}, \beta_{i+1}, \ldots, \beta_n$.

Now, let $\boldsymbol{x}$ be any vector in V. Since $\{\boldsymbol{x}_1, \boldsymbol{x}_2, \ldots, \boldsymbol{x}_n\}$ spans V, there are scalars, $\alpha_1, \ldots, \alpha_n$, such that

$$\boldsymbol{x} = \alpha_1\boldsymbol{x}_1 + \cdots + \alpha_{i-1}\boldsymbol{x}_{i-1} + \alpha_i\boldsymbol{x}_i + \alpha_{i+1}\boldsymbol{x}_{i+1} + \cdots + \alpha_n\boldsymbol{x}_n$$

By substituting the above expression for $\boldsymbol{x}_i$ into the linear combination which equals $\boldsymbol{x}$, we see that $\boldsymbol{x}$ may be expressed as a linear combination of $\boldsymbol{x}_1, \ldots, \boldsymbol{x}_{i-1}, \boldsymbol{x}_{i+1}, \ldots, \boldsymbol{x}_n$.

Thus, every vector in V is a linear combination of $\boldsymbol{x}_1, \ldots, \boldsymbol{x}_{i-1}, \boldsymbol{x}_{i+1}, \ldots, \boldsymbol{x}_n$, showing that this set spans V. ■

Using this, we may show the following theorem.

Theorem 1 Let V be a vector space spanned by finitely many vectors. Then V has a finite basis.

Proof Suppose $\{\boldsymbol{x}_1, \boldsymbol{x}_2, \ldots, \boldsymbol{x}_n\}$ spans V. If $\{\boldsymbol{x}_1, \boldsymbol{x}_2, \ldots, \boldsymbol{x}_n\}$ is a linearly independent set, it is a basis, and we are finished. So suppose it is not linearly independent. By the preceding lemma, we may delete a vector in the set and obtain a smaller set which still spans V. If this new set is linearly independent, we are finished, since we then have a finite basis. If it is not linearly independent, we may obtain a yet smaller set which spans V. By repeating this process it must eventually come to pass that we obtain a linearly independent set which spans V, demonstrating our theorem. ∎

Actually our proof yielded the stronger statement: If $\{\boldsymbol{x}_1, \ldots, \boldsymbol{x}_n\}$ spans a vector space V, some subset of $\{\boldsymbol{x}_1, \ldots, \boldsymbol{x}_n\}$ is a basis for V.

Let H be a subspace of $\boldsymbol{R}^3$ consisting of those vectors lying on some fixed plane which passes through the origin. Let us determine the number of vectors in some basis for H. In §4.5 we saw that any set of three coplanar vectors in $\boldsymbol{R}^3$ is necessarily linearly dependent. Thus, a basis for H, since it must consist of linearly independent vectors, must have at most two vectors. Since the vectors in any basis for H spans a plane, no basis for H can have only one vector. Thus, any basis for H must have exactly two vectors.

This remarkable and useful fact, namely, that any two bases have the same number of vectors, is true in any finite-dimensional vector space. This follows immediately from the next result.

Theorem 2 Let V be a finite-dimensional vector space and $\{\boldsymbol{x}_1, \boldsymbol{x}_2, \ldots, \boldsymbol{x}_n\}$ be a basis for V with n elements. If $\{\boldsymbol{y}_1, \boldsymbol{y}_2, \ldots, \boldsymbol{y}_m\}$ is a set of m linearly independent vectors in V, then $m \leq n$.

We may restate this: If V is a finite-dimensional space having a basis of n elements and $\{\boldsymbol{y}_1, \ldots, \boldsymbol{y}_m\}$ is a set of vectors in V with $m > n$, then the vectors $\boldsymbol{y}_1, \boldsymbol{y}_2, \ldots, \boldsymbol{y}_m$ are linearly dependent. We prove the second statement.

Proof Let $\{\boldsymbol{x}_1, \boldsymbol{x}_2, \ldots, \boldsymbol{x}_n\}$ be a basis for V. In order to show that $\{\boldsymbol{y}_1, \boldsymbol{y}_2, \ldots, \boldsymbol{y}_m\}$ is a linearly dependent set, we must exhibit scalars $\beta_1, \beta_2, \ldots, \beta_m$ not all zero, such that

$$\sum_{j=1}^{m} \beta_j \boldsymbol{y}_j = \beta_1 \boldsymbol{y}_1 + \beta_2 \boldsymbol{y}_2 + \cdots + \beta_m \boldsymbol{y}_m = \boldsymbol{0}$$

Since $\{\boldsymbol{x}_1, \boldsymbol{x}_2, \ldots, \boldsymbol{x}_n\}$ is a basis for V, there are scalars α_{ij}, such that

$$\boldsymbol{y}_j = \sum_{i=1}^{n} \alpha_{ij} \boldsymbol{x}_i$$

Thus, we wish to find $\beta_1, \beta_2, \ldots, \beta_m$, such that

$$\sum_{j=1}^{m} \beta_j \left(\sum_{i=1}^{n} \alpha_{ij} \boldsymbol{x}_i \right) = \boldsymbol{0}$$

or

$$\sum_{i=1}^{n}\left(\sum_{j=1}^{m}\alpha_{ij}\beta_j\right)\boldsymbol{x}_i = \boldsymbol{0}$$

Thus, if we can find a nontrivial solution to the system of homogeneous equations

$$\begin{aligned}\alpha_{11}\beta_1 + \cdots + \alpha_{1m}\beta_m &= 0\\ \alpha_{21}\beta_1 + \cdots + \alpha_{2m}\beta_m &= 0\\ &\vdots\\ \alpha_{n1}\beta_1 + \cdots + \alpha_{nm}\beta_m &= 0\end{aligned}$$

our theorem will be demonstrated. Since $m > n$, the number of unknowns in the system is greater than the number of equations, hence, we know from §1.4 that a nontrivial solution exists. ∎

Immediately we obtain the next result.

Theorem 3 In a finite-dimensional space any two bases have the same number of elements.

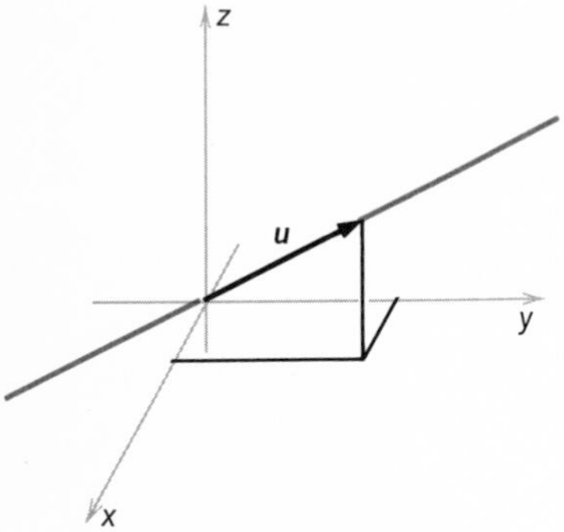

FIGURE 7-1

Proof Let $\{\boldsymbol{x}_1, \boldsymbol{x}_2, \ldots, \boldsymbol{x}_m\}$ and $\{\boldsymbol{y}_1, \boldsymbol{y}_2, \ldots, \boldsymbol{y}_n\}$ be bases for V, a finite-dimensional space. Since $\{\boldsymbol{y}_1, \boldsymbol{y}_2, \ldots, \boldsymbol{y}_n\}$ is a linearly independent set and $\{\boldsymbol{x}_1, \boldsymbol{x}_2, \ldots, \boldsymbol{x}_m\}$ is a basis, by the preceding theorem $m \geq n$. Since $\{\boldsymbol{x}_1, \boldsymbol{x}_2, \ldots, \boldsymbol{x}_m\}$ is a linearly independent set and $\{\boldsymbol{y}_1, \boldsymbol{y}_2, \ldots, \boldsymbol{y}_n\}$ is a basis, we also have $n \geq m$. Thus, $m = n$. ∎

In $\boldsymbol{R}^n$ we exhibited the standard basis $\{\boldsymbol{e}_1, \boldsymbol{e}_2, \ldots, \boldsymbol{e}_n\}$ with n elements. By Theorem 3, we see that any other basis whatsoever must also have n elements.

If V is a finite-dimensional space, we define the **dimension** of V to be the number of elements in some basis for V. By Theorem 3, this number is independent of the basis we choose. The dimension of V is often denoted by dim V. According to this definition dim $\boldsymbol{R}^n = n$.

We have seen in $\boldsymbol{R}^3$ that the vectors on any line passing through the origin all consist of scalar multiples of a single vector. Thus, if H denotes a subspace of $\boldsymbol{R}^3$ consisting solely of vectors on some line passing through the origin, we see that dim $H = 1$. Thus, our definition of dimension makes precise the statement that a line is a "one-dimensional object." (See Figure 7-1.)

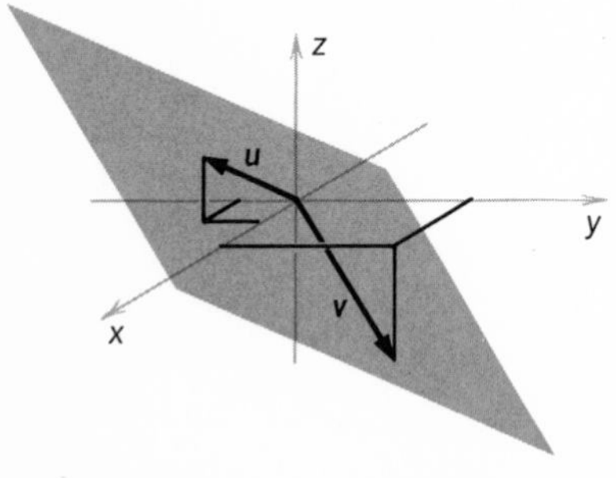

FIGURE 7-2

If, in a similar manner, we let K denote the vectors which lie on some plane passing through the origin, then we have seen that K has a basis of two vectors, and so dim $K = 2$. Thus, we see why it makes perfectly good sense to say that a plane is a "two-dimensional object." (See Figure 7-2.)

Example 1 Let P_n be the space of polynomials of degree less than or equal to n. We have seen that $1, x, x^2, \ldots, x^n$ is a basis for P_n. Since the set $\{1, x, \ldots, x^i, \ldots, x^n\}$ has $n + 1$ elements, we see that $\dim P_n = n + 1$.

Example 2 Consider the 2×2 matrices with real coefficients, M_{22}. Observe that the matrices

$$\begin{bmatrix} 1 & 0 \\ 0 & 0 \end{bmatrix}, \quad \begin{bmatrix} 0 & 1 \\ 0 & 0 \end{bmatrix}, \quad \begin{bmatrix} 0 & 0 \\ 1 & 0 \end{bmatrix}, \quad \begin{bmatrix} 0 & 0 \\ 0 & 1 \end{bmatrix}$$

form a basis.

To see this, note that

$$\begin{bmatrix} a & b \\ c & d \end{bmatrix} = \begin{bmatrix} a & 0 \\ 0 & 0 \end{bmatrix} + \begin{bmatrix} 0 & b \\ 0 & 0 \end{bmatrix} + \begin{bmatrix} 0 & 0 \\ c & 0 \end{bmatrix} + \begin{bmatrix} 0 & 0 \\ 0 & d \end{bmatrix}$$

$$= a\begin{bmatrix} 1 & 0 \\ 0 & 0 \end{bmatrix} + b\begin{bmatrix} 0 & 1 \\ 0 & 0 \end{bmatrix} + c\begin{bmatrix} 0 & 0 \\ 1 & 0 \end{bmatrix} + d\begin{bmatrix} 0 & 0 \\ 0 & 1 \end{bmatrix}$$

Thus, the given set spans the space of 2×2 matrices. If

$$a\begin{bmatrix} 1 & 0 \\ 0 & 0 \end{bmatrix} + b\begin{bmatrix} 0 & 1 \\ 0 & 0 \end{bmatrix} + c\begin{bmatrix} 0 & 0 \\ 1 & 0 \end{bmatrix} + d\begin{bmatrix} 0 & 0 \\ 0 & 1 \end{bmatrix} = \begin{bmatrix} 0 & 0 \\ 0 & 0 \end{bmatrix}$$

then

$$\begin{bmatrix} a & b \\ c & d \end{bmatrix} = \begin{bmatrix} 0 & 0 \\ 0 & 0 \end{bmatrix}$$

So $a = b = c = d = 0$.

Thus, the given set of matrices is linearly independent. Since the space of 2×2 matrices with real coefficients admits a basis with four elements, it is of dimension 4.

Example 3 Consider the system of homogeneous equations

$$\begin{aligned} x_1 + 2x_2 - x_3 + x_4 &= 0 \\ x_1 - x_2 + x_3 + 2x_4 &= 0 \end{aligned}$$

In matrix notation

$$\begin{bmatrix} 1 & 2 & -1 & 1 \\ 1 & -1 & 1 & 2 \end{bmatrix}\begin{bmatrix} x_1 \\ x_2 \\ x_3 \\ x_4 \end{bmatrix} = \begin{bmatrix} 0 \\ 0 \end{bmatrix}$$

We have seen in Example 2, §4.3 that the set of solutions form a subspace of $\boldsymbol{R}^4$. Let us find a basis for this subspace and calculate its dimension.

We apply Gaussian elimination.

$$\begin{aligned} x_1 + 2x_2 - x_3 + x_4 &= 0 \\ x_1 - x_2 + x_3 + 2x_4 &= 0 \end{aligned}$$

Use x_1 and the first equation.

$$\downarrow$$

$$\begin{aligned} x_1 + 2x_2 - x_3 + x_4 &= 0 \\ - 3x_2 + 2x_3 + x_4 &= 0 \end{aligned}$$

Use x_4 and the second equation.

$$\downarrow$$

$$\begin{aligned} x_1 + 5x_2 - 3x_3 \qquad &= 0 \\ - 3x_2 + 2x_3 + x_4 &= 0 \end{aligned}$$

The procedure comes to a halt since there are no unused equations.

If we choose $x_2 = c$, $x_3 = d$, where c and d are arbitrary numbers, and then choose

$$\begin{aligned} x_1 &= -5c + 3d \\ x_4 &= 3c - 2d \end{aligned}$$

we see that any solution to the system is of the form

$$\begin{bmatrix} -5c + 3d \\ c \\ d \\ 3c - 2d \end{bmatrix} = c \begin{bmatrix} -5 \\ 1 \\ 0 \\ 3 \end{bmatrix} + d \begin{bmatrix} 3 \\ 0 \\ 1 \\ -2 \end{bmatrix}$$

Thus, we see that the vectors

$$\begin{bmatrix} -5 \\ 1 \\ 0 \\ 3 \end{bmatrix} \quad \text{and} \quad \begin{bmatrix} 3 \\ 0 \\ 1 \\ -2 \end{bmatrix}$$

span the space of solutions. Since the two vectors are clearly linearly independent, it follows that taken together they provide a basis for the space of solutions. Since the space of solutions is a two-dimensional vector space, we see precisely what it means to say that the solutions form a "two-parameter family."

EXERCISES

1. Calculate the dimension of the following subspaces of the space of 2×2 matrices.
   (a) The diagonal matrices.
   (b) The symmetric matrices.
   (c) The skew-symmetric matrices.
   (d) The matrices of the form

$$\begin{bmatrix} a & b \\ -b & a \end{bmatrix}$$

   (e) The matrices of the form

$$\begin{bmatrix} 0 & a \\ 0 & b \end{bmatrix}$$

   (f) The upper triangular matrices.
   (g) The matrices of the form

$$\begin{bmatrix} a & b \\ -b & c \end{bmatrix}$$

2. Calculate the dimension of the following subspaces of P_3.
   (a) $\{f \mid f(0) = 0\}$ (b) $\{f \mid f(1) = 0\}$
   (c) $\{f \mid f(0) = f'(0) = 0\}$ (d) $\{f \mid f'(0) + f(0) = 0\}$
3. Let $i_1 < i_2 < \cdots < i_n$ be integers. Calculate the dimension of the subspace of $\boldsymbol{R}^m$ which consists of all vectors such that i_1th, i_2th, . . . , i_nth components vanish. ($m \geq n$.)
4. Find a basis and the dimension of the space of solutions to each system of linear homogeneous equations.
   (a) $x + y + z = 0$
   $x - y + z = 0$
   (b) $x + y + z = 0$
   $x + 2y + 3z = 0$
   (c) $x_1 + x_2 + x_3 - x_4 = 0$
   $x_1 - x_2 + x_3 - x_4 = 0$
   (d) $x_1 + x_2 + 2x_3 - 4x_4 = 0$
   $x_1 + x_2 - x_3 - x_4 = 0$
   $2x_1 + 2x_2 - x_3 - 3x_4 = 0$
   (e) $x_1 + x_2 - 3x_3 + x_4 + x_5 = 0$
   $x_1 - x_2 + x_3 + x_4 + x_5 = 0$
   $x_1 + 2x_2 + 3x_3 + 2x_4 + x_5 = 0$
5. Show that the complex numbers with the usual addition and multiplication by scalars form a two-dimensional vector space over the reals.

6. What is the dimension of the space of $n \times n$ diagonal matrices?
7. If f is a polynomial of degree n, show that $f, f', f'', \ldots, f^{(n)}$ form a basis for P_n.
8. Show that the dimension of the space of $n \times m$ matrices is nm.
9. Let H_e be the subspace of P_n consisting of the even polynomials, and H_o be the subspace of odd polynomials. What is the dimension of H_o and H_e?
10. What is the dimension of the space of upper triangular $n \times n$ matrices?
11. Show that the dimension of the space of $n \times n$ symmetric matrices is $\frac{1}{2}n(n + 1)$, while the dimension of the space of skew-symmetric matrices is $\frac{1}{2}n(n - 1)$.
12. Let V be an n-dimensional vector space. If V is spanned by $\{\boldsymbol{x}_1, \boldsymbol{x}_2, \ldots, \boldsymbol{x}_n\}$, a set of n vectors, show that $\{\boldsymbol{x}_1, \boldsymbol{x}_2, \ldots, \boldsymbol{x}_n\}$ are linearly independent.
13. Show that the space of polynomials in two variables of degree less than or equal to n, i.e., all polynomials of the form

$$\sum_{\substack{0 \le i+j \le n \\ 0 \le i \\ 0 \le j}} a_{ij} x^i y^j$$

form a vector space under ordinary addition and scalar multiplication, and show that the dimension of this space is $\frac{1}{2}(n + 1)(n + 2)$.
14. If $\boldsymbol{x}$, $\boldsymbol{y}$, and $\boldsymbol{z}$ form a basis for a three-dimensional vector space, show also that $\boldsymbol{x} + \boldsymbol{y}$, $\boldsymbol{y} + \boldsymbol{z}$, $\boldsymbol{x} + \boldsymbol{z}$ form a basis as well.
15. If $f_0, f_1, \ldots, f_{n+1}$ are polynomials of degree less than or equal to n, show that there are scalars, $\alpha_0, \alpha_1, \ldots, \alpha_{n+1}$, not all zero, such that $\alpha_0 f_0 + \alpha_1 f_1 + \cdots + \alpha_{n+1} f_{n+1} = 0$.
16. Show that any $n \times n$ matrix with real coefficients satisfies a polynomial equation $f(A) = 0$, where $f(x)$ is a nonzero polynomial with real coefficients. [Hint: Show that the matrices I_n, A, $A^2, \ldots, A^{n^2}$, must be linearly dependent.]

8 ADDITIONAL PROPERTIES OF FINITE-DIMENSIONAL SPACES

In this section we note some of the consequences of theorems in previous sections.

Theorem 1 Let V be a vector space and suppose $\boldsymbol{x}_1, \boldsymbol{x}_2, \ldots, \boldsymbol{x}_m$ span V. Then $\dim V \le m$.

Proof By Theorem 1, §4.7, some subset of $\boldsymbol{x}_1, \boldsymbol{x}_2, \ldots, \boldsymbol{x}_m$ is a basis for V. But $\dim V$ is the number of elements in this subset which is a basis. Consequently $\dim V \le m$. ∎

Preliminary to our next results, we have

Lemma Let V be a vector space and H be a subspace of V. Let $\{\boldsymbol{x}_1, \boldsymbol{x}_2, \ldots, \boldsymbol{x}_n\}$ be a basis for H and suppose $\boldsymbol{y}$ does not belong to H. Then, the set $\{\boldsymbol{y}, \boldsymbol{x}_1, \boldsymbol{x}_2, \ldots, \boldsymbol{x}_n\}$ is linearly independent.

Proof We show that if $\{\boldsymbol{y}, \boldsymbol{x}_1, \boldsymbol{x}_2, \ldots, \boldsymbol{x}_m\}$ is linearly dependent, then $\boldsymbol{y}$ belongs to H, proving our assertion.

So suppose $\{\boldsymbol{y}, \boldsymbol{x}_1, \boldsymbol{x}_2, \ldots, \boldsymbol{x}_n\}$ is linearly dependent. Then, there are scalars $\beta, \alpha_1, \alpha_2, \ldots, \alpha_n$, not all zero, such that

$$\beta\boldsymbol{y} + \alpha_1\boldsymbol{x}_1 + \alpha_2\boldsymbol{x}_2 + \cdots + \alpha_n\boldsymbol{x}_n = \boldsymbol{0}$$

Now, β is not zero. For if $\beta = 0$, $\alpha_1\boldsymbol{x}_1 + \alpha_2\boldsymbol{x}_2 + \cdots + \alpha_n\boldsymbol{x}_n = \boldsymbol{0}$. Since not all of the scalars $\alpha_1, \alpha_2, \ldots, \alpha_n$ are zero, it follows that $\boldsymbol{x}_1, \ldots, \boldsymbol{x}_n$ are linearly dependent. But by hypothesis, $\boldsymbol{x}_1, \boldsymbol{x}_2, \ldots, \boldsymbol{x}_n$ are linearly independent. Consequently, $\beta \neq 0$.

Since $\beta \neq 0$, $\boldsymbol{y} = -(\beta^{-1}\alpha_1)\boldsymbol{x}_1 + \cdots + (-\beta^{-1}\alpha_n)\boldsymbol{x}_n$. Thus, $\boldsymbol{y}$ is a linear combination of vectors in H and so belongs to H. ∎

Let H be a subspace of $\boldsymbol{R}^3$ consisting of those vectors lying on some fixed plane through the origin. We saw in §4.4 that the linear combinations of two noncollinear vectors in H fill out H. Restated in algebraic terminology, two linearly independent vectors in H span H and therefore form a basis for H. The generalization of this result to arbitrary finite-dimensional spaces is formulated in the next theorem.

Theorem 2 Let V be a vector space of dimension n. If $\{\boldsymbol{x}_1, \boldsymbol{x}_2, \ldots, \boldsymbol{x}_n\}$ is a set of n linearly independent vectors in V, then $\{\boldsymbol{x}_1, \boldsymbol{x}_2, \ldots, \boldsymbol{x}_n\}$ is a basis for V.

Proof Let $L = \text{sp}(\boldsymbol{x}_1, \boldsymbol{x}_2, \ldots, \boldsymbol{x}_n)$. We desire to show $L = V$.

If $L \neq V$, there is some vector, say $\boldsymbol{y}$, in V, such that $\boldsymbol{y} \notin L$. By the lemma, the vectors in the set $\{\boldsymbol{y}, \boldsymbol{x}_1, \boldsymbol{x}_2, \ldots, \boldsymbol{x}_n\}$ are linearly independent. But the set $\{\boldsymbol{y}, \boldsymbol{x}_1, \ldots, \boldsymbol{x}_n\}$ has $n + 1$ elements, and by Theorem 2 of §4.7, we know that in any n-dimensional vector space, we can find at most n linearly independent vectors. Thus, it was false to assume that $L \neq V$, and so $L = V$. ∎

We can interpret this theorem in $\boldsymbol{R}^3$. If $\boldsymbol{x}_1$, $\boldsymbol{x}_2$, and $\boldsymbol{x}_3$ are any three linearly independent vectors in $\boldsymbol{R}^3$, then $\boldsymbol{x}_1, \boldsymbol{x}_2, \boldsymbol{x}_3$ form a basis for $\boldsymbol{R}^3$. In geometric terminology, any three noncoplaner (i.e., linearly independent) vectors in $\boldsymbol{R}^3$ span $\boldsymbol{R}^3$. Figure 8-1 illustrates this principle. The reader may find it a worthwhile exercise to demonstrate the same result from geometric considerations by using a procedure analogous to that of constructing Cartesian coordinates in three-dimensional space. (See §2.2.)

We have another intuitively reasonable result.

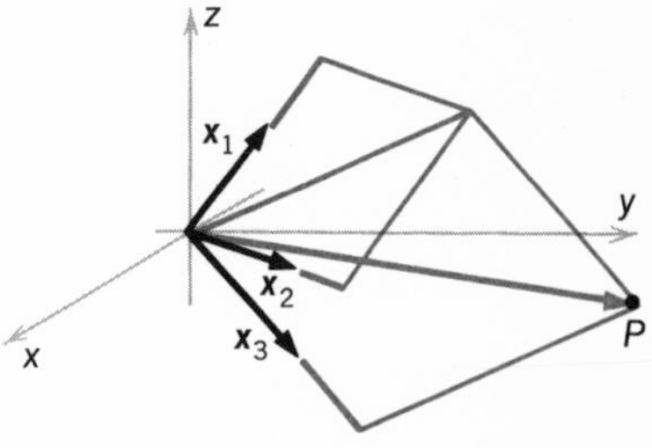

FIGURE 8-1

Theorem 3 Let V be a finite-dimensional vector space. If H is a subspace of V, H is finite dimensional and $\dim H \leq \dim V$.

Proof If $H = \{\mathbf{0}\}$, the zero subspace, then $\dim H = 0 \leq \dim V$. If H is not the zero subspace, it has some nonzero element, say $\boldsymbol{x}_1$. Consider $\text{sp}(\boldsymbol{x}_1)$. If $\text{sp}(\boldsymbol{x}_1) = H$, H is finite dimensional. If $\text{sp}(\boldsymbol{x}_1) \neq H$, then there is some vector $\boldsymbol{x}_2$ in H but not in $\text{sp}(\boldsymbol{x}_1)$. By the lemma of this section $\{\boldsymbol{x}_1, \boldsymbol{x}_2\}$ is a linearly independent set. If $H = \text{sp}(\boldsymbol{x}_1, \boldsymbol{x}_2)$, H is finite dimensional. If $H \neq \text{sp}(\boldsymbol{x}_1, \boldsymbol{x}_2)$, there is some vector $\boldsymbol{x}_3$ in H but not in $\text{sp}(\boldsymbol{x}_1, \boldsymbol{x}_2)$. By the same lemma $\{\boldsymbol{x}_1, \boldsymbol{x}_2, \boldsymbol{x}_3\}$ is a linearly independent set.

By continuing this process we obtain sets of linearly dependent vectors $\{\boldsymbol{x}_1, \boldsymbol{x}_2, \ldots, \boldsymbol{x}_i\}$. As long as $\text{sp}(\boldsymbol{x}_1, \boldsymbol{x}_2, \ldots, \boldsymbol{x}_i)$ is properly contained in H, we can find a vector $\boldsymbol{x}_{i+1}$ in H but not in $\text{sp}(\boldsymbol{x}_1, \boldsymbol{x}_2, \ldots, \boldsymbol{x}_i)$. By the lemma we then obtain a larger set of independent vectors $\{\boldsymbol{x}_1, \boldsymbol{x}_2, \ldots, \boldsymbol{x}_i, \boldsymbol{x}_{i+1}\}$. However, we know that V is a finite-dimensional space and that any linearly independent set of vectors in V has at most $\dim V$ elements.

Thus, eventually we must have $\text{sp}(\boldsymbol{x}_1, \boldsymbol{x}_2, \ldots, \boldsymbol{x}_i) = H$, and so H is finite dimensional. Choosing a basis for H, we know, by definition of dimension, that it contains $\dim H$ elements. By Theorem 2, §4.7, $\dim H \leq \dim V$. ∎

Note that the bulk of the previous theorem consisted in demonstrating that H was finite dimensional. As soon as we obtained finite dimensionality, the fact that $\dim H \leq \dim V$ was immediately apparent.

As an application of these theorems, let us determine all subspaces of $\boldsymbol{R}^3$. We know, by Theorem 3, that any such subspace H is finite dimensional and satisfies $0 \leq \dim H \leq 3$.

If $\dim H = 0$, $H = \{\mathbf{0}\}$, the zero subspace.

If $\dim H = 3$, we choose a basis for H with three elements. By Theorem 2, this basis, having three elements, is also a basis for $\boldsymbol{R}^3$. Thus, $H = \boldsymbol{R}^3$.

From these two facts, we see that any proper subspace H of $\boldsymbol{R}^3$ must satisfy $1 \leq \dim H \leq 2$.

If $\dim H = 1$, H consists of all scalar multiples of a single vector, or in other words, H consists of all vectors lying on some line through the origin.

If $\dim H = 2$, H consists of all linear combinations of two noncollinear vectors, or, as we have seen earlier, H consists of all vectors lying on some plane which passes through the origin.

Thus, we see, as is intuitively plausible, that all proper subspaces of $\boldsymbol{R}^3$ are formed by lines or planes passing through the origin.

EXERCISES

1. If H is a subspace of the space of 2×2 matrices, show that $\dim H = 0$, 1, 2, 3, or 4.

2. Show that the polynomials $1, x - \alpha, (x - \alpha)^2, \ldots, (x - \alpha)^n$ form a basis for P_n, if α is any real number.
3. If f is a polynomial of degree n and g is a polynomial of degree less than or equal to n, show there are constants $\alpha_0, \alpha_1, \ldots, \alpha_n$, such that $g = \alpha_0 f + \alpha_1 f' + \alpha_2 f'' + \cdots + \alpha_n f^{(n)}$.
4. If $\boldsymbol{x}$ is a nonzero vector in a finite-dimensional vector space, show that $\boldsymbol{x}$ belongs to some basis.
5. If a vector space V is spanned by $\boldsymbol{x}_1, \boldsymbol{x}_2, \ldots, \boldsymbol{x}_k$, show that a set of $k + 1$ vectors in V is linearly dependent.
6. Find a set of $n + 1$ vectors in $\boldsymbol{R}^n$ such that any subset of n elements is linearly independent.
7. Let $\{\boldsymbol{x}_1, \boldsymbol{x}_2, \ldots, \boldsymbol{x}_n\}$ be a set of independent vectors in a vector space V. If $\{\boldsymbol{x}_1, \boldsymbol{x}_2, \ldots, \boldsymbol{x}_n\}$ is contained in no larger set of independent vectors in V, show that $\{\boldsymbol{x}_1, \boldsymbol{x}_2, \ldots, \boldsymbol{x}_n\}$ is a basis for V.
8. If V and W are subspaces of a vector space, $V \subset W$, and $\dim V = \dim W$, show that $V = W$.
9. Show that any independent set $\{\boldsymbol{x}_1, \ldots, \boldsymbol{x}_m\}$ of vectors in a finite dimensional space V can be extended to a basis for V, i.e., there are vectors $\{\boldsymbol{y}_1, \ldots, \boldsymbol{y}_n\}$, such that $\{\boldsymbol{x}_1, \ldots, \boldsymbol{x}_m, \boldsymbol{y}_1, \ldots, \boldsymbol{y}_n\}$ is a basis for V.
10. Let S be a set of vectors all of whose subsets of $k + 1$ vectors are linearly dependent but some subset of k vectors is linearly independent. Show that each vector in S is a linear combination of those k vectors.
11. If V and W are subspaces of a vector space, with $V \subset W$ and $\dim V \leq l \leq \dim W$, show that there is a subspace U, such that $V \subset U \subset W$ and $\dim U = l$.
12. If $f_0, f_1, \ldots, f_n$ are polynomials in P_n, such that $\deg f_i = i$, show that $f_0, f_1, \ldots, f_n$ form a basis for P_n.
13. If $\{\boldsymbol{x}_1, \boldsymbol{x}_2, \ldots, \boldsymbol{x}_n\}$ spans V, but no smaller subset spans V, show that $\{\boldsymbol{x}_1, \boldsymbol{x}_2, \ldots, \boldsymbol{x}_n\}$ is a basis for V.
14. Let $\alpha_0, \alpha_1, \ldots, \alpha_n$ be $n + 1$ distinct real numbers. Consider the polynomials $p_i(x) = (x - \alpha_0) \ldots (x - \alpha_{i-1})(x - \alpha_{i+1}) \ldots (x - \alpha_n)$. Show that $p_0, p_1, \ldots, p_n$ form a basis for P_n.
15. If H is a subspace of a finite-dimensional vector space V, show that there is a subspace K, such that

$$H \cap K = \boldsymbol{0}$$

$$H + K = V$$

16. Let V be a finite-dimensional space.
    (a) Let $H_0 \leq H_1 \leq H_2 \leq H_3 \leq \cdots$ be an increasing collection of subspaces of V. Show that we must eventually have $H_n = H_{n+1} = H_{n+2} = \cdots$.
    (b) Let $H_0 \geq H_1 \geq H_2 \geq H_3 \geq \cdots$ be a decreasing family of subspaces of V. Show that we must eventually have $H_n = H_{n+1} = H_{n+2} = \cdots$.
17. Let V be a vector space and $\boldsymbol{x}_1, \ldots, \boldsymbol{x}_n$ be a collection of vectors in V. Suppose $\boldsymbol{y}_1, \boldsymbol{y}_2, \ldots, \boldsymbol{y}_n$ belongs to $\text{sp}(\boldsymbol{x}_1, \boldsymbol{x}_2, \ldots, \boldsymbol{x}_n)$. Let μ be the maximum

number of linearly independent vectors in $\{\boldsymbol{x}_1, \boldsymbol{x}_2, \ldots, \boldsymbol{x}_n\}$ and ν be the maximum number of linearly independent vectors in $\{\boldsymbol{y}_1, \ldots, \boldsymbol{y}_n\}$. Show that $\nu \leq \mu$.

18. If $\boldsymbol{x}_1, \boldsymbol{x}_2, \ldots, \boldsymbol{x}_n$ are linearly independent vectors in a vector space V and if $\boldsymbol{x} \notin \text{sp}(\boldsymbol{x}_1, \boldsymbol{x}_2, \ldots, \boldsymbol{x}_n)$, show that $\{\boldsymbol{x}_1 + \boldsymbol{x}, \boldsymbol{x}_2 + \boldsymbol{x}, \ldots, \boldsymbol{x}_n + \boldsymbol{x}\}$ are also linearly independent in V.
19. Let S be the set of vectors in $\boldsymbol{R}^n$ which have exactly two nonzero components and these nonzero components are both 1. Show that S is a linearly independent set of vectors if and only if $n \leq 3$.
20. If a is a complex number which is not real, show that a and a^2 form a basis for the complexes, regarded as a vector space over the reals.

9 CHANGE OF COORDINATES

There are many problems in mathematics and the sciences for which the standard coordinate basis $\{\boldsymbol{e}_1, \boldsymbol{e}_2, \ldots, \boldsymbol{e}_n\}$ is not the most convenient for dealing with a particular problem. For example, in studying the conic sections in plane analytic geometry, it is often convenient to bring the equations representing the conics into a standard form by means of a rotation of axes. It is to facilitate such change of coordinates that this section is presented.

Thus, let V be an n-dimensional vector space and $\{\boldsymbol{x}_1, \boldsymbol{x}_2, \ldots, \boldsymbol{x}_n\}$ be a basis for V. As we have seen, given any $\boldsymbol{x}$ in V, there are unique scalars $\alpha_1, \alpha_2, \ldots, \alpha_n$, such that $\boldsymbol{x} = \alpha_1\boldsymbol{x}_1 + \alpha_2\boldsymbol{x}_2 + \cdots + \alpha_n\boldsymbol{x}_n$. This fact enables us to "coordinatize" the space relative to the given basis. This procedure is analogous to the by now familiar method of "coordinatizing" $\boldsymbol{R}^n$ by means of the standard basis $\{\boldsymbol{e}_1, \boldsymbol{e}_2, \ldots, \boldsymbol{e}_n\}$. For if $\boldsymbol{x}$ is a vector in $\boldsymbol{R}^n$,

$$\begin{bmatrix} \alpha_1 \\ \alpha_2 \\ \vdots \\ \alpha_n \end{bmatrix} = \alpha_1 \begin{bmatrix} 1 \\ 0 \\ 0 \\ \vdots \\ 0 \end{bmatrix} + \alpha_2 \begin{bmatrix} 0 \\ 1 \\ 0 \\ \vdots \\ 0 \end{bmatrix} + \cdots + \alpha_n \begin{bmatrix} 0 \\ 0 \\ 0 \\ \vdots \\ 1 \end{bmatrix}$$

$$= \alpha_1\boldsymbol{e}_1 + \alpha_2\boldsymbol{e}_2 + \cdots + \alpha_n\boldsymbol{e}_n$$

In this case the ith coordinate of the vector $\boldsymbol{x}$ is just the coefficient of $\boldsymbol{e}_i$ in the linear expression for $\boldsymbol{x}$ relative to the basis $\{\boldsymbol{e}_1, \boldsymbol{e}_2, \ldots, \boldsymbol{e}_n\}$.

Thus, if $\mathcal{B} = \{\boldsymbol{x}_1, \boldsymbol{x}_2, \ldots, \boldsymbol{x}_n\}$ is a basis for V, and if $\boldsymbol{x}$ is a vector in V and $\boldsymbol{x} = \alpha_1\boldsymbol{x}_1 + \alpha_2\boldsymbol{x}_2 + \cdots + \alpha_n\boldsymbol{x}_n$, we say that α_i is the ith coordinate of $\boldsymbol{x}$ relative to the basis $\mathcal{B}$. We may thus associate with the vector $\boldsymbol{x}$ its coordinate n-tuple relative to the basis $\mathcal{B}$:

$$\begin{bmatrix} \alpha_1 \\ \alpha_2 \\ \vdots \\ \alpha_n \end{bmatrix} \underset{\mathcal{B}}{\leftrightarrow} \boldsymbol{x}$$

The $\mathcal{B}$ below the arrow indicates that $\alpha_1, \alpha_2, \ldots, \alpha_n$ are coordinates of $\boldsymbol{x}$ relative to $\mathcal{B}$. Since the expression

$$\boldsymbol{x} = \alpha_1\boldsymbol{x}_1 + \alpha_2\boldsymbol{x}_2 + \cdots + \alpha_n\boldsymbol{x}_n$$

for $\boldsymbol{x}$ relative to the basis $\mathcal{B}$ is unique, it is clear that if

$$\begin{bmatrix} \alpha_1 \\ \alpha_2 \\ \vdots \\ \alpha_n \end{bmatrix} \underset{\mathcal{B}}{\leftrightarrow} \boldsymbol{x} \quad \text{and} \quad \begin{bmatrix} \beta_1 \\ \beta_2 \\ \vdots \\ \beta_n \end{bmatrix} \underset{\mathcal{B}}{\leftrightarrow} \boldsymbol{x}$$

we must have

$$\alpha_1 = \beta_1, \qquad \alpha_2 = \beta_2, \qquad \ldots, \qquad \alpha_n = \beta_n$$

Moreover, it is clear that given any n-tuple

$$\begin{bmatrix} \gamma_1 \\ \gamma_2 \\ \vdots \\ \gamma_n \end{bmatrix}$$

there is a vector, namely $\boldsymbol{y} = \gamma_1\boldsymbol{x}_1 + \cdots + \gamma_n\boldsymbol{x}_n$, such that

$$\begin{bmatrix} \gamma_1 \\ \gamma_2 \\ \vdots \\ \gamma_n \end{bmatrix} \underset{\mathcal{B}}{\leftrightarrow} \boldsymbol{y}$$

Coordinate systems may be convenient tools for calculation, since if

$$\boldsymbol{x} \leftrightarrow \begin{bmatrix} \alpha_1 \\ \alpha_2 \\ \vdots \\ \alpha_n \end{bmatrix} \quad \text{and} \quad \boldsymbol{y} \leftrightarrow \begin{bmatrix} \beta_1 \\ \beta_2 \\ \vdots \\ \beta_n \end{bmatrix}$$

then

$$\alpha \boldsymbol{x} + \beta \boldsymbol{y} \leftrightarrow \begin{bmatrix} \alpha\alpha_1 + \beta\beta_1 \\ \alpha\alpha_2 + \beta\beta_2 \\ \vdots \\ \alpha\alpha_n + \beta\beta_n \end{bmatrix}$$

In the following we denote the standard basis $\{\boldsymbol{e}_1, \boldsymbol{e}_2, \ldots, \boldsymbol{e}_n\}$ for $\boldsymbol{R}^n$ by $\mathcal{E}$.

As an example, let us calculate the coordinates of a vector

$$\boldsymbol{x} = \begin{bmatrix} \alpha_1 \\ \alpha_2 \end{bmatrix}$$

relative to basis $\{\boldsymbol{x}_1, \boldsymbol{x}_2\} = \mathcal{B}$, where

$$\boldsymbol{x}_1 = \begin{bmatrix} 1 \\ 0 \end{bmatrix}, \qquad \boldsymbol{x}_2 = \begin{bmatrix} 1 \\ 1 \end{bmatrix}$$

Observe that

$$\boldsymbol{e}_1 = \boldsymbol{x}_1, \qquad \boldsymbol{e}_2 = \boldsymbol{x}_2 - \boldsymbol{x}_1$$

Thus,

$$\boldsymbol{x} = \begin{bmatrix} \alpha_1 \\ \alpha_2 \end{bmatrix}_{\mathcal{E}} = \alpha_1\boldsymbol{e}_1 + \alpha_2\boldsymbol{e}_2 = \alpha_1\boldsymbol{x}_1 + \alpha_2(\boldsymbol{x}_2 - \boldsymbol{x}_1) = (\alpha_1 - \alpha_2)\boldsymbol{x}_1 + \alpha_2\boldsymbol{x}_2$$

Thus,

$$\boldsymbol{x} \leftrightarrow \begin{bmatrix} \alpha_1 - \alpha_2 \\ \alpha_2 \end{bmatrix}_{\mathcal{B}}$$

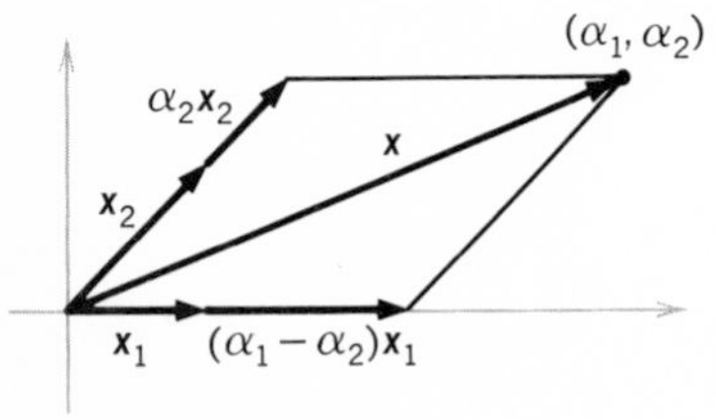

FIGURE 9-1

A geometric means of visualizing and determining the coordinates of the vector $\boldsymbol{x}$ is by the procedure described in §4.4 and depicted in Figure 9-1.

There is a useful rule for relating the coordinates of a vector with respect to one basis, say $\mathcal{B} = \{\boldsymbol{x}_1, \boldsymbol{x}_2, \ldots, \boldsymbol{x}_n\}$, to the coordinates with respect to another basis, say $\mathcal{B}' = \{\boldsymbol{x}_1', \boldsymbol{x}_2', \ldots, \boldsymbol{x}_n'\}$.

Relative to the basis $\mathcal{B}$, we have

$$\boldsymbol{x} \underset{\mathcal{B}}{\leftrightarrow} \begin{bmatrix} \alpha_1 \\ \alpha_2 \\ \vdots \\ \alpha_n \end{bmatrix}$$

and relative to $\mathcal{B}'$, we have

$$\boldsymbol{x} \underset{\mathcal{B}'}{\leftrightarrow} \begin{bmatrix} \alpha_1' \\ \alpha_2' \\ \vdots \\ \alpha_n' \end{bmatrix}$$

Suppose

$$\boldsymbol{x}_j' = \sum_{i=1}^{n} a_{ij}\boldsymbol{x}_i = a_{1j}\boldsymbol{x}_1 + a_{2j}\boldsymbol{x}_2 + \cdots + a_{nj}\boldsymbol{x}_n$$

We know that such numbers a_{ij} exist since $\{\boldsymbol{x}_1, \boldsymbol{x}_2, \ldots, \boldsymbol{x}_n\}$ is a basis for V.

Let A be the $n \times n$ matrix, $A = [a_{ij}]_{(nn)}$. We claim

$$\begin{bmatrix} \alpha_1 \\ \alpha_2 \\ \vdots \\ \alpha_n \end{bmatrix} = A \begin{bmatrix} \alpha_1' \\ \alpha_2' \\ \vdots \\ \alpha_n' \end{bmatrix}$$

To prove this, observe

$$\begin{aligned} \boldsymbol{x} &= \sum_{j=1}^{n} \alpha_j'\boldsymbol{x}_j' \\ &= \sum_{j=1}^{n} \alpha_j' \left(\sum_{i=1}^{n} a_{ij}\boldsymbol{x}_i\right) \\ &= \sum_{j=1}^{n} \sum_{i=1}^{n} a_{ij}\alpha_j'\boldsymbol{x}_i \\ &= \sum_{i=1}^{n} \left(\sum_{j=1}^{n} a_{ij}\alpha_j'\right)\boldsymbol{x}_i \\ &= \sum_{i=1}^{n} \alpha_i\boldsymbol{x}_i \end{aligned}$$

Thus, it follows that

$$\sum_{j=1}^{n} a_{ij}\alpha_j' = \alpha_i, \quad \text{for } i = 1, 2, \ldots, n$$

If we interpret this result using matrix multiplication, the desired assertion follows.

For example, in $\boldsymbol{R}^3$, let

$$\boldsymbol{x}_1 = \begin{bmatrix} 1 \\ 0 \\ 0 \end{bmatrix}, \qquad \boldsymbol{x}_2 = \begin{bmatrix} 0 \\ 1 \\ 0 \end{bmatrix}, \qquad \boldsymbol{x}_3 = \begin{bmatrix} 0 \\ 0 \\ 1 \end{bmatrix}$$

be the standard basis, and let

$$\boldsymbol{x}_1' = \begin{bmatrix} 1 \\ 0 \\ 0 \end{bmatrix}, \qquad \boldsymbol{x}_2' = \begin{bmatrix} 1 \\ 1 \\ 0 \end{bmatrix}, \qquad \boldsymbol{x}_3' = \begin{bmatrix} 1 \\ 1 \\ 1 \end{bmatrix}$$

be another basis.

Suppose $\boldsymbol{x}$ is a vector in $\boldsymbol{R}^3$,

$$\boldsymbol{x} = \begin{bmatrix} \alpha_1 \\ \alpha_2 \\ \alpha_3 \end{bmatrix} = \alpha_1\boldsymbol{x}_1 + \alpha_2\boldsymbol{x}_2 + \alpha_3\boldsymbol{x}_3$$

In this case the coordinates $\alpha_1, \alpha_2, \alpha_3, \alpha_1', \alpha_2'$, and α_3', are related by

$$\begin{bmatrix} \alpha_1 \\ \alpha_2 \\ \alpha_3 \end{bmatrix} = \begin{bmatrix} 1 & 1 & 1 \\ 0 & 1 & 1 \\ 0 & 0 & 1 \end{bmatrix} \begin{bmatrix} \alpha_1' \\ \alpha_2' \\ \alpha_3' \end{bmatrix}$$

Note that the matrix A has in its jth column the coordinates of the vector $\boldsymbol{x}_j'$ with respect to the basis $\{\boldsymbol{x}_1, \boldsymbol{x}_2, \ldots, \boldsymbol{x}_n\}$.

A matrix obtained by a change of coordinates has one very engaging property: It is invertible.

To see this let $\mathcal{B} = \{\boldsymbol{x}_1, \boldsymbol{x}_2, \ldots, \boldsymbol{x}_n\}$ be a basis and $\mathcal{B}' = \{\boldsymbol{x}_1', \boldsymbol{x}_2', \ldots, \boldsymbol{x}_n'\}$ be a second basis. Let

$$\begin{bmatrix} \alpha_1' \\ \alpha_2' \\ \vdots \\ \alpha_n' \end{bmatrix} \underset{\mathcal{B}'}{\leftrightarrow} \boldsymbol{x} \underset{\mathcal{B}}{\leftrightarrow} \begin{bmatrix} \alpha_1 \\ \alpha_2 \\ \vdots \\ \alpha_n \end{bmatrix}$$

be coordinate n-tuples of the vector $\boldsymbol{x}$ relative to the bases $\mathcal{B}$ and $\mathcal{B}'$.

Let $\boldsymbol{x}_j' = \Sigma_{j=1}^{n} a_{ij}\boldsymbol{x}_i$ be the expression of $\boldsymbol{x}_j'$ relative to the basis $\mathcal{B}$, $A = [a_{ij}]_{(nn)}$ and $\boldsymbol{x}_j = \Sigma_{i=1}^{n} b_{ij}\boldsymbol{x}_j'$ be the expression of $\boldsymbol{x}_j$ relative to the basis $\mathcal{B}'$, $B = [b_{ij}]_{(nn)}$.

We saw above that

$$\begin{bmatrix} \alpha_1 \\ \alpha_2 \\ \vdots \\ \alpha_n \end{bmatrix} = A \begin{bmatrix} \alpha_1' \\ \alpha_2' \\ \vdots \\ \alpha_n' \end{bmatrix}$$

and

$$\begin{bmatrix} \alpha_1' \\ \alpha_2' \\ \vdots \\ \alpha_n' \end{bmatrix} = B \begin{bmatrix} \alpha_1 \\ \alpha_2 \\ \vdots \\ \alpha_n \end{bmatrix}$$

Therefore, for all n-tuples

$$\begin{bmatrix} \alpha_1 \\ \alpha_2 \\ \vdots \\ \alpha_n \end{bmatrix} = AB \begin{bmatrix} \alpha_1 \\ \alpha_2 \\ \vdots \\ \alpha_n \end{bmatrix}$$

and

$$\begin{bmatrix} \alpha_1' \\ \alpha_2' \\ \vdots \\ \alpha_n' \end{bmatrix} = BA \begin{bmatrix} \alpha_1' \\ \alpha_2' \\ \vdots \\ \alpha_n' \end{bmatrix}$$

By the lemma of §3.7, it follows that

$$AB = I_n \quad \text{and} \quad BA = I_n$$

Thus, the matrix A is invertible and its inverse is the matrix B.

NOTE Again we mention that the ith column of A is just the coordinate n-tuple of $\boldsymbol{x}_i'$ relative to the basis $\mathcal{B}$. The ith column of B is the coordinate n-tuple of $\boldsymbol{x}_i$ relative to the basis $\mathcal{B}'$.

Example 1 Let P_4 be, as usual, the space of polynomials in the variable x, of degree less than or equal to 4.

Consider the two bases for P_4: $1, x, x^2, x^3, x^4$ and $1, 1 + x, (1 + x)^2, (1 + x)^3, (1 + x)^4$. Suppose we have f in P_4. Then,

$$\begin{aligned} f &= \alpha_0 + \alpha_1 x + \alpha_2 x^2 + \alpha_3 x^3 + \alpha_4 x^4 \\ &= \beta_0 + \beta_1(1 + x) + \beta_2(1 + x)^2 + \beta_3(1 + x)^3 + \beta_4(1 + x)^4 \end{aligned}$$

If we form the matrix A, having as its columns the coordinates of $1, 1 + x, (1 + x)^2, (1 + x)^3, (1 + x)^4$ relative to the basis $1, x, x^2, x^3, x^4$, we obtain

$$A = \begin{bmatrix} 1 & 1 & 1 & 1 & 1 \\ 0 & 1 & 2 & 3 & 4 \\ 0 & 0 & 1 & 3 & 6 \\ 0 & 0 & 0 & 1 & 4 \\ 0 & 0 & 0 & 0 & 1 \end{bmatrix}$$

Thus,

$$\begin{bmatrix} \alpha_0 \\ \alpha_1 \\ \alpha_2 \\ \alpha_3 \\ \alpha_4 \end{bmatrix} = \begin{bmatrix} 1 & 1 & 1 & 1 & 1 \\ 0 & 1 & 2 & 3 & 4 \\ 0 & 0 & 1 & 3 & 6 \\ 0 & 0 & 0 & 1 & 4 \\ 0 & 0 & 0 & 0 & 1 \end{bmatrix} \begin{bmatrix} \beta_0 \\ \beta_1 \\ \beta_2 \\ \beta_3 \\ \beta_4 \end{bmatrix}$$

If we form the matrix B, having as its columns the coordinates of $1, x, x^2, x^3, x^4$, relative to the basis $1, 1 + x, (1 + x)^2, (1 + x)^3, (1 + x)^4$, by observing that $x^k = ((1 + x) - 1)^k$, we see

$$B = \begin{bmatrix} 1 & -1 & 1 & -1 & 1 \\ 0 & 1 & -2 & 3 & -4 \\ 0 & 0 & 1 & -3 & 6 \\ 0 & 0 & 0 & 1 & -4 \\ 0 & 0 & 0 & 0 & 1 \end{bmatrix}$$

These matrices afford a convenient means of passing from the expression of a polynomial relative to one basis to its expression relative to another basis. As an auxiliary result we see

$$\begin{bmatrix} 1 & 1 & 1 & 1 & 1 \\ 0 & 1 & 2 & 3 & 4 \\ 0 & 0 & 1 & 3 & 6 \\ 0 & 0 & 0 & 1 & 4 \\ 0 & 0 & 0 & 0 & 1 \end{bmatrix}^{-1} = \begin{bmatrix} 1 & -1 & 1 & -1 & 1 \\ 0 & 1 & -2 & 3 & -4 \\ 0 & 0 & 1 & -3 & 6 \\ 0 & 0 & 0 & 1 & -4 \\ 0 & 0 & 0 & 0 & 1 \end{bmatrix}$$

Example 2 Using the methods of this section we derive the standard formulas for rotation of axes encountered in plane analytic geometry.

Let

$$\boldsymbol{i} = \begin{bmatrix} 1 \\ 0 \end{bmatrix}, \qquad \boldsymbol{j} = \begin{bmatrix} 0 \\ 1 \end{bmatrix}$$

be the standard basis for $\boldsymbol{R}^2$.

Let

$$\boldsymbol{i}_\theta = (\cos\theta)\boldsymbol{i} + (\sin\theta)\boldsymbol{j}$$
$$\boldsymbol{j}_\theta = -(\sin\theta)\boldsymbol{i} + (\cos\theta)\boldsymbol{j}$$

be the vectors obtained by rotating $\boldsymbol{i}$ and $\boldsymbol{j}$ by θ degrees, respectively.

We wish to find a relationship between the coordinates of a point determined by the $\{\boldsymbol{i}, \boldsymbol{j}\}$ basis and its coordinates as determined by the $\{\boldsymbol{i}_\theta, \boldsymbol{j}_\theta\}$ basis. Note that the matrix $[\boldsymbol{i}_\theta, \boldsymbol{j}_\theta]$ is

$$\begin{bmatrix} \cos\theta & -\sin\theta \\ \sin\theta & \cos\theta \end{bmatrix}$$

Let $\boldsymbol{v} = x\boldsymbol{i} + y\boldsymbol{j}$ be the linear expression for the vector $\boldsymbol{v}$ relative to the $\{\boldsymbol{i}, \boldsymbol{j}\}$ basis, and $\boldsymbol{v} = x_\theta \boldsymbol{i}_\theta + y_\theta \boldsymbol{j}_\theta$ be the expression relative to the $\{\boldsymbol{i}_\theta, \boldsymbol{j}_\theta\}$ basis. Then, by an earlier result, we have

$$\begin{bmatrix} x \\ y \end{bmatrix} = \begin{bmatrix} \cos\theta & -\sin\theta \\ \sin\theta & \cos\theta \end{bmatrix} \begin{bmatrix} x_\theta \\ y_\theta \end{bmatrix}$$

Since the $\{\boldsymbol{i}, \boldsymbol{j}\}$ basis may be obtained by rotating the $\{\boldsymbol{i}_\theta, \boldsymbol{j}_\theta\}$ basis by $-\theta$ degrees, it follows that

$$\begin{bmatrix} x_\theta \\ y_\theta \end{bmatrix} = \begin{bmatrix} \cos(-\theta) & -\sin(-\theta) \\ \sin(-\theta) & \cos(-\theta) \end{bmatrix} \begin{bmatrix} x \\ y \end{bmatrix}$$

$$= \begin{bmatrix} \cos\theta & \sin\theta \\ -\sin\theta & \cos\theta \end{bmatrix} \begin{bmatrix} x \\ y \end{bmatrix}$$

Writing out the two matrix equations explicitly, we obtain

$$x = (\cos\theta)x_\theta - (\sin\theta)y_\theta$$
$$y = (\sin\theta)x_\theta + (\cos\theta)y_\theta$$
$$x_\theta = (\cos\theta)x + (\sin\theta)y$$
$$y_\theta = (-\sin\theta)x + (\cos\theta)y$$

which are, of course, the standard formulas for rotation of axes in analytic geometry.

Our previous results imply (as the reader may verify by other means) that

$$\begin{bmatrix} \cos\theta & -\sin\theta \\ \sin\theta & \cos\theta \end{bmatrix}^{-1} = \begin{bmatrix} \cos\theta & \sin\theta \\ -\sin\theta & \cos\theta \end{bmatrix}$$

As another application of the ideas of this section, we have

Theorem 1 Let $\boldsymbol{x}_1, \boldsymbol{x}_2, \ldots, \boldsymbol{x}_n$ be vectors in $\boldsymbol{R}^n$ (or $\boldsymbol{C}^n$). Let $A = [\boldsymbol{x}_1, \boldsymbol{x}_2, \ldots, \boldsymbol{x}_n]$ be the matrix whose jth column is the vector $\boldsymbol{x}_j$. Then, the following are equivalent:

(i) The vectors $\boldsymbol{x}_1, \boldsymbol{x}_2, \ldots, \boldsymbol{x}_n$ are linearly independent.
(ii) The matrix A is invertible.
(iii) $\det A \neq 0$.

Proof We already know from chapter 3 that (ii) and (iii) are equivalent. Thus, it will suffice to show that (i) and (ii) are equivalent.

First, suppose the vectors $\boldsymbol{x}_1, \boldsymbol{x}_2, \ldots, \boldsymbol{x}_n$ are linearly independent. By Theorem 2, §4.8, we know that $\boldsymbol{x}_1, \boldsymbol{x}_2, \ldots, \boldsymbol{x}_n$ form a basis for $\boldsymbol{R}^n$. Thus, the matrix A represents a change of coordinates from the standard basis $\{\boldsymbol{e}_1, \boldsymbol{e}_2, \ldots, \boldsymbol{e}_n\}$ to the basis $\{\boldsymbol{x}_1, \boldsymbol{x}_2, \ldots, \boldsymbol{x}_n\}$, and we have seen above that any such matrix is invertible. Thus (i) implies (ii).

Secondly, suppose the vectors $\boldsymbol{x}_1, \boldsymbol{x}_2, \ldots, \boldsymbol{x}_n$ are linearly dependent. Then, one of the vectors, say $\boldsymbol{x}_1$, is a linear combination of the remaining vectors

$$\boldsymbol{x}_1 = \alpha_2\boldsymbol{x}_2 + \alpha_3\boldsymbol{x}_3 + \cdots + \alpha_n\boldsymbol{x}_n$$

Then,

$$\begin{aligned}\det A &= \det[\boldsymbol{x}_1, \boldsymbol{x}_2, \ldots, \boldsymbol{x}_n] \\ &= \det[\boldsymbol{x}_1 - \alpha_2\boldsymbol{x}_2 - \alpha_3\boldsymbol{x}_3 - \alpha_n\boldsymbol{x}_n, \boldsymbol{x}_2, \ldots, \boldsymbol{x}_n] \quad \text{[by (D5)]} \\ &= \det[\boldsymbol{0}, \boldsymbol{x}_2, \ldots, \boldsymbol{x}_n] \\ &= 0 \quad \text{[by (D7)]}\end{aligned}$$

Thus, the matrix A is not invertible. From this we see that (ii) implies (i). ∎

This theorem provides a convenient means for testing whether a given set of vectors forms a basis.

Example 3 Do the vectors

$$\begin{bmatrix}1\\2\\4\\8\end{bmatrix}, \quad \begin{bmatrix}1\\0\\1\\7\end{bmatrix}, \quad \begin{bmatrix}6\\1\\0\\1\end{bmatrix}, \quad \begin{bmatrix}-1\\-1\\0\\0\end{bmatrix}$$

form a basis for $\boldsymbol{R}^4$? By the previous theorem, it is only necessary to calculate the determinant

$$\begin{vmatrix} 1 & 1 & 6 & -1 \\ 2 & 0 & 1 & -1 \\ 4 & 1 & 0 & 0 \\ 8 & 7 & 1 & 0 \end{vmatrix} = \begin{vmatrix} 1 & 0 & 0 & 0 \\ 2 & -2 & -11 & 1 \\ 4 & -3 & -24 & 4 \\ 8 & -1 & -47 & 8 \end{vmatrix} = \begin{vmatrix} 1 & 0 & 0 & 0 \\ 0 & 0 & 0 & 1 \\ -4 & 5 & 20 & 4 \\ -8 & 15 & 41 & 8 \end{vmatrix}$$

$$= \begin{vmatrix} 5 & 20 \\ 15 & 41 \end{vmatrix} = 205 - 300 = -95 \neq 0$$

Thus, the vectors do indeed form a basis.

Example 4 Are the vectors

$$\begin{bmatrix} 1 \\ 3 \\ -1 \end{bmatrix}, \quad \begin{bmatrix} 2 \\ 1 \\ 3 \end{bmatrix}, \quad \begin{bmatrix} 3 \\ 4 \\ 2 \end{bmatrix}$$

linearly independent? We consider the determinant

$$\begin{vmatrix} 1 & 2 & 3 \\ 3 & 1 & 4 \\ -1 & 3 & 2 \end{vmatrix} = \begin{vmatrix} 1 & 0 & 0 \\ 3 & -5 & -5 \\ -1 & 5 & 5 \end{vmatrix} = \begin{vmatrix} -5 & -5 \\ 5 & 5 \end{vmatrix} = 0$$

Thus, the vectors are linearly dependent.

Actually, the method of proof of Theorem 1 yields the stronger result:

Let V be an n-dimensional vector space and let $\mathbf{x}_1, \mathbf{x}_2, \ldots, \mathbf{x}_n$ be a basis for V. Let $\mathbf{x}'_1, \mathbf{x}'_2, \ldots, \mathbf{x}'_n$ be n vectors in V, and suppose $\mathbf{x}'_j = \sum_{j=1}^{n} \alpha_{ij}\mathbf{x}_j$ is the expression for $\mathbf{x}'_j$ relative to the basis $\mathbf{x}_1, \mathbf{x}_2, \ldots, \mathbf{x}_n$.

Then, the following are equivalent statements.

(1) $\mathbf{x}'_1, \mathbf{x}'_2, \mathbf{x}'_3, \ldots, \mathbf{x}'_n$ are linearly independent.

(2)

$$A = \begin{bmatrix} \alpha_{11} & \alpha_{12} & \cdots & \alpha_{1n} \\ \alpha_{21} & \alpha_{22} & \cdots & \alpha_{2n} \\ & & \vdots & \\ \alpha_{n1} & \alpha_{n2} & \cdots & \alpha_{nn} \end{bmatrix} \text{ is invertible.}$$

(3) $\det A \neq 0$.

Notice that the matrix A is obtained by placing the coordinates of the vector $\boldsymbol{x}_j'$ relative to the basis $\boldsymbol{x}_1, \boldsymbol{x}_2, \ldots, \boldsymbol{x}_n$ down the jth column of A.

Example 5 For what values of α do the matrices

$$\begin{bmatrix} \alpha & 1 \\ 1 & 0 \end{bmatrix}, \quad \begin{bmatrix} 1 & \alpha \\ 0 & 1 \end{bmatrix}, \quad \begin{bmatrix} 1 & 0 \\ \alpha & 1 \end{bmatrix}, \quad \begin{bmatrix} 0 & 1 \\ 1 & \alpha \end{bmatrix}$$

form a basis for M_{22}, the space of 2×2 matrices?

Taking as basis for M_{22} the matrices

$$\begin{bmatrix} 1 & 0 \\ 0 & 0 \end{bmatrix}, \quad \begin{bmatrix} 0 & 1 \\ 0 & 0 \end{bmatrix}, \quad \begin{bmatrix} 0 & 0 \\ 1 & 0 \end{bmatrix}, \quad \begin{bmatrix} 0 & 0 \\ 0 & 1 \end{bmatrix}$$

we see that the above matrices form a basis if and only if

$$\begin{vmatrix} \alpha & 1 & 1 & 0 \\ 1 & \alpha & 0 & 1 \\ 1 & 0 & \alpha & 1 \\ 0 & 1 & 1 & \alpha \end{vmatrix} \neq 0$$

Since

$$\begin{vmatrix} \alpha & 1 & 1 & 0 \\ 1 & \alpha & 0 & 1 \\ 1 & 0 & \alpha & 1 \\ 0 & 1 & 1 & \alpha \end{vmatrix} = \begin{vmatrix} \alpha & 1-\alpha^2 & 1 & -\alpha \\ 1 & 0 & 0 & 0 \\ 1 & -\alpha & \alpha & 0 \\ 1 & 1 & 1 & \alpha \end{vmatrix} = - \begin{vmatrix} 1-\alpha^2 & 1 & -\alpha \\ -\alpha & \alpha & 0 \\ 1 & 1 & \alpha \end{vmatrix} = - \begin{vmatrix} 2-\alpha^2 & 1 & -\alpha \\ 0 & \alpha & 0 \\ 2 & 1 & \alpha \end{vmatrix}$$

$$= (-\alpha) \begin{vmatrix} 2-\alpha^2 & -\alpha \\ 2 & \alpha \end{vmatrix} = (-\alpha)(2\alpha - \alpha^3 + 2\alpha) = -\alpha^2(4 - \alpha^2)$$

we see that the matrices

$$\begin{bmatrix} \alpha & 1 \\ 1 & 0 \end{bmatrix}, \quad \begin{bmatrix} 1 & \alpha \\ 0 & 1 \end{bmatrix}, \quad \begin{bmatrix} 1 & 0 \\ \alpha & 1 \end{bmatrix}, \quad \begin{bmatrix} 0 & 1 \\ 1 & \alpha \end{bmatrix}$$

form a basis for M_{22} if and only if $\alpha \neq 0, 2, -2$.

Theorem 2 Let

$$\begin{aligned} a_{11}x_1 + a_{12}x_2 + \cdots + a_{1n}x_n &= 0 \\ a_{21}x_1 + a_{22}x_2 + \cdots + a_{2n}x_n &= 0 \\ &\vdots \\ a_{n1}x_1 + a_{n2}x_2 + \cdots + a_{nn}x_n &= 0 \end{aligned}$$

be a system of n homogeneous linear equations in n variables. The system admits a nontrivial solution, i.e., a solution in which not all the x_i's are 0, if and only if

$$\begin{vmatrix} a_{11} & a_{12} & \cdots & a_{1n} \\ a_{21} & a_{22} & \cdots & a_{2n} \\ & \vdots & & \\ a_{n1} & a_{n2} & \cdots & a_{nn} \end{vmatrix} = 0$$

Proof We let

$$A = \begin{bmatrix} a_{11} & a_{12} & \cdots & a_{1n} \\ a_{21} & a_{22} & \cdots & a_{2n} \\ & \vdots & & \\ a_{n1} & a_{n2} & \cdots & a_{nn} \end{bmatrix} = [A_1, A_2, \ldots, A_n]$$

where A_j is the jth column of the matrix A. If $\det A = 0$, by Theorem 1, we know that the columns of the matrix A are linearly dependent. Thus we have $\alpha_1 A_1 + \alpha_2 A_2 + \cdots + \alpha_n A_n = 0$, for scalars $\alpha_1, \alpha_2, \ldots, \alpha_n$ not all zero. But

$$\begin{bmatrix} a_{11} & a_{12} & \cdots & a_{1n} \\ a_{21} & a_{22} & \cdots & a_{2n} \\ & \vdots & & \\ a_{n1} & a_{n2} & \cdots & a_{nn} \end{bmatrix} \begin{bmatrix} \alpha_1 \\ \alpha_2 \\ \vdots \\ \alpha_n \end{bmatrix} = \begin{bmatrix} \alpha_1 a_{11} + \alpha_2 a_{12} + \cdots + \alpha_n a_{1n} \\ \alpha_1 a_{21} + \alpha_2 a_{22} + \cdots + \alpha_n a_{2n} \\ \vdots \\ \alpha_1 a_{n1} + \alpha_2 a_{n2} + \cdots + \alpha_n a_{nn} \end{bmatrix}$$

$$= [\alpha_1 A_1 + \alpha_2 A_2 + \cdots + \alpha_n A_n]$$

$$= \begin{bmatrix} 0 \\ 0 \\ \vdots \\ 0 \end{bmatrix}$$

Thus, $x_1 = \alpha_1, x_2 = \alpha_2, \ldots, x_n = \alpha_n$ is a nontrivial solution to the system.

If, on the other hand, $\det A \neq 0$, then A^{-1} exists. So if $A\boldsymbol{x} = 0$, then $A^{-1}(A\boldsymbol{x}) = \boldsymbol{x} = \mathbf{0}$. Thus, no nontrivial solution exists. ∎

Example 6 Let (x_1, y_1), (x_2, y_2), and (x_3, y_3) be three points in the plane. These points are collinear if and only if

$$\Delta = \begin{vmatrix} 1 & x_1 & y_1 \\ 1 & x_2 & y_2 \\ 1 & x_3 & y_3 \end{vmatrix} = 0$$

For, if the points are collinear, they all satisfy an equation

$$Ax + By + C = 0$$

where A, B, and C are not all zero.

Thus, it follows that

$$\begin{bmatrix} 1 & x_1 & y_1 \\ 1 & x_2 & y_2 \\ 1 & x_3 & y_3 \end{bmatrix} \begin{bmatrix} C \\ A \\ B \end{bmatrix} = \begin{bmatrix} 0 \\ 0 \\ 0 \end{bmatrix}$$

and so

$$\begin{vmatrix} 1 & x_1 & y_1 \\ 1 & x_2 & y_2 \\ 1 & x_3 & y_3 \end{vmatrix} = 0$$

If, on the other hand $\Delta = 0$, there are numbers C, A, B, not all zero, such that

$$\begin{bmatrix} 1 & x_1 & y_1 \\ 1 & x_2 & y_2 \\ 1 & x_3 & y_3 \end{bmatrix} \begin{bmatrix} C \\ A \\ B \end{bmatrix} = \begin{bmatrix} 0 \\ 0 \\ 0 \end{bmatrix}$$

Thus, the points (x_1, y_1), (x_2, y_2), and (x_3, y_3) all lie on the line $Ax + By + C = 0$.

EXERCISES

1. Determine the coordinates of the vector

$$\begin{bmatrix} x \\ y \end{bmatrix}$$

in $\boldsymbol{R}^2$, relative to each of the following bases for $\boldsymbol{R}^2$.

(a) $\begin{bmatrix} 1 \\ 0 \end{bmatrix}, \quad \begin{bmatrix} 1 \\ 1 \end{bmatrix}$ (b) $\begin{bmatrix} 1 \\ 1 \end{bmatrix}, \quad \begin{bmatrix} 1 \\ -1 \end{bmatrix}$

(c) $\begin{bmatrix} 0 \\ 1 \end{bmatrix}, \quad \begin{bmatrix} -1 \\ -1 \end{bmatrix}$ (d) $\begin{bmatrix} 2 \\ 1 \end{bmatrix}, \quad \begin{bmatrix} 5 \\ 3 \end{bmatrix}$

2. Determine the coordinates of the vector

$$\begin{bmatrix} x \\ y \\ z \end{bmatrix}$$

relative to each of the following bases for $\boldsymbol{R}^3$.

(a) $\begin{bmatrix} 1 \\ 0 \\ 0 \end{bmatrix}, \quad \begin{bmatrix} 1 \\ 2 \\ 1 \end{bmatrix}, \quad \begin{bmatrix} 1 \\ 5 \\ 3 \end{bmatrix}$ (b) $\begin{bmatrix} 1 \\ 1 \\ 0 \end{bmatrix}, \quad \begin{bmatrix} 0 \\ 1 \\ 1 \end{bmatrix}, \quad \begin{bmatrix} 1 \\ 0 \\ 1 \end{bmatrix}$

(c) $\begin{bmatrix} 1 \\ 3 \\ 0 \end{bmatrix}, \quad \begin{bmatrix} 2 \\ 1 \\ 3 \end{bmatrix}, \quad \begin{bmatrix} -1 \\ 1 \\ 2 \end{bmatrix}$ (d) $\begin{bmatrix} 1 \\ 1 \\ -1 \end{bmatrix}, \quad \begin{bmatrix} 1 \\ -1 \\ 1 \end{bmatrix}, \quad \begin{bmatrix} -1 \\ 1 \\ 1 \end{bmatrix}$

3. Determine which of the following sets of vectors are independent in the indicated vector spaces.

(a) $\begin{bmatrix} -1 & 1 \\ 1 & 1 \end{bmatrix}, \quad \begin{bmatrix} 1 & -1 \\ 1 & 1 \end{bmatrix}, \quad \begin{bmatrix} 1 & 1 \\ -1 & 1 \end{bmatrix}, \quad \begin{bmatrix} 1 & 1 \\ 1 & -1 \end{bmatrix}$
in the space of 2×2 matrices.

(b) $\begin{bmatrix} 1 & -1 \\ 0 & 1 \end{bmatrix}, \quad \begin{bmatrix} -1 & 1 \\ 0 & 1 \end{bmatrix}, \quad \begin{bmatrix} 1 & 1 \\ 0 & -1 \end{bmatrix}$
in the space of 2×2 upper triangular matrices.

(c) $1 + x + x^2 + x^3, \quad 1 - x + x^2 + x^3, \quad 1 + x - x^2 + x^3,$
$1 + x + x^2 - x^3,$ in P_3.

(d) $\begin{bmatrix} 1 & 1 \\ -1 & -1 \end{bmatrix}, \quad \begin{bmatrix} 1 & -1 \\ 1 & -1 \end{bmatrix}, \quad \begin{bmatrix} -1 & 1 \\ 1 & -1 \end{bmatrix}, \quad \begin{bmatrix} 1 & -1 \\ 0 & 0 \end{bmatrix}$
in the space of 2×2 matrices.

(e) $\begin{bmatrix} 1 & 1 \\ 0 & 1 \\ 0 & 0 \end{bmatrix}, \quad \begin{bmatrix} 1 & 1 \\ 1 & 0 \\ 0 & 0 \end{bmatrix}, \quad \begin{bmatrix} 0 & 1 \\ 1 & 1 \\ 0 & 0 \end{bmatrix}, \quad \begin{bmatrix} 1 & 0 \\ 1 & 1 \\ 0 & 0 \end{bmatrix}$
in the space of 3×2 matrices.

(f) $\begin{bmatrix}1 & 1\\ 1 & -1\end{bmatrix}, \quad \begin{bmatrix}1 & 0\\ 0 & -1\end{bmatrix}, \quad \begin{bmatrix}3 & 2\\ 2 & -3\end{bmatrix}$
in the space of 2×2 symmetric matrices.

4. If

$$\begin{bmatrix}a\\ b\end{bmatrix} \neq 0$$

is a vector in $\boldsymbol{R}^2$, give a formula for another vector

$$\begin{bmatrix}c\\ d\end{bmatrix}$$

in $\boldsymbol{R}^2$, such that

$$\begin{bmatrix}a\\ b\end{bmatrix}, \quad \begin{bmatrix}c\\ d\end{bmatrix}$$

is a basis for $\boldsymbol{R}^2$.

5. For what values of λ do the following sets of vectors form a basis for $\boldsymbol{R}^3$?

(a) $\begin{bmatrix}1\\ \lambda\\ 0\end{bmatrix}, \quad \begin{bmatrix}\lambda\\ 1\\ \lambda\end{bmatrix}, \quad \begin{bmatrix}0\\ \lambda\\ 1\end{bmatrix}$ (b) $\begin{bmatrix}1\\ 0\\ \lambda\end{bmatrix}, \quad \begin{bmatrix}0\\ 1\\ 0\end{bmatrix}, \quad \begin{bmatrix}\lambda\\ 0\\ 1\end{bmatrix}$

(c) $\begin{bmatrix}1\\ 0\\ 0\end{bmatrix}, \quad \begin{bmatrix}\lambda\\ 1\\ 0\end{bmatrix}, \quad \begin{bmatrix}\lambda^2\\ \lambda\\ 1\end{bmatrix}$ (d) $\begin{bmatrix}0\\ \lambda\\ \lambda\end{bmatrix}, \quad \begin{bmatrix}\lambda\\ 0\\ \lambda\end{bmatrix}, \quad \begin{bmatrix}\lambda\\ \lambda\\ 0\end{bmatrix}$

6. For what real values of λ do the following systems of equations admit nontrivial solutions?

(a)
$$\begin{aligned} \lambda x + y + z &= 0\\ x + \lambda y + z &= 0\\ x + y + \lambda z &= 0 \end{aligned}$$

(b)
$$\begin{aligned} x + y + z &= 0\\ \lambda x + y + z &= 0\\ \lambda^2 x + y + z &= 0 \end{aligned}$$

(c)
$$\begin{aligned} x + \lambda y + z &= 0\\ \lambda x + y + z &= 0\\ \lambda x + \lambda y + z &= 0 \end{aligned}$$

7. Let (x_1, y_1), (x_2, y_2), (x_3, y_3), (x_4, y_4) be the coordinates of four points in the plane. Show that the points lie on a line or circle if and only if the determinant

$$\begin{vmatrix} x_1^2+y_1^2 & x_1 & y_1 & 1 \\ x_2^2+y_2^2 & x_2 & y_2 & 1 \\ x_3^2+y_3^2 & x_3 & y_3 & 1 \\ x_4^2+y_4^2 & x_4 & y_4 & 1 \end{vmatrix}$$

is zero.

8. Let

$$\boldsymbol{a} = \begin{bmatrix} \alpha_1 \\ \alpha_2 \\ \alpha_3 \end{bmatrix} \quad \text{and} \quad \boldsymbol{b} = \begin{bmatrix} \beta_1 \\ \beta_2 \\ \beta_3 \end{bmatrix}$$

be two linearly independent vectors in $\boldsymbol{R}^3$. Show that the vector

$$\begin{bmatrix} x \\ y \\ z \end{bmatrix}$$

lies in the subspace spanned by the vectors $\boldsymbol{a}$ and $\boldsymbol{b}$ if and only if

$$\begin{vmatrix} \alpha_1 & \beta_1 & x \\ \alpha_2 & \beta_2 & y \\ \alpha_3 & \beta_3 & z \end{vmatrix} = 0$$

9. In P_3 determine the matrices associated with the change from the basis 1, x, x^2, x^3 to the basis 1, $1 + x$, $(1 + x)^2$, $(1 + x)^3$.

10. If $\boldsymbol{x}_1, \boldsymbol{x}_2, \ldots, \boldsymbol{x}_n$ form a basis for the vector space V, show that the following sets of vectors also form a basis for V.
   (a) $\boldsymbol{x}_1, \boldsymbol{x}_1 + \boldsymbol{x}_2, \boldsymbol{x}_1 + \boldsymbol{x}_2 + \boldsymbol{x}_3, \ldots, \boldsymbol{x}_1 + \boldsymbol{x}_2 + \boldsymbol{x}_3 + \cdots + \boldsymbol{x}_n$
   (b) $\boldsymbol{x}_1 + \boldsymbol{x}_2 + \cdots + \boldsymbol{x}_n, \boldsymbol{x}_1 - \boldsymbol{x}_2 + \boldsymbol{x}_3 + \cdots + \boldsymbol{x}_n,$
   $\boldsymbol{x}_1 + \boldsymbol{x}_2 - \boldsymbol{x}_3 + \cdots + \boldsymbol{x}_n, \ldots, \boldsymbol{x}_1 + \boldsymbol{x}_2 + \boldsymbol{x}_3 + \cdots + (-\boldsymbol{x}_n)$
   (c) $\boldsymbol{x}_1, \boldsymbol{x}_1 + \boldsymbol{x}_2, \boldsymbol{x}_1 + \boldsymbol{x}_3, \ldots, \boldsymbol{x}_1 + \boldsymbol{x}_n$

11. Show that any set of three distinct vectors whose endpoints lie on the parametrized parabolic curve

$$x(t) = 1, \qquad y(t) = t, \qquad z(t) = t^2$$

form a basis for $\boldsymbol{R}^3$.

12. Let

$$\begin{bmatrix} a_1 \\ a_2 \\ a_3 \\ \vdots \\ a_n \end{bmatrix}, \quad \begin{bmatrix} b_1 \\ b_2 \\ b_3 \\ \vdots \\ b_n \end{bmatrix}, \quad \begin{bmatrix} c_1 \\ c_2 \\ c_3 \\ \vdots \\ c_n \end{bmatrix}$$

be three vectors in $\boldsymbol{R}^n$. If

$$\begin{vmatrix} a_1 & b_1 & c_1 \\ a_2 & b_2 & c_2 \\ a_3 & b_3 & c_3 \end{vmatrix} \neq 0$$

show that the three vectors are linearly independent.

13. If p_0, p_1, p_2 are independent polynomials in P_2 and x_0, x_1, and x_2 are distinct real numbers, show that the vectors

$$\begin{bmatrix} p_0(x_0) \\ p_1(x_0) \\ p_2(x_0) \end{bmatrix}, \quad \begin{bmatrix} p_0(x_1) \\ p_1(x_1) \\ p_2(x_1) \end{bmatrix}, \quad \begin{bmatrix} p_0(x_2) \\ p_1(x_2) \\ p_2(x_2) \end{bmatrix}$$

are linearly independent.

14. If $\boldsymbol{x}_1, \boldsymbol{x}_2, \ldots, \boldsymbol{x}_n$ is a basis for a vector space V, show that the vectors $\boldsymbol{x}_1 - \boldsymbol{x}, \boldsymbol{x}_2 - \boldsymbol{x}, \ldots, \boldsymbol{x}_n - \boldsymbol{x}$ form a basis for V if and only if $\boldsymbol{x}$ cannot be expressed in the form

$$\boldsymbol{x} = \alpha_1\boldsymbol{x}_1 + \alpha_2\boldsymbol{x}_2 + \cdots + \alpha_n\boldsymbol{x}_n$$

with

$$\alpha_1 + \alpha_2 + \cdots + \alpha_n = 1$$

15. If $\boldsymbol{x}_1, \boldsymbol{x}_2, \ldots, \boldsymbol{x}_m$ are vectors in $\boldsymbol{R}^n$, they may also be regarded as vectors in $\boldsymbol{C}^n$. Show that they are linearly independent over $\boldsymbol{R}$ if and only if they are linearly independent over $\boldsymbol{C}$.

16. If

$$a_1x + b_1y + c_1 = 0$$
$$a_2x + b_2y + c_2 = 0$$
$$a_3x + b_3y + c_3 = 0$$

are the equations of three lines in the plane, show that if

$$\begin{vmatrix} a_1 & b_1 & c_1 \\ a_2 & b_2 & c_2 \\ a_3 & b_3 & c_3 \end{vmatrix} = 0$$

the lines intersect at a point or are parallel. Prove the converse.

17. If $\boldsymbol{x}_1, \boldsymbol{x}_2, \ldots, \boldsymbol{x}_n$ is a basis for a vector space V, show that $\boldsymbol{x}_1 + \boldsymbol{x}_2$, $\boldsymbol{x}_2 + \boldsymbol{x}_3$, $\boldsymbol{x}_3 + \boldsymbol{x}_4, \ldots, \boldsymbol{x}_{n-1} + \boldsymbol{x}_n$, $\boldsymbol{x}_n + \boldsymbol{x}_1$ is a basis for V if and only if V is of odd dimension.

18. Give a determinant criterion which guarantees that five points in the plane lie on some line or conic section. [Hint: See exercise 7 above.]

19. Show that the necessary and sufficient condition that two straight lines

$$x = a_1t + b_1, \qquad y = a_2t + b_2, \qquad z = a_3t + b_3$$

$$x' = c_1t' + d_1, \qquad y = c_2t' + d_2, \qquad z = c_3t' + d_3$$

with t, t' parameters, either intersect or are parallel is

$$\begin{vmatrix} a_1 & c_1 & b_1 - d_1 \\ a_2 & c_2 & b_2 - d_2 \\ a_3 & c_3 & b_3 - d_3 \end{vmatrix} = 0$$

5

LINEAR TRANSFORMATIONS

1 DEFINITION OF LINEAR TRANSFORMATION

In studying vector spaces it is of interest to consider those functions from one vector space to another which preserve the algebraic structure. In the following definition we make precise what is meant by preserving the algebraic structure.

Definition Let V and W be vector spaces and T be a function from V into W with the properties

(i) $T(\boldsymbol{x} + \boldsymbol{y}) = T(\boldsymbol{x}) + T(\boldsymbol{y})$, for all vectors $\boldsymbol{x}$ and $\boldsymbol{y}$ in V.

(ii) $T(\alpha\boldsymbol{x}) = \alpha T(\boldsymbol{x})$, for all vectors $\boldsymbol{x}$ in V and all scalars α.

Then, T is said to be a **linear transformation** from the vector space V to the vector space W.

Often we write $T : V \rightarrow W$ to indicate that T is a linear transformation from the vector space V into the vector space W.

As a first example let us consider the function T_θ from $\boldsymbol{R}^2$ to $\boldsymbol{R}^2$, which assigns to every vector $\boldsymbol{v}$ in the plane the vector $T_\theta(\boldsymbol{v})$ obtained by rotating the vector $\boldsymbol{v}$ θ degrees in the counterclockwise direction. (See Figure 1-1.)

Let r be the length of the vector $\boldsymbol{v}$ and ψ be the angle the vector $\boldsymbol{v}$ makes with the x axis. If the vector $\boldsymbol{v}$ ends at the point (x, y) in the plane, we have $\boldsymbol{v} = x\boldsymbol{i} + y\boldsymbol{j}$.

By trigonometry, it follows that $x = r \cos \psi$ and $y = r \sin \psi$. Thus, $\boldsymbol{v} = (r \cos \psi)\boldsymbol{i} + (r \sin \psi)\boldsymbol{j}$.

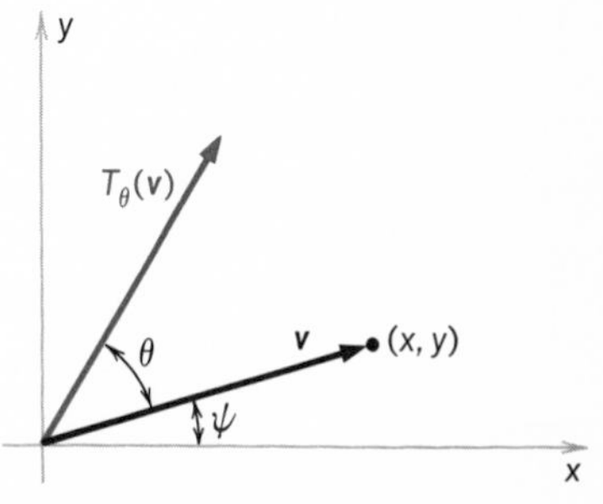

FIGURE 1-1

Since the vector $T_\theta(\boldsymbol{v})$ is obtained by rotating $\boldsymbol{v}$ by θ degrees, $T_\theta(\boldsymbol{v})$ has the same length as $\boldsymbol{v}$, namely r, and makes an angle of $\theta + \psi$ degrees with the x axis. Thus,

$$\begin{aligned} T_\theta(\boldsymbol{v}) &= (r\cos(\theta+\psi))\boldsymbol{i} + (r\sin(\theta+\psi))\boldsymbol{j} \\ &= (r\cos\theta\cos\psi - r\sin\theta\sin\psi)\boldsymbol{i} \\ &\quad + (r\cos\theta\sin\psi + r\sin\theta\cos\psi)\boldsymbol{j} \\ &= (x\cos\theta - y\sin\theta)\boldsymbol{i} + (y\cos\theta + x\sin\theta)\boldsymbol{j} \end{aligned}$$

In terms of matrices and column vectors,

$$T_\theta\begin{bmatrix} x \\ y \end{bmatrix} = \begin{bmatrix} \cos\theta & -\sin\theta \\ \sin\theta & \cos\theta \end{bmatrix}\begin{bmatrix} x \\ y \end{bmatrix}$$

If A_θ denotes the matrix

$$A_\theta = \begin{bmatrix} \cos\theta & -\sin\theta \\ \sin\theta & \cos\theta \end{bmatrix}$$

it follows that $T_\theta(\boldsymbol{v}) = A_\theta\boldsymbol{v}$.

From the last equation we can show that T_θ is linear. If $\boldsymbol{v}_1$ and $\boldsymbol{v}_2$ are vectors in $\boldsymbol{R}^2$, we have

$$\begin{aligned} T_\theta(\boldsymbol{v}_1 + \boldsymbol{v}_2) &= A_\theta(\boldsymbol{v}_1 + \boldsymbol{v}_2) \\ &= A_\theta\boldsymbol{v}_1 + A_\theta\boldsymbol{v}_2 \\ &= T_\theta(\boldsymbol{v}_1) + T_\theta(\boldsymbol{v}_2) \end{aligned}$$

and if α is a scalar and $\boldsymbol{v}$ is a vector in $\boldsymbol{R}^2$,

$$\begin{aligned} T_\theta(\alpha\boldsymbol{v}) &= A_\theta(\alpha\boldsymbol{v}) \\ &= \alpha A_\theta\boldsymbol{v} \\ &= \alpha T_\theta(\boldsymbol{v}) \end{aligned}$$

Thus, having verified conditions (i) and (ii) in the definition of a linear transformation, it follows that T_θ is a linear transformation from the vector space $\boldsymbol{R}^2$ into itself. In this case the linear transformation may be represented by multiplying the vector $\boldsymbol{v}$ by a suitable matrix. Shortly, we shall see that any linear transformation of $\boldsymbol{R}^2$ into itself, indeed any linear transformation from $\boldsymbol{R}^n$ to $\boldsymbol{R}^m$, may be obtained in this straightforward manner.

The verification that the function T_θ is a linear transformation was of a mixed algebraic and geometric nature. That is, the geometric definition of T_θ was used to derive an algebraic expression representing T_θ. This algebraic expression was then used to demonstrate the linearity of the function T_θ. It is possible to reach the same conclusion from purely geometric considerations. Observe that if A and B are points in the plane and C is the endpoint of the parallelogram with adjacent sides OA and OB, and if we denote by A', B', and C' the points obtained by rotating A, B, and C, respectively, by θ degrees, then C' is the endpoint of the diagonal of the parallelogram with adjacent sides OA' and OB'. (See Figure 1-2.) Thus, if $\boldsymbol{u}$ is the vector ending at the point A and $\boldsymbol{v}$ is the vector ending at the point B, we see that $\boldsymbol{u} + \boldsymbol{v}$ ends at the point C, $T_\theta(\boldsymbol{u})$ ends at A', $T_\theta(\boldsymbol{v})$ ends at B', and $T_\theta(\boldsymbol{u} + \boldsymbol{v})$ ends at C'. Since the vector $T_\theta(\boldsymbol{u}) + T_\theta(\boldsymbol{v})$ also ends at C', we obtain $T_\theta(\boldsymbol{u} + \boldsymbol{v}) = T_\theta(\boldsymbol{u}) + T_\theta(\boldsymbol{v})$. That $T_\theta(\alpha\boldsymbol{u}) = \alpha T_\theta(\boldsymbol{u})$ may be demonstrated in a similar manner.

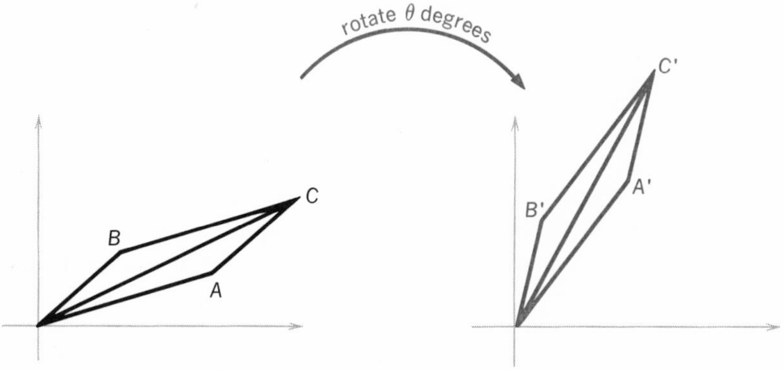

FIGURE 1-2

In our first proof that the function T_θ is linear, we relied on the fact that T_θ could be obtained by multiplication by a suitable matrix. In the next example we show that any transformation obtained in this manner, i.e., by matrix multiplication, is a linear transformation.

Example 1 Let A be an $m \times n$ matrix with real entries. Consider the function T_A from $\boldsymbol{R}^n$ into $\boldsymbol{R}^m$ defined by

$$T_A(\boldsymbol{x}) = A\boldsymbol{x}$$

where $\boldsymbol{x}$ is an n-vector

If $\boldsymbol{x}$ is an n-vector, the product $A\boldsymbol{x}$ is, of course, defined and is an m-vector. Hence, T_A defines a function from $\boldsymbol{R}^n$ into $\boldsymbol{R}^m$.

To verify linearity, observe that if $\boldsymbol{x}$ and $\boldsymbol{y}$ belong to $\boldsymbol{R}^n$ and α is a scalar, we have

$$\begin{aligned} T_A(\boldsymbol{x}+\boldsymbol{y}) &= A(\boldsymbol{x}+\boldsymbol{y}) && \text{(by definition of } T_A\text{)} \\ &= A\boldsymbol{x} + A\boldsymbol{y} && \text{(by the distributive law of matrix multiplication)} \\ &= T_A(\boldsymbol{x}) + T_A(\boldsymbol{y}) && \text{(by definition of } T_A\text{)} \end{aligned}$$

and

$$\begin{aligned} T_A(\alpha\boldsymbol{x}) &= A(\alpha\boldsymbol{x}) \\ &= \alpha A\boldsymbol{x} \\ &= \alpha T_A(\boldsymbol{x}) \end{aligned}$$

For example, the function T, from $\boldsymbol{R}^3$ to $\boldsymbol{R}^2$, defined by

$$T\left(\begin{bmatrix} x \\ y \\ z \end{bmatrix}\right) = \begin{bmatrix} 1 & -1 & 1 \\ 0 & 1 & 1 \end{bmatrix}\begin{bmatrix} x \\ y \\ z \end{bmatrix} = \begin{bmatrix} x - y + z \\ y + z \end{bmatrix}$$

is linear.

Example 2 If A is an $m \times n$ matrix with complex entries, the function T_A, defined by $T_A(\boldsymbol{x}) = A\boldsymbol{x}$, for $\boldsymbol{x}$ belonging to $\boldsymbol{C}^n$, defines a linear transformation from $\boldsymbol{C}^n$ to $\boldsymbol{C}^m$.

Example 3 Let M_{mn} denote the space of $m \times n$ matrices with real entries and M_{nm} the space of $n \times m$ matrices with real entries. Consider the function T from M_{mn} to M_{nm} defined by $T(A) = A^T$, where A is an $m \times n$ matrix.

Since the transpose of an $m \times n$ matrix is an $n \times m$ matrix, T is a well-defined function from M_{mn} to M_{nm}. If A and B are $m \times n$ matrices and α is a scalar,

$$\begin{aligned} T(A+B) &= (A+B)^T && \text{(by definition of } T\text{)} \\ &= A^T + B^T && (\S 2.7) \\ &= T(A) + T(B) && \text{(by definition of } T\text{)} \end{aligned}$$

and

$$\begin{aligned} T(\alpha A) &= (\alpha A)^T && \text{(by definition of } T\text{)} \\ &= \alpha A^T && (\S 2.7) \\ &= \alpha T(A) && \text{(by definition of } T\text{)} \end{aligned}$$

It follows that T is a linear transformation.

Example 4 Let D be the function from P_n into P_n defined by $D(f) = f'$, where f' denotes the derivative of f. Since the derivative of a polynomial of degree less than or equal to n is a polynomial of degree less than or equal to n, D is a well-defined function from P_n into P_n.

To demonstrate linearity, observe that

$$\begin{aligned} D(f + g) &= (f + g)' && \text{(by definition of } D) \\ &= f' + g' && \text{(by Calculus)} \\ &= D(f) + D(g) && \text{(by definition of } D) \end{aligned}$$

and

$$\begin{aligned} D(\alpha f) &= (\alpha f)' && \text{(by definition of } D) \\ &= \alpha f' && \text{(by Calculus)} \\ &= \alpha D(f) && \text{(by definition of } D) \end{aligned}$$

Linear transformations are important because of the wide variety of situations in which functions satisfying the linearity conditions are encountered.

If $\boldsymbol{x}$ is a vector in a vector space V and T is a linear transformation from V into a vector space W, and if $T(\boldsymbol{x}) = \boldsymbol{y}$, we say that the vector $\boldsymbol{y}$ is the **image** under T of the vector $\boldsymbol{x}$. We may also say that $\boldsymbol{x}$ goes into $\boldsymbol{y}$ under T, or that $\boldsymbol{x}$ is sent into $\boldsymbol{y}$ by T. Thus, for example, the image of $1 + x + 3x^2$ under the differentiation operator D of Example 4 is $1 + 6x$.

In many cases a linear transformation of a vector space into itself is called a **linear operator.**

Let $T : \boldsymbol{R}^2 \to \boldsymbol{R}^2$ be a linear operator on $\boldsymbol{R}^2$. Using conditions (i) and (ii) in the definition of a linear transformation we can show how to determine the image of an arbitrary vector in $\boldsymbol{R}^2$ as soon as the images under T of the basis vectors $\boldsymbol{i}$ and $\boldsymbol{j}$ are known.

Suppose

$$T(\boldsymbol{i}) = \boldsymbol{v} \text{ and } T(\boldsymbol{j}) = \boldsymbol{w}$$

Then

$$\begin{aligned} T(x\boldsymbol{i} + y\boldsymbol{j}) &= T(x\boldsymbol{i}) + T(y\boldsymbol{j}) \\ &= xT(\boldsymbol{i}) + yT(\boldsymbol{j}) \\ &= x\boldsymbol{v} + y\boldsymbol{w} \end{aligned}$$

(See Figure 1-3.)

Thus, the T image of the vector $x\boldsymbol{i} + y\boldsymbol{j}$ is completely determined by $\boldsymbol{v}$ and $\boldsymbol{w}$, the images of the basis vectors $\boldsymbol{i}$ and $\boldsymbol{j}$.

As another illustration of the geometric nature of linear transformations we show that any linear transformation carries a line into a point or line.

To see this, let $\boldsymbol{r}(t) = \boldsymbol{a} + t\boldsymbol{b}$ be the parametric equation of a line, where $\boldsymbol{a}$ and $\boldsymbol{b}$ are vectors in $\boldsymbol{R}^3$ and t is a real parameter. If T is a linear

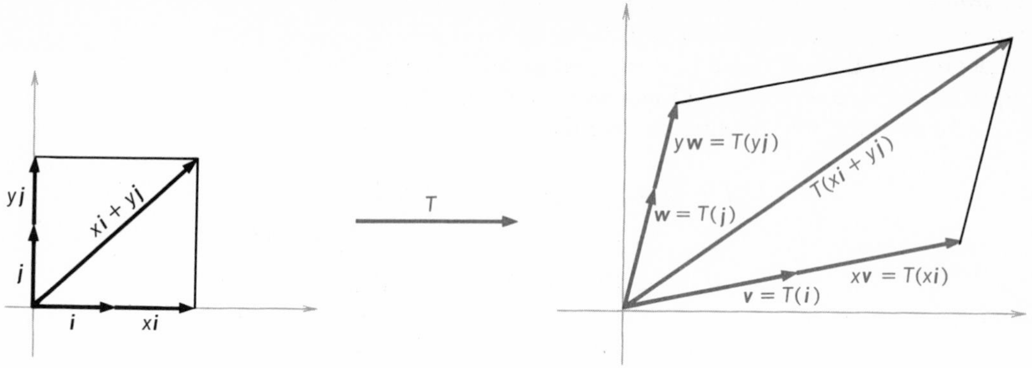

FIGURE 1-3

transformation on $\boldsymbol{R}^3$, let $\boldsymbol{s}(t) = T(\boldsymbol{r}(t))$. Then, using the linearity of T,

$$\begin{aligned} \boldsymbol{s}(t) &= T(\boldsymbol{r}(t)) \\ &= T(\boldsymbol{a} + t\boldsymbol{b}) \\ &= T(\boldsymbol{a}) + T(t\boldsymbol{b}) \\ &= T(\boldsymbol{a}) + tT(\boldsymbol{b}) \end{aligned}$$

If $T(\boldsymbol{b}) \neq \boldsymbol{0}$, $\boldsymbol{s}(t)$ is the equation of the line through the point $T(\boldsymbol{a})$ in the direction of the vector $T(\boldsymbol{b})$.

If $T(\boldsymbol{b}) = \boldsymbol{0}$, $\boldsymbol{s}(t) = T(\boldsymbol{a})$ is constant, and so the line $\boldsymbol{r}(t)$ is carried into a point by T.

The following example illustrates a case where both situations may occur.

Consider the linear transformation of $\boldsymbol{R}^2$ into itself associated with the matrix

$$\begin{bmatrix} 1 & 0 \\ 0 & 0 \end{bmatrix}$$

that is,

$$T\left(\begin{bmatrix} x \\ y \end{bmatrix}\right) = \begin{bmatrix} 1 & 0 \\ 0 & 0 \end{bmatrix}\begin{bmatrix} x \\ y \end{bmatrix} = \begin{bmatrix} x \\ 0 \end{bmatrix}$$

The parametric form of the line $x + y = 1$ is

$$\boldsymbol{r}(t) = t\boldsymbol{i} + (1 - t)\boldsymbol{j}$$

Its image is

$$\boldsymbol{s}(t) = T(t\boldsymbol{i} + (1 - t)\boldsymbol{j}) = t\boldsymbol{i}$$

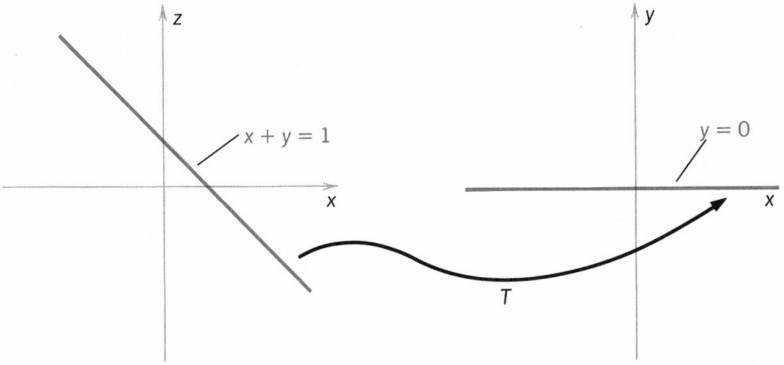

FIGURE 1-4

which is the parametric form of the line $y = 0$. (See Figure 1-4.)

On the other hand, consider the line $x = 1$. Its parametric form is

$$\boldsymbol{r}(t) = \boldsymbol{i} + t\boldsymbol{j}$$

which is carried into

$$\boldsymbol{s}(t) = T(\boldsymbol{r}(t)) = T(\boldsymbol{i} + t\boldsymbol{j}) = \boldsymbol{i}$$

which is the point $(1, 0)$. (See Figure 1-5.)

The reader may verify that the linear transformation T carries each line parallel to the y axis into a single point, while all other lines are carried into the line $y = 0$.

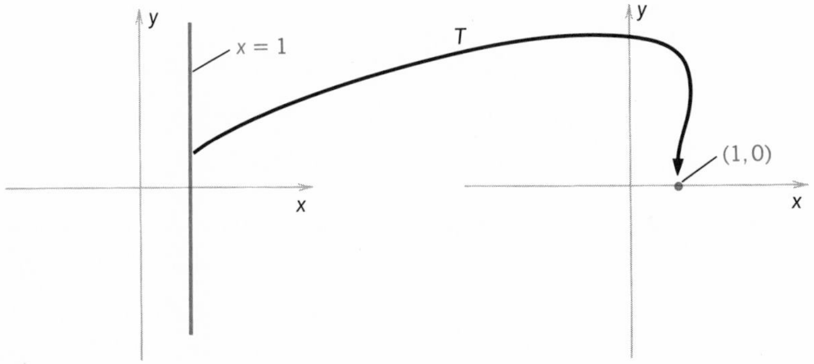

FIGURE 1-5

EXERCISES

1. Determine which of the following functions from $\boldsymbol{R}^2$ to $\boldsymbol{R}^2$ are linear.

   (a) $T\left(\begin{bmatrix} x \\ y \end{bmatrix}\right) = \begin{bmatrix} y \\ x \end{bmatrix}$ (b) $T\left(\begin{bmatrix} x \\ y \end{bmatrix}\right) = \begin{bmatrix} x+1 \\ y \end{bmatrix}$

   (c) $T\left(\begin{bmatrix} x \\ y \end{bmatrix}\right) = \begin{bmatrix} x^2 \\ y \end{bmatrix}$ (d) $T\left(\begin{bmatrix} x \\ y \end{bmatrix}\right) = \begin{bmatrix} x+y \\ x-y \end{bmatrix}$

   (e) $T\left(\begin{bmatrix} x \\ y \end{bmatrix}\right) = \begin{bmatrix} 2x-3y \\ y \end{bmatrix}$ (f) $T\left(\begin{bmatrix} x \\ y \end{bmatrix}\right) = \begin{bmatrix} x \\ (x^2+y^2)^{1/2} \end{bmatrix}$

2. Let R be the function from $\boldsymbol{R}^2$ to $\boldsymbol{R}^2$ which reflects points about the x axis.
   (a) Show that R is linear.
   (b) Show that

$$R\left(\begin{bmatrix} x \\ y \end{bmatrix}\right) = \begin{bmatrix} x \\ -y \end{bmatrix}$$

3. Show that the following functions from $\boldsymbol{R}^3$ to $\boldsymbol{R}^3$ are linear.

   (a)
$$T\left(\begin{bmatrix} x \\ y \\ z \end{bmatrix}\right) = \begin{bmatrix} x+3y+2z \\ x+y \\ 2x+y+z \end{bmatrix}$$

   (b) $$T\left(\begin{bmatrix} x \\ y \\ z \end{bmatrix}\right) = \begin{bmatrix} x-y+z \\ x+z \\ x+y \end{bmatrix}$$

   (c) $$T\left(\begin{bmatrix} x \\ y \\ z \end{bmatrix}\right) = \begin{bmatrix} 4x+3y \\ x-y \\ x+2y+z \end{bmatrix}$$

4. Let R_θ be the function which assigns to each vector $\boldsymbol{v}$ the vector $R_\theta(\boldsymbol{v})$, obtained by rotating the vector $\boldsymbol{v}$ through an angle of θ degrees about the z axis. The endpoint of $\boldsymbol{v}$ is to remain in the same plane perpendicular to the z axis throughout the rotation.
   (a) Show that R_θ is linear and that

$$R_\theta\left(\begin{bmatrix} x \\ y \\ z \end{bmatrix}\right) = \begin{bmatrix} \cos\theta & -\sin\theta & 0 \\ \sin\theta & \cos\theta & 0 \\ 0 & 0 & 1 \end{bmatrix}\begin{bmatrix} x \\ y \\ z \end{bmatrix}$$

   (b) If in the above instructions we replace the z axis by the y axis throughout, show that

$$R_\theta\left(\begin{bmatrix} x \\ y \\ z \end{bmatrix}\right) = \begin{bmatrix} \cos\theta & 0 & -\sin\theta \\ 0 & 1 & 0 \\ \sin\theta & 0 & \cos\theta \end{bmatrix}\begin{bmatrix} x \\ y \\ z \end{bmatrix}$$

5. If A is an $m \times n$ matrix, consider the function T from M_{np} into M_{mp},

$$T(B) = AB, \quad \text{with } B \text{ in } M_{np}$$

Show that T is linear.

6. Let P be the function from $\boldsymbol{R}^n$ into $\boldsymbol{R}^n$ defined by

$$P\left(\begin{bmatrix} x_1 \\ x_2 \\ \vdots \\ x_n \end{bmatrix}\right) = \begin{bmatrix} 0 \\ x_2 \\ \vdots \\ x_n \end{bmatrix}$$

Show that P is linear.

7. Give an example of a function T from $\boldsymbol{R}^3$ to $\boldsymbol{R}^3$ that satisfies $T(\alpha\boldsymbol{v}) = \alpha T(\boldsymbol{v})$ for all $\boldsymbol{v}$ in $\boldsymbol{R}^3$ and all scalars α, which is not linear.

8. If B is an invertible $n \times n$ matrix, show that the function from M_{nn} into itself defined by $T(A) = BAB^{-1}$ is a linear transformation.

9. If B is a fixed $n \times n$ matrix, show that the following functions from M_{nn} into M_{nn} are linear.
   (a) $T(A) = AB - BA$ (b) $T(A) = AB + BA$
   (c) $T(A) = AB - B^2A$

10. If V is a vector space and α is a fixed scalar, show that the function $T: V \rightarrow V$, defined by $T(\boldsymbol{x}) = \alpha\boldsymbol{x}$ is linear. Interpret this function geometrically if $V = \boldsymbol{R}^2$.

11. Let V be a vector space and $\boldsymbol{x}_1, \boldsymbol{x}_2, \ldots, \boldsymbol{x}_n$ be a basis for V. Then, if $\boldsymbol{x}$ is a vector in V, there are unique scalars $\alpha_1, \alpha_2, \alpha_3, \ldots, \alpha_n$, such that $\boldsymbol{x} = \alpha_1\boldsymbol{x}_1 + \alpha_2\boldsymbol{x}_2 + \cdots + \alpha_n\boldsymbol{x}_n$. Consider the function $f(\boldsymbol{x}) = \alpha_1$, with α_1 the first coordinate of $\boldsymbol{x}$ relative to the basis $\boldsymbol{x}_1, \boldsymbol{x}_2, \ldots, \boldsymbol{x}_n$. Show that f is linear.

12. Let $f: V \rightarrow \boldsymbol{R}$ and $g: V \rightarrow \boldsymbol{R}$ be linear transformations from a vector space V to $\boldsymbol{R}$. Show that

$$T(\boldsymbol{x}) = \begin{bmatrix} f(\boldsymbol{x}) \\ g(\boldsymbol{x}) \end{bmatrix}$$

is a linear function from V into $\boldsymbol{R}^2$. Generalize to the case of n functions.

13. Let P_n be the space of polynomials of degree less than or equal to n in a variable x with real coefficients. Show that the following functions from P_n into P_n are linear transformations.
   (a) $(T(f))(x) = f(x + \alpha)$, where α is a fixed real number. In other words, T carries $f(x)$ into $f(x + \alpha)$.
   (b) $T(f) = a_0f + (b_0 + b_1x)f' + (c_0 + c_1x + c_2x^2)f''$, where f', f'' are first and second derivatives of f, respectively, and $a_0, b_0, b_1, c_0, c_1, c_2$ are real numbers.
   (c) $(T(f))(x) = \int_0^x tf'(t)\,dt$.
   (d) $(T(f))(x) = f(\alpha x)$, α a real number.
   (e) $T(f) = f(\alpha)$, α a real number.

14. Is the function $f(A) = \det A$ from M_{nn} into the reals, a linear transformation?

15. Let A be an invertible 2×2 matrix,

$$A = \begin{bmatrix} a & c \\ b & d \end{bmatrix}$$

and let T_A be the linear transformation of $\boldsymbol{R}^2$ into itself,

$$T_A\left(\begin{bmatrix} x \\ y \end{bmatrix}\right) = A\begin{bmatrix} x \\ y \end{bmatrix}$$

Show that the vectors which lie in the unit square

$$S = \{(x, y) \mid 0 \leq x \leq 1 \quad \text{and} \quad 0 \leq y \leq 1\}$$

all go into some parallelogram whose adjacent sides are the directed line segments represented by the vectors

$$\begin{bmatrix} a \\ b \end{bmatrix} \quad \text{and} \quad \begin{bmatrix} c \\ d \end{bmatrix}$$

If A is not invertible and nonzero, show that the points of S go into some line segment.

16. Let A be an invertible 2×2 matrix. (a) Show that the function $T_A(\boldsymbol{x}) = A\boldsymbol{x}$ from $\boldsymbol{R}^2$ into itself carries lines into lines. (b) Show that lines through the origin go into lines through the origin.

17. Let $T : M_{33} \to M_{22}$ be the function

$$T\left(\begin{bmatrix} a_{11} & a_{12} & a_{13} \\ a_{21} & a_{22} & a_{23} \\ a_{31} & a_{32} & a_{33} \end{bmatrix}\right) = \begin{bmatrix} a_{11} & a_{12} \\ a_{21} & a_{22} \end{bmatrix}$$

Show that T is linear.

18. Let P be the space of all polynomials with real coefficients in a variable x. Show that the functions S and T, from P to P, defined by

$$S(f) = xf \quad \text{and} \quad (T(f))(x) = \int_0^x f(t)\, dt$$

are linear transformations from P into itself.

19. Let T be a function from V into V, a real vector space, such that
(1) $T(\boldsymbol{x} + \boldsymbol{y}) = T(\boldsymbol{x}) + T(\boldsymbol{y})$
(2) $T(\alpha\boldsymbol{x}) = \alpha T(\boldsymbol{x})$, if $\alpha \geq 0$
Show that T is a linear transformation.

2 ADDITIONAL PROPERTIES OF LINEAR TRANSFORMATIONS

In this section we call attention to some additional important properties of linear transformations.

Proposition Let V and W be vector spaces, $T: V \to W$ a linear transformation. Then,

(i) $T(\mathbf{0}) = \mathbf{0}$
(ii) $T(\alpha\boldsymbol{x} + \beta\boldsymbol{y}) = \alpha T(\boldsymbol{x}) + \beta T(\boldsymbol{y})$
(iii) $T(\Sigma_{i=1}^{n}\alpha_i\boldsymbol{x}_i) = \Sigma_{i=1}^{n}\alpha_i T(\boldsymbol{x}_i)$

Proof (i) $T(\mathbf{0}) = T(0 \cdot \boldsymbol{x}) = 0 \cdot T(\boldsymbol{x}) = \mathbf{0}$

(ii)
$$T(\alpha\boldsymbol{x} + \beta\boldsymbol{y}) = T(\alpha\boldsymbol{x}) + T(\beta\boldsymbol{y}) = \alpha T(\boldsymbol{x}) + \beta T(\boldsymbol{y})$$

(iii) The proof of (iii) is by induction on n. If $n = 1$, (iii) clearly holds by condition (ii) in the definition of a linear transformation. If $n > 1$, we have

$$T\left(\sum_{i=1}^{n}\alpha_i\boldsymbol{x}_i\right) = T\left(\sum_{i=1}^{n-1}\alpha_i\boldsymbol{x}_i + \alpha_n\boldsymbol{x}_n\right) = T\left(\sum_{i=1}^{n-1}\alpha_i\boldsymbol{x}_i\right) + T(\alpha_n\boldsymbol{x}_n)$$

By the induction hypothesis

$$T\left(\sum_{i=1}^{n-1}\alpha_i\boldsymbol{x}_i\right) = \sum_{i=1}^{n-1}\alpha_i T(\boldsymbol{x}_i)$$

Thus,

$$T\left(\sum_{i=1}^{n}\alpha_i\boldsymbol{x}_i\right) = \sum_{i=1}^{n-1}\alpha_i T(\boldsymbol{x}_i) + \alpha_n T(\boldsymbol{x}_n) = \sum_{i=1}^{n}\alpha_i T(\boldsymbol{x}_i)$$

There are two important linear transformations which can be defined for any vector space V. The first is the function N defined on V by $N(\boldsymbol{x}) = \mathbf{0}$, for all $\boldsymbol{x}$ in V. Since $N(\boldsymbol{x} + \boldsymbol{y}) = \mathbf{0}$ and $N(\boldsymbol{x}) = N(\boldsymbol{y}) = \mathbf{0}$, we see that $N(\boldsymbol{x} + \boldsymbol{y}) = N(\boldsymbol{x}) + N(\boldsymbol{y})$. Also, $N(\alpha\boldsymbol{x}) = \boldsymbol{0}$ and $N(\boldsymbol{x}) = \boldsymbol{0}$, so $N(\alpha\boldsymbol{x}) = \alpha N(\boldsymbol{x})$. Thus, N is linear. Appropriately enough, N is called the **zero transformation.**

Another linear transformation defined in any vector space V is the function I_V, where $I_V(\boldsymbol{x}) = \boldsymbol{x}$, for all $\boldsymbol{x}$ in V. I_V is linear, since $I_V(\boldsymbol{x} + \boldsymbol{y}) = \boldsymbol{x} + \boldsymbol{y}$, by definition of I_V, and $I_V(\boldsymbol{x}) = \boldsymbol{x}$ and $I_V(\boldsymbol{y}) = \boldsymbol{y}$, thus $I_V(\boldsymbol{x} + \boldsymbol{y}) = I_V(\boldsymbol{x}) + I_V(\boldsymbol{y})$. Also, $I_V(\alpha\boldsymbol{x}) = \alpha\boldsymbol{x} = \alpha I_V(\boldsymbol{x})$. I_V is called the **identity transformation** on the vector space V. When the vector space V is clear from context, I_V is often simply denoted by I.

Let T_1 and T_2 be two linear transformations from a vector space V into a vector space W. Under what circumstances are T_1 and T_2 equal? We say that T_1 and T_2 are equal if they are equal as functions. In other words, T_1 and T_2 are equal if and only if $T_1(\boldsymbol{x}) = T_2(\boldsymbol{x})$ for all vectors $\boldsymbol{x}$ in V. If T_1 and T_2 are equal, we write $T_1 = T_2$.

A general method of constructing linear transformations is indicated in the following theorem.

Theorem Let V be a vector space with $\boldsymbol{x}_1, \boldsymbol{x}_2, \ldots, \boldsymbol{x}_n$ a basis for V. Let W be another vector space and $\boldsymbol{y}_1, \boldsymbol{y}_2, \ldots, \boldsymbol{y}_n$ be n arbitrary vectors in W. Then, there is one and only one linear transformation T from V into W such that $T(\boldsymbol{x}_i) = \boldsymbol{y}_i$, $i = 1, 2, \ldots, n$.

Proof First, we show that such a linear transformation exists.

Since $\boldsymbol{x}_1, \boldsymbol{x}_2, \ldots, \boldsymbol{x}_n$ is a basis, if $\boldsymbol{x}$ is a vector in V there are unique scalars $\alpha_1, \alpha_2, \ldots, \alpha_n$ such that $\boldsymbol{x} = \alpha_1\boldsymbol{x}_1 + \alpha_2\boldsymbol{x}_2 + \cdots + \alpha_n\boldsymbol{x}_n$.

We define $T(\boldsymbol{x})$ by $T(\boldsymbol{x}) = \alpha_1\boldsymbol{y}_1 + \alpha_2\boldsymbol{y}_2 + \cdots + \alpha_n\boldsymbol{y}_n$. Since the scalars $\alpha_1, \alpha_2, \ldots, \alpha_n$ are uniquely determined by $\boldsymbol{x}$, the function T is well-defined.

We prove T is linear. Suppose $\boldsymbol{x}$ and $\boldsymbol{y}$ are vectors in V, then for suitable scalars $\alpha_1, \alpha_2, \ldots, \alpha_n, \beta_1, \beta_2, \ldots, \beta_n$, we have

$$\boldsymbol{x} = \alpha_1\boldsymbol{x}_1 + \alpha_2\boldsymbol{x}_2 + \cdots + \alpha_n\boldsymbol{x}_n$$

and

$$\boldsymbol{y} = \beta_1\boldsymbol{x}_1 + \beta_2\boldsymbol{x}_2 + \cdots + \beta_n\boldsymbol{x}_n$$

So $\boldsymbol{x} + \boldsymbol{y} = (\alpha_1 + \beta_1)\boldsymbol{x}_1 + (\alpha_2 + \beta_2)\boldsymbol{x}_2 + \cdots + (\alpha_n + \beta_n)\boldsymbol{x}_n$.

By definition of T, we have

$$T(\boldsymbol{x}) = \alpha_1\boldsymbol{y}_1 + \alpha_2\boldsymbol{y}_2 + \cdots + \alpha_n\boldsymbol{y}_n$$

$$T(\boldsymbol{y}) = \beta_1\boldsymbol{y}_1 + \beta_2\boldsymbol{y}_2 + \cdots + \beta_n\boldsymbol{y}_n$$

and

$$T(\boldsymbol{x} + \boldsymbol{y}) = (\alpha_1 + \beta_1)\boldsymbol{y}_1 + (\alpha_2 + \beta_2)\boldsymbol{y}_2 + \cdots + (\alpha_n + \beta_n)\boldsymbol{y}_n$$

Thus, $T(\boldsymbol{x} + \boldsymbol{y}) = T(\boldsymbol{x}) + T(\boldsymbol{y})$.

If $\boldsymbol{x}$ belongs to V and α is a scalar, $\boldsymbol{x} = \alpha_1\boldsymbol{x}_1 + \alpha_2\boldsymbol{x}_2 + \cdots + \alpha_n\boldsymbol{x}_n$ for suitable scalars $\alpha_1, \alpha_2, \ldots, \alpha_n$. Then

$$\alpha\boldsymbol{x} = (\alpha\alpha_1)\boldsymbol{x}_1 + (\alpha\alpha_2)\boldsymbol{x}_2 + \cdots + (\alpha\alpha_n)\boldsymbol{x}_n$$

Thus, by definition of T,

$$T(\boldsymbol{x}) = \alpha_1\boldsymbol{y}_1 + \alpha_2\boldsymbol{y}_2 + \cdots + \alpha_n\boldsymbol{y}_n$$
$$T(\alpha\boldsymbol{x}) = (\alpha\alpha_1)\boldsymbol{y}_1 + (\alpha\alpha_2)\boldsymbol{y}_2 + \cdots + (\alpha\alpha_n)\boldsymbol{y}_n$$

Thus, $T(\alpha\boldsymbol{x}) = \alpha T(\boldsymbol{x})$.

Hence we can conclude that the function T is linear. By definition of T, $T(\boldsymbol{x}_i) = \boldsymbol{y}_i$, and so the first half of the theorem is proved.

The second half of the theorem asserts that there is only one linear transformation such that $T(\boldsymbol{x}_1) = \boldsymbol{y}_1, T(\boldsymbol{x}_2) = \boldsymbol{y}_2, \ldots, T(\boldsymbol{x}_n) = \boldsymbol{y}_n$.

To prove this suppose S and T are two linear operators such that $T(\boldsymbol{x}_i) = \boldsymbol{y}_i$ and $S(\boldsymbol{x}_i) = \boldsymbol{y}_i$ for $i = 1, 2, \ldots, n$. We desire to show that $T = S$, i.e., that $T(\boldsymbol{x}) = S(\boldsymbol{x})$ for all $\boldsymbol{x}$ in V.

If $\boldsymbol{x}$ belongs to V, there are scalars $\alpha_1, \alpha_2, \ldots, \alpha_n$ such that $\boldsymbol{x} = \alpha_1\boldsymbol{x}_1 + \alpha_2\boldsymbol{x}_2 + \cdots + \alpha_n\boldsymbol{x}_n$. Then

$$T(\boldsymbol{x}) = \alpha_1T(\boldsymbol{x}_1) + \alpha_2T(\boldsymbol{x}_2) + \cdots + \alpha_nT(\boldsymbol{x}_n) = \alpha_1\boldsymbol{y}_1 + \alpha_2\boldsymbol{y}_2 + \cdots + \alpha_n\boldsymbol{y}_n$$

and

$$S(\boldsymbol{x}) = \alpha_1S(\boldsymbol{x}_1) + \alpha_2S(\boldsymbol{x}_2) + \cdots + \alpha_nS(\boldsymbol{x}_n) = \alpha_1\boldsymbol{y}_1 + \alpha_2\boldsymbol{y}_2 + \cdots + \alpha_n\boldsymbol{y}_n$$

Thus, $T(\boldsymbol{x}) = S(\boldsymbol{x})$. Since $\boldsymbol{x}$ was an arbitrary vector, it follows that $T = S$. ▧

Notice that the theorem has two parts. The first part enables us to construct linear transformations by suitably transforming the basis vectors, while the images of the remaining vectors are determined by linearity. For example, the zero transformation might have been constructed by specifying $\boldsymbol{y}_1 = \mathbf{0}, \boldsymbol{y}_2 = \mathbf{0}, \ldots, \boldsymbol{y}_n = \mathbf{0}$. The identity transformation might be constructed by specifying $V = W$ and $\boldsymbol{y}_1 = \boldsymbol{x}_1$, $\boldsymbol{y}_2 = \boldsymbol{x}_2, \ldots, \boldsymbol{y}_n = \boldsymbol{x}_n$. On P_3, we may construct a linear transformation by insisting that, under T,

$$1 \to 0, \qquad x \to 1, \qquad x^2 \to 2x, \qquad x^3 \to 3x^2$$

The resulting linear operator is just the differentiation operator of Example 4, §5.1.

The second half of the theorem has an alternative formulation. If T and S are linear transformations from a vector space V into a vector space W and T and S agree for the vectors in some basis for V, then T

and S agree on all of V. This statement generalizes to arbitrary finite-dimensional spaces the fact we demonstrated in the last section. There we saw that a linear transformation from $\boldsymbol{R}^2$ into $\boldsymbol{R}^2$ is completely determined by the images of the basis vectors $\boldsymbol{i}$ and $\boldsymbol{j}$.

We are now in a position to generalize a result from the previous section and show that any linear transformation from $\boldsymbol{R}^n$ into $\boldsymbol{R}^m$ may be induced by multiplication by a suitable matrix.

Theorem 2 Let $T : \boldsymbol{R}^n \to \boldsymbol{R}^m$ be a linear transformation. Then, there is an $m \times n$ matrix A, with real entries, such that $T(\boldsymbol{x}) = A\boldsymbol{x}$ for all $\boldsymbol{x}$ in $\boldsymbol{R}^n$.

Proof Let $\boldsymbol{e}_1, \boldsymbol{e}_2, \ldots, \boldsymbol{e}_n$ denote the standard basis for $\boldsymbol{R}^n$ and $\boldsymbol{e}_1', \boldsymbol{e}_2', \ldots, \boldsymbol{e}_m'$ denote the standard basis for $\boldsymbol{R}^m$.

Since $\boldsymbol{e}_1', \boldsymbol{e}_2', \ldots, \boldsymbol{e}_m'$ is a basis for $\boldsymbol{R}^m$, there are scalars a_{ij} such that

$$T(\boldsymbol{e}_j) = \sum_{i=1}^{m} a_{ij}\boldsymbol{e}_i'$$

If

$$\boldsymbol{x} = \begin{bmatrix} x_1 \\ x_2 \\ \vdots \\ x_n \end{bmatrix} = \sum_{j=1}^{n} x_j\boldsymbol{e}_j$$

is a vector in $\boldsymbol{R}^n$, we have

$$\begin{aligned} T(\boldsymbol{x}) &= T\left(\sum_{j=1}^{n} x_j\boldsymbol{e}_j\right) \\ &= \sum_{j=1}^{n} x_jT(\boldsymbol{e}_j) \\ &= \sum_{j=1}^{n} x_j\left(\sum_{i=1}^{m} a_{ij}\boldsymbol{e}_i'\right) \\ &= \sum_{j=1}^{n}\sum_{i=1}^{m} a_{ij}x_j\boldsymbol{e}_i' \\ &= \sum_{i=1}^{m}\left(\sum_{j=1}^{n} a_{ij}x_j\right)\boldsymbol{e}_i' \end{aligned}$$

Thus, the ith component of the vector $T(\boldsymbol{x})$ is just $\Sigma_{j=1}^{n} a_{ij}x_j$.

If we let $A = [a_{ij}]_{(mn)}$, then the ith component of the product $A\boldsymbol{x}$ is precisely $\Sigma_{j=1}^{n} a_{ij}x_i$.

Thus, for $i = 1, 2, \ldots, m$ the vectors $A\boldsymbol{x}$ and $T(\boldsymbol{x})$ have the same ith components, and so $T(\boldsymbol{x}) = A\boldsymbol{x}$. Hence, the linear transformation T is precisely that induced by multiplication by the matrix A. ∎

If A is an $m \times n$ matrix, we often denote the linear transformation from $\boldsymbol{R}^n$ to $\boldsymbol{R}^m$ induced by A simply by T_A. It is important to note that the

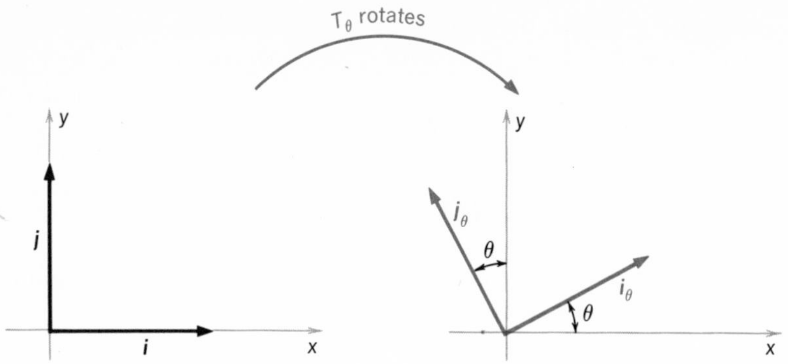

FIGURE 2-1

matrix A has as its jth column the vector $T(\boldsymbol{e}_j)$. Thus, if T is a linear operator from $\boldsymbol{R}^n$ to $\boldsymbol{R}^m$, in order to calculate its associated matrix we need only calculate the image of the basis vectors $\boldsymbol{e}_1, \boldsymbol{e}_2, \ldots, \boldsymbol{e}_n$ and form the matrix $A = [T(\boldsymbol{e}_1), T(\boldsymbol{e}_2), \ldots, T(\boldsymbol{e}_n)]$. It is also clear that the matrix A is uniquely determined by the image under T of the basis vectors $\boldsymbol{e}_1, \boldsymbol{e}_2, \ldots, \boldsymbol{e}_n$.

As an example, recall the linear transformation T_θ which rotates the vectors in the plane by θ degrees. (See Figure 2-1.)

Since

$$\boldsymbol{i}_\theta = T_\theta(\boldsymbol{i}) = (\cos\theta)\boldsymbol{i} + (\sin\theta)\boldsymbol{j}$$

$$\boldsymbol{j}_\theta = T_\theta(\boldsymbol{j}) = -(\sin\theta)\boldsymbol{i} + (\cos\theta)\boldsymbol{j}$$

we see that the matrix associated with T_θ is

$$A_\theta = \begin{bmatrix} \cos\theta & -\sin\theta \\ \sin\theta & \cos\theta \end{bmatrix}$$

This formula is obtained by writing the vector $\boldsymbol{i}_\theta$ in the first and $\boldsymbol{j}_\theta$ in the second columns of the matrix A_θ.

As another example, we consider the following projection transformation. Let l be a line in the plane through the origin which makes an angle of θ degrees with the x axis. If $\boldsymbol{v}$ is a vector in the plane ending at the point R, let $P_\theta(\boldsymbol{v})$ be the vector along the line l ending at the point S, where S is obtained by dropping a perpendicular from the point R to the line l. (See Figure 2-2.)

The linearity of the geometrically defined function P_θ may be demonstrated in a manner completely analogous to the proof, in §2.2 that $\boldsymbol{v}(x, y) + \boldsymbol{v}(x', y') = \boldsymbol{v}(x + x', y + y')$, that is, that the geometric and algebraic definitions of vector addition are equivalent. Indeed, if $\theta = 0$, i.e., the line l is the x axis, the proofs are completely identical, for in this case $P_\theta(\boldsymbol{v}(x, y)) = x\boldsymbol{i}$.

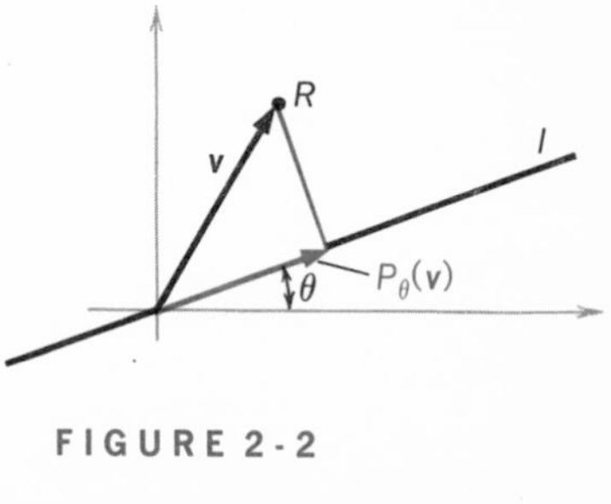

FIGURE 2-2

Having obtained the linearity of the function P_θ from geometric considerations, it follows that there is some matrix, say B_θ, such that $P_\theta(\boldsymbol{v}) = B_\theta \boldsymbol{v}$.

We complete our determination of the function P_θ by finding the matrix B_θ. In order to obtain the matrix B_θ it is only necessary to determine $P_\theta(\boldsymbol{i})$ and $P_\theta(\boldsymbol{j})$. (See Figure 2-3.)

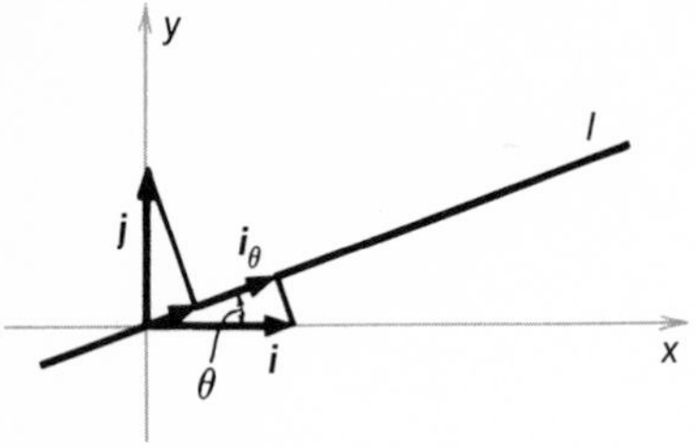

FIGURE 2-3

If $\boldsymbol{i}_\theta$ is the vector of unit length in the direction of the line l, it is clear that $\boldsymbol{i}_\theta = (\cos\theta)\boldsymbol{i} + (\sin\theta)\boldsymbol{j}$.

From trigonometry the length of the vector $P_\theta(\boldsymbol{i})$ is $\cos\theta$ and the length of $P_\theta(\boldsymbol{j})$ is $\sin\theta$. Thus,

$$P_\theta(\boldsymbol{i}) = (\cos\theta)\boldsymbol{i}_\theta = (\cos^2\theta)\boldsymbol{i} + (\cos\theta\sin\theta)\boldsymbol{j}$$

$$P_\theta(\boldsymbol{j}) = (\sin\theta)\boldsymbol{i}_\theta = (\sin\theta\cos\theta)\boldsymbol{i} + (\sin^2\theta)\boldsymbol{j}$$

If

$$B_\theta = \begin{bmatrix} \cos^2\theta & \cos\theta\sin\theta \\ \cos\theta\sin\theta & \sin^2\theta \end{bmatrix}$$

we see that

$$P_\theta\left(\begin{bmatrix} x \\ y \end{bmatrix}\right) = \begin{bmatrix} \cos^2\theta & \cos\theta\sin\theta \\ \cos\theta\sin\theta & \sin^2\theta \end{bmatrix}\begin{bmatrix} x \\ y \end{bmatrix}$$

EXERCISES

1. Find all linear transformations from $\boldsymbol{R}^2$ into itself which carry vectors on the line $x = 0$ into vectors on the line $x = 0$ and vectors on the line $y = 0$ into vectors on the line $y = 0$.
2. If

$$\begin{bmatrix} a \\ b \end{bmatrix} \quad \text{and} \quad \begin{bmatrix} c \\ d \end{bmatrix}$$

are vectors in $\boldsymbol{R}^2$, give a matrix representation of the linear transformation which carries

$$\begin{bmatrix} 1 \\ 0 \end{bmatrix} \to \begin{bmatrix} a \\ b \end{bmatrix} \quad \text{and} \quad \begin{bmatrix} 0 \\ 1 \end{bmatrix} \to \begin{bmatrix} c \\ d \end{bmatrix}$$

3. Let P be the function from $\boldsymbol{R}^3$ into $\boldsymbol{R}^3$ defined as follows: If $\boldsymbol{v}$ is a vector in $\boldsymbol{R}^3$, $P(\boldsymbol{v})$ is obtained by dropping a perpendicular from the endpoint of $\boldsymbol{v}$ to the xy plane. The directed line segment from the origin to the point where the perpendicular intersects the xy plane represents the vector $P(\boldsymbol{v})$.
(a) Show that P is linear.

(b) Show that

$$P\left(\begin{bmatrix} x \\ y \\ z \end{bmatrix}\right) = \begin{bmatrix} x \\ y \\ 0 \end{bmatrix}$$

(c) Find a matrix A, such that

$$P\left(\begin{bmatrix} x \\ y \\ z \end{bmatrix}\right) = A\begin{bmatrix} x \\ y \\ z \end{bmatrix}$$

4. If T is a linear transformation between two vector spaces V and W, and $\boldsymbol{x}_1, \boldsymbol{x}_2, \ldots, \boldsymbol{x}_n$ are vectors in V such that $T(\boldsymbol{x}_1), T(\boldsymbol{x}_2), \ldots, T(\boldsymbol{x}_n)$ are linearly independent, show that the $\boldsymbol{x}_i$'s are linearly independent in V.

5. Suppose $T : V \to W$ and $S : V \to W$ are linear transformations. If $\boldsymbol{x}_1, \boldsymbol{x}_2, \ldots, \boldsymbol{x}_n$ are vectors in V such that sp $(\boldsymbol{x}_1, \boldsymbol{x}_2, \ldots, \boldsymbol{x}_n) = V$, and if

$$T(\boldsymbol{x}_1) = S(\boldsymbol{x}_1),\ T(\boldsymbol{x}_2) = S(\boldsymbol{x}_2), \ldots, T(\boldsymbol{x}_n) = S(\boldsymbol{x}_n)$$

show that $T = S$.

6. Find a linear transformation T from $\boldsymbol{R}^2$ to $\boldsymbol{R}^2$ such that

$$T\left(\begin{bmatrix} 1 \\ 1 \end{bmatrix}\right) = \begin{bmatrix} -1 \\ 0 \end{bmatrix}, \quad T\left(\begin{bmatrix} 1 \\ -2 \end{bmatrix}\right) = \begin{bmatrix} 1 \\ 3 \end{bmatrix}$$

7. Find all linear transformations from $\boldsymbol{R}^2$ into $\boldsymbol{R}^2$ which
(a) Carry the line $x = 0$ into the line $x = 0$.
(b) Carry the line $y = 0$ into the line $y = 0$.
(c) Carry the line $x = y$ into the line $x = y$.

8. If H is a subspace of a vector space V, and $T : V \to W$ is a linear transformation, let K denote the subset of W each of whose elements is the image of some vector in H, i.e., $K = T(H) = \{\boldsymbol{w} : \boldsymbol{w} \in W \text{ and } \boldsymbol{w} = T(\boldsymbol{h})$ for some $\boldsymbol{h}$ in $H\}$. Show that K is a subspace of W.

9. Give an example of a linear transformation T from $\boldsymbol{R}^2$ into $\boldsymbol{R}^2$ such that $\boldsymbol{x}_1$ and $\boldsymbol{x}_2$ are linearly independent, but $T(\boldsymbol{x}_1)$ and $T(\boldsymbol{x}_2)$ are linearly dependent.

10. Let V be a two-dimensional vector space and let $\boldsymbol{x}_1, \boldsymbol{x}_2, \boldsymbol{x}_3$ be vectors in V, any two of which are linearly independent. Let $T : V \to V$ be a linear transformation such that

$$T(\boldsymbol{x}_1) = \alpha_1\boldsymbol{x}_1, \qquad T(\boldsymbol{x}_2) = \alpha_2\boldsymbol{x}_2, \qquad T(\boldsymbol{x}_3) = \alpha_3\boldsymbol{x}_3$$

for scalars α_1, α_2, and α_3. Show that there is a scalar α such that $T(\boldsymbol{x}) = \alpha\boldsymbol{x}$ for all $\boldsymbol{x}$ in V. Interpret geometrically if $V = \boldsymbol{R}^2$.

11. Let

$$\begin{bmatrix} x_1 \\ x_2 \end{bmatrix} \text{ and } \begin{bmatrix} x_1' \\ x_2' \end{bmatrix}$$

be linearly independent vectors in $\boldsymbol{R}^2$. Let T be a linear transformation from $\boldsymbol{R}^2$ into $\boldsymbol{R}^2$ such that

$$T\left(\begin{bmatrix} x_1 \\ x_2 \end{bmatrix}\right) = \begin{bmatrix} y_1 \\ y_2 \end{bmatrix}, \qquad T\left(\begin{bmatrix} x_1' \\ x_2' \end{bmatrix}\right) = \begin{bmatrix} y_1' \\ y_2' \end{bmatrix}$$

Show that $T(\boldsymbol{x}) = A\boldsymbol{x}$, where A is the 2×2 matrix

$$A = \begin{bmatrix} y_1 & y_1' \\ y_2 & y_2' \end{bmatrix} \begin{bmatrix} x_1 & x_1' \\ x_2 & x_2' \end{bmatrix}^{-1}$$

12. Let T be a linear transformation of $\boldsymbol{R}^3$ into itself. Show that T carries any plane through the origin into a plane through the origin, a line through the origin, or the origin itself. Give an example of each of the three cases.
13. Let $\boldsymbol{a} = a_1\boldsymbol{i} + a_2\boldsymbol{j} + a_3\boldsymbol{k}$ be a vector in $\boldsymbol{R}^3$ of length 1, i.e., $a_1^2 + a_2^2 + a_3^2 = 1$. Let P be the function from $\boldsymbol{R}^3$ into $\boldsymbol{R}^3$ obtained as follows: If $\boldsymbol{v}$ is a vector in $\boldsymbol{R}^3$, drop a perpendicular from the endpoint of $\boldsymbol{v}$ to the line determined by the scalar multiples of the vector $\boldsymbol{a}$. This perpendicular intersects the line determined by $\boldsymbol{a}$ at some point $\alpha\boldsymbol{a}$. Then, $P(\boldsymbol{v})$ is defined to be $\alpha\boldsymbol{a}$.
(a) Show that P is linear.
(b) Show that

$$P\left(\begin{bmatrix} x \\ y \\ z \end{bmatrix}\right) = \begin{bmatrix} a_1^2 & a_2a_1 & a_3a_1 \\ a_1a_2 & a_2^2 & a_3a_2 \\ a_1a_3 & a_2a_3 & a_3^2 \end{bmatrix} \begin{bmatrix} x \\ y \\ z \end{bmatrix}$$

14. If $f : \boldsymbol{R}^n \to \boldsymbol{R}$ is a linear transformation, show that there are scalars a_1, $a_2, \ldots, a_n$ such that

$$f\left(\begin{bmatrix} x_1 \\ x_2 \\ \vdots \\ x_n \end{bmatrix}\right) = a_1x_1 + a_2x_2 + \cdots + a_nx_n$$

15. If $\boldsymbol{e}_1, \boldsymbol{e}_2, \ldots, \boldsymbol{e}_n$ is the standard basis for $\boldsymbol{R}^n$, obtain the matrix representation of the linear operator T, where
(a) $T(\boldsymbol{e}_1) = \boldsymbol{e}_2$, $T(\boldsymbol{e}_2) = \boldsymbol{e}_3, \ldots, T(\boldsymbol{e}_{n-1}) = \boldsymbol{e}_n$, $T(\boldsymbol{e}_n) = \boldsymbol{0}$
(b) $T(\boldsymbol{e}_1) = \boldsymbol{e}_2$, $T(\boldsymbol{e}_2) = \boldsymbol{e}_3, \ldots, T(\boldsymbol{e}_{n-1}) = \boldsymbol{e}_n$, $T(\boldsymbol{e}_n) = \boldsymbol{e}_1$
(c) $T(\boldsymbol{e}_1) = \boldsymbol{e}_1$, $T(\boldsymbol{e}_2) = \boldsymbol{e}_2 + \boldsymbol{e}_1, \ldots, T(\boldsymbol{e}_{n-1}) = \boldsymbol{e}_{n-1} + \boldsymbol{e}_{n-2}$, $T(\boldsymbol{e}_n) = \boldsymbol{e}_n + \boldsymbol{e}_{n-1}$
16. Find the matrix representation of the linear transformation from $\boldsymbol{R}^n$ into $\boldsymbol{R}^n$ which carries the vector $\boldsymbol{x}$ into the vector $\alpha\boldsymbol{x}$ where α is a fixed scalar.

3 RANGE SPACE

Let T be a linear transformation from a vector space V into a vector space W. We wish to study those vectors in W, each of which is the image of some vector in V. Let us call this collection of vectors the range space of the linear transformation T. We denote this set of vectors by R_T. Thus, $R_T = \{\boldsymbol{y} \mid \boldsymbol{y} \in W \text{ and for some } \boldsymbol{x} \text{ in } V, \boldsymbol{y} = T(\boldsymbol{x})\}$. In order to justify this terminology we prove the following:

Theorem 1 Let $T : V \to W$ be a linear transformation of a vector space V into a vector space W. Then, the set of vectors $R_T = \{\boldsymbol{y} \mid \boldsymbol{y} \in W$ and $\boldsymbol{y} = T(\boldsymbol{x})$, for some $\boldsymbol{x}$ in $V\}$ is a subspace of W.

Proof Suppose $\boldsymbol{y}_1$ and $\boldsymbol{y}_2$ are vectors belonging to R_T. By definition of R_T, there are vectors $\boldsymbol{x}_1$ and $\boldsymbol{x}_2$ in V such that $T(\boldsymbol{x}_1) = \boldsymbol{y}_1$ and $T(\boldsymbol{x}_2) = \boldsymbol{y}_2$. By the linearity of T, $T(\boldsymbol{x}_1 + \boldsymbol{x}_2) = T(\boldsymbol{x}_1) + T(\boldsymbol{x}_2) = \boldsymbol{y}_1 + \boldsymbol{y}_2$, and so since $\boldsymbol{y}_1 + \boldsymbol{y}_2$ is the image under T of some vector in V, namely, $\boldsymbol{x}_1 + \boldsymbol{x}_2$, $\boldsymbol{y}_1 + \boldsymbol{y}_2$ belongs to R_T.

Next, suppose $\boldsymbol{y}$ belongs to R_T and α is a scalar. Since $\boldsymbol{y}$ is in R_T, there is some vector $\boldsymbol{x}$ in V such that $T(\boldsymbol{x}) = \boldsymbol{y}$. By the linearity of T, $T(\alpha\boldsymbol{x}) = \alpha T(\boldsymbol{x}) = \alpha\boldsymbol{y}$. Since $\alpha\boldsymbol{y}$ is the T image of some vector in V, namely, $\alpha\boldsymbol{x}$, we see that $\alpha\boldsymbol{y}$ belongs to R_T. Since the subset R_T is closed under the algebraic operations of addition and scalar multiplication, it follows that R_T is a subspace of W. ∎

For example, the range space of N, the zero transformation, is precisely the zero subspace. If $I_v : V \to V$ is the identity operator on the vector space V, the range space is V, the whole space.

Using the fact that R_T is a subspace of the vector space W, and Theorem 3, §4.8, we can immediately conclude that $\dim R_T \leq \dim W$.

The quantity $\dim R_T$ is of such importance that it is given a special name. It is called the rank of T and is denoted by $r(T)$.

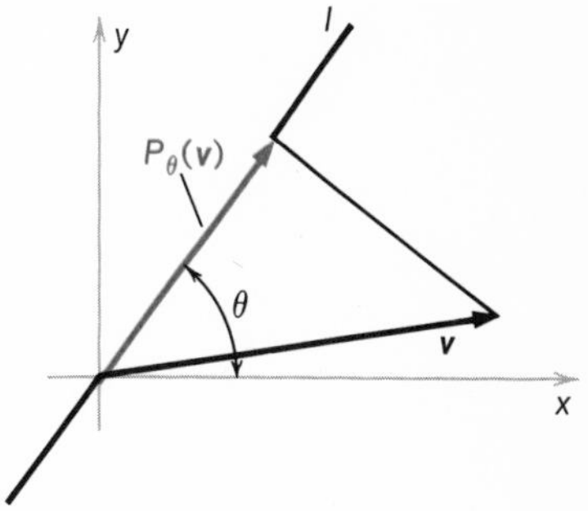

FIGURE 3-1

Example 1 In the last example of §5.2 we defined a projection transformation P_θ from $\boldsymbol{R}^2$ into itself. $P_\theta(\boldsymbol{v})$ is the vector obtained by projecting $\boldsymbol{v}$ perpendicularly on the line making an angle θ with the x axis. (See Figure 3-1.) From the definition of P_θ, it immediately follows that any vector in the range space of P_θ is a scalar multiple of the vector $\boldsymbol{i}_\theta = (\cos\theta)\boldsymbol{i} + (\sin\theta)\boldsymbol{j}$. Thus, R_{P_θ} is the one-dimensional subspace of $\boldsymbol{R}^2$ spanned by the vector $\boldsymbol{i}_\theta$. Moreover, $r(P_\theta) = 1$.

Example 2 Let

$$\begin{aligned}
a_{11}x_1 + a_{12}x_2 + \cdots + a_{1n}x_n &= y_1 \\
a_{21}x_1 + a_{22}x_2 + \cdots + a_{2n}x_n &= y_2 \\
&\vdots \\
a_{m1}x_1 + a_{m2}x_2 + \cdots + a_{mn}x_n &= y_m
\end{aligned}$$

be a system of m linear equations in n variables. Using matrix notation, with $A = [a_{ij}]_{(mn)}$, $\boldsymbol{x} = [x_i]_{(n1)}$, $\boldsymbol{y} = [y_j]_{(m1)}$, the system of equations becomes $A\boldsymbol{x} = \boldsymbol{y}$.

Forming the linear operator $T_A : \boldsymbol{R}^n \to \boldsymbol{R}^m$ by defining $T_A(\boldsymbol{x}) = A\boldsymbol{x}$, we see that the range space of T_A consists precisely of those vectors $\boldsymbol{y}$ in $\boldsymbol{R}^m$ for which there is some vector $\boldsymbol{x}$ in $\boldsymbol{R}^n$, with $A\boldsymbol{x} = \boldsymbol{y}$. Or, in other words, the range space contains precisely those vectors $\boldsymbol{y}$, such that the system $A\boldsymbol{x} = \boldsymbol{y}$ is solvable.

The procedure of Gaussian elimination discussed in chapter 1 provides a computational means of determining the range space of the linear transformation associated with a matrix. If

$$A = \begin{bmatrix} 1 & -1 & 0 \\ 3 & 1 & 7 \\ 4 & 0 & 7 \end{bmatrix}$$

we can consider the linear transformation from $\boldsymbol{R}^3$ to $\boldsymbol{R}^3$, defined by $T_A(\boldsymbol{x}) = A\boldsymbol{x}$. In order to obtain the range space of T_A, we must find those column vectors with components u, v, w for which the system of equations

$$\begin{aligned} x - y &= u \\ 3x + y + 7z &= v \\ 4x + 7z &= w \end{aligned}$$

is solvable.

We follow the elimination procedure

$$\begin{aligned} x - y &= u \\ 3x + y + 7z &= v \\ 4x + 7z &= w \end{aligned}$$

Use x and the first equation.

$$\downarrow$$

$$\begin{aligned} x - y &= u \\ 4y + 7z &= v - 3u \\ 4y + 7z &= w - 4u \end{aligned}$$

Use z and the second equation.

$$\downarrow$$

$$\begin{aligned} x - y &= u \\ 4y + 7z &= v - 3u \\ 0 &= w - v - u \end{aligned}$$

The procedure comes to a halt, since in the only unused equation, the third, all variables have coefficient 0. Thus in order that the system be satisfied, it is necessary to have $w - v - u = 0$. On the other hand, if we let $y = 0$, and find x and z using the first two equations, we see that the system indeed admits a solution. Thus, the range space of the linear transformation T_A consists exactly of those vectors for which $w = v + u$.

In order to determine $r(T_A)$, we must determine the dimension of this subspace. It is clear that the vectors

$$\boldsymbol{a} = \begin{bmatrix} 1 \\ 0 \\ 1 \end{bmatrix} \quad \text{and} \quad \boldsymbol{b} = \begin{bmatrix} 0 \\ 1 \\ 1 \end{bmatrix}$$

are both members of R_{T_A}.

If $\alpha\boldsymbol{a} + \beta\boldsymbol{b} = \boldsymbol{0}$, we have

$$\alpha \begin{bmatrix} 1 \\ 0 \\ 1 \end{bmatrix} + \beta \begin{bmatrix} 0 \\ 1 \\ 1 \end{bmatrix} = \begin{bmatrix} \alpha \\ \beta \\ \alpha + \beta \end{bmatrix} = \begin{bmatrix} 0 \\ 0 \\ 0 \end{bmatrix}$$

Thus, $\alpha = 0$ and $\beta = 0$. From this it is evident that $\boldsymbol{a}$ and $\boldsymbol{b}$ are linearly independent.

Suppose there is another vector in R_{T_A} with components c_1, c_2, c_3. Since $c_3 = c_1 + c_2$, we have

$$\begin{bmatrix} c_1 \\ c_2 \\ c_3 \end{bmatrix} = \begin{bmatrix} c_1 \\ c_2 \\ c_1 + c_2 \end{bmatrix} = c_1 \begin{bmatrix} 1 \\ 0 \\ 1 \end{bmatrix} + c_2 \begin{bmatrix} 0 \\ 1 \\ 1 \end{bmatrix}$$
$$= c_1\boldsymbol{a} + c_2\boldsymbol{b}$$

Thus, the vectors $\boldsymbol{a}$ and $\boldsymbol{b}$ span R_{T_A}. Since it was already shown that $\boldsymbol{a}$ and $\boldsymbol{b}$ are linearly independent, it follows that the vectors $\boldsymbol{a}$ and $\boldsymbol{b}$ form a basis for R_{T_A}. Thus, $\dim R_{T_A} = 2$, and so $r(T_A) = 2$.

In subsequent sections we shall develop other procedures for determining the rank and range space of a linear transformation associated with a matrix. As further explication of Example 2 above we state and prove Theorem 2.

Theorem 2 Let A be an $m \times n$ matrix and T_A be the linear transformation from $\boldsymbol{R}^n$ into $\boldsymbol{R}^m$ induced by A. Then, the columns of the matrix A span the range space of T_A.

Proof Let $\boldsymbol{e}_1, \boldsymbol{e}_2, \ldots, \boldsymbol{e}_n$ be the standard basis for $\boldsymbol{R}^n$. If $\boldsymbol{x}$ is in $\boldsymbol{R}^n$, there are scalars $\alpha_1, \alpha_2, \ldots, \alpha_n$ such that $\boldsymbol{x} = \alpha_1\boldsymbol{e}_1 + \alpha_2\boldsymbol{e}_2 + \cdots + \alpha_n\boldsymbol{e}_n$. Thus,

$$\begin{aligned} T_A(\boldsymbol{x}) &= T_A(\alpha_1\boldsymbol{e}_1 + \alpha_2\boldsymbol{e}_2 + \cdots + \alpha_n\boldsymbol{e}_n) \\ &= \alpha_1 T_A(\boldsymbol{e}_1) + \alpha_2 T_A(\boldsymbol{e}_2) + \cdots + \alpha_n T_A(\boldsymbol{e}_n) \\ &= \alpha_1(A\boldsymbol{e}_1) + \alpha_2(A\boldsymbol{e}_2) + \cdots + \alpha_n(A\boldsymbol{e}_n) \end{aligned}$$

But Ae_i is just the ith column of the matrix A. Since every vector in R_{T_A} can be expressed as a linear combination of $Ae_1, Ae_2, \ldots, Ae_n$, we have the desired result. ∎

For example, if A is the 2×4 matrix

$$\begin{bmatrix} -1 & 0 & -1 & 3 \\ 1 & 2 & 0 & 4 \end{bmatrix}$$

our theorem tells us that the vectors

$$\begin{bmatrix} -1 \\ 1 \end{bmatrix}, \quad \begin{bmatrix} 0 \\ 2 \end{bmatrix}, \quad \begin{bmatrix} -1 \\ 0 \end{bmatrix}, \quad \begin{bmatrix} 3 \\ 4 \end{bmatrix}$$

span the range of T_A. Since the vectors

$$\begin{bmatrix} -1 \\ 1 \end{bmatrix} \text{ and } \begin{bmatrix} 0 \\ 2 \end{bmatrix}$$

form a basis for $\boldsymbol{R}^2$, we see that $R_{T_A} = \boldsymbol{R}^2$, and $r(T_A) = 2$.

Example 3 Let A be an invertible $n \times n$ matrix. If T_A is the linear operator on $\boldsymbol{R}^n$ induced by A, $R_{T_A} = \boldsymbol{R}^n$.

For under these circumstances, the equation $T_A(\boldsymbol{x}) = \boldsymbol{y}$, i.e., $A\boldsymbol{x} = \boldsymbol{y}$ is always solvable. Indeed, its solution is $\boldsymbol{x} = A^{-1}\boldsymbol{y}$.

Example 4 Let S be the function from M_{nn} to M_{nn}, defined by $S(A) = A + A^T$.

First, S is linear, since if A and B are $n \times n$ matrices and α is a scalar then

$$\begin{aligned} S(A + B) &= A + B + (A + B)^T \\ &= A + B + A^T + B^T \\ &= A + A^T + B + B^T \\ &= S(A) + S(B) \\ S(\alpha A) &= \alpha A + (\alpha A)^T \\ &= \alpha(A + A^T) \\ &= \alpha S(A) \end{aligned}$$

We claim that the range of S is precisely the space of symmetric matrices. Since $(A + A^T)^T = A^T + A = A + A^T$, if B belongs to R_S, we have $B = A + A^T$ for some A, and so $B = B^T$. Thus, any matrix in R_S is symmetric.

Next, we show that any symmetric matrix belongs to R_S. If B is symmetric,

$$S(\tfrac{1}{2}B) = \tfrac{1}{2}B + (\tfrac{1}{2}B)^T$$
$$= (\tfrac{1}{2} + \tfrac{1}{2})B = B$$

Thus, B belongs to R_S.

Since all matrices in R_S are symmetric and all symmetric matrices belong to R_S, R_S is exactly the space of symmetric matrices. Since the space of symmetric matrices is of dimension $\frac{1}{2}n(n+1)$ (§4.7 Exercise 11) we see that $r(S) = \frac{1}{2}n(n+1)$.

Given a linear transformation $T : V \to W$, if $R_T = W$, T is said to be **onto**. For example, I_V, the identity operator on some vector space V, is onto. In Example 3 above, the linear transformation from $\boldsymbol{R}^n$ into $\boldsymbol{R}^n$ induced by an invertible matrix was seen to be onto. If A is an $m \times n$ matrix, the linear transformation T_A from $\boldsymbol{R}^n$ to $\boldsymbol{R}^m$ induced by the matrix A is onto if and only if the systems of equations $A\boldsymbol{x} = \boldsymbol{y}$ is always solvable.

EXERCISES

1. Find a basis for the range space and the rank of the linear transformations induced by the following matrices.

(a) $\begin{bmatrix} -1 & 1 & 3 \\ 0 & 2 & 0 \end{bmatrix}$ (b) $\begin{bmatrix} 1 & 0 \\ 3 & 1 \\ 0 & 2 \end{bmatrix}$

(c) $\begin{bmatrix} 0 & 1 & 2 \\ 1 & 2 & 5 \\ 0 & -1 & -2 \end{bmatrix}$ (d) $\begin{bmatrix} 1 & 2 & 3 \\ 0 & 0 & 0 \\ -3 & 1 & -2 \\ 1 & 3 & 4 \end{bmatrix}$

(e) $\begin{bmatrix} 1 & -1 & 0 & 3 \\ 0 & 3 & 2 & -1 \\ 2 & 7 & 1 & -2 \end{bmatrix}$

2. Let $D : P_n \to P_n$ be the differentiation operator on P_n. Show that $R_D = P_{n-1}$, and $r(D) = n$.

3. If $T : V \to W$ is a linear transformation and if $R_T = 0$, show that T is the zero operator.

4. Let A be an $n \times n$ matrix. If $T : M_{nn} \to M_{nn}$ is the linear transformation defined by $T(B) = AB$, show that T is onto if and only if A is invertible.

5. Let T be a linear transformation between two vector spaces V and W. If $\boldsymbol{x}_1, \boldsymbol{x}_2, \ldots, \boldsymbol{x}_n$ are vectors in V such that $\text{sp}(\boldsymbol{x}_1, \boldsymbol{x}_2, \ldots, \boldsymbol{x}_n) = V$, show that $T(\boldsymbol{x}_1), T(\boldsymbol{x}_2), \ldots, T(\boldsymbol{x}_n)$ span R_T.

6. Let $T : M_{nn} \to M_{nn}$ be the linear operator defined on the space of $n \times n$ matrices by $T(A) = A^T$. Show that T is onto.

7. Calculate a basis for the range space and the rank of the following linear operators on P_n.
   (a) $T(f) = xf'$, f' is the derivative of f.
   (b) $(T(f))(x) = \int_0^x tf''(t)\,dt$, f'' is the second derivative of f.
   (c) $(T(f))(x) = f(x + 1)$.

8. Let V be a vector space, and let $\boldsymbol{x}_1, \boldsymbol{x}_2, \ldots, \boldsymbol{x}_n$ be a basis for V. Let T be the function from V into $\boldsymbol{R}^n$ which carries the vector $\boldsymbol{x}$ in V into its coordinate n-tuple in $\boldsymbol{R}^n$ relative to the basis $\boldsymbol{x}_1, \boldsymbol{x}_2, \ldots, \boldsymbol{x}_n$. Show that T is a linear transformation from V onto $\boldsymbol{R}^n$.

9. Let A be a 2×2 diagonal matrix which is not a scalar multiple of the identity. Let $C : M_{22} \to M_{22}$ be the linear operator defined by $C(B) = AB - BA$. Show that the matrices

$$\begin{bmatrix} 0 & 1 \\ 0 & 0 \end{bmatrix} \quad \text{and} \quad \begin{bmatrix} 0 & 0 \\ 1 & 0 \end{bmatrix}$$

form a basis for R_C.

10. Let $\boldsymbol{x} \neq \boldsymbol{0}$ be a fixed vector in $\boldsymbol{R}^n$. Show that the function from the space of $n \times n$ matrices into $\boldsymbol{R}^n$ defined by $P(A) = A\boldsymbol{x}$ for each $n \times n$ matrix A is a linear transformation onto $\boldsymbol{R}^n$.

11. Let $S : M_{nn} \to M_{nn}$ be the linear operator defined by $S(A) = A - A^T$. Show that R_S consists precisely of the skew-symmetric matrices, i.e., those matrices such that $B^T = -B$.

12. Let D be a diagonal $n \times n$ matrix. Let $T(\boldsymbol{x}) = D\boldsymbol{x}$ be the linear transformation from $\boldsymbol{R}^n$ to $\boldsymbol{R}^n$ induced by D. Show that the rank of D is precisely the number of nonzero entries on the diagonal of D.

13. Let A and B be $n \times n$ matrices with B invertible. Let $T(\boldsymbol{x}) = A\boldsymbol{x}$, $S(\boldsymbol{x}) = (AB)\boldsymbol{x}$ be the linear transformations of $\boldsymbol{R}^n$ induced by A and AB, respectively. Show that T and S have the same range space.

14. Let $T : \boldsymbol{R}^n \to \boldsymbol{R}^n$ be a linear operation of rank 1. Show that there are scalars $a_1, a_2, \ldots, a_n, b_1, b_2, \ldots, b_n$ such that $T(\boldsymbol{x}) = A\boldsymbol{x}$, where

$$A = \begin{bmatrix} b_1a_1 & b_2a_1 & \ldots & b_na_1 \\ b_1a_2 & b_2a_2 & \ldots & b_na_2 \\ & & \vdots & \\ b_1a_n & b_2a_n & \ldots & b_na_n \end{bmatrix}$$

Show that conversely any matrix of this sort induces a linear transformation from $\boldsymbol{R}^n$ into itself of rank 1.

15. Let $T : \boldsymbol{R}^n \to \boldsymbol{R}^n$ be a linear transformation of rank r. Suppose $\boldsymbol{x}_1$, $\boldsymbol{x}_2, \ldots, \boldsymbol{x}_r$ is a basis for $\boldsymbol{R}_T$. If $T(e_j) = a_{j1}\boldsymbol{x}_1 + a_{j2}\boldsymbol{x}_2 + \cdots + a_{jr}\boldsymbol{x}_r$, show that

$$T(\boldsymbol{x}) = [\boldsymbol{x}_1, \boldsymbol{x}_2, \ldots, \boldsymbol{x}_r]\begin{bmatrix} a_{11} & \ldots & a_{n1} \\ & \vdots & \\ a_{1r} & \ldots & a_{nr} \end{bmatrix}\boldsymbol{x}$$

where $[\boldsymbol{x}_1, \boldsymbol{x}_2, \ldots, \boldsymbol{x}_r]$ is the $n \times r$ matrix whose columns are $\boldsymbol{x}_1, \boldsymbol{x}_2, \ldots, \boldsymbol{x}_n$ successively.

This shows that a matrix of rank r can be expressed as a product of an $n \times r$ matrix and an $r \times n$ matrix. Why does this imply exercise 14 above?

16. Let A and B be $n \times n$ matrices T_A, T_B, T_{A+B} the linear transformations of $\boldsymbol{R}^n$ induced by the matrices A, B, and $A + B$, respectively. Show that

$$r(T_A) + r(T_B) \geq r(T_{A+B})$$

17. Find a linear transformation which maps the space of 3×3 matrices onto the space of 2×2 matrices.

18. Let A be a fixed $n \times n$ matrix. Let T be the function from P_n into M_{nn}, which sends the polynomial $f(x)$ into the matrix $f(A)$, i.e., if $f(x) = \alpha_0 + \alpha_1 x + \cdots + \alpha_n x^n$, then $T(f) = \alpha_0 I_n + \alpha_1 A + \cdots + \alpha_n A^n$. Show that T is a linear transformation from P_n into M_{nn}. If $n > 1$, why is T not onto?

4 NULLSPACE

Let T be a linear transformation from a vector space V into a vector space W. There is a subspace of V associated with the linear transformation T which is, in some sense, complementary to the range space discussed in the previous section. We denote by N_T the subset of V which consists of those vectors $\boldsymbol{x}$ such that $T(\boldsymbol{x}) = \boldsymbol{0}$. In other words, the elements of N_T are just those vectors sent into $\boldsymbol{0}$ by the linear transformation T. N_T is called the **nullspace** of T. This terminology is reasonable, in view of the following:

Theorem 1 Let $T : V \to W$ be a linear transformation from a vector space V into a vector space W. Then, $N_T = \{\boldsymbol{x} \mid \boldsymbol{x} \in V \text{ and } T(\boldsymbol{x}) = \boldsymbol{0}\}$ is a subspace of V.

Proof Suppose $\boldsymbol{x}$ and $\boldsymbol{y}$ belong to N_T. Then $T(\boldsymbol{x} + \boldsymbol{y}) = T(\boldsymbol{x}) + T(\boldsymbol{y}) = \boldsymbol{0} + \boldsymbol{0} = \boldsymbol{0}$. Thus, $\boldsymbol{x} + \boldsymbol{y}$ belongs to N_T. If $\boldsymbol{x}$ is in N_T and α is a scalar, $T(\alpha\boldsymbol{x}) = \alpha T(\boldsymbol{x}) = \alpha \cdot \boldsymbol{0} = \boldsymbol{0}$, and so $\alpha\boldsymbol{x}$ belongs to N_T. Thus, N_T is a subspace of V. ∎

If N is the zero operator from V to W, since N sends all vectors in V into $\mathbf{0}$, we see that $N_T = V$. If I_V is the identity operator on V and $I(\boldsymbol{x}) = \mathbf{0}$, it follows that $\boldsymbol{x} = \mathbf{0}$ since $I(\boldsymbol{x}) = \boldsymbol{x}$. Thus $N_{I_V} = 0$, the zero subspace.

The quantity $\dim N_T$ is, like $\dim R_T$, of interest. It is called the **nullity** of T and is denoted by $n(T)$.

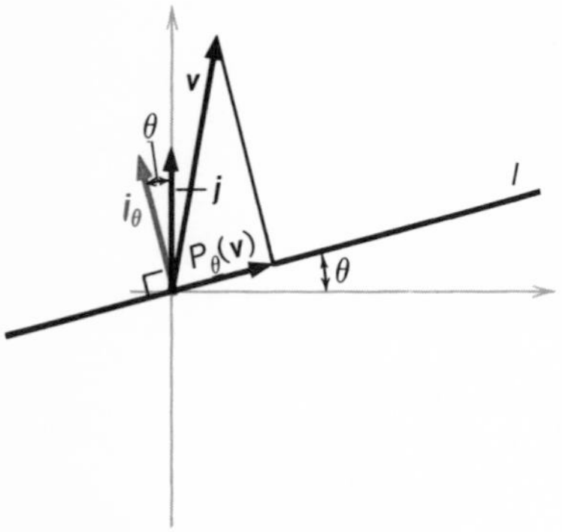

FIGURE 4-1

Example 1 Let P_θ be the projection transformation of $\boldsymbol{R}^2$ defined in §5.2. Recall that $P_\theta(\boldsymbol{v})$ is obtained by dropping a perpendicular from the endpoint of $\boldsymbol{v}$ to the line l making an angle of θ degrees with the x axis. $P_\theta(\boldsymbol{v})$ is then the vector ending at the point where the perpendicular from the endpoint of $\boldsymbol{v}$ intersects l. (See Figure 4-1.) It is clear from this geometric definition that a vector $\boldsymbol{v}$ belongs to the nullspace of P_θ when and only when it is perpendicular to the vector $\boldsymbol{i}_\theta = (\cos\theta)\boldsymbol{i} + (\sin\theta)\boldsymbol{j}$, or in other words if and only if $\boldsymbol{v}$ is a scalar multiple of the vector $\boldsymbol{j}_\theta = -(\sin\theta)\boldsymbol{i} + (\cos\theta)\boldsymbol{j}$. Thus, the nullspace of P_θ is spanned by the vector $\boldsymbol{j}_\theta$, and so $n(P_\theta) = 1$.

Example 2 Let A be an $m \times n$ matrix. Let T_A be the linear transformation from $\boldsymbol{R}^n$ to $\boldsymbol{R}^m$ induced by the matrix A. A vector $\boldsymbol{x}$ belongs to the nullspace of T_A if and only if $A\boldsymbol{x} = \mathbf{0}$, hence we see that the nullspace of T_A is just the set of solutions to the system of homogeneous linear equations written in matrix notation as $A\boldsymbol{x} = \mathbf{0}$.

Using the procedure of Gaussian elimination, it is possible to obtain a basis for the nullspace of a linear operator induced by a matrix. For example, if T is the linear transformation from $\boldsymbol{R}^5$ into $\boldsymbol{R}^3$ defined by

$$T\begin{bmatrix} x_1 \\ x_2 \\ x_3 \\ x_4 \\ x_5 \end{bmatrix} = \begin{bmatrix} 1 & -1 & 3 & 2 & 0 \\ 4 & 1 & -2 & 0 & 1 \\ 1 & 3 & 1 & 0 & 2 \end{bmatrix}\begin{bmatrix} x_1 \\ x_2 \\ x_3 \\ x_4 \\ x_5 \end{bmatrix}$$

in order to find its nullspace we must solve the following system of equations:

$$\begin{aligned} x_1 - x_2 + 3x_3 + 2x_4 \qquad &= 0 \\ 4x_1 + x_2 - 2x_3 \qquad + x_5 &= 0 \\ x_1 + 3x_2 + x_3 \qquad + 2x_5 &= 0 \end{aligned}$$

Use x_5 and equation two.

$$\downarrow$$

$$\begin{aligned} x_1 - x_2 + 3x_3 + 2x_4 \qquad &= 0 \\ 4x_1 + x_2 - 2x_3 \qquad + x_5 &= 0 \\ -7x_1 + x_2 + 5x_3 \qquad &= 0 \end{aligned}$$

Use x_2 and equation three.

$$\begin{aligned}
-6x_1 \quad\quad &+ 8x_3 + 2x_4 &= 0\\
11x_1 \quad\quad &- 7x_3 \quad\quad + x_5 &= 0\\
-7x_1 + x_2 &+ 5x_3 &= 0
\end{aligned}$$

We pretend to use x_4 and equation one, thereby producing a system in which all equations have been used.

Letting $x_1 = c$ and $x_3 = d$, we see that the most general solution is of the form

$$\begin{bmatrix} c \\ 7c - 5d \\ d \\ 3c - 4d \\ -11c + 7d \end{bmatrix} = c\begin{bmatrix} 1 \\ 7 \\ 0 \\ 3 \\ -11 \end{bmatrix} + d\begin{bmatrix} 0 \\ -5 \\ 1 \\ -4 \\ 7 \end{bmatrix}$$

Since the vectors on the right above are clearly linearly independent and span N_T, we see that they form a basis for N_T. Since N_T admits a basis of two vectors, $n(T) = 2$.

Example 3 Let $D : P_n \to P_n$ be the differentiation operator on the space of polynomials of degree less than or equal to n, $D(f) = f'$. If a polynomial f belongs to N_D, then $f' = 0$. By calculus or by using the standard basis $1, x, \ldots, x^n$ for P_n, we see that f must be a constant, i.e., f is a scalar multiple of the polynomial 1. Thus, 1 forms a basis for N_D, and so $n(D) = 1$.

Example 4 Consider the transformation on the space of 2×2 matrices defined by

$$T(A) = AC - CA, \qquad \text{where } C = \begin{bmatrix} 1 & 1 \\ 0 & 1 \end{bmatrix}$$

T is linear since if A and B are 2×2 matrices and α is a scalar

$$\begin{aligned}
T(A + B) &= (A + B)C - C(A + B)\\
&= AC + BC - CA - CB\\
&= AC - CA + BC - CB\\
&= T(A) + T(B)\\
T(\alpha A) &= (\alpha A)C - C(\alpha A)\\
&= \alpha(AC - CA)\\
&= \alpha T(A)
\end{aligned}$$

It is immediately obvious that $N_T \neq 0$, since $T(I_2) = I_2C - CI_2 = C - C = 0$. To calculate a basis for N_T, let

$$\begin{bmatrix} a & b \\ c & d \end{bmatrix}$$

be a matrix in N_T. Then

$$\begin{bmatrix} a & b \\ c & d \end{bmatrix}\begin{bmatrix} 1 & 1 \\ 0 & 1 \end{bmatrix} - \begin{bmatrix} 1 & 1 \\ 0 & 1 \end{bmatrix}\begin{bmatrix} a & b \\ c & d \end{bmatrix} = \begin{bmatrix} 0 & 0 \\ 0 & 0 \end{bmatrix}$$

or

$$\begin{bmatrix} a & a+b \\ c & c+d \end{bmatrix} - \begin{bmatrix} a+c & b+d \\ c & d \end{bmatrix} = \begin{bmatrix} 0 & 0 \\ 0 & 0 \end{bmatrix}$$

so

$$\begin{aligned} a &= a + c \\ a + b &= b + d \\ c &= c \\ c + d &= d \end{aligned}$$

Thus, we obtain $c = 0$ and $a = d$. (Notice that our choice of b is arbitrary.) Thus, any matrix in N_T is of the form

$$\begin{bmatrix} a & b \\ 0 & a \end{bmatrix}$$

It is also easy to show that all matrices of this form belong to N_T. Thus, a basis for N_T is formed by the matrices

$$\begin{bmatrix} 1 & 0 \\ 0 & 1 \end{bmatrix}, \quad \begin{bmatrix} 0 & 1 \\ 0 & 0 \end{bmatrix}$$

and we also see $n(T) = 2$. Note that in this case N_T consists precisely of those matrices which commute with the matrix C.

A linear transformation $T: V \to W$ is said to be **one-one** if $T(\boldsymbol{x}_1) = T(\boldsymbol{x}_2)$ implies $\boldsymbol{x}_1 = \boldsymbol{x}_2$. Stated otherwise, each vector of W is the image of at most one vector in V.

For example, if A is an invertible $n \times n$ matrix, the linear operator defined on $\boldsymbol{R}^n$ by $T_A(\boldsymbol{x}) = A\boldsymbol{x}$ is one-one. Indeed, if $A\boldsymbol{x}_1 = A\boldsymbol{x}_2$, multiplying both sides of the forgoing equation by A^{-1}, we have $A^{-1}(A\boldsymbol{x}_1) = A^{-1}(A\boldsymbol{x}_2)$, or $\boldsymbol{x}_1 = \boldsymbol{x}_2$.

On the other hand, as an example of a linear transformation which is not one-one, consider the differentiation operator D on P_n. We have $D(1) = D(0) = 0$. But, certainly $1 \neq 0$.

It is possible to characterize one-one transformations in terms of null space.

Theorem 2 Let $T : V \rightarrow W$ be a linear transformation between two vector spaces V and W. Then, T is one-one if and only if $N_T = \mathbf{0}$, i.e., $n(T) = 0$.

Proof First, we suppose T is one-one. Let $\boldsymbol{x}$ be a vector in N_T. Then, $T(\boldsymbol{x}) = \mathbf{0}$. Since $T(\mathbf{0}) = \mathbf{0}$, it follows that the vectors $\mathbf{0}$ and $\boldsymbol{x}$ have the same image in W. Since T is one-one, these vectors must be equal, i.e., $\boldsymbol{x} = \mathbf{0}$. Hence we have shown that any vector in N_T is zero, and it follows that N_T is the zero subspace.

If, on the other hand, N_T is the zero subspace, suppose $T(\boldsymbol{x}_1) = T(\boldsymbol{x}_2)$. Then, $T(\boldsymbol{x}_1 - \boldsymbol{x}_2) = T(\boldsymbol{x}_1) - T(\boldsymbol{x}_2) = \mathbf{0}$. But, since $N_T = 0$, $\boldsymbol{x}_1 - \boldsymbol{x}_2 = \mathbf{0}$, or $\boldsymbol{x}_1 = \boldsymbol{x}_2$. Since $T(\boldsymbol{x}_1) = T(\boldsymbol{x}_2)$ implies $\boldsymbol{x}_1 = \boldsymbol{x}_2$, T is one-one. ∎

Suppose the linear transformation induced by an $m \times n$ matrix A is one-one. We see that the system of equations $A\boldsymbol{x} = \boldsymbol{y}$ is not always solvable; however, when it is solvable, the solutions are unique. It is worthwhile to contrast the case of a linear transformation being one-one with the case of a linear transformation being onto. In the latter situation the system $A\boldsymbol{x} = \boldsymbol{y}$ is always solvable, but the solution need not be unique.

There is a fundamental theorem connecting the rank and nullity of a linear transformation.

Theorem 3 Let $T : V \rightarrow W$ be a linear transformation between two finite-dimensional vector spaces. Then, $\dim R_T + \dim N_T = \dim V$, or, $r(T) + n(T) = \dim V$.

Proof Let $\boldsymbol{x}_1, \boldsymbol{x}_2, \ldots, \boldsymbol{x}_m$ be a basis for N_T and $\boldsymbol{y}_1, \boldsymbol{y}_2, \ldots, \boldsymbol{y}_n$ be a basis for R_T. Then $\dim N_T = m$ and $\dim R_T = n$. Since $\boldsymbol{y}_1, \boldsymbol{y}_2, \ldots, \boldsymbol{y}_n$ belong to R_T, there are vectors $\boldsymbol{z}_1, \boldsymbol{z}_2, \ldots, \boldsymbol{z}_n$ in V such that

$$T(\boldsymbol{z}_1) = \boldsymbol{y}_1, \qquad T(\boldsymbol{z}_2) = \boldsymbol{y}_2, \qquad \ldots, \qquad T(\boldsymbol{z}_n) = \boldsymbol{y}_n$$

We wish to show that $\boldsymbol{x}_1, \boldsymbol{x}_2, \ldots, \boldsymbol{x}_m, \boldsymbol{z}_1, \boldsymbol{z}_2, \ldots, \boldsymbol{z}_n$ form a basis for V. First, the vectors $\boldsymbol{x}_1, \boldsymbol{x}_2, \ldots, \boldsymbol{x}_m, \boldsymbol{z}_1, \boldsymbol{z}_2, \ldots, \boldsymbol{z}_n$ are linearly independent. For suppose we have

$$\alpha_1\boldsymbol{x}_1 + \alpha_2\boldsymbol{x}_2 + \cdots + \alpha_m\boldsymbol{x}_m + \beta_1\boldsymbol{z}_1 + \beta_2\boldsymbol{z}_2 + \cdots + \beta_n\boldsymbol{z}_n = \mathbf{0}$$

Then, applying T,

$$\alpha_1 T(x_1) + \alpha_2 T(x_2) + \cdots + \alpha_m T(x_m) + \beta_1 T(z_1) + \beta_2 T(z_2)$$
$$+ \cdots + \beta_n T(z_n) = 0$$

Since $x_1, x_2, \ldots, x_m$ belong to N_T, $T(x_1) = T(x_2) = \cdots = 0$. So,

$$\beta_1 T(z_1) + \beta_2 T(z_2) + \cdots + \beta_n T(z_n) = 0$$

But $T(z_1) = y_1$, $T(z_2) = y_2, \ldots, T(z_n) = y_n$ is by hypothesis a basis for R_T. Thus, $\beta_1 = \beta_2 = \cdots = \beta_n = 0$, and so $\alpha_1 x_1 + \alpha_2 x_2 + \cdots + \alpha_m x_m = 0$. Since $x_1, x_2, \ldots, x_m$ is a basis for N_T, we have $\alpha_1 = \alpha_2 = \cdots = \alpha_m = 0$. It therefore follows that the vectors $x_1, x_2, \ldots, x_n, z_1, z_2, \ldots, z_n$ are linearly independent.

Secondly, the vectors $x_1, x_2, \ldots, x_m, z_1, z_2, \ldots, z_n$ span V. For suppose x is a vector in V. Then $T(x)$ is a vector in R_T, and since $y_1, y_2, \ldots, y_n$ is a basis for R_T, there are scalars $\beta_1, \beta_2, \ldots, \beta_n$ such that $T(x) = \beta_1 y_1 + \beta_2 y_2 + \cdots + \beta_n y_n$.

Consider the vector $a = x - \beta_1 z_1 - \beta_2 z_2 - \cdots - \beta_n z_n$. Note that

$$T(a) = T(x) - \beta_1 T(z_1) - \cdots - \beta_n T(z_n)$$
$$= \beta_1 y_1 + \beta_2 y_2 + \cdots + \beta_n y_n - \beta_1 y_1 - \beta_2 y_2 - \cdots - \beta_n y_n$$
$$= 0$$

Thus, $a \in N_T$, and since $x_1, x_2, \ldots, x_m$ is a basis for N_T, there are scalars $\alpha_1, \alpha_2, \ldots, \alpha_m$ such that $a = \alpha_1 x_1 + \alpha_2 x_2 + \cdots + \alpha_m x_m$. Rewriting $a = x - \beta_1 z_1 - \beta_2 z_2 - \cdots - \beta_n z_n = \alpha_1 x_1 + \alpha_2 x_2 + \cdots + \alpha_m x_m$, we see that $x = \alpha_1 x_1 + \alpha_2 x_2 + \cdots + \alpha_m x_m + \beta_1 z_1 + \beta_2 z_2 + \cdots + \beta_n z_n$.

Thus, the vectors $x_1, x_2, \ldots, x_n, z_1, z_2, \ldots, z_n$ span V, and since their linear independence was demonstrated above, we see that $x_1, x_2, \ldots, x_m, z_1, z_2, \ldots, z_n$ form a basis for V. Thus,

$$\dim V = m + n$$
$$= \dim N_T + \dim R_T \ \blacksquare$$

Corollary 1 If $T: V \to W$ is a linear transformation, then $\dim R_T \leq \dim V$.

Proof From the above theorem, we know that $\dim V = \dim R_T + \dim N_T \geq \dim R_T$. $\blacksquare$

Now suppose that $\dim W > \dim V$. Since $\dim V \geq \dim R_T$, we must have $\dim W > \dim R_T$. Thus, if $T : V \to W$ is a linear transformation with $\dim V < \dim W$, T is not onto.

Example 5 Let

$$
\begin{aligned}
a_{11}x_1 + a_{12}x_2 + \cdots + a_{1n}x_n &= y_1 \\
a_{21}x_1 + a_{22}x_2 + \cdots + a_{2n}x_n &= y_2 \\
&\vdots \\
a_{m1}x_1 + a_{m2}x_2 + \cdots + a_{mn}x_n &= y_m
\end{aligned}
$$

be a system of m equations in n variables. If $m > n$, i.e., the number of equations is greater than the number of unknowns, by Corollary 1, it is possible to choose values of $y_1, y_2, \ldots, y_m$ so that the system will not be solvable.

Corollary 2 If $T: V \to W$ is a linear transformation, and $\dim W < \dim V$, then $N_T \neq \mathbf{0}$, i.e., $n(T) > 0$.

Proof We know that $\dim N_T + \dim R_T = \dim V$. Since R_T is a subspace of W, $\dim R_T \leq \dim W < \dim V$. Thus, $\dim N_T = \dim V - \dim R_T > 0$. ∎

Example 6 Let

$$
\begin{aligned}
a_{11}x_1 + a_{12}x_2 + \cdots + a_{1n}x_n &= 0 \\
a_{21}x_1 + a_{22}x_2 + \cdots + a_{2n}x_n &= 0 \\
&\vdots \\
a_{m1}x_1 + a_{m2}x_2 + \cdots + a_{mn}x_n &= 0
\end{aligned}
$$

be a system of m homogeneous linear equations in n variables, where $m < n$, i.e., the number of equations is less than the number of unknowns. As usual, let $A = [a_{ij}]_{(mn)}$ and consider the linear transformation $T_A : \mathbf{R}^n \to \mathbf{R}^m$, associated with the matrix A. Since $\dim \mathbf{R}^m < \dim \mathbf{R}^n$, by Corollary 2, N_{T_A} must contain some vector other than the zero vector. This is merely a restatement of the fact we proved in chapter 1, namely, that the above system of equations admits a nontrivial solution.

Corollary 3 Let $T : V \to W$ be a linear transformation and suppose $\dim V = \dim W$. Then, T is one-one if and only if T is onto.

Proof By Theorem 3, $\dim R_T + \dim N_T = \dim V = \dim W$. Thus, $\dim N_T = \dim W - \dim R_T$. So, $\dim N_T = 0$ (i.e., T is one-one) if and only if $\dim W = \dim R_T$ (i.e., T is onto). ∎

Example 7 Let A be an $n \times n$ matrix. Consider the system of linear equations $A\boldsymbol{x} = \boldsymbol{y}$, where $\boldsymbol{x}$ and $\boldsymbol{y}$ are n-vectors. By Corollary 3, we see that the system $A\boldsymbol{x} = \boldsymbol{y}$ is solvable regardless of the value of $\boldsymbol{y}$ if and only if the system $A\boldsymbol{x} = \boldsymbol{0}$ admits only the trivial solution $\boldsymbol{x} = \boldsymbol{0}$.

EXERCISES

1. Calculate a basis for the nullspace and the nullity of the linear transformation associated with the following matrices.

(a) $\begin{bmatrix} -1 & 2 \\ 3 & -6 \end{bmatrix}$ (b) $\begin{bmatrix} 1 & 0 & 3 \\ -1 & 2 & 1 \end{bmatrix}$ (c) $\begin{bmatrix} 0 & 1 & 7 \\ -1 & 2 & 3 \\ -1 & 3 & 10 \end{bmatrix}$

(d) $\begin{bmatrix} 1 & 3 & -1 & 2 \\ 3 & -4 & 0 & 1 \\ 5 & 2 & -2 & 5 \\ 4 & -1 & -1 & 3 \end{bmatrix}$ (e) $\begin{bmatrix} 0 & 1 & 3 & 1 & 4 \\ 1 & 0 & 2 & -1 & -1 \\ 1 & 0 & 1 & -1 & -2 \end{bmatrix}$

2. Let $a_1, a_2, \ldots, a_n$ be n real numbers not all of which are zero. Show that the function

$$f\left(\begin{bmatrix} x_1 \\ x_2 \\ \vdots \\ x_n \end{bmatrix}\right) = a_1x_1 + a_2x_2 + \cdots + a_nx_n$$

is linear from $\boldsymbol{R}^n$ to $\boldsymbol{R}$. Show that its nullspace is of dimension $n - 1$.

3. Let D^k be the linear transformation from P_n into P_n which carries each polynomial into its kth derivative. Show that the nullspace of D^k is of dimension k.

4. Let T be the linear operator on P_2 defined by $T(f) = f - xf' + (a - x^2)f''$, where a is some number. Show that the polynomial x spans N_T.

5. Find a linear transformation T from $\boldsymbol{R}^2$ into itself such that $R_T = N_T$.

6. Let D be an $n \times n$ diagonal matrix and $T_D(\boldsymbol{x}) = D\boldsymbol{x}$ be the linear operator on $\boldsymbol{R}^n$ induced by D. Show that the nullity of T_D is just the number of zeroes on the diagonal of D.

7. Let V be a vector space and $\boldsymbol{x}_1, \boldsymbol{x}_2, \ldots, \boldsymbol{x}_n$ be a basis for V. Let $T: V \to V$ be the linear transformation such that

$$T(\boldsymbol{x}_1) = \boldsymbol{x}_2, \qquad T(\boldsymbol{x}_2) = \boldsymbol{x}_3, \qquad \ldots, \qquad T(\boldsymbol{x}_{n-1}) = \boldsymbol{x}_n, \qquad T(\boldsymbol{x}_n) = \boldsymbol{0}$$

What is $n(T)$?

8. Let $T: V \to V$ be a linear transformation with the property that $N_T = V$. Show that T is the zero operator.

9. If A is an $n \times n$ matrix and there is some nonzero n-vector $\boldsymbol{x}$ such that $A\boldsymbol{x} = \boldsymbol{0}$, show that $r(A) < n$. Prove the converse.

10. Let A be a fixed $n \times n$ matrix. Suppose the nullspace of the operator on $\boldsymbol{R}^n$ induced by A is of dimension k. Let T be the linear transformation from the space of $n \times n$ matrices into itself defined by $T(B) = AB$.
    (a) Show that B belongs to N_T if and only if each column of B belongs to N_A.
    (b) Show that $n(T) = nk$.
    (c) Show that $r(T) = nr(A)$.
11. Let A be an $m \times n$ matrix. Let $T_A : \boldsymbol{R}^n \to \boldsymbol{R}^m$ be the linear transformation $T_A(\boldsymbol{x}) = A\boldsymbol{x}$, where $\boldsymbol{x}$ belongs to $\boldsymbol{R}^n$. Prove the following statements.
    (a) T_A is onto if and only if the columns of A span $\boldsymbol{R}^m$.
    (b) T_A is one-one if and only if the columns of A are linearly independent.
    (c) If T_A is one-one and onto, $m = n$.
12. Let T be a linear transformation from a vector space V to a vector space W. Let $\boldsymbol{y}$ be a vector in W. If the equation $T(\boldsymbol{x}) = \boldsymbol{y}$ is uniquely solvable for a vector $\boldsymbol{x}$ in V, show that T is one-one.
13. Let $T : P_n \to P_n$ be the linear operator on the space of polynomials of degree less than or equal to n in a variable x defined by $T(f) = f + xf'$. Show that T is one-one and onto.
14. Give an example of linear transformation on $\boldsymbol{R}^3$ of rank 1; of rank 2.
15. Let T be a linear transformation on $\boldsymbol{R}^2$, $T \neq 0$ and T not onto. What are $r(T)$ and $n(T)$?
16. Let $T : V \to W$ be a linear transformation with the property that whenever $\boldsymbol{x}_1, \boldsymbol{x}_2, \ldots, \boldsymbol{x}_n$ are linearly independent in V, $T(\boldsymbol{x}_1), T(\boldsymbol{x}_2), \ldots, T(\boldsymbol{x}_n)$ are linearly independent in W. Show that T is one-one. Show that if T is one-one, it has this property.
17. Let $T : P_n \to P_n$ be the function which carries the polynomial $f(x)$ into the polynomial $\frac{1}{2}(f(x) + f(-x))$.
    (a) Show that T is linear.
    (b) Show that R_T consists precisely of the even polynomials in P_n.
    (c) Show that N_T consists precisely of the odd polynomials in P_n.
18. Let $T : \boldsymbol{R}^n \to \boldsymbol{R}^n$ be the linear operator

$$T\begin{bmatrix} x_1 \\ x_2 \\ x_3 \\ \vdots \\ x_n \end{bmatrix} = \begin{bmatrix} 1-n & 1 & 1 & \cdots & 1 \\ 1 & 1-n & 1 & \cdots & 1 \\ 1 & 1 & 1-n & \cdots & 1 \\ & & \vdots & & \\ 1 & 1 & 1 & \cdots & 1-n \end{bmatrix}\begin{bmatrix} x_1 \\ x_2 \\ x_3 \\ \vdots \\ x_n \end{bmatrix}$$

    Show that N_T is spanned by the column n-vector all of whose entries are 1. What is $r(T)$?
19. Let A be a fixed $n \times n$ matrix. Let T be the linear transformation from M_{nn} into M_{nn} defined by $T(B) = AB - BA$. Show that T is not one-one.
20. If f is a polynomial of degree less than or equal to n, show that there is a polynomial g of degree less than or equal to n such that $g + g' = f$.
21. Let T be a linear transformation between two vector spaces V and W.

Let $\boldsymbol{x}_1, \boldsymbol{x}_2, \ldots, \boldsymbol{x}_n$ and $\boldsymbol{y}_1, \boldsymbol{y}_2, \ldots, \boldsymbol{y}_m$ be vectors in V. If $\mathrm{sp}(\boldsymbol{x}_1, \boldsymbol{x}_2, \ldots, \boldsymbol{x}_n) = N_T$ and $\mathrm{sp}(T(\boldsymbol{y}_1), T(\boldsymbol{y}_2), \ldots, T(\boldsymbol{y}_m)) = R_T$, show that $\mathrm{sp}(\boldsymbol{x}_1, \boldsymbol{x}_2, \ldots, \boldsymbol{x}_n, \boldsymbol{y}_1, \boldsymbol{y}_2, \ldots, \boldsymbol{y}_m) = V$.

22. Let $T : V \to V$ be a linear operator on a vector space V. If $R_T = N_T$, show that $\dim V$ is even.

23. If A is an $m \times n$ matrix, show that there is an $n \times n$ matrix B, $B \neq 0$, such that $AB = 0$ if and only if $r(A) < n$.

24. If A is an $m \times n$ matrix, show that there is an $m \times m$ matrix B, $B \neq 0$, such that $BA = 0$ if and only if $r(A) < m$.

25. If A is an upper triangular $n \times n$ matrix, show that the nullity of A is just the number of zeroes on the diagonal of A.

26. Let A and B be $n \times n$ matrices with B invertible. Let S and T be the linear transformations on $\boldsymbol{R}^n$ defined by $S(\boldsymbol{x}) = A\boldsymbol{x}$, $T(\boldsymbol{x}) = (BA)\boldsymbol{x}$. Show that S and T have the same nullspace.

27. Let a, b, and c be real numbers. If a, b, and c are not all equal, show that the linear transformation T defined by

$$T\left(\begin{bmatrix} x \\ y \\ z \end{bmatrix}\right) = \begin{bmatrix} 1 & a & b+c \\ 1 & b & a+c \\ 1 & c & a+b \end{bmatrix}\begin{bmatrix} x \\ y \\ z \end{bmatrix}$$

is of rank 2 and its nullspace is spanned by the vector

$$\begin{bmatrix} a+b+c \\ -1 \\ -1 \end{bmatrix}$$

28. Show that the k points $(x_1, y_1), (x_2, y_2), \ldots, (x_k, y_k)$ in the plane lie on some straight line if and only if the rank of

$$\begin{bmatrix} x_1 & y_1 & 1 \\ x_2 & y_2 & 1 \\ & \vdots & \\ x_k & y_k & 1 \end{bmatrix}$$

is less than or equal to 2.

29. Let H be the subset of the $n \times n$ matrices consisting of those matrices A such that $A\boldsymbol{x} = \boldsymbol{0}$, where $\boldsymbol{x}$ is a fixed nonzero n-vector. Show that H is a subspace of the $n \times n$ matrices of dimension $n^2 - n$.

5 RANK AND ELEMENTARY MATRICES

If A is an $m \times n$ matrix, A induces a linear transformation $T_A : \boldsymbol{R}^n \to \boldsymbol{R}^m$ by defining $T_A(\boldsymbol{x}) = A\boldsymbol{x}$. We defined the rank of T_A as dimension of the range space of T_A. By the **rank of the matrix** A we mean simply the rank of the linear transformation T_A. We also call the range space of T_A simply the range of A.

If we denote $A_1, A_2, \ldots, A_n$ the columns of the matrix A, we saw in §5.3 that the range space of T_A is spanned by the m-vectors, $A_1, A_2, \ldots, A_n$. For this reason the range space of T_A is sometimes called the **column space** of the matrix A. Thus, the rank of the matrix A might well be defined to be the dimension of the span of the m-vectors $A_1, A_2, \ldots, A_n$. By Theorem 1 of §4.7, we see that the rank of A is the maximum number of linearly independent vectors in the collection $A_1, A_2, \ldots, A_n$. In this section we simplify the problem of computing the rank of a matrix by providing an algorithm for its calculation.

The following theorem is important for our development of a process for calculating rank.

Theorem 1 Let A be an $m \times n$ matrix, B be an $m \times m$ matrix, and C be an $n \times n$ matrix. Then, $r(BA) \leq r(A)$ and $r(AC) \leq r(A)$.

Proof First, $r(BA) \leq r(A)$.

Let $\boldsymbol{y}_1, \boldsymbol{y}_2, \ldots, \boldsymbol{y}_{r(A)}$ be a basis for the range of A. Thus, if $\boldsymbol{x}$ belongs to $\boldsymbol{R}^n$, there are scalars $\alpha_1, \alpha_2, \ldots, \alpha_{r(A)}$ such that

$$A\boldsymbol{x} = \alpha_1\boldsymbol{y}_1 + \alpha_2\boldsymbol{y}_2 + \cdots + \alpha_{r(A)}\boldsymbol{y}_{r(A)}$$

So,

$$\begin{aligned}(BA)(\boldsymbol{x}) &= B(A\boldsymbol{x})\\ &= B(\alpha_1\boldsymbol{y}_1 + \alpha_2\boldsymbol{y}_2 + \cdots + \alpha_{r(A)}\boldsymbol{y}_{r(A)})\\ &= \alpha_1 B\boldsymbol{y}_1 + \alpha_2 B\boldsymbol{y}_2 + \cdots + \alpha_{r(A)} B\boldsymbol{y}_{r(A)}\end{aligned}$$

Thus, the vectors $B\boldsymbol{y}_1, B\boldsymbol{y}_2, \ldots, B\boldsymbol{y}_{r(A)}$ span the range space of BA. By Theorem 1, §4.8, we see that $\dim R_{BA} \leq \dim R_A$. Thus, $r(BA) \leq r(A)$.

Next, we show that $r(AC) \leq r(A)$. We claim that the range space of AC is contained in the range space of A. Suppose $\boldsymbol{y}$ belongs to the range of AC, $\boldsymbol{y} = (AC)(\boldsymbol{x}) = A(C\boldsymbol{x})$. Thus, $\boldsymbol{y}$ is the A image of a vector, namely $C\boldsymbol{x}$, and so $\boldsymbol{y}$ belongs to the range of A. Thus, the range space of AC is a subspace of the range space of A, and so, by Theorem 3, §4.8, $\dim R_{AC} \leq \dim R_A$, or $r(AC) \leq r(A)$. ∎

Immediately from Theorem 1 we get the following:

Theorem 2 Let A be an $m \times n$ matrix, B be an invertible $m \times m$ matrix, and C be an invertible $n \times n$ matrix. Then,

$$r(BA) = r(A) \quad \text{and} \quad r(AC) = r(A)$$

Proof By Theorem 1, $r(BA) \leq r(A)$. Again, by Theorem 1,

$$r(B^{-1}(BA)) \leq r(BA)$$

or

$$r(A) \leq r(BA)$$

Thus,

$$r(A) = r(BA)$$

The proof that $r(AC) = r(A)$ is similar. ∎

For example, the matrix

$$A = \begin{bmatrix} 1 & 0 & -1 \\ 0 & 1 & -1 \\ 0 & 0 & 1 \end{bmatrix}$$

is invertible, since $\det A = 1 \neq 0$.

Thus, if B is the matrix

$$\begin{bmatrix} 1 & 0 & 1 \\ 3 & 1 & 4 \\ 7 & 2 & 9 \end{bmatrix}$$

$r(B) = r(BA)$. From

$$\begin{bmatrix} 1 & 0 & 1 \\ 3 & 1 & 4 \\ 7 & 2 & 9 \end{bmatrix}\begin{bmatrix} 1 & 0 & -1 \\ 0 & 1 & -1 \\ 0 & 0 & 1 \end{bmatrix} = \begin{bmatrix} 1 & 0 & 0 \\ 3 & 1 & 0 \\ 7 & 2 & 0 \end{bmatrix}$$

and from the fact that the matrix on the right has exactly two linearly independent columns, we see that $r(B) = 2$.

Theorem 2 may be restated: Multiplication of a matrix by an invertible matrix does not change its rank. Our next goal is to find a family of invertible matrices of a simple enough character to enable us to reduce a given matrix by a series of matrix multiplications to one whose rank is easily computed. One such family is the collection of elementary matrices described below.

Let E be a matrix obtained from the identity matrix by one of the three operations:

(1) Interchanging two columns of the identity matrix.

(2) Multiplying a column of the identity matrix by a nonzero scalar.

(3) Adding a scalar multiple of one column of the identity matrix to another column.

Then, E is said to be an **elementary matrix.** For example,

$$E = \begin{bmatrix} 0 & 0 & 1 \\ 0 & 1 & 0 \\ 1 & 0 & 0 \end{bmatrix}$$

is obtained by interchanging the first and third columns of the identity

$$I_3 = \begin{bmatrix} 1 & 0 & 0 \\ 0 & 1 & 0 \\ 0 & 0 & 1 \end{bmatrix}$$

Therefore, E is an elementary matrix.

The matrix

$$E = \begin{bmatrix} 1 & 0 & 0 \\ 0 & 2 & 0 \\ 0 & 0 & 1 \end{bmatrix}$$

is obtained by multiplying the second column of the identity matrix by 2. Thus, E is an elementary matrix.

Since the matrix

$$E = \begin{bmatrix} 1 & 0 & -7 \\ 0 & 1 & 0 \\ 0 & 0 & 1 \end{bmatrix}$$

is obtained by adding -7 times the first column of I_3 to its third column, E is an elementary matrix.

Elementary matrices are important mainly because they enable us to present a simple description of the result of multiplying an elementary matrix and another matrix. First, we describe what happens to an $m \times n$ matrix A when we postmultiply it by an $n \times n$ elementary matrix E, that is, when we form the product AE.

In what follows we make extensive use of the fact that if A is an $m \times n$ matrix and $\boldsymbol{e}_i$ is the ith vector in the standard basis for $\boldsymbol{R}^n$, then $A\boldsymbol{e}_i$ is just the ith column of the matrix A, which we denote by A_i.

Proposition 1 Suppose that E is an elementary matrix obtained by interchanging columns i and j, $i < j$, in the identity matrix I_n. Then the product AE is the matrix obtained from A by interchanging columns i and j in the matrix A, all other columns remaining the same.

Proof Since E is obtained from I_n by interchanging columns i and j, the ith column of E is $\boldsymbol{e}_j$, the jth column of E is $\boldsymbol{e}_i$, and the kth column of E, for $k \neq i$ or j, is $\boldsymbol{e}_k$. Thus, $E\boldsymbol{e}_i = \boldsymbol{e}_j$, $E\boldsymbol{e}_j = \boldsymbol{e}_i$, and $E\boldsymbol{e}_k = \boldsymbol{e}_k$ for $k \neq i$ or j. Then,

$$(AE)\boldsymbol{e}_i = A(E\boldsymbol{e}_i) = A\boldsymbol{e}_j = A_j$$
$$(AE)\boldsymbol{e}_j = A(E\boldsymbol{e}_j) = A\boldsymbol{e}_i = A_i$$
$$(AE)\boldsymbol{e}_k = A(E\boldsymbol{e}_k) = A\boldsymbol{e}_k = A_k$$

for $k \neq i$ or j.

From this, it is clear that the ith column of AE is A_j, the jth column of AE is A_i, and the kth column of AE is A_k, if $k \neq i$ or j. Stated otherwise, AE is obtained from A by interchange of columns i and j.

For example, if

$$A = \begin{bmatrix} a_1 & b_1 & c_1 \\ a_2 & b_2 & c_2 \\ a_3 & b_3 & c_3 \end{bmatrix}$$

and

$$E = \begin{bmatrix} 1 & 0 & 0 \\ 0 & 0 & 1 \\ 0 & 1 & 0 \end{bmatrix}$$

is obtained by interchanging the second and third columns of the identity matrix,

$$AE = \begin{bmatrix} a_1 & b_1 & c_1 \\ a_2 & b_2 & c_2 \\ a_3 & b_3 & c_3 \end{bmatrix} \begin{bmatrix} 1 & 0 & 0 \\ 0 & 0 & 1 \\ 0 & 1 & 0 \end{bmatrix} = \begin{bmatrix} a_1 & c_1 & b_1 \\ a_2 & c_2 & b_2 \\ a_3 & c_3 & b_3 \end{bmatrix}$$

is obtained by interchanging the second and third columns of the matrix A.

Proposition 2 Suppose the elementary matrix E is obtained by multiplying the ith column of the identity matrix by a nonzero scalar α. The product AE is the matrix obtained by multiplying the ith column of the matrix A by the scalar α, all other columns remaining the same.

Proof It is clear that the ith column of E is $\alpha\boldsymbol{e}_i$, while the kth column of E, for $k \neq i$ is $\boldsymbol{e}_k$. Thus, $E\boldsymbol{e}_i = \alpha\boldsymbol{e}_i$ and $E\boldsymbol{e}_k = \boldsymbol{e}_k$, for $k \neq i$.

$$(AE)\boldsymbol{e}_k = A(E\boldsymbol{e}_k) = A\boldsymbol{e}_k = A_k, \quad \text{for } k \neq i$$

and

$$(AE)\boldsymbol{e}_i = A(E\boldsymbol{e}_i) = A(\alpha \boldsymbol{e}_i) = \alpha A\boldsymbol{e}_i = \alpha A_i$$

From this, we see that the kth column of AE is A_k, if $k \neq i$. The ith column of AE is αA_i, that is, AE is obtained by multiplying the ith column of A by α. ∎

For example, if

$$A = \begin{bmatrix} a_1 & b_1 & c_1 \\ a_2 & b_2 & c_2 \\ a_3 & b_3 & c_3 \end{bmatrix}$$

and

$$E = \begin{bmatrix} 1 & 0 & 0 \\ 0 & \alpha & 0 \\ 0 & 0 & 1 \end{bmatrix}$$

is obtained by multiplying the second column of the identity matrix by the scalar α,

$$AE = \begin{bmatrix} a_1 & b_1 & c_1 \\ a_2 & b_2 & c_2 \\ a_3 & b_3 & c_3 \end{bmatrix} \begin{bmatrix} 1 & 0 & 0 \\ 0 & \alpha & 0 \\ 0 & 0 & 1 \end{bmatrix} = \begin{bmatrix} a_1 & \alpha b_1 & c_1 \\ a_2 & \alpha b_2 & c_2 \\ a_3 & \alpha b_3 & c_3 \end{bmatrix}$$

is obtained by multiplying the second column of the matrix A by the scalar α.

Proposition 3 Suppose that the elementary matrix E is obtained from I_n by adding α times the jth column of I_n to the ith column of I_n, for $i \neq j$. Then, the matrix AE is obtained from the matrix A by adding α times the jth column of A to the ith column of A, all other columns remaining the same.

Proof It is clear that the kth column of E, for $k \neq i$, is $\boldsymbol{e}_k$ and that the ith column of E is $\boldsymbol{e}_i + \alpha \boldsymbol{e}_j$. Thus,

$$E\boldsymbol{e}_k = \boldsymbol{e}_k, \quad \text{if } k \neq i$$

$$E\boldsymbol{e}_i = \boldsymbol{e}_i + \alpha \boldsymbol{e}_j$$

Therefore,

$$(AE)\boldsymbol{e}_k = A(E\boldsymbol{e}_k) = A\boldsymbol{e}_k = A_k, \quad \text{for } k \neq i$$

$$(AE)\boldsymbol{e}_i = A(E\boldsymbol{e}_i) = A(\boldsymbol{e}_i + \alpha \boldsymbol{e}_j) = A_i + \alpha A_j$$

Hence we see that the kth column of the matrix AE, for $k \neq i$, is A_k, and the ith column of AE is $A_i + \alpha A_j$. So AE can be obtained from A by adding α times the jth column of A to the ith column of A. ∎

For example, if

$$A = \begin{bmatrix} a_1 & b_1 & c_1 \\ a_2 & b_2 & c_2 \\ a_3 & b_3 & c_3 \end{bmatrix}$$

and

$$E = \begin{bmatrix} 1 & 0 & \alpha \\ 0 & 1 & 0 \\ 0 & 0 & 1 \end{bmatrix}$$

is obtained by adding α times the first column of the identity matrix to the third column of the identity,

$$AE = \begin{bmatrix} a_1 & b_1 & c_1 \\ a_2 & b_2 & c_2 \\ a_3 & b_3 & c_3 \end{bmatrix}\begin{bmatrix} 1 & 0 & \alpha \\ 0 & 1 & 0 \\ 0 & 0 & 1 \end{bmatrix} = \begin{bmatrix} a_1 & b_1 & c_1 + \alpha a_1 \\ a_2 & b_2 & c_2 + \alpha a_2 \\ a_3 & b_3 & c_3 + \alpha a_3 \end{bmatrix}$$

is obtained by adding α times the first column of A to the third column of A.

We may combine propositions 1, 2, and 3 into one rule: Postmultiplication of A by an elementary matrix E performs an operation on A which is the same type as that operation used to obtain E from the identity matrix.

As a consequence of Propositions 1, 2, and 3 we have

Theorem 3 Any elementary matrix is invertible and its inverse is a matrix of the same type.

Proof (i) Let E be an elementary matrix obtained by interchanging columns of i and j of the identity for $i < j$. Then, EE, the result of postmultiplying E by E, is a matrix obtained by interchanging the ith and jth columns of E. Thus, $EE = I_n$. So E is invertible and its inverse is E, again an elementary matrix.

(ii) Let E be an elementary matrix obtained by multiplying the ith column of the identity matrix by the nonzero scalar α. Then, E^{-1} is just the matrix obtained by multiplying the ith column of the identity matrix by α^{-1}.

(iii) Let E be an elementary matrix obtained by adding α times the jth column of the identity matrix to the ith column of the identity matrix. Then, E^{-1} is the matrix obtained by adding $-\alpha$ times the jth column of the identity matrix to the ith column of the identity matrix. ∎

For example,

$$\begin{bmatrix} 1 & 0 & 0 \\ 0 & 0 & 1 \\ 0 & 1 & 0 \end{bmatrix}^{-1} = \begin{bmatrix} 1 & 0 & 0 \\ 0 & 0 & 1 \\ 0 & 1 & 0 \end{bmatrix}$$

$$\begin{bmatrix} 1 & 0 & 0 \\ 0 & \alpha & 0 \\ 0 & 0 & 1 \end{bmatrix}^{-1} = \begin{bmatrix} 1 & 0 & 0 \\ 0 & \alpha^{-1} & 0 \\ 0 & 0 & 1 \end{bmatrix}, \qquad \alpha \neq 0$$

$$\begin{bmatrix} 1 & 0 & \alpha \\ 0 & 1 & 0 \\ 0 & 0 & 1 \end{bmatrix}^{-1} = \begin{bmatrix} 1 & 0 & -\alpha \\ 0 & 1 & 0 \\ 0 & 0 & 1 \end{bmatrix}$$

In a similar way it is possible to determine the effect of premultiplying an $m \times n$ matrix A by an $m \times m$ elementary matrix E, that is, forming the product EA. Note that the matrix obtained by interchanging two rows of the identity matrix, multiplying some row by a nonzero scalar, or adding a scalar multiple of one row to another is again elementary.

(i′) If E is an elementary matrix obtained from the identity matrix by interchanging rows i and j, EA is obtained from A by interchanging rows i and j.

(ii′) If E is an elementary matrix obtained from the identity matrix by multiplying the ith row of the identity matrix by the nonzero scalar α, EA is obtained from A by multiplying the ith row of A by the scalar α.

(iii′) If E is an elementary matrix obtained by adding α times the jth row of the identity matrix to the ith row of the identity matrix, for $i \neq j$, then the matrix EA is obtained by adding α times the jth row of A to the ith row of A.

As examples, we see

$$\begin{bmatrix} 1 & 0 & 0 \\ 0 & 0 & 1 \\ 0 & 1 & 0 \end{bmatrix} \begin{bmatrix} a_1 & b_1 & c_1 \\ a_2 & b_2 & c_2 \\ a_3 & b_3 & c_3 \end{bmatrix} = \begin{bmatrix} a_1 & b_1 & c_1 \\ a_3 & b_3 & c_3 \\ a_2 & b_2 & c_2 \end{bmatrix}$$

$$\begin{bmatrix} 1 & 0 & 0 \\ 0 & \alpha & 0 \\ 0 & 0 & 1 \end{bmatrix} \begin{bmatrix} a_1 & b_1 & c_1 \\ a_2 & b_2 & c_2 \\ a_3 & b_3 & c_3 \end{bmatrix} = \begin{bmatrix} a_1 & b_1 & c_1 \\ \alpha a_2 & \alpha b_2 & \alpha c_2 \\ a_3 & b_3 & c_3 \end{bmatrix}$$

$$\begin{bmatrix} 1 & 0 & \alpha \\ 0 & 1 & 0 \\ 0 & 0 & 1 \end{bmatrix}\begin{bmatrix} a_1 & b_1 & c_1 \\ a_2 & b_2 & c_2 \\ a_3 & b_3 & c_3 \end{bmatrix} = \begin{bmatrix} a_1 + \alpha a_3 & b_1 + \alpha b_3 & c_1 + \alpha c_3 \\ a_2 & b_2 & c_2 \\ a_3 & b_3 & c_3 \end{bmatrix}$$

If A is a matrix we sometimes use the word **line** to denote either a row or a column of A.

If A is a matrix, we define an **elementary operation** on A to be any of the following three transformations.

(1) Interchanging two parallel lines of A.
(2) Multiplication of a line of A by a nonzero constant.
(3) Addition of a scalar multiple of a line of A to another parallel line of A.

We have seen above that any elementary operation may be performed by premultiplying or postmultiplying A by a suitable elementary matrix. The object of introducing elementary operations is to facilitate the process of calculating the rank of a matrix. The significance of elementary operations in this procedure stems for the most part from the following theorem.

Theorem 4 Let A and A' be $m \times n$ matrices and suppose that A' is obtained from A by an elementary operation. Then, $r(A) = r(A')$.

Proof Since A' is obtained from A by means of an elementary operation, we have $A' = EA$ or $A' = AE$, where E is a suitable elementary matrix. Since any elementary matrix is invertible by Theorem 3 above, and since by Theorem 2, multiplication of A by an invertible matrix does not change its rank, we see that $r(A') = r(A)$. ∎

If the matrix B can be obtained from the matrix A by a sequence of elementary operations, we say that the matrices A and B are **equivalent** and write $A \sim B$. By Theorem 4, $A \sim B$ implies $r(A) = r(B)$. Note the following properties of equivalence:

(1) $A \sim A$.
(2) $A \sim B$ implies $B \sim A$.
(3) $A \sim B$ and $B \sim C$ implies $A \sim C$.

The first property is obvious. To prove (2), note that if $A \sim B$, there are elementary matrices $E_1, E_2, \ldots, E_m, F_1, F_2, \ldots, F_n$, such that $B = E_1E_2 \cdots E_mAF_1F_2 \cdots F_n$. Therefore, $A = E_m^{-1}E_{m-1}^{-1} \cdots E_1^{-1}BF_n^{-1}F_{n-1}^{-1} \cdots F_1^{-1}$. Since the inverse of an elementary matrix is again an elementary matrix by Theorem 3, we see that A can be obtained from B by elementary operations and so $B \sim A$. Property (3) says that if B is obtained from A by elementary operations and C is obtained from B by elementary operations, then C can be obtained from A by elementary operations.

For example,

$$\begin{bmatrix} 1 & -2 & -1 \\ 0 & 1 & 1 \\ 3 & 2 & 5 \end{bmatrix}$$

Add twice the first column to the second column.

$$\downarrow$$

$$\sim \begin{bmatrix} 1 & 0 & -1 \\ 0 & 1 & 1 \\ 3 & 8 & 5 \end{bmatrix}$$

Add the first column to the third column.

$$\downarrow$$

$$\sim \begin{bmatrix} 1 & 0 & 0 \\ 0 & 1 & 1 \\ 3 & 8 & 8 \end{bmatrix}$$

Add -1 times the second column to the third column.

$$\downarrow$$

$$\sim \begin{bmatrix} 1 & 0 & 0 \\ 0 & 1 & 0 \\ 3 & 8 & 0 \end{bmatrix}$$

Add -3 times the first row to the third row.

$$\downarrow$$

$$\sim \begin{bmatrix} 1 & 0 & 0 \\ 0 & 1 & 0 \\ 0 & 8 & 0 \end{bmatrix}$$

Add -8 times the second row to the third row.

$$\downarrow$$

$$\begin{bmatrix} 1 & 0 & 0 \\ 0 & 1 & 0 \\ 0 & 0 & 0 \end{bmatrix}$$

Since a basis for the range space of the final matrix above is

$$\begin{bmatrix}1\\0\\0\end{bmatrix}, \quad \begin{bmatrix}0\\1\\0\end{bmatrix}$$

we see that the rank of that matrix is 2. Since equivalent matrices have the same rank, the rank of the first, and indeed of all intervening matrices, is 2.

Using a procedure analogous to Gaussian elimination and suggested by the previous example, it is possible to show that any matrix is equivalent to a matrix of one of the following types:

$$\left[\begin{array}{c}I_r\\ \hline 0\end{array}\right], \quad \left[\begin{array}{c|c}I_r & 0\end{array}\right], \quad \left[\begin{array}{c|c}I_r & 0\\ \hline 0 & 0\end{array}\right], \quad I_r$$

I_r is the identity matrix of order r and 0 stands for a block of 0's. This equivalence may be effected by the following procedure:

(1) Using interchanges of columns and rows, obtain a nonzero element (preferably a 1) in the first row and column.

(2) Divide the first column by this element.

(3) By adding appropriate multiples of the first column to each of the remaining columns and then appropriate multiples of the first row to each of the remaining rows, obtain an equivalent matrix of the following form

$$\left[\begin{array}{c|cccc}1 & 0 & 0 & \cdots & 0\\ \hline 0 & a_{22} & a_{23} & \cdots & a_{2n}\\ & & \vdots & & \\ 0 & a_{m2} & a_{m3} & \cdots & a_{mn}\end{array}\right]$$

(4) Repeat the procedure on the submatrix

$$\begin{bmatrix}a_{22} & a_{23} & \cdots & a_{2n}\\ & \vdots & & \\ a_{m2} & a_{m3} & \cdots & a_{mn}\end{bmatrix}$$

(5) As asserted, one of the following types of matrices is obtained

$$\left[\begin{array}{c}I_r\\ \hline 0\end{array}\right], \quad \left[\begin{array}{c|c}I_r & 0\end{array}\right], \quad \left[\begin{array}{c|c}I_r & 0\\ \hline 0 & 0\end{array}\right], \quad I_r$$

Since each of the last matrices admits the vectors $e_1, e_2, \ldots, e_r$ as a basis for its range space, it is clear that each of the last matrices is of rank r. Matrices of the type depicted in (5) are said to be of **normal form.** Thus, every matrix is equivalent to some matrix of normal form.

Example 1 Calculate the rank of the matrix:

$$\begin{bmatrix} 3 & 1 & 3 & 7 \\ -1 & -3 & -1 & -5 \\ 7 & 0 & 2 & 9 \\ 0 & 1 & 3 & 4 \\ 2 & 0 & 1 & 3 \end{bmatrix}$$

Interchange columns one and two.

$$\downarrow$$

$$\sim\begin{bmatrix} 1 & 3 & 3 & 7 \\ -3 & -1 & -1 & -5 \\ 0 & 7 & 2 & 9 \\ 1 & 0 & 3 & 4 \\ 0 & 2 & 1 & 3 \end{bmatrix}$$

Add suitable multiples of column one to each remaining column.

$$\downarrow$$

$$\sim\begin{bmatrix} 1 & 0 & 0 & 0 \\ -3 & 8 & 8 & 16 \\ 0 & 7 & 2 & 9 \\ 1 & -3 & 0 & -3 \\ 0 & 2 & 1 & 3 \end{bmatrix}$$

Add suitable multiples of row one to each remaining row.

$$\downarrow$$

$$\sim\begin{bmatrix} 1 & 0 & 0 & 0 \\ 0 & 8 & 8 & 16 \\ 0 & 7 & 2 & 9 \\ 0 & -3 & 0 & -3 \\ 0 & 2 & 1 & 3 \end{bmatrix}$$

Interchange columns two and three, rows two and five.

$$\downarrow$$

$$\sim\begin{bmatrix}1&0&0&0\\0&1&2&3\\0&2&7&9\\0&0&-3&-3\\0&8&8&16\end{bmatrix}$$

Add suitable multiples of column two to each remaining column. Then, add suitable multiples of row two to each remaining row.

$$\downarrow$$

$$\sim\begin{bmatrix}1&0&0&0\\0&1&0&0\\0&0&3&3\\0&0&-3&-3\\0&0&-8&-8\end{bmatrix}$$

Add (-1) times column three to column four.

$$\downarrow$$

$$\sim\begin{bmatrix}1&0&0&0\\0&1&0&0\\0&0&3&0\\0&0&-3&0\\0&0&-8&0\end{bmatrix}$$

Multiply row three by $\frac{1}{3}$ and add suitable multiples to each remaining row.

$$\downarrow$$

$$\sim\begin{bmatrix}1&0&0&0\\0&1&0&0\\0&0&1&0\\0&0&0&0\\0&0&0&0\end{bmatrix}$$

Thus, the rank of the initial matrix is 3.

The computational procedure used to calculate the rank of a matrix may be used to solve other problems involving dimension.

Example 2 Calculate the dimension of the subspace of solutions to the system of linear homogeneous equations

$$\begin{aligned} x_1 + 2x_2 - 3x_3 + x_4 + x_5 &= 0 \\ -3x_1 + x_2 + 7x_3 - x_4 + x_5 &= 0 \\ -2x_1 + 3x_2 + 4x_3 \qquad + 2x_5 &= 0 \end{aligned}$$

In other words, we wish to calculate the dimension of the nullspace of the linear transformation $T : \boldsymbol{R}^5 \to \boldsymbol{R}^3$. We can write

$$T\begin{bmatrix} x_1 \\ x_2 \\ x_3 \\ x_4 \\ x_5 \end{bmatrix} = \begin{bmatrix} 1 & 2 & -3 & 1 & 1 \\ -3 & 1 & 7 & -1 & 1 \\ -2 & 3 & 4 & 0 & 2 \end{bmatrix}\begin{bmatrix} x_1 \\ x_2 \\ x_3 \\ x_4 \\ x_5 \end{bmatrix}$$

Since $\dim R_T + \dim N_T = 5$, in order to determine $\dim N_T$ we need only determine the rank of T, which is the rank of the matrix

$$\begin{bmatrix} 1 & 2 & -3 & 1 & 1 \\ -3 & 1 & 7 & -1 & 1 \\ -2 & 3 & 4 & 0 & 2 \end{bmatrix}$$

Add suitable multiples of the first column to each of the remaining columns.

$$\sim \begin{bmatrix} 1 & 0 & 0 & 0 & 0 \\ -3 & 7 & -2 & 2 & 4 \\ 2 & 7 & -2 & 2 & 4 \end{bmatrix}$$

Add suitable multiples of the first row to each remaining row.

$$\sim \begin{bmatrix} 1 & 0 & 0 & 0 & 0 \\ 0 & 7 & -2 & 2 & 4 \\ 0 & 7 & -2 & 2 & 4 \end{bmatrix}$$

Add (-1) times the second row to the third row.

$$\sim \begin{bmatrix} 1 & 0 & 0 & 0 & 0 \\ 0 & 7 & -2 & 2 & 4 \\ 0 & 0 & 0 & 0 & 0 \end{bmatrix}$$

Multiply the second column by $\frac{1}{7}$ and add suitable multiples to the remaining columns.

$$\downarrow$$

$$\sim \begin{bmatrix} 1 & 0 & 0 & 0 & 0 \\ 0 & 1 & 0 & 0 & 0 \\ 0 & 0 & 0 & 0 & 0 \end{bmatrix}$$

Thus, $r(T) = 2$. So $\dim N_T = 3$.

Example 3 What is the dimension of the subspace of $\boldsymbol{R}^4$ spanned by the vectors

$$\begin{bmatrix} 1 \\ -1 \\ 1 \\ -1 \end{bmatrix}, \quad \begin{bmatrix} 0 \\ 1 \\ 3 \\ 2 \end{bmatrix}, \quad \begin{bmatrix} 1 \\ 1 \\ 7 \\ 3 \end{bmatrix}$$

Observe that the subspace of $\boldsymbol{R}^4$ spanned by these vectors is precisely the column space of the matrix

$$A = \begin{bmatrix} 1 & 0 & 1 \\ -1 & 1 & 1 \\ 1 & 3 & 7 \\ -1 & 2 & 3 \end{bmatrix}$$

Hence, it suffices to determine $r(A)$.

$$\begin{bmatrix} 1 & 0 & 1 \\ -1 & 1 & 1 \\ 1 & 3 & 7 \\ -1 & 2 & 3 \end{bmatrix} \sim \begin{bmatrix} 1 & 0 & 0 \\ -1 & 1 & 2 \\ 1 & 3 & 6 \\ -1 & 2 & 4 \end{bmatrix} \sim \begin{bmatrix} 1 & 0 & 0 \\ 0 & 1 & 2 \\ 0 & 3 & 6 \\ 0 & 2 & 4 \end{bmatrix}$$

$$\sim \begin{bmatrix} 1 & 0 & 0 \\ 0 & 1 & 0 \\ 0 & 3 & 0 \\ 0 & 2 & 0 \end{bmatrix} \sim \begin{bmatrix} 1 & 0 & 0 \\ 0 & 1 & 0 \\ 0 & 0 & 0 \\ 0 & 0 & 0 \end{bmatrix}$$

Thus, the given collection of vectors spans a space of dimension 2.

Using the procedure of reduction to normal form we obtain Theorem 5.

Theorem 5 Let A be an $m \times n$ matrix of rank r and N be the $m \times n$ matrix of rank r in normal form. Then, there are $m \times m$ elementary matrices $E_1, E_2, \ldots, E_k$ and $n \times n$ elementary matrices $F_1, F_2, \ldots, F_l$ such that $A = E_1E_2 \ldots E_kNF_1F_2 \ldots F_l$.

Proof We know that $A \sim N$. Thus, there are $m \times m$ elementary matrices $G_1, G_2, \ldots, G_k$ and $n \times n$ elementary matrices $H_1, H_2, \ldots, H_l$ such that

$$N = G_1G_2 \ldots G_kAH_1H_2 \ldots H_l$$

or

$$A = G_k^{-1} \ldots G_1^{-1}NH_l^{-1} \ldots H_1^{-1}$$

Since the inverse of an elementary matrix is elementary by Theorem 3, renaming $G_k^{-1}, \ldots, G_1^{-1}, H_l^{-1}, \ldots, H_1^{-1}$, we obtain the desired result. ∎

Corollary 1 Any two matrices of the same rank are equivalent.

Proof If A and B are $m \times n$ matrices of rank r, then $A = E_1 \ldots E_kNF_1F_2 \ldots F_l$, $B = G_1 \ldots G_pNH_1 \ldots H_q$, for elementary matrices E_i, F_i, G_i, H_i. Thus, $G_1 \ldots G_pE_k^{-1} \ldots E_1^{-1}AF_l^{-1} \ldots F_1^{-1}H_1 \ldots H_q = B$. So, A and B are equivalent. ∎

Corollary 2 Any invertible $n \times n$ matrix can be factored into the product of elementary matrices.

Proof Since A is invertible, it is of rank n, and its normal form is just the matrix I_n. Thus, there are elementary matrices $E_1, E_2, \ldots, E_k, F_1, F_2, \ldots, F_l$ such that $A = (E_1E_2 \ldots E_k)I_n(F_1F_2 \ldots F_l)$. ∎

Corollary 2 is often useful in proving results about invertible matrices.

By keeping careful track of the operations used to reduce a matrix to normal form it is possible to calculate the inverse of the matrix, when it exists.

Example 4 Find the inverse of the matrix

$$\begin{bmatrix} 1 & 3 & 4 \\ -3 & 1 & -2 \\ 2 & 0 & 3 \end{bmatrix}$$

Add -3 times the first column to the second column.

$$\downarrow$$

$$\sim \begin{bmatrix} 1 & 0 & 4 \\ -3 & 10 & -2 \\ 2 & -6 & 3 \end{bmatrix}$$

Add -4 times the first column to the third column.

$$\sim \begin{bmatrix} 1 & 0 & 0 \\ -3 & 10 & 10 \\ 2 & -6 & -5 \end{bmatrix}$$

Add -1 times the second column to the third column.

$$\sim \begin{bmatrix} 1 & 0 & 0 \\ -3 & 10 & 0 \\ 2 & -6 & 1 \end{bmatrix}$$

Add -2 times the third column to the first column.

$$\sim \begin{bmatrix} 1 & 0 & 0 \\ -3 & 10 & 0 \\ 0 & -6 & 1 \end{bmatrix}$$

Add 6 times the third column to the second column.

$$\sim \begin{bmatrix} 1 & 0 & 0 \\ -3 & 10 & 0 \\ 0 & 0 & 1 \end{bmatrix}$$

Multiply the second column by $\frac{1}{10}$.

$$\sim \begin{bmatrix} 1 & 0 & 0 \\ -3 & 1 & 0 \\ 0 & 0 & 1 \end{bmatrix}$$

Add 3 times the second column to the first column.

$$\sim \begin{bmatrix} 1 & 0 & 0 \\ 0 & 1 & 0 \\ 0 & 0 & 1 \end{bmatrix}$$

Now, if we consider the elementary matrix required to effect each step in the operation, we have

$$E_1 = \begin{bmatrix} 1 & -3 & 0 \\ 0 & 1 & 0 \\ 0 & 0 & 1 \end{bmatrix}, \qquad E_2 = \begin{bmatrix} 1 & 0 & -4 \\ 0 & 1 & 0 \\ 0 & 0 & 1 \end{bmatrix}$$

$$E_3 = \begin{bmatrix} 1 & 0 & 0 \\ 0 & 1 & -1 \\ 0 & 0 & 1 \end{bmatrix}, \qquad E_4 = \begin{bmatrix} 1 & 0 & 0 \\ 0 & 1 & 0 \\ -2 & 0 & 1 \end{bmatrix}$$

$$E_5 = \begin{bmatrix} 1 & 0 & 0 \\ 0 & 1 & 0 \\ 0 & 6 & 1 \end{bmatrix}, \qquad E_6 = \begin{bmatrix} 1 & 0 & 0 \\ 0 & \frac{1}{10} & 0 \\ 0 & 0 & 1 \end{bmatrix}$$

$$E_7 = \begin{bmatrix} 1 & 0 & 0 \\ 3 & 1 & 0 \\ 0 & 0 & 1 \end{bmatrix}$$

and

$$\begin{bmatrix} 1 & 3 & 4 \\ -3 & 1 & -2 \\ 2 & 0 & 3 \end{bmatrix} E_1E_2E_3E_4E_5E_6E_7 = I_3$$

Thus,

$$\begin{bmatrix} 1 & 3 & 4 \\ -3 & 1 & -2 \\ 2 & 0 & 3 \end{bmatrix}^{-1} = E_1E_2E_3E_4E_5E_6E_7$$

What is the matrix $E_1E_2E_3E_4E_5E_6E_7$? It is the matrix obtained by applying steps 1 through 7 above to the identity matrix. Doing this,

$$\begin{bmatrix} 1 & 0 & 0 \\ 0 & 1 & 0 \\ 0 & 0 & 1 \end{bmatrix} \sim \begin{bmatrix} 1 & -3 & 0 \\ 0 & 1 & 0 \\ 0 & 0 & 1 \end{bmatrix} \sim \begin{bmatrix} 1 & -3 & -4 \\ 0 & 1 & 0 \\ 0 & 0 & 1 \end{bmatrix}$$

$$\sim \begin{bmatrix} 1 & -3 & -1 \\ 0 & 1 & -1 \\ 0 & 0 & 1 \end{bmatrix} \sim \begin{bmatrix} 3 & -3 & -1 \\ 2 & 1 & -1 \\ -2 & 0 & 1 \end{bmatrix} \sim \begin{bmatrix} 3 & -9 & -1 \\ 2 & -5 & -1 \\ -2 & 6 & 1 \end{bmatrix}$$

$$\sim \begin{bmatrix} 3 & -\frac{9}{10} & -1 \\ 2 & -\frac{1}{2} & -1 \\ -2 & \frac{3}{5} & 1 \end{bmatrix} \sim \begin{bmatrix} \frac{3}{10} & -\frac{9}{10} & -1 \\ \frac{1}{2} & -\frac{1}{2} & -1 \\ -\frac{1}{5} & \frac{3}{5} & 1 \end{bmatrix}$$

We conclude that

$$\begin{bmatrix} 1 & 3 & 4 \\ -3 & 1 & -2 \\ 2 & 0 & 3 \end{bmatrix}^{-1} = \begin{bmatrix} \frac{3}{10} & -\frac{9}{10} & -1 \\ \frac{1}{2} & -\frac{1}{2} & -1 \\ -\frac{1}{5} & \frac{3}{5} & 1 \end{bmatrix}$$

The general principle may be stated thus: Keep track of the column operations required to reduce the given matrix to the identity matrix. Then perform the same operations in the same order on the identity. The resulting matrix is the inverse of the original matrix.

We provide an alternative definition of rank which is sometimes useful. If A is a matrix, by a **submatrix** of A, we mean a matrix obtained by deleting a certain number of rows and columns of A.

For example, if

$$A = \begin{bmatrix} 1 & 0 & 3 & 1 \\ -3 & 1 & 4 & 2 \\ 7 & 2 & 3 & 1 \end{bmatrix}$$

we see that

$$\begin{bmatrix} 1 & 3 \\ -3 & 4 \\ 7 & 3 \end{bmatrix} \quad \text{and} \quad \begin{bmatrix} 1 & 1 \\ -3 & 2 \end{bmatrix}$$

are submatrices of A. The first is obtained by deleting the second and fourth columns of A. The second is obtained by deleting the second and third columns, as well as the third row, of A.

If A is a matrix, by a **minor** of order r of A, we mean the determinant of some $r \times r$ submatrix of A.

For example, if

$$A = \begin{bmatrix} 1 & -1 & 2 \\ 0 & 3 & 7 \end{bmatrix}$$

then

$$\begin{vmatrix} 1 & -1 \\ 0 & 3 \end{vmatrix} = 3, \quad \begin{vmatrix} 1 & 2 \\ 0 & 7 \end{vmatrix} = 7, \quad \text{and} \begin{vmatrix} -1 & 2 \\ 3 & 7 \end{vmatrix} = -13$$

are second-order minors of A. If

$$A = \begin{bmatrix} 1 & -3 & 7 & 0 \\ 1 & 2 & 0 & 0 \\ 1 & 0 & 2 & 1 \end{bmatrix}$$

then

$$\begin{vmatrix} 1 & 7 & 0 \\ 1 & 0 & 0 \\ 1 & 2 & 1 \end{vmatrix} = -7 \quad \text{and} \quad \begin{vmatrix} -3 & 0 \\ 2 & 0 \end{vmatrix} = 0$$

are minors of order 3 and 2 of the matrix A, respectively.

If A is an $m \times n$ matrix we define an integer $p(A)$, depending on A, as the largest nonvanishing minor of the matrix A.

For example, in the matrix

$$A = \begin{bmatrix} 1 & 1 & 0 \\ 0 & 0 & 0 \\ 0 & 0 & 0 \end{bmatrix}$$

$p(A) = 1$. Indeed, inspection of A shows that all minors of order 2 and 3 vanish. Since A has nonzero entries, it has nonvanishing minors of order 1.

In the matrix

$$A = \begin{bmatrix} 1 & 3 & 4 \\ 2 & 1 & 3 \\ 0 & 2 & 2 \end{bmatrix}$$

$p(A) = 2$. Since the third column is the sum of the first and second columns, $\det A = 0$. But A has at least one nonzero minor of order 2, namely,

$$\begin{vmatrix} 1 & 3 \\ 2 & 1 \end{vmatrix} = -5 \neq 0$$

By calculating the rank of each matrix in the preceding examples, it can be seen that $r(A) = p(A)$. It is our goal to prove that this relation is always true.

If A is a matrix in normal form then $p(A) = r = r(A)$. For A certainly has a nonvanishing minor of order r, namely I_r, while any larger minor of A has a line of zeroes, and thus is necessarily zero.

As a preliminary step in the proof of the relation $p(A) = r(A)$, we prove the following lemma.

Lemma If the matrix B is obtained from A by an elementary operation, then $p(B) \leq p(A)$.

Proof (i) Suppose B is obtained from A by interchange of parallel lines. Then, any minor of B can be obtained from a minor of A by interchanging parallel lines. Since this merely changes the sign of the minor, $p(B) = p(A)$.

(ii) If B is obtained from A by multiplication of some line by a nonzero scalar, say α, then any minor of B is either a minor of A or can be obtained from a minor of A by multiplying by the nonzero scalar α. Thus, $p(B) = p(A)$.

(iii) Suppose B is obtained from A by the addition of some multiple of a line of A to another parallel line of A. Suppose $m > p(A)$ and consider a minor of B of order m. Either the minor of B of order m is a minor of A, or it is the sum of two minors of A. In either case, it is clear that the given minor is zero. Thus, all minors of B of order m are zero, and so $p(B) \leq p(A)$. ∎

We are now ready to prove the final result.

Theorem 6 If A is an $m \times n$ matrix, $p(A) = r(A)$.

Proof Suppose $A \sim B$. Since B can, by definition, be obtained from A by means of elementary transformations, $p(B) \leq p(A)$. However, $A \sim B$ implies $B \sim A$, so we also see that $p(A) \leq p(B)$. Thus, $A \sim B$ implies $p(A) = p(B)$.

We are now finished, for we have seen that $A \sim B$ implies $p(A) = p(B)$ and that if N is a matrix in normal form, then $p(N) = r(N)$. Since any martix A is equivalent to a matrix in normal form N

$$p(A) = p(N)$$

$$r(A) = r(N)$$

and

$$p(N) = r(N)$$

Thus, $p(A) = r(A)$. ∎

There are many examples in which this definition of rank is useful.

Example 5 Let A be an $n \times n$ matrix, with adjunct $\mathcal{A}$. If $\mathcal{A}$ is the zero matrix, by the definition of the adjunct, all cofactors of A are 0. Thus, all minors of A of order n and $n - 1$ are 0. By definition of $p(A)$, we see that $p(A) \leq n - 2$. Since $p(A) = r(A)$, we obtain $r(A) \leq n - 2$.

On the other hand, suppose $r(A) \leq n - 2$. Since $p(A) = r(A)$, all minors of order $n - 1$ are zero. Thus $\mathcal{A}$, the adjunct of A, is 0. We have shown, therefore, that the adjunct $\mathcal{A}$ of an $n \times n$ matrix is zero if and only if $r(A) \leq n - 2$.

EXERCISES

1. Find the rank of the following matrices.

(a) $\begin{bmatrix} 1 & -3 & -2 \\ 2 & 1 & 3 \\ 0 & 3 & 3 \end{bmatrix}$ (b) $\begin{bmatrix} 1 & 3 & 5 & 8 \\ -2 & 1 & -3 & -2 \\ 0 & 7 & 7 & 14 \end{bmatrix}$

(c) $\begin{bmatrix} -1 & 7 & 0 \\ -3 & 0 & 1 \\ 0 & -1 & 2 \\ 2 & 3 & 0 \end{bmatrix}$ (d) $\begin{bmatrix} 1 & 7 \\ -3 & 5 \\ 0 & 6 \\ 7 & 1 \end{bmatrix}$

(e) $\begin{bmatrix} 4 & 1 & 0 \\ 0 & 0 & 1 \\ 7 & -1 & 3 \\ 3 & 2 & 7 \end{bmatrix}$ (f) $\begin{bmatrix} -1 & 0 & 2 & 3 \\ 3 & 7 & 1 & 4 \end{bmatrix}$

(g) $\begin{bmatrix} 1 & 0 & 1 \\ -4 & 2 & -2 \\ 3 & -3 & 0 \\ 1 & 1 & 2 \end{bmatrix}$

2. Determine the dimension of the solution space for each system of homogeneous linear equations.

(a) $\begin{aligned} -7x + 2y - 3z &= 0 \\ x + 3y + 5z &= 0 \\ -2x - y - 4z &= 0 \end{aligned}$ (b) $\begin{aligned} 3x_1 - 5x_2 + 4x_3 - x_4 &= 0 \\ x_1 - x_2 + 3x_3 + 2x_4 &= 0 \\ x_1 + x_2 - 3x_3 + 3x_4 &= 0 \end{aligned}$

3. Invert the following elementary matrices.

(a) $\begin{bmatrix} 1 & \alpha & 0 \\ 0 & 1 & 0 \\ 0 & 0 & 1 \end{bmatrix}$ (b) $\begin{bmatrix} 1 & 0 & 0 & 0 \\ 0 & 0 & 1 & 0 \\ 0 & 1 & 0 & 0 \\ 0 & 0 & 0 & 1 \end{bmatrix}$

(c) $\begin{bmatrix} \alpha & 0 & 0 \\ 0 & 1 & 0 \\ 0 & 0 & 1 \end{bmatrix}, \alpha \neq 0$ (d) $\begin{bmatrix} 1 & 0 & 0 \\ 0 & 1 & 0 \\ \alpha & 0 & 1 \end{bmatrix}$

4. Express as a product of elementary matrices.

(a) $\begin{bmatrix} 1 & 0 & 1 \\ 3 & 1 & 4 \\ -1 & 2 & 3 \end{bmatrix}$ (b) $\begin{bmatrix} -1 & -2 \\ 3 & 5 \end{bmatrix}$

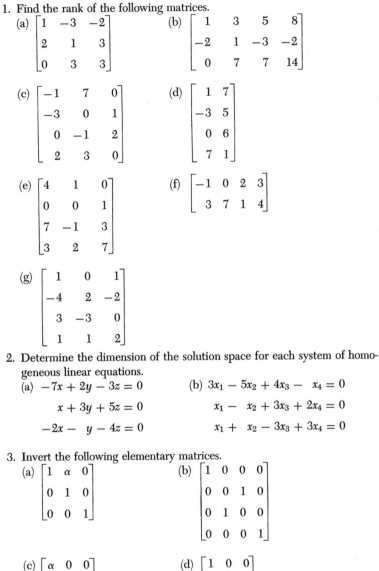

(c) $\begin{bmatrix} 1 & 0 & 3 \\ 0 & 2 & 2 \\ -1 & 1 & 0 \end{bmatrix}$ (d) $\begin{bmatrix} 1 & 3 \\ 2 & 7 \end{bmatrix}$

5. Suppose the $n \times n$ matrix A is postmultiplied by the matrix

$$G = \begin{bmatrix} 1 & 0 & 0 & \dots & 0 & 1 \\ 0 & 1 & 0 & \dots & 0 & 1 \\ 0 & 0 & 1 & \dots & 0 & 1 \\ & & & \vdots & & \\ 0 & 0 & 0 & \dots & 1 & 1 \\ 0 & 0 & 0 & \dots & 0 & 1 \end{bmatrix}$$

that is, the product AG is formed. Show that AG is obtained from A by adding all columns of A except the last to the last column of A, and that all other columns of AG are the same as those of A.

6. If A is an $n \times n$ matrix, show that there are invertible matrices B and C such that
   (a) AB is lower triangular. (b) CA is upper triangular.
7. If A is an $m \times n$ matrix and B is an $n \times p$ matrix, show that $r(AB) \leq \min(r(A), r(B))$.
8. If A is an $m \times n$ matrix, show that there is an invertible $m \times m$ matrix B and an invertible $n \times n$ matrix C such that BAC is a matrix in normal form.
9. Prove in two different ways that premultiplication of a matrix A by an elementary matrix E performs on A an elementary operation of the type described in the text.
   (a) Use the fact that the transpose of an elementary matrix is an elementary matrix and $(EA)^T = A^TE^T$.
   (b) Use a standard basis for the space of row n-vectors.
10. Show that if A is an $m \times n$ matrix, $r(A) = r(A^T)$.
11. If all minors of order r of a matrix are 0, show that all larger minors are 0.
12. Let A be the $2n + 1$ matrix

$$A = \begin{bmatrix} 1 & 0 & 0 & \dots & 0 & 0 & 1 \\ 0 & 1 & 0 & \dots & 0 & 1 & 0 \\ 0 & 0 & 1 & \dots & 1 & 0 & 0 \\ & & & \vdots & & & \\ 0 & 0 & 1 & \dots & 1 & 0 & 0 \\ 0 & 1 & 0 & \dots & 0 & 1 & 0 \\ 1 & 0 & 0 & \dots & 0 & 0 & 1 \end{bmatrix}$$

Show that $r(A) = n + 1$.

13. Let A be a matrix of order $2n$ obtained from the identity matrix by interchanging rows 1 and 2, rows 3 and 4, . . . , rows $2n - 1$ and $2n$. Show that $A^{-1} = A$.

14. Using elementary operations, invert the following matrices.

(a) $\begin{bmatrix} 1 & 0 & 0 & 1 \\ 0 & 1 & 0 & 1 \\ 0 & 0 & 1 & 1 \\ 0 & 0 & 0 & 1 \end{bmatrix}$ (b) $\begin{bmatrix} 1 & 0 & 0 & 1 \\ 1 & 1 & 0 & 1 \\ 1 & 0 & 1 & 1 \\ 0 & 0 & 0 & 1 \end{bmatrix}$

(c) $\begin{bmatrix} 1 & 0 & 0 & \alpha \\ 0 & 1 & \alpha & 0 \\ 0 & 0 & 1 & 0 \\ 0 & 0 & 0 & 1 \end{bmatrix}$ (d) $\begin{bmatrix} 0 & 0 & 0 & 1 \\ 1 & 0 & 0 & 0 \\ 0 & 1 & 0 & 0 \\ 0 & 0 & 1 & 0 \end{bmatrix}$

15. Let A be an $m \times n$ matrix, H be the subspace of $\boldsymbol{R}^n$ spanned by the columns of A, and K be the subspace of $\boldsymbol{R}^m$ spanned by the rows of A. Show $\dim H = \dim K$. [Hint: Use exercise 10.]

16. If A is an $m \times n$ matrix of rank $r(A)$, and $0 \leq k \leq r(A)$, show that there is an $m \times m$ matrix B such that $r(BA) = k$, and an $n \times n$ matrix C such that $r(AC) = k$.

17. Determine the dimension of the subspace of $\boldsymbol{R}^n$ spanned by the following vectors.

(a) $\begin{bmatrix} -1 \\ 3 \\ 2 \end{bmatrix}, \begin{bmatrix} 1 \\ 7 \\ 3 \end{bmatrix}, \begin{bmatrix} 1 \\ 0 \\ 2 \end{bmatrix}$ (b) $\begin{bmatrix} -1 \\ 0 \\ 2 \\ 1 \end{bmatrix}, \begin{bmatrix} 3 \\ 1 \\ -2 \\ -3 \end{bmatrix}, \begin{bmatrix} 2 \\ 1 \\ 0 \\ -2 \end{bmatrix}$

(c) $\begin{bmatrix} 1 \\ -5 \\ 0 \\ 3 \end{bmatrix}, \begin{bmatrix} 1 \\ 0 \\ -3 \\ 2 \end{bmatrix}, \begin{bmatrix} 0 \\ 5 \\ 3 \\ 5 \end{bmatrix}, \begin{bmatrix} 1 \\ 5 \\ 6 \\ 1 \end{bmatrix}$

18. If A and B are $m \times n$ matrices, show that the following are equivalent statements.
 (a) $r(A) = r(B)$.
 (b) There exists an $m \times m$ matrix C and an $n \times n$ matrix D, both invertible, such that $A = CBD$.

19. If A and B are $m \times n$ matrices, show that the following statements are equivalent.
 (a) The range spaces of A and B are the same subspaces of $\boldsymbol{R}^m$.
 (b) There exists an invertible $n \times n$ matrix C such that $A = BC$.

20. If A is an $m \times n$ matrix with the property that r of its columns are linearly independent but any $r + 1$ columns are linearly dependent, show that $r(A) = r$.

21. Let A be an $n \times n$ matrix with adjunct $\mathcal{A}$. If $r(A) = n - 1$, show that $r(\mathcal{A}) = 1$. If $r(A) < n - 1$, show that $r(\mathcal{A}) = 0$.

22. Show that any rectangular matrix of rank r is the sum of r matrices of rank 1.

23. Show that any invertible 2×2 matrix can be expressed as a product of matrices of the form

$$\begin{bmatrix} 1 & t \\ 0 & 1 \end{bmatrix}, \quad \begin{bmatrix} 1 & 0 \\ t & 1 \end{bmatrix}, \quad \begin{bmatrix} t & 0 \\ 0 & 1 \end{bmatrix}, \quad \begin{bmatrix} 1 & 0 \\ 0 & t \end{bmatrix}, \quad \begin{bmatrix} 0 & 1 \\ 1 & 0 \end{bmatrix}$$

24. Consider the system of linear equations

$$\begin{aligned} a_{11}x_1 + a_{12}x_2 + \cdots + a_{1n}x_n &= y_1 \\ a_{21}x_1 + a_{22}x_2 + \cdots + a_{2n}x_n &= y_2 \\ &\vdots \\ a_{m1}x_1 + a_{m2}x_2 + \cdots + a_{mn}x_n &= y_m \end{aligned}$$

Show that the system admits a solution if and only if the rank of the matrix

$$\begin{bmatrix} a_{11} & a_{12} & \cdots & a_{1n} \\ a_{21} & a_{22} & \cdots & a_{2n} \\ & & \vdots & \\ a_{m1} & a_{m2} & \cdots & a_{mn} \end{bmatrix}$$

is the same as that of the matrix

$$\begin{bmatrix} a_{11} & a_{12} & \cdots & a_{1n} & y_1 \\ a_{21} & a_{22} & \cdots & a_{2n} & y_2 \\ & & \vdots & & \\ a_{m1} & a_{m2} & \cdots & a_{mn} & y_m \end{bmatrix}$$

25. Let $A_ix + B_iy + C_i = 0$, $i = 1, 2, \ldots, k$ be a set of k lines. Show that the lines intersect at a point or are parallel if and only if the rank of the matrix

$$\begin{bmatrix} A_1 & B_1 & C_1 \\ A_2 & B_2 & C_2 \\ & \vdots & \\ A_k & B_k & C_k \end{bmatrix}$$

is less than or equal to 2.

26. Show that the k points $(x_1, y_1), (x_2, y_2), (x_3, y_3), \ldots, (x_k, y_k)$ lie on some line or circle in the plane if and only if the rank of the matrix

$$\begin{bmatrix} x_1^2 + y_1^2 & x_1 & y_1 & 1 \\ x_2^2 + y_2^2 & x_2 & y_2 & 1 \\ & \vdots & & \\ x_k^2 + y_k^2 & x_k & y_k & 1 \end{bmatrix}$$

is less than or equal to 3.

27. If A and B are $n \times n$ matrices and C and D are invertible $n \times n$ matrices, show that $AB = BA = 0$ if and only if

$$(CAD)(D^{-1}BC^{-1}) = (D^{-1}BC^{-1})(CAD) = 0$$

28. If A is a noninvertible $n \times n$ matrix, show that there is a nonzero $n \times n$ matrix B such that $AB = BA = 0$.
29. If A' is a submatrix of the rectangular matrix A, show that $r(A') \leq r(A)$.
30. If the matrix E is obtained by rearranging the columns of the identity matrix, show that AE may be obtained from A by rearranging the columns of A in the same way.
31. Determine the rank of the matrix

$$A = \begin{bmatrix} 1-t & 0 & 1-t \\ 0 & 1+t & 1+t \\ 1-t & 1+t & 2 \end{bmatrix}$$

as a function of t.

32. Determine the rank of the matrix

$$A = \begin{bmatrix} 1 & x & x \\ x & 1 & x \\ x & x & 1 \end{bmatrix}$$

as a function of x.

38. Show that the matrix

$$A = \begin{bmatrix} 0 & z & y \\ z & 0 & x \\ y & x & 0 \end{bmatrix}$$

is of rank 3 if $xyz \neq 0$, rank 2 if $xyz = 0$ but one of the numbers x, y, z is nonzero.

6 ISOMORPHISM

Let $T : V \to W$ be a linear transformation from a vector space V to a vector space W. If T is both one-one and onto, T is said to be an **isomorphism**. Therefore, T is an isomorphism if and only if $R_T = W$ and $N_T = 0$.

Example 1 Let A be an $n \times n$ matrix. Consider the linear transformation $T_A : \boldsymbol{R}^n \to \boldsymbol{R}^n$, defined by $T_A(\boldsymbol{x}) = A\boldsymbol{x}$. By Theorem 2, §4.9, T_A is one-one if and only if det $A = 0$. If det $A \neq 0$, A is invertible, and as we have seen in Example 3, §5.3, T_A is onto. Thus, T_A is an isomorphism if and only if A is invertible.

Example 2 In chapter 2 we gave two definitions of a vector in $\boldsymbol{R}^3$. The algebraic definition regarded vectors as columns of numbers. The other definition was geometric and regarded vectors as directed line segments beginning at the origin. With our algebraic definition of vectors, we gave an algebraic definition of vector addition.

$$\begin{bmatrix} x \\ y \\ z \end{bmatrix} + \begin{bmatrix} x' \\ y' \\ z' \end{bmatrix} = \begin{bmatrix} x + x' \\ y + y' \\ z + z' \end{bmatrix}$$

and scalar multiplication

$$\alpha \begin{bmatrix} x \\ y \\ z \end{bmatrix} = \begin{bmatrix} \alpha x \\ \alpha y \\ \alpha z \end{bmatrix}$$

With our geometric definition of vectors, the definition of vector addition became the parallelogram law and scalar multiplication was thought of as signed magnification of length.

Let us now consider the transformation T from the space of column vectors into the space of directed line segments beginning at the origin.

$$T\left(\begin{bmatrix} x \\ y \\ z \end{bmatrix}\right) = \boldsymbol{v}_{(x,y,z)}$$

In other words, T assigns to the column vector with components x, y, z the directed line segment from the origin to the point (x, y, z). Because of the correspondence between triples of real numbers and points in space, the transformation T is one-one and onto. In §2.2, we saw that

$$\boldsymbol{v}(x, y, z) + \boldsymbol{v}(x', y', z') = \boldsymbol{v}(x + x', y + y', z + z')$$

and

$$\alpha\boldsymbol{v}(x, y, z) = \boldsymbol{v}(\alpha x, \alpha y, \alpha z)$$

We can reinterpret this using the transformation T, defined above, as

$$T\left(\begin{bmatrix} x \\ y \\ z \end{bmatrix}\right) + T\left(\begin{bmatrix} x' \\ y' \\ z' \end{bmatrix}\right) = T\left(\begin{bmatrix} x \\ y \\ z \end{bmatrix} + \begin{bmatrix} x' \\ y' \\ z' \end{bmatrix}\right)$$

and

$$T\left(\alpha \begin{bmatrix} x \\ y \\ z \end{bmatrix}\right) = \alpha T\left(\begin{bmatrix} x \\ y \\ z \end{bmatrix}\right)$$

In other words, T defines a linear transformation from an algebraic system into a geometric system. Because T is one-one and onto, we see that T is an isomorphism.

As a result of this correspondence between the algebraically and geometrically defined vectors, we agreed to regard the two as equivalent. Because of the correspondence between algebraically and geometrically defined operations, we agreed to regard them as equivalent. This correspondence helps to illuminate the essential meaning of the concept of isomorphism.

The word "isomorphism" is derived from the Greek "iso = same" and "morphos = form." It is appropriate to ask, why, in this case, the two spaces have the "same form." The two spaces have the same form in the sense that T induces a one-one correspondence which renders the two definitions of addition and scalar multiplication equivalent. Using this equivalence, we may prove results in whichever space is the more convenient and use the isomorphism T to transfer the results to the other space, thereby exhibiting a correspondence between algebraic and geometric statements.

For example, by virtue of the isomorphism T, the geometric statement "If $\boldsymbol{v}_1$ and $\boldsymbol{v}_2$ are two noncollinear directed line segments in the plane beginning at the origin, then every directed line segment in the plane which begins at the origin is the diagonal of some parallelogram with adjacent sides lying on the lines determined by extending $\boldsymbol{v}_1$ and $\boldsymbol{v}_2$ respectively" is equivalent to the algebraic statement "If $\boldsymbol{v}_1$ and $\boldsymbol{v}_2$ are linearly independent vectors in $\boldsymbol{R}^2$, their linear combinations span $\boldsymbol{R}^2$." See Figure 6-1.

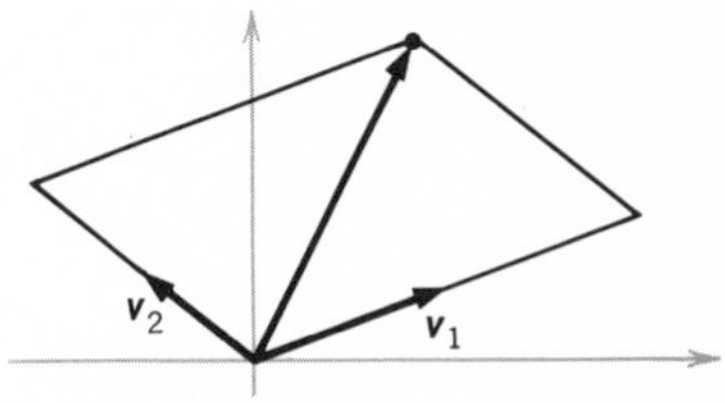

FIGURE 6-1

Example 3 Consider the linear transformation T from $\boldsymbol{R}^3$ into P_2, which sends $\boldsymbol{e}_1$ into 1, $\boldsymbol{e}_2$ into x, $\boldsymbol{e}_3$ into x^2. By the theorem in §5.2 such a linear transformation does exist. By the definition of T we have

$$T\left(\begin{bmatrix}\alpha\\ \beta\\ \gamma\end{bmatrix}\right) = \alpha + \beta x + \gamma x^2$$

T is clearly one-one and onto. Thus, T induces an isomorphism of $\boldsymbol{R}^3$ and P_2.

In the case of an isomorphism $T : V \to W$ between two vector spaces, T renders the algebraic operations in V equivalent to those of W. The next theorem helps make this statement more precise.

Theorem 1 Let $T : V \to W$ be an isomorphism of two vector spaces. Then,

(i) If $\boldsymbol{x}_1, \boldsymbol{x}_2, \ldots, \boldsymbol{x}_n$ are linearly independent in V, $T(\boldsymbol{x}_1), T(\boldsymbol{x}_2), \ldots, T(\boldsymbol{x}_n)$ are linearly independent in W.

(ii) If $\boldsymbol{x}_1, \boldsymbol{x}_2, \ldots, \boldsymbol{x}_n$ span V, $T(\boldsymbol{x}_1), T(\boldsymbol{x}_2), \ldots, T(\boldsymbol{x}_n)$ span W.

(iii) If $\boldsymbol{x}_1, \boldsymbol{x}_2, \ldots, \boldsymbol{x}_n$ form a basis for V, $T(\boldsymbol{x}_1), T(\boldsymbol{x}_2), \ldots, T(\boldsymbol{x}_n)$ form a basis for W.

(iv) $\dim V = \dim W$.

Proof (i) Suppose $\boldsymbol{x}_1, \boldsymbol{x}_2, \ldots, \boldsymbol{x}_n$ are independent in V. If $\alpha_1 T(\boldsymbol{x}_1) + \alpha_2 T(\boldsymbol{x}_2) + \cdots + \alpha_n T(\boldsymbol{x}_n) = \boldsymbol{0}$ in W, then by linearity of T, $T(\alpha_1\boldsymbol{x}_1 + \alpha_2\boldsymbol{x}_2 + \cdots + \alpha_n\boldsymbol{x}_n) = \boldsymbol{0}$. Because T is one-one, we conclude that $\alpha_1\boldsymbol{x}_1 + \alpha_2\boldsymbol{x}_2 + \cdots + \alpha_n\boldsymbol{x}_n = \boldsymbol{0}$. Since $\boldsymbol{x}_1, \boldsymbol{x}_2, \ldots, \boldsymbol{x}_n$ are independent in V, $\alpha_1 = \alpha_2 = \cdots = \alpha_n = 0$. Thus, $\alpha_1 T(\boldsymbol{x}_1) + \alpha_2 T(\boldsymbol{x}_2) + \cdots + \alpha_n T(\boldsymbol{x}_n) = \boldsymbol{0}$ implies that $\alpha_1 = \alpha_2 = \cdots = \alpha_n = 0$, so $T(\boldsymbol{x}_1), T(\boldsymbol{x}_2), \ldots, T(\boldsymbol{x}_n)$ are linearly independent.

(ii) Suppose $\boldsymbol{x}_1, \boldsymbol{x}_2, \ldots, \boldsymbol{x}_n$ span V and $\boldsymbol{y}$ is a vector in W. Since T is onto, there is a vector $\boldsymbol{x}$ in V, such that $T(\boldsymbol{x}) = \boldsymbol{y}$. Since $\boldsymbol{x}_1, \boldsymbol{x}_2, \ldots, \boldsymbol{x}_n$ span V, there are scalars $\alpha_1, \alpha_2, \ldots, \alpha_n$ such that $\boldsymbol{x} = \alpha_1\boldsymbol{x}_1 + \alpha_2\boldsymbol{x}_2 + \cdots + \alpha_n\boldsymbol{x}_n$. Thus, $\boldsymbol{y} = T(\boldsymbol{x}) = T(\alpha_1\boldsymbol{x}_1 + \cdots + \alpha_n\boldsymbol{x}_n) = \alpha_1 T(\boldsymbol{x}_1) + \cdots + \alpha_n T(\boldsymbol{x}_n)$. Thus, if $\boldsymbol{y}$ belongs to W, $\boldsymbol{y}$ is a linear combination of $T(\boldsymbol{x}_1), T(\boldsymbol{x}_2), \ldots, T(\boldsymbol{x}_n)$. Therefore, $T(\boldsymbol{x}_1), T(\boldsymbol{x}_2), \ldots, T(\boldsymbol{x}_n)$ span W.

(iii) Suppose $\boldsymbol{x}_1, \boldsymbol{x}_2, \ldots, \boldsymbol{x}_n$ is a basis for V. By (a), $T(\boldsymbol{x}_1), T(\boldsymbol{x}_2), \ldots, T(\boldsymbol{x}_n)$ are linearly independent in W. By (b), $T(\boldsymbol{x}_1), T(\boldsymbol{x}_2), \ldots, T(\boldsymbol{x}_n)$ span W. Thus, $T(\boldsymbol{x}_1), T(\boldsymbol{x}_2), \ldots, T(\boldsymbol{x}_n)$ forms a basis for W.

(iv) If $\dim V = n$, then there is some basis, say $\boldsymbol{x}_1, \boldsymbol{x}_2, \ldots, \boldsymbol{x}_n$ for V with n elements. Since $T(\boldsymbol{x}_1), T(\boldsymbol{x}_2), \ldots, T(\boldsymbol{x}_n)$ is a basis for W with n elements, we see that $\dim V = \dim W$. ∎

Suppose $T : V \to W$ is a linear transformation between two finite dimensional spaces V and W and $\dim V = \dim W$. By Theorem 3 of §5.4, $\dim R_T + \dim N_T = \dim V$. Thus, $\dim N_T = \dim V - \dim R_T$. We see, therefore, that T is one-one (i.e., $\dim N_T = 0$), if and only if T is onto (i.e., $\dim V = \dim R_T$). Thus, the following statements are equivalent:

(1) T is an isomorphism.

(2) T is one-one.

(3) T is onto.

Example 4 Consider the transformation from P_n to P_n defined by $H(f) = f + f'$, where f' denotes the derivative of f.

H is linear, since

$$\begin{aligned} H(f_1 + f_2) &= (f_1 + f_2) + (f_1 + f_2)' \\ &= f_1 + f_2 + f_1' + f_2' \\ &= f_1 + f_1' + f_2 + f_2' \\ &= H(f_1) + H(f_2) \end{aligned}$$

$$\begin{aligned} H(\alpha f) &= \alpha f + (\alpha f)' \\ &= \alpha f + \alpha f' \\ &= \alpha(f + f') \\ &= \alpha H(f) \end{aligned}$$

Since H is a linear transformation from an $(n + 1)$-dimensional vector space into itself, in order to show that H is an isomorphism, we need only show that it is one-one.

Let us suppose $H(f) = 0$, then $f + f' = 0$, or, $f' = -f$. If f is a nonzero polynomial of degree k, f' is a polynomial of degree $k - 1$, and so f cannot possibly equal $-f'$. Thus, $f = 0$ and we conclude that H is one-one. By the above, H is an isomorphism.

Since H is onto, we see that any polynomial g of degree less than or equal to n may be expressed in the form $g = f + f'$, where f is a polynomial of degree less than or equal to n.

By using Theorem 1 above, we see that if $f_0, f_1, \ldots, f_n$ is a basis for P_n, then another basis is $f_0 + f_0', f_1 + f_1', \ldots, f_n + f_n'$. For example, if $1, x, x^2, \ldots, x^n$ is the usual basis for P_n, then, $1, x + 1, x^2 + 2x, \ldots, x^n + nx^{n-1}$ is also a basis for P_n.

Previously, we saw that if there exists an isomorphism $T : V \to W$ between two vector spaces V and W, then V and W have the same dimension. Interestingly enough, the converse of this theorem is also true.

Theorem 2 Let V and W be two n-dimensional real vector spaces. Then, V and W are isomorphic.

Proof Take $\boldsymbol{x}_1, \boldsymbol{x}_2, \ldots, \boldsymbol{x}_n$ as a basis for V and $\boldsymbol{y}_1, \boldsymbol{y}_2, \ldots, \boldsymbol{y}_n$ as a basis for W. By the theorem of §5.2, there is a linear transformation T from V to W, such that $T(\boldsymbol{x}_i) = \boldsymbol{y}_i$, $i = 1, 2, \ldots, n$.

We claim T is an isomorphism. First we show that T is one-one. Suppose there is some $\boldsymbol{x}$ in V such that $T(\boldsymbol{x}) = \boldsymbol{0}$. Since $\boldsymbol{x}_1, \boldsymbol{x}_2, \ldots, \boldsymbol{x}_n$ is a basis for V, there are scalars $\alpha_1, \alpha_2, \ldots, \alpha_n$ such that

$$\boldsymbol{x} = \alpha_1\boldsymbol{x}_1 + \alpha_2\boldsymbol{x}_2 + \cdots + \alpha_n\boldsymbol{x}_n$$

Then,

$$\begin{aligned} T(\boldsymbol{x}) &= T(\alpha_1\boldsymbol{x}_1 + \alpha_2\boldsymbol{x}_2 + \cdots + \alpha_n\boldsymbol{x}_n) \\ &= \alpha_1 T(\boldsymbol{x}_1) + \alpha_2 T(\boldsymbol{x}_2) + \cdots + \alpha_n T(\boldsymbol{x}_n) = \boldsymbol{0} \end{aligned}$$

or $\alpha_1\boldsymbol{y}_1 + \alpha_2\boldsymbol{y}_2 + \cdots + \alpha_n\boldsymbol{y}_n = \boldsymbol{0}$. Since $\boldsymbol{y}_1, \boldsymbol{y}_2, \ldots, \boldsymbol{y}_n$ is a basis for W, we must have $\alpha_1 = \alpha_2 = \cdots = \alpha_n = 0$. Therefore,

$$\boldsymbol{x} = \alpha_1\boldsymbol{x}_1 + \cdots + \alpha_n\boldsymbol{x}_n = 0 \cdot \boldsymbol{x}_1 + \cdots + 0 \cdot \boldsymbol{x}_n = \boldsymbol{0}$$

Thus, we see that if $T(\boldsymbol{x}) = \boldsymbol{0}$, then $\boldsymbol{x} = \boldsymbol{0}$, or $N_T = \boldsymbol{0}$.

Next we show T is onto. Assume that $\boldsymbol{y}$ belongs to W. Then there are scalars $\beta_1, \beta_2, \ldots, \beta_n$ such that $\boldsymbol{y} = \beta_1\boldsymbol{y}_1 + \beta_2\boldsymbol{y}_2 + \cdots + \beta_n\boldsymbol{y}_n$. Let $\boldsymbol{x} = \beta_1\boldsymbol{x}_1 + \beta_2\boldsymbol{x}_2 + \cdots + \beta_n\boldsymbol{x}_n$. Then,

$$\begin{aligned} T(\boldsymbol{x}) &= T(\beta_1\boldsymbol{x}_1 + \beta_2\boldsymbol{x}_2 + \cdots + \beta_n\boldsymbol{x}_n) \\ &= \beta_1 T(\boldsymbol{x}_1) + \beta_2 T(\boldsymbol{x}_2) + \cdots + \beta_n T(\boldsymbol{x}_n) \\ &= \beta_1\boldsymbol{y}_1 + \beta_2\boldsymbol{y}_2 + \cdots + \beta_n\boldsymbol{y}_n = \boldsymbol{y} \end{aligned}$$

Thus, given $\boldsymbol{y} \in W$, there is an $\boldsymbol{x}$ in V such that $T(\boldsymbol{x}) = \boldsymbol{y}$. So $R_T = W$ and T is onto. ∎

Corollary If V is an n-dimensional real vector space, V is isomorphic to $\boldsymbol{R}^n$.

This corollary means that there is an isomorphism from V to $\boldsymbol{R}^n$. It is in this sense that $\boldsymbol{R}^n$ is the model for real finite-dimensional spaces. Using a proof similar to that of Theorem 2 we can conclude that any two complex n-dimensional vector spaces are isomorphic, and any complex n-dimensional vector space is isomorphic to $\boldsymbol{C}^n$.

We have already seen that the space P_n is a vector space of dimension $n + 1$. Therefore, P_n and $\boldsymbol{R}^{n+1}$ are isomorphic. In a similar manner we see that the space of 2×2 real matrices, which is of dimension 4, is isomorphic to $\boldsymbol{R}^4$.

EXERCISES

1. Which of the following matrices induce isomorphisms of $\boldsymbol{R}^n$?
   (a) $\begin{bmatrix} -1 & 0 & 3 \\ 1 & 2 & 0 \\ 0 & 2 & 3 \end{bmatrix}$ (b) $\begin{bmatrix} 7 & 5 \\ 4 & 3 \end{bmatrix}$

   (c) $\begin{bmatrix} 1 & 1 & 3 \\ 3 & 2 & 5 \\ 7 & 5 & 12 \end{bmatrix}$ (d) $\begin{bmatrix} 1 & 5 \\ 3 & 0 \end{bmatrix}$
2. Let V be a vector space with basis $\boldsymbol{x}_1, \boldsymbol{x}_2, \ldots, \boldsymbol{x}_n$. Let T be the function from V to $\boldsymbol{R}^n$ which assigns to each vector $\boldsymbol{x}$ its coordinate n-tuple relative to the basis $\boldsymbol{x}_1, \boldsymbol{x}_2, \ldots, \boldsymbol{x}_n$. Show that T is an isomorphism of V and $\boldsymbol{R}^n$.
3. Let P_e be the vector space of even polynomials in a variable x of degree less than or equal to $2n$ and let P_o be the space of odd polynomials of degree less than or equal to $2n + 1$. Show that the function $T : P_e \rightarrow P_o$ defined by $T(f) = xf$ is an isomorphism of P_e onto P_o.
4. Show that the function from the space of $n \times n$ matrices into itself which carries A into A^T is an isomorphism.

5. Let T be the mapping from P_n into P_n defined by

$$T(f) = f + \alpha_1 f' + \alpha_2 f'' + \cdots + \alpha_n f^{(n)}, \qquad \alpha_1, \alpha_2, \ldots, \alpha_n, \text{ real numbers}$$

Show that T is an isomorphism.

6. Let A be an invertible $n \times n$ matrix. Show that the following linear transformations from the space of $n \times n$ matrices into itself are isomorphisms.
(a) $T(B) = AB$ (b) $T(B) = BA$
(c) $T(B) = ABA$ (d) $T(B) = ABA^{-1}$

7. Let $T: V \rightarrow W$ be a linear transformation between two finite-dimensional vector spaces. Show that the following statements are equivalent.
(a) T is an isomorphism.
(b) If $\boldsymbol{x}_1, \boldsymbol{x}_2, \ldots, \boldsymbol{x}_n$ are linearly independent in V, then $T(\boldsymbol{x}_1), T(\boldsymbol{x}_2), \ldots, T(\boldsymbol{x}_n)$ are linearly independent in W, and if $\boldsymbol{y}_1, \boldsymbol{y}_2, \ldots, \boldsymbol{y}_n$ span V, then $T(\boldsymbol{y}_1), T(\boldsymbol{y}_2), \ldots, T(\boldsymbol{y}_n)$ span W.
(c) If $\boldsymbol{x}_1, \boldsymbol{x}_2, \ldots, \boldsymbol{x}_n$ is a basis for V, then $T(\boldsymbol{x}_1), T(\boldsymbol{x}_2), \ldots, T(\boldsymbol{x}_n)$ is a basis for W.

8. Show that the linear transformation of P_n which carries $f(x)$ into $f(x + \alpha)$, with α real, is an isomorphism of P_n.

9. Let T be an isomorphism of $\boldsymbol{R}^3$. Show that T carries planes through the origin into planes through the origin and lines through the origin into lines through the origin.

10. Construct an isomorphism between the space of 3×4 matrices and the space of 2×6 matrices. When, in general, it is possible to find an isomorphism between the space of $k \times l$ matrices and the space of $m \times n$ matrices?

11. Show that the space of upper triangular matrices is isomorphic to the space of lower triangular matrices.

12. Show that the function from the complex numbers (regarded here as a real vector space) to the space of 2×2 matrices of the form

$$\begin{bmatrix} a & b \\ -b & a \end{bmatrix}$$

defined by

$$T(a + bi) = \begin{bmatrix} a & b \\ -b & a \end{bmatrix}$$

is an isomorphism. Show that this isomorphism preserves products, i.e., $T(a + bi)T(c + di) = T((a + bi)(c + di))$.

13. Let V be a vector space and T be a fixed vector in V. Define new operations on V by

$$\boldsymbol{x} \oplus \boldsymbol{y} = \boldsymbol{x} + \boldsymbol{y} + \boldsymbol{t}$$
$$\alpha * \boldsymbol{x} = \alpha\boldsymbol{x} + (1 - \alpha)\boldsymbol{t}$$

Show that V with the operations $\oplus$ and $*$ is a vector space. Show that the function $T(\boldsymbol{x}) = \boldsymbol{x} + \boldsymbol{t}$ which carries V, regarded as a vector space under $+$ and $\cdot$, into V, regarded as a vector space under $\oplus$ and $*$ induces an isomorphism.

14. For what values of λ is the function from P_n into P_n defined by $T(f) = \lambda f - xf'$ an isomorphism of P_n.
15. Let $\boldsymbol{x}_1, \boldsymbol{x}_2, \boldsymbol{x}_3$ be one set of linearly independent vectors in $\boldsymbol{R}^3$ and let $\boldsymbol{y}_1, \boldsymbol{y}_2, \boldsymbol{y}_3$ be another set of linearly independent vectors in $\boldsymbol{R}^3$. Suppose T is a linear operator on $\boldsymbol{R}^3$ such that $T(\boldsymbol{x}_i) = \boldsymbol{y}_i$, $i = 1, 2, 3$.
    (a) Show that T is an isomorphism.
    (b) Show that $T(\boldsymbol{x}) = A\boldsymbol{x}$, where A is the 3×3 matrix,

$$[\boldsymbol{y}_1, \boldsymbol{y}_2, \boldsymbol{y}_3][\boldsymbol{x}_1, \boldsymbol{x}_2, \boldsymbol{x}_3]^{-1}$$

    where $[\boldsymbol{y}_1, \boldsymbol{y}_2, \boldsymbol{y}_3]$ is the 3×3 matrix with columns $\boldsymbol{y}_1, \boldsymbol{y}_2, \boldsymbol{y}_3$ and $[\boldsymbol{x}_1, \boldsymbol{x}_2, \boldsymbol{x}_3]$ is the 3×3 matrix with columns $\boldsymbol{x}_1, \boldsymbol{x}_2, \boldsymbol{x}_3$.
16. Which of the following pairs of spaces are isomorphic?
    (a) 3×3 skew-symmetric matrices with real entries and $\boldsymbol{R}^3$.
    (b) $n \times n$ diagonal matrices with real entries and P_n.
    (c) $(n - 1) \times (n - 1)$ symmetric matrices with real entries and $n \times n$ skew-symmetric matrices with real entries.
17. (a) If $\boldsymbol{a}$ and $\boldsymbol{b}$ are linearly independent vectors in the plane, show that the triangle with one vertex at the origin and adjacent sides $\boldsymbol{a}$ and $\boldsymbol{b}$ consists exactly of the vectors of the form $\alpha\boldsymbol{a} + \beta\boldsymbol{b}$ where $\alpha \geq 0$, $\beta \geq 0$, and $\alpha + \beta \leq 1$.
    (b) If $\boldsymbol{a}$ and $\boldsymbol{b}$ are linearly independent vectors in the plane and $\boldsymbol{c}$ is another vector in the plane, show that the vectors which end within the triangle having one vertex at the endpoint of $\boldsymbol{c}$ and sides parallel to $\boldsymbol{a}$ and $\boldsymbol{b}$ are precisely the vectors of the form

$$\boldsymbol{c} + \alpha\boldsymbol{a} + \beta\boldsymbol{b}, \qquad \alpha \geq 0, \qquad \beta \geq 0, \qquad \alpha + \beta \leq 1$$

    (c) Vectors $\boldsymbol{x}$, $\boldsymbol{y}$, and $\boldsymbol{z}$ in the plane are said to be **point-wise independent** if and only if $\alpha\boldsymbol{x} + \beta\boldsymbol{y} + \gamma\boldsymbol{z} = \boldsymbol{0}$ and $\alpha + \beta + \gamma = 0$ imply that $\alpha = 0$, $\beta = 0$, and $\gamma = 0$. Show that $\boldsymbol{x}$, $\boldsymbol{y}$, and $\boldsymbol{z}$ are pointwise independent if and only if $\boldsymbol{x} - \boldsymbol{y}$ and $\boldsymbol{y} - \boldsymbol{z}$ are linearly independent.
    (d) Show that the vectors $\boldsymbol{x}$, $\boldsymbol{y}$, and $\boldsymbol{z}$ are pointwise independent if and only if their endpoints are noncollinear.
    (e) If $\boldsymbol{x}$, $\boldsymbol{y}$, and $\boldsymbol{z}$ are pointwise independent vectors, show that the vectors which end in the triangle having as vertices the endpoints of $\boldsymbol{x}$, $\boldsymbol{y}$, and $\boldsymbol{z}$ are precisely those vectors of the form $\alpha\boldsymbol{x} + \beta\boldsymbol{y} + \gamma\boldsymbol{z}$, where $\alpha \geq 0$, $\beta \geq 0$, $\gamma \geq 0$, and $\alpha + \beta + \gamma = 1$.
    (f) Show that any linear isomorphism of the plane carries triangles onto triangles.

7 ALGEBRA OF LINEAR TRANSFORMATIONS

If V and W are two vector spaces, let $L(V, W)$ denote the family of all linear transformations from V into W. There are certain natural algebraic operations defined in $L(V, W)$ which we study in this section.

Suppose T_1 and T_2 are two members of $L(V, W)$, that is, T_1 and T_2 are linear transformations from the vector space V into the vector space W. We define the **sum** of T_1 and T_2, denoted by $T_1 + T_2$, to be $(T_1 + T_2)(\boldsymbol{x}) = T_1(\boldsymbol{x}) + T_2(\boldsymbol{x})$.

In order for this definition to be of interest, the sum should again be a linear transformation. To see that this is indeed the case, let $\boldsymbol{x}$ and $\boldsymbol{y}$ belong to V, then

$$\begin{aligned}(T_1 + T_2)(\boldsymbol{x} + \boldsymbol{y}) &= T_1(\boldsymbol{x} + \boldsymbol{y}) + T_2(\boldsymbol{x} + \boldsymbol{y}) && \text{(by definition of } T_1 + T_2\text{)}\\ &= T_1(\boldsymbol{x}) + T_1(\boldsymbol{y}) + T_2(\boldsymbol{x}) + T_2(\boldsymbol{y}) && \text{(by linearity of } T_1, T_2\text{)}\\ &= T_1(\boldsymbol{x}) + T_2(\boldsymbol{x}) + T_1(\boldsymbol{y}) + T_2(\boldsymbol{y}) && \text{[by (V1)]}\\ &= (T_1 + T_2)(\boldsymbol{x}) + (T_1 + T_2)(\boldsymbol{y}) && \text{(by definition of } T_1 + T_2\text{)}\end{aligned}$$

Similarly, if $\boldsymbol{x}$ belongs to V and α is a scalar, we have

$$\begin{aligned}(T_1 + T_2)(\alpha\boldsymbol{x}) &= T_1(\alpha\boldsymbol{x}) + T_2(\alpha\boldsymbol{x}) && \text{(by definition of } T_1 + T_2\text{)}\\ &= \alpha T_1(\boldsymbol{x}) + \alpha T_2(\boldsymbol{x}) && \text{(by linearity of } T_1, T_2\text{)}\\ &= \alpha(T_1(\boldsymbol{x}) + T_2(\boldsymbol{x})) && \text{[by (V6)]}\\ &= \alpha(T_1 + T_2)(\boldsymbol{x}) && \text{(by definition of } T_1 + T_2\text{)}\end{aligned}$$

There is also a natural operation of scalar multiplication on $L(V, W)$. If T belongs to $L(V, W)$ we define the **scalar multiple** of T by the scalar α, denoted by αT, to be $(\alpha T)(\boldsymbol{x}) = \alpha(T(\boldsymbol{x}))$.

αT is again a linear transformation, since if $\boldsymbol{x}$ and $\boldsymbol{y}$ belong to V,

$$\begin{aligned}(\alpha T)(\boldsymbol{x} + \boldsymbol{y}) &= \alpha(T(\boldsymbol{x} + \boldsymbol{y})) && \text{(by definition of } \alpha T\text{)}\\ &= \alpha(T(\boldsymbol{x}) + T(\boldsymbol{y})) && \text{(by linearity of } T\text{)}\\ &= \alpha T(\boldsymbol{x}) + \alpha T(\boldsymbol{y}) && \text{[by (V6)]}\\ &= (\alpha T)(\boldsymbol{x}) + (\alpha T)(\boldsymbol{y}) && \text{(by definition of } \alpha T\text{)}\end{aligned}$$

Moreover, if $\boldsymbol{x}$ is a vector in V and β is a scalar, we have

$$\begin{aligned}(\alpha T)(\beta\boldsymbol{x}) &= \alpha(T(\beta\boldsymbol{x})) && \text{(by definition of } \alpha T\text{)}\\ &= \alpha(\beta T(\boldsymbol{x})) && \text{(by linearity of } T\text{)}\\ &= (\alpha\beta)(T(\boldsymbol{x})) && \text{[by (V7)]}\\ &= \beta(\alpha T(\boldsymbol{x})) && \text{[by (V7)]}\\ &= \beta(\alpha T)(\boldsymbol{x}) && \text{(by definition of } \alpha T\text{)}\end{aligned}$$

In view of the choice of notation, it is not surprising that the following theorem is valid.

Theorem 1 $L(V, W)$, with the addition and scalar multiplication defined above, is a vector space.

Proof Let T_1, T_2, T_3 be members of $L(V, W)$, α, β be scalars, and $\boldsymbol{x}$ be an arbitrary vector in V.

(V1)

$$\begin{aligned} (T_1 + T_2)(\boldsymbol{x}) &= T_1(\boldsymbol{x}) + T_2(\boldsymbol{x}) && \text{(by definition of } T_1 + T_2) \\ &= T_2(\boldsymbol{x}) + T_1(\boldsymbol{x}) && \text{[using (V1) in } W] \\ &= (T_2 + T_1)(\boldsymbol{x}) && \text{(by definition of } T_2 + T_1) \end{aligned}$$

(V2)

$$\begin{aligned} ((T_1 + T_2) + T_3)(\boldsymbol{x}) &= (T_1 + T_2)(\boldsymbol{x}) + T_3(\boldsymbol{x}) \\ & \qquad \text{[by definition of } (T_1 + T_2) + T_3] \\ &= (T_1(\boldsymbol{x}) + T_2(\boldsymbol{x})) + T_3(\boldsymbol{x}) \\ & \qquad \text{(by definition of } T_1 + T_2) \\ &= T_1(\boldsymbol{x}) + (T_2(\boldsymbol{x}) + T_3(\boldsymbol{x})) && \text{[using (V2) in } W] \\ &= T_1(\boldsymbol{x}) + (T_2 + T_3)(\boldsymbol{x}) && \text{(by definition of } T_2 + T_3) \\ &= (T_1 + (T_2 + T_3))(\boldsymbol{x}) \\ & \qquad \text{[by definition of } T_1 + (T_2 + T_3)] \end{aligned}$$

(V3)

Let 0 denote the zero transformation from V into W, i.e., the operator sending all vectors of V into $\mathbf{0}$. Then,

$$\begin{aligned} (0 + T_1)(\boldsymbol{x}) &= 0(\boldsymbol{x}) + T_1(\boldsymbol{x}) && \text{(by definition of } 0 + T_1) \\ &= \mathbf{0} + T_1(\boldsymbol{x}) && \text{(by definition of } 0) \\ &= T_1(\boldsymbol{x}) && \text{[using (V3) in } W] \end{aligned}$$

(V4)

For T in $L(V, W)$, define the negative of T by $(-T)(\boldsymbol{x}) = -(T(\boldsymbol{x}))$. Then,

$$\begin{aligned} (T + (-T))(\boldsymbol{x}) &= T(\boldsymbol{x}) + (-T)(\boldsymbol{x}) && \text{[by definition of } T + (-T)] \\ &= T(\boldsymbol{x}) - T(\boldsymbol{x}) && \text{[by definition of } (-T)(\boldsymbol{x})] \\ &= \mathbf{0} && \text{[using (V4) in } W] \\ &= 0(\boldsymbol{x}) && \text{(by definition of the zero operator)} \end{aligned}$$

(V5)

$$\begin{aligned} ((\alpha + \beta)T)(\boldsymbol{x}) &= (\alpha + \beta)T(\boldsymbol{x}) && [\text{by definition of } (\alpha + \beta)T] \\ &= \alpha T(\boldsymbol{x}) + \beta T(\boldsymbol{x}) && [\text{using (V5) in } W] \\ &= (\alpha T)(\boldsymbol{x}) + (\beta T)(\boldsymbol{x}) && (\text{by definition of } \alpha T \text{ and } \beta T) \\ &= (\alpha T + \beta T)(\boldsymbol{x}) && (\text{by definition of } \alpha T + \beta T) \end{aligned}$$

(V6)

$$\begin{aligned} \alpha(T_1 + T_2)(\boldsymbol{x}) &= \alpha((T_1 + T_2)(\boldsymbol{x})) && [\text{by definition of } \alpha(T_1 + T_2)] \\ &= \alpha(T_1(\boldsymbol{x}) + T_2(\boldsymbol{x})) && (\text{by definition of } T_1 + T_2) \\ &= \alpha T_1(\boldsymbol{x}) + \alpha T_2(\boldsymbol{x}) && [\text{using (V6) in } W] \\ &= (\alpha T_1)(\boldsymbol{x}) + (\alpha T_2)(\boldsymbol{x}) && (\text{by definition of } \alpha T_1 \text{ and } \alpha T_2) \\ &= (\alpha T_1 + \alpha T_2)(\boldsymbol{x}) && (\text{by definition of } \alpha T_1 + \alpha T_2) \end{aligned}$$

(V7)

$$\begin{aligned} ((\alpha\beta)T)(\boldsymbol{x}) &= (\alpha\beta)(T(\boldsymbol{x})) && [\text{by definition of } (\alpha\beta)T] \\ &= \alpha(\beta(T(\boldsymbol{x}))) && [\text{using (V7) in } W] \\ &= \alpha((\beta T)(\boldsymbol{x})) && (\text{by definition of } \beta T) \\ &= (\alpha(\beta T))(\boldsymbol{x}) && [\text{by definition of } \alpha(\beta T)] \end{aligned}$$

(V8)

$$\begin{aligned} (1 \cdot T)(\boldsymbol{x}) &= 1 \cdot T(\boldsymbol{x}) && (\text{by definition of } 1 \cdot T) \\ &= T(\boldsymbol{x}) && [\text{using (V8) in } W] \end{aligned}$$

Having verified that (V1)–(V8) hold in $L(V, W)$, we see that $L(V, W)$ is a vector space. ▨

Example 1 Let A be an $m \times n$ matrix with real entries and T_A be the linear transformation from $\boldsymbol{R}^n$ into $\boldsymbol{R}^m$ defined by $T_A(\boldsymbol{x}) = A\boldsymbol{x}$, $\boldsymbol{x} \in \boldsymbol{R}^n$. If B is another $m \times n$ matrix,

$$\begin{aligned} (T_A + T_B)(\boldsymbol{x}) &= T_A(\boldsymbol{x}) + T_B(\boldsymbol{x}) \\ &= A\boldsymbol{x} + B\boldsymbol{x} \\ &= (A + B)\boldsymbol{x} \\ &= T_{A+B}(\boldsymbol{x}) \end{aligned}$$

Since $\boldsymbol{x}$ was an arbitrary vector, we see that $T_A + T_B = T_{A+B}$, i.e., addition of transformations corresponds to addition of matrices. Likewise, it is not difficult to demonstrate that $\alpha T_A = T_{\alpha A}$, which means scalar multiplication of linear transformations corresponds to scalar multiplication of matrices.

Let L denote the function from M_{mn}, the space of $m \times n$ matrices with real entries, into $L(\boldsymbol{R}^n, \boldsymbol{R}^m)$, defined by $L(A) = T_A$. In other words, L carries the matrix A into the linear transformation induced by multiplication by A. Since

$$\begin{aligned} L(A + B) &= T_{A+B} \\ &= T_A + T_B \\ &= L(A) + L(B) \end{aligned}$$

and

$$\begin{aligned} L(\alpha A) &= T_{\alpha A} \\ &= \alpha T_A \\ &= \alpha L(A) \end{aligned}$$

we see that L is linear.

We saw in Theorem 2, §5.2 that any linear transformation from $\boldsymbol{R}^n$ into $\boldsymbol{R}^m$ is induced by multiplication by a suitable $m \times n$ matrix. That is, given a linear transformation $T : \boldsymbol{R}^n \to \boldsymbol{R}^m$, there is some matrix A, such that $T = T_A$. In other words, the function L from M_{mn} into $L(\boldsymbol{R}^n, \boldsymbol{R}^m)$ is onto.

The function L is also one-one. For if $L(A) = 0$, then $T_A(\boldsymbol{x}) = A\boldsymbol{x} = \boldsymbol{0}$, for all vectors $\boldsymbol{x}$ in $\boldsymbol{R}^n$. By letting $\boldsymbol{x} = \boldsymbol{e}_1, \boldsymbol{e}_2, \ldots, \boldsymbol{e}_n$, we see that all columns of A are zero, and so $A = 0$.

Thus, the function L defined above induces an isomorphism from the space of $m \times n$ matrices into the space of linear transformations from $\boldsymbol{R}^n$ to $\boldsymbol{R}^m$.

Just as addition and scalar multiplication of linear transformations are analogous to the operations with matrices, there is also a method of combining linear transformations which corresponds to matrix multiplication.

Let $T : U \to V$ and $S : V \to W$ be linear transformations, with U, V, and W vector spaces. We define the **composition** of S and T, denoted by $S \circ T$, or more often simply ST, by

$$(S \circ T)(\boldsymbol{x}) = S(T(\boldsymbol{x})), \quad \text{for } \boldsymbol{x} \text{ in } U$$

It seems reasonable to expect that the composition of two linear transformations would again be linear. This is indeed the case, since if $\boldsymbol{x}_1$ and $\boldsymbol{x}_2$ are vectors in U, we have

$$
\begin{aligned}
(S \circ T)(\boldsymbol{x}_1 + \boldsymbol{x}_2) &= S(T(\boldsymbol{x}_1 + \boldsymbol{x}_2)) && \text{(by definition of } S \circ T\text{)} \\
&= S(T(\boldsymbol{x}_1) + T(\boldsymbol{x}_2)) && \text{(by linearity of } T\text{)} \\
&= S(T(\boldsymbol{x}_1)) + S(T(\boldsymbol{x}_2)) && \text{(by linearity of } S\text{)} \\
&= (S \circ T)(\boldsymbol{x}_1) + (S \circ T)(\boldsymbol{x}_2) && \text{(by definition of } S \circ T\text{)}
\end{aligned}
$$

Similarly, we have, with $\boldsymbol{x}$ in U and α a scalar,

$$
\begin{aligned}
(S \circ T)(\alpha\boldsymbol{x}) &= S(T(\alpha\boldsymbol{x})) && \text{(by definition of } S \circ T\text{)} \\
&= S(\alpha T(\boldsymbol{x})) && \text{(by linearity of } T\text{)} \\
&= \alpha S(T(\boldsymbol{x})) && \text{(by linearity of } S\text{)} \\
&= \alpha(S \circ T)(\boldsymbol{x}) && \text{(by definition of } S \circ T\text{)}
\end{aligned}
$$

Since we have just verified conditions (i) and (ii) in the definition of a linear transformation, it follows that $S \circ T$ is a linear transformation from U to W.

Example 2 Our purpose here is to show precisely how composition of linear transformations corresponds to multiplication of matrices.

Suppose $T_A : \boldsymbol{R}^p \to \boldsymbol{R}^n$ and $T_B : \boldsymbol{R}^n \to \boldsymbol{R}^m$ are linear transformations induced by the matrices A, an $n \times p$ matrix, and B, an $m \times n$ matrix, respectively. That is $T_A(\boldsymbol{x}) = A\boldsymbol{x}$, for $\boldsymbol{x}$ in $\boldsymbol{R}^p$, and $T_B(\boldsymbol{y}) = B\boldsymbol{y}$, for $\boldsymbol{y}$ in $\boldsymbol{R}^n$.

Then, for all $\boldsymbol{x}$ in $\boldsymbol{R}^p$

$$
\begin{aligned}
(T_B \circ T_A)(\boldsymbol{x}) &= T_B(T_A(\boldsymbol{x})) && \text{(by definition of } T_B \circ T_A\text{)} \\
&= T_B(A\boldsymbol{x}) && \text{(by definition of } T_A\text{)} \\
&= B(A\boldsymbol{x}) && \text{(by definition of } T_B\text{)} \\
&= (BA)(\boldsymbol{x}) && \text{(by associativity of matrix multiplication)}
\end{aligned}
$$

Since $\boldsymbol{x}$ was arbitrary $T_B \circ T_A = T_{BA}$, where T_{BA} is the linear transformation induced by the $m \times p$ matrix BA.

This natural correspondence between composition of linear functions and multiplication of matrices often enables us to give a geometric interpretation of a matrix product. For example, consider the linear operator T_θ of $\boldsymbol{R}^2$ into itself, discussed in §5.1. $T_\theta(\boldsymbol{v})$ was defined to be the vector obtained by rotating a given vector $\boldsymbol{v}$ by θ degrees. If θ_1 and θ_2 are two numbers, what is the composition $T_{\theta_1} \circ T_{\theta_2}$?

If $\boldsymbol{v}$ is a vector in the plane, $T_{\theta_2}(\boldsymbol{v})$ is obtained by rotating $\boldsymbol{v}$ by θ_2 degrees. $T_{\theta_1}(T_{\theta_2}(\boldsymbol{v}))$ is then found by rotating $T_{\theta_2}(\boldsymbol{v})$ by θ_1 degrees. All together then, $T_{\theta_1}(T_{\theta_2}(\boldsymbol{v}))$ is that vector resulting from rotating the vector $\boldsymbol{v}$ by $\theta_1 + \theta_2$ degrees. In other words, $T_{\theta_1+\theta_2} = T_{\theta_1} \circ T_{\theta_2}$.

Since

$$T_\theta\left(\begin{bmatrix} x \\ y \end{bmatrix}\right) = \begin{bmatrix} \cos\theta & -\sin\theta \\ \sin\theta & \cos\theta \end{bmatrix}\begin{bmatrix} x \\ y \end{bmatrix}$$

we see that

$$\begin{bmatrix} \cos\theta_1 & -\sin\theta_1 \\ \sin\theta_1 & \cos\theta_1 \end{bmatrix}\begin{bmatrix} \cos\theta_2 & -\sin\theta_2 \\ \sin\theta_2 & \cos\theta_2 \end{bmatrix} = \begin{bmatrix} \cos(\theta_1+\theta_2) & -\sin(\theta_1+\theta_2) \\ \sin(\theta_1+\theta_2) & \cos(\theta_1+\theta_2) \end{bmatrix}$$

This result may, of course, be obtained by an explicit multiplication of the two matrices in question.

The identity transformation plays a role in composition of linear transformations analogous to that played by the identity matrix in matrix multiplication.

Theorem 2 Let $T : V \to W$ be a linear transformation between two vector spaces V and W. Let $I_V : V \to V$ and $I_W : W \to W$ be the identity transformations on V and W, respectively. Then,

$$T \circ I_V = T \quad \text{and} \quad I_W \circ T = T$$

Proof Let $\boldsymbol{x}$ be a vector in V. Then,

$$\begin{aligned} (T \circ I_V)(\boldsymbol{x}) &= T(I_V(\boldsymbol{x})) && \text{(by definition of } T \circ I_V) \\ &= T(\boldsymbol{x}) && \text{(by definition of } I_V) \end{aligned}$$

Since this holds for all $\boldsymbol{x}$, we have $T \circ I_V = T$. The proof of $I_W \circ T = T$ is similar. ∎

Next we verify that composition of linear transformations satisfies the associative law.

Theorem 3 Let $T_1 : V_1 \to V_2$, $T_2 : V_2 \to V_3$, and $T_3 : V_3 \to V_4$ be linear transformations, where V_1, V_2, V_3, V_4 are vector spaces. Then,

$$T_3 \circ (T_2 \circ T_1) = (T_3 \circ T_2) \circ T_1$$

Proof Let $\boldsymbol{x}$ be any vector in V. Then,

$$\begin{aligned} (T_3 \circ (T_2 \circ T_1))(\boldsymbol{x}) &= T_3((T_2 \circ T_1)(\boldsymbol{x})) && [\text{by definition of } T_3 \circ (T_2 \circ T_1)] \\ &= T_3(T_2(T_1(\boldsymbol{x}))) && (\text{by definition of } T_2 \circ T_1) \\ &= (T_3 \circ T_2)(T_1(\boldsymbol{x})) && (\text{by definition of } T_3 \circ T_2) \\ &= ((T_3 \circ T_2) \circ T_1)(\boldsymbol{x}) && [\text{by definition of } (T_3 \circ T_2) \circ T_1] \end{aligned}$$

Since for all $\boldsymbol{x}$ in V, $(T_3 \circ (T_2 \circ T_1))(\boldsymbol{x}) = ((T_3 \circ T_2) \circ T_1)(\boldsymbol{x})$, we have proved that $T_3 \circ (T_2 \circ T_1) = (T_3 \circ T_2) \circ T_1$. ∎

Theorem 3 implies that the multiplication of matrices is associative. For if A_1, A_2, and A_3 are matrices of appropriate order, we have, by Theorem 3,

$$T_{A_3} \circ (T_{A_2} \circ T_{A_1}) = (T_{A_3} \circ T_{A_2}) \circ T_{A_1}$$

Using $T_{AB} = T_A \circ T_B$, repeatedly,

$$T_{A_3} \circ T_{A_2A_1} = T_{A_3A_2} \circ T_{A_1}$$
$$T_{A_3(A_2A_1)} = T_{(A_3A_2)A_1}$$

Since $T_A = T_B$ implies $A = B$, we have $A_3(A_2A_1) = (A_3A_2)A_1$, a result proved in chapter 2 by an entirely different method.

In view of the close connection between matrix multiplication and composition of linear transformations, it is not surprising that the following distributive laws are also valid.

Theorem 4 Let $T_2 : V_2 \to V_3$, $T_2' : V_2 \to V_3$, $T_1 : V_1 \to V_2$, and $T_3 : V_3 \to V_4$ be linear transformations on vector spaces V_1, V_2, V_3, V_4. Then,

$$(T_2 + T_2') \circ T_1 = T_2 \circ T_1 + T_2' \circ T_1$$

and

$$T_3 \circ (T_2 + T_2') = T_3 \circ T_2 + T_3 \circ T_2'$$

Proof Let $\boldsymbol{x}$ be any vector in V_1. Then,

$$\begin{aligned}
&((T_2 + T_2') \circ T_1)(\boldsymbol{x}) \\
&\quad = (T_2 + T_2')(T_1(\boldsymbol{x})) && [\text{by definition of } (T_2 + T_2') \circ T_1] \\
&\quad = T_2(T_1(\boldsymbol{x})) + T_2'(T_1(\boldsymbol{x})) && (\text{by definition of } T_2 + T_2') \\
&\quad = (T_2 \circ T_1)(\boldsymbol{x}) + (T_2' \circ T_1)(\boldsymbol{x}) && (\text{by definition of } T_2 \circ T_1 \text{ and } T_2' \circ T_1) \\
&\quad = (T_2 \circ T_1 + T_2' \circ T_1)(\boldsymbol{x}) && (\text{by definition of } T_2 \circ T_1 + T_2' \circ T_1)
\end{aligned}$$

Since the above equality holds for all $\boldsymbol{x}$ in V_1, we have shown that

$$(T_2 + T_2') \circ T_1 = T_2 \circ T_1 + T_2' \circ T_1$$

The other distributive law may be proved analogously. ∎

If $T : V \to V$ is a linear operator on V, a vector space, we define powers of V in the usual way:

$$T^2(x) = T(T(x))$$
$$T^3(x) = T(T^2(x))$$
$$\vdots$$
$$T^n(x) = T(T^{n-1}(x))$$

Because of the associative law T^n is just T composed with itself n times, regardless of the order in which these compositions take place.

Example 3 Let D be the differentiation operator on P_n, $D(f) = f'$. Then,

$$\begin{aligned} D^2(f) &= D(D(f)) \\ &= D(f') \\ &= f'' \end{aligned}$$

Thus, D^2 is the linear operator which carries f into its second derivative. In general we see that $D^k(f) = f^{(k)}$, i.e., D^k carries f into its kth derivative. Since the $(n + 1)$st derivative of any polynomial of degree less than or equal to n is 0, we see that $D^{n+1} = 0$. Operators of this sort, i.e., those having the property that some power of the operator is 0, are said to be **nilpotent**.

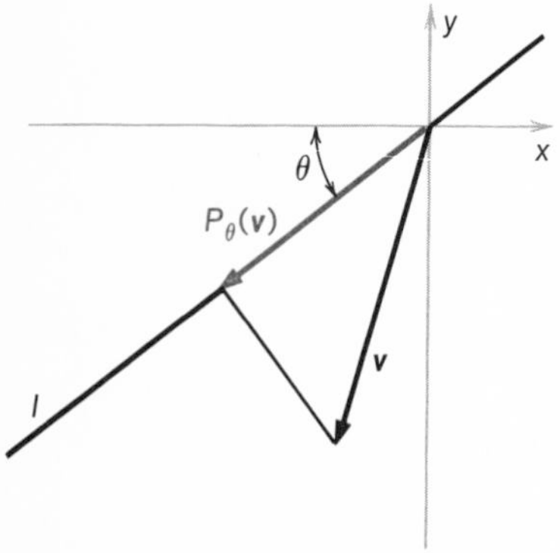

FIGURE 7-1

Example 4 Let P_θ be the projection transformation on $\mathbf{R}^2$ (see §5.2). Recall that $P_\theta(\mathbf{v})$ is the vector which lies on a line l making an angle θ with the x axis and $P_\theta(\mathbf{v})$ obtained by dropping a perpendicular from the endpoint of $\mathbf{v}$ to l. See Figure 7-1. If $\mathbf{u}$ is any vector lying on l, we see that $P_\theta(\mathbf{u}) = \mathbf{u}$. Thus, since $P_\theta(\mathbf{v})$ lies on l, $P_\theta(P_\theta(\mathbf{v})) = P_\theta(\mathbf{v})$, or $P_\theta^2(\mathbf{v}) = P_\theta(\mathbf{v})$. Since this is true for all vectors $\mathbf{v}$ in $\mathbf{R}^2$, we obtain $P_\theta^2 = P_\theta$.

Operators of this sort, that is, those operators which equal their square, are said to be **idempotent**.

In view of the important role played by the inverse in the study of matrix multiplication, it is appropriate to ask if there is a corresponding entity for linear transformations.

Definition Let $T : V \to W$ be a linear transformation between two vector spaces V and W. If there is a linear transformation $S : W \to V$ such that $T \circ S = I_W$ and $S \circ T = I_V$, then T is said to be **invertible** and S is said to be its **inverse**.

Note that if T has an inverse, it has at most one inverse. Suppose that S_1 and S_2 are two linear transformations from W to V, which satisfy the condition of the inverse in the definition above. Then

$$\begin{aligned} S_1 &= S_1 \circ I_W && \text{(by Theorem 2)} \\ &= S_1 \circ (T \circ S_2) && \text{(since } T \circ S_2 = I_W\text{, by hypothesis)} \\ &= (S_1 \circ T) \circ S_2 && \text{(by Theorem 3)} \\ &= I_V \circ S_2 && \text{(since } S_1 \circ T = I_V\text{, by hypothesis)} \\ &= S_2 && \text{(by Theorem 2)} \end{aligned}$$

Thus, if the inverse exists, it is unique. The inverse of a transformation T is generally denoted by T^{-1}.

Example 5 Let A be an invertible $n \times n$ matrix and let T be the linear operator on $\boldsymbol{R}^n$, defined by $T_A(\boldsymbol{x}) = A\boldsymbol{x}$. We also have the operation $T_{A^{-1}}$ on $\boldsymbol{R}^n$, defined by $T_{A^{-1}}(\boldsymbol{x}) = A^{-1}\boldsymbol{x}$.

Calculating,

$$\begin{aligned} (T_A \circ T_{A^{-1}})(\boldsymbol{x}) &= T_A(T_{A^{-1}}(\boldsymbol{x})) && \text{(by definition of } T_A \circ T_{A^{-1}}) \\ &= T_A(A^{-1}\boldsymbol{x}) && \text{(by definition of } T_{A^{-1}}) \\ &= A(A^{-1}\boldsymbol{x}) && \text{(by definition of } T_A) \\ &= \boldsymbol{x} \end{aligned}$$

Hence we have $T_A \circ T_{A^{-1}} = I$. Similarly, it can be shown that $T_{A^{-1}} \circ T_A = I$. Thus, T_A is invertible and $(T_A)^{-1} = T_{A^{-1}}$.

Example 6 Let $S : P_n \to P_n$ be the linear operator which carries the polynomial $f(x)$ into the polynomial $f(x + 1)$, or $(S(f))(x) = f(x + 1)$. Thus, x is transformed into $x + 1$, x^2 into $(x + 1)^2$, and so on.

Now, if we let $T : P_n \to P_n$ be the linear operator which carries $f(x)$ into $f(x - 1)$, or $(T(f))(x) = f(x - 1)$, then S is invertible and $S^{-1} = T$.

For we have

$$\begin{aligned} ((S \circ T)(f))(x) &= (S(T(f)))(x) && \text{(by definition of } S \circ T) \\ &= (T(f))(x + 1) && \text{(by definition of } S) \\ &= f((x + 1) - 1) && \text{(by definition of } T) \\ &= f(x) \end{aligned}$$

Thus, for all f, $(S \circ T)f = f$, so $S \circ T = I$. In a like manner, it can be shown that $T \circ S = I$.

The concept of the inverse of a linear operator is closely related to the concept of isomorphism. This relationship is made explicit in the following theorem.

Theorem 5 Let $T : V \to W$ be a linear transformation between two vector spaces V and W. Then, T is invertible if and only if T is an isomorphism.

Proof Suppose T is invertible and $T^{-1} : W \to V$ is its inverse. If $T(\boldsymbol{x}) = \boldsymbol{0}$, then, $T^{-1}(T(\boldsymbol{x})) = T^{-1}(\boldsymbol{0}) = \boldsymbol{0}$. But $T^{-1}(T(\boldsymbol{x})) = \boldsymbol{x}$. Thus, $\boldsymbol{x} = \boldsymbol{0}$. Since $T(\boldsymbol{x}) = \boldsymbol{0}$ implies $\boldsymbol{x} = \boldsymbol{0}$, we see that $N_T = \boldsymbol{0}$. Next, we show T is onto. If $\boldsymbol{y}$ is a vector in W, let $\boldsymbol{x} = T^{-1}(\boldsymbol{y})$. Then $T(\boldsymbol{x}) = T(T^{-1}(\boldsymbol{y})) = \boldsymbol{y}$. Thus, $\boldsymbol{y}$ is the image under T of the vector $\boldsymbol{x}$. From this, it follows that T is onto. Since T is one-one and onto, T is an isomorphism.

Next suppose T is an isomorphism. We must construct an inverse of T.

Since T is onto, given $\boldsymbol{y}$ in W there is some vector $\boldsymbol{x}$ in V such that $T(\boldsymbol{x}) = \boldsymbol{y}$. We define $S(\boldsymbol{y}) = \boldsymbol{x}$, then S is well-defined, since if $T(\boldsymbol{x}_1) = \boldsymbol{y}$ and $T(\boldsymbol{x}_2) = \boldsymbol{y}$, the assumption that T is one-one implies $\boldsymbol{x}_1 = \boldsymbol{x}_2$. Thus, S is a function from W to V.

Next we must show that S is linear. Let $\boldsymbol{y}_1$ and $\boldsymbol{y}_2$ be vectors in W. Then, there are vectors $\boldsymbol{x}_1$ and $\boldsymbol{x}_2$ in V such that $T(\boldsymbol{x}_1) = \boldsymbol{y}_1$ and $T(\boldsymbol{x}_2) = \boldsymbol{y}_2$. Then, $T(\boldsymbol{x}_1 + \boldsymbol{x}_2) = \boldsymbol{y}_1 + \boldsymbol{y}_2$ by linearity of T. Hence, $S(\boldsymbol{y}_1) = \boldsymbol{x}_1$, $S(\boldsymbol{y}_2) = \boldsymbol{x}_2$, and $S(\boldsymbol{y}_1 + \boldsymbol{y}_2) = \boldsymbol{x}_1 + \boldsymbol{x}_2$, by definition of S. Therefore, $S(\boldsymbol{y}_1 + \boldsymbol{y}_2) = S(\boldsymbol{y}_1) + S(\boldsymbol{y}_2)$.

Next, suppose $\boldsymbol{y}$ belongs to W and α is a scalar. Since T is onto, $T(\boldsymbol{x}) = \boldsymbol{y}$, for some $\boldsymbol{x}$ in V. Then $T(\alpha\boldsymbol{x}) = \alpha T(\boldsymbol{x}) = \alpha\boldsymbol{y}$ by linearity of T. By definition of S, $S(\boldsymbol{y}) = \boldsymbol{x}$ and $S(\alpha\boldsymbol{y}) = \alpha\boldsymbol{x}$. Thus, $S(\alpha\boldsymbol{y}) = \alpha S(\boldsymbol{y})$. So it follows that S is linear.

Now we claim $S \circ T = I_V$ and $T \circ S = I_W$. Pick any vector $\boldsymbol{x}$ in V and let $T(\boldsymbol{x}) = \boldsymbol{y}$. By definition of S, $S(\boldsymbol{y}) = \boldsymbol{x}$. Thus,

$$\begin{aligned}
(S \circ T)(\boldsymbol{x}) &= S(T(\boldsymbol{x})) && \text{(by definition of } S \circ T\text{)} \\
&= S(\boldsymbol{y}) && [\text{since } \boldsymbol{y} = T(\boldsymbol{x})] \\
&= \boldsymbol{x} && \text{(by definition of } S\text{)}
\end{aligned}$$

Therefore we have shown that $S \circ T = I_V$.

Next, pick any vector $\boldsymbol{y}$ in W. Since T is onto there is some $\boldsymbol{x}$ in V such that $T(\boldsymbol{x}) = \boldsymbol{y}$. By definition of S, $S(\boldsymbol{y}) = \boldsymbol{x}$.

Then,

$$\begin{aligned}
(T \circ S)(\boldsymbol{y}) &= T(S(\boldsymbol{y})) && \text{(by definition of } T \circ S\text{)} \\
&= T(\boldsymbol{x}) && \text{(by definition of } S\text{)} \\
&= \boldsymbol{y} && [\text{since } T(\boldsymbol{x}) = \boldsymbol{y}]
\end{aligned}$$

It follows that $T \circ S = I_W$.

Thus, we have demonstrated that any isomorphism has an inverse. ∎

In some types of problems the procedure just described can be used to find the inverse of a given transformation. Indeed, if $T : V \rightarrow W$ is a linear isomorphism between V and W, and $\boldsymbol{x}_1, \boldsymbol{x}_2, \ldots, \boldsymbol{x}_n$ is a basis for V, it follows by Theorem 1, §5.6 that $T(\boldsymbol{x}_1) = \boldsymbol{y}_1, T(\boldsymbol{x}_2) = \boldsymbol{y}_2, \ldots, T(\boldsymbol{x}_n) = \boldsymbol{y}_n$ is a basis for W. Thus, T^{-1} is the linear function defined by $T^{-1}(\boldsymbol{y}_1) = \boldsymbol{x}_1$, $T^{-1}(\boldsymbol{y}_2) = \boldsymbol{x}_2, \ldots, T^{-1}(\boldsymbol{y}_n) = \boldsymbol{x}_n$. An example may help to clarify the method.

Example 7 Consider the linear transformation $T(f) = f + f'$ from P_4 to P_4 defined in Example 4, §5.6. We saw that T is an isomorphism of P_4 with itself. We calculate the images of the basis vectors $1, x, x^2, x^3, x^4$.

$$
\begin{aligned}
T(1) &= 1 \\
T(x) &= 1 + x \\
T(x^2) &= 2x + x^2 \\
T(x^3) &= 3x^2 + x^3 \\
T(x^4) &= 4x^3 + x^4
\end{aligned}
$$

Thus,

$$
\begin{aligned}
T^{-1}(1) &= 1 \\
T^{-1}(1 + x) &= x \\
T^{-1}(2x + x^2) &= x^2 \\
T^{-1}(3x^2 + x^3) &= x^3 \\
T^{-1}(4x^3 + x^4) &= x^4
\end{aligned}
$$

Using the linearity of T^{-1},

$$
\begin{aligned}
T^{-1}(1) &= 1 \\
T^{-1}(x) &= T^{-1}((x + 1) - 1) = x - 1 \\
T^{-1}(x^2) &= T^{-1}((x^2 + 2x) - 2x) \\
&= x^2 - 2(x - 1) = x^2 - 2x + 2 \\
T^{-1}(x^3) &= T^{-1}((x^3 + 3x^2) - 3x^2) \\
&= x^3 - 3(x^2 - 2x + 2) \\
&= x^3 - 3x^2 + 6x - 6 \\
T^{-1}(x^4) &= T^{-1}((x^4 + 4x^3) - 4x^3) \\
&= x^4 - 4(x^3 - 3x^2 + 6x - 6) \\
&= x^4 - 4x^3 + 12x^2 - 24x + 24
\end{aligned}
$$

Since we know the T^{-1} image of each basis element, using linearity we can determine $T^{-1}(\alpha_0 + \alpha_1 x + \alpha_2 x^2 + \alpha_3 x^3 + \alpha_4 x^4)$.

EXERCISES

1. Let S, T, and U be the linear transformation of $\mathbf{R}^2$ defined by

$$S\left(\begin{bmatrix} x \\ y \end{bmatrix}\right) = \begin{bmatrix} 1 & 1 \\ 0 & 1 \end{bmatrix}\begin{bmatrix} x \\ y \end{bmatrix}$$

$$T\left(\begin{bmatrix} x \\ y \end{bmatrix}\right) = \begin{bmatrix} 1 & 0 \\ 0 & -1 \end{bmatrix}\begin{bmatrix} x \\ y \end{bmatrix}$$

$$U\left(\begin{bmatrix} x \\ y \end{bmatrix}\right) = \begin{bmatrix} 0 & 1 \\ 1 & 0 \end{bmatrix}\begin{bmatrix} x \\ y \end{bmatrix}$$

Calculate the matrices associated with the following linear transformations.
(a) $S \circ T$ (b) $T \circ S$
(c) S^2 (d) S^3
(e) S^n, n a positive integer (f) S^{-1}
(g) $(S \circ T) \circ U$ (h) $S \circ (T \circ U)$
(i) T^n, n any integer (j) $S \circ T - T \circ S$
(k) $T \circ U - U \circ T$ (l) $U^2 - I_2$
(m) $S^2 - 2S + I$ (n) $S \circ U - U \circ S$

2. Let S, T, U be the linear operators on P_3 defined by

$$S(f) = f + f'$$

$$T(f) = f + xf' + x^2 f''$$

$$U(f) = f''$$

Calculate the following linear operators.
(a) $S \circ U$ (b) $U \circ S$
(c) $U \circ T$ (d) $T \circ U$
(e) $T \circ U - U \circ T$ (f) U^2
(g) S^2 (h) S^3
(i) S^n if n is any positive integer (j) $T \circ S$
(k) $S \circ T$

3. Which of the following matrices induce invertible linear operators on $\mathbf{R}^n$? Determine the inverse if it exists.

(a) $\begin{bmatrix} 1 & 3 \\ 2 & 7 \end{bmatrix}$ (b) $\begin{bmatrix} 3 & -6 \\ -2 & 4 \end{bmatrix}$

(c) $\begin{bmatrix} 1 & 1 & -2 \\ -7 & 3 & 0 \\ 0 & 1 & 1 \end{bmatrix}$ (d) $\begin{bmatrix} 1 & -3 & 2 \\ 4 & 1 & 0 \\ 5 & -2 & 2 \end{bmatrix}$

4. Let V be a vector space and $\boldsymbol{x}_1, \boldsymbol{x}_2, \ldots, \boldsymbol{x}_n$ be a basis for V. Let T be the linear operator on V, such that $T(\boldsymbol{x}_1) = \boldsymbol{x}_2$, $T(\boldsymbol{x}_2) = \boldsymbol{x}_3, \ldots, T(\boldsymbol{x}_{n-1}) = \boldsymbol{x}_n$, $T(\boldsymbol{x}_n) = \boldsymbol{0}$. Show that T is nilpotent.

5. Let A be an $n \times n$ matrix. Let $T_1(A) = \frac{1}{2}(A + A^T)$, $T_2(A) = \frac{1}{2}(A - A^T)$.
   (a) Show that $T_1^2 = T_1$, i.e., T_1 is idempotent.
   (b) Show that $T_2^2 = T_2$, i.e., T_2 is idempotent.
   (c) Show that $T_1T_2 = T_2T_1 = 0$.

6. Let $T : V \to V$ be a linear operator on a vector space V such that $T^2 = I_V$. Show that
   (a) The operator $T_1 = \frac{1}{2}(1 + T)$ satisfies $T_1^2 = T_1$, i.e., T_1 is idempotent.
   (b) The operator $T_2 = \frac{1}{2}(1 - T)$ satisfies $T_2^2 = T_2$, i.e., T_2 is idempotent.
   (c) $T_1T_2 = T_2T_1 = 0$, $T_1 + T_2 = I_V$.
   (d) $R_{T_1} = N_{T_2}$, $N_{T_2} = R_{T_1}$.
   (e) If $V = P_n$, the space of polynomials of degree less than or equal to n in a variable x, and $T(f(x)) = f(-x)$, show that $T^2 = I_{P_n}$. What are the operators T_1 and T_2 of (a) and (b)? What polynomials belong to R_{T_1} and R_{T_2}?
   (f) If $V = P_n$, as in part (e), and $(T(f))(x) = f(1 - x)$, show that $T^2 = I_{P_n}$.

7. If V is a vector space and $\boldsymbol{x}_1, \boldsymbol{x}_2, \ldots, \boldsymbol{x}_n$ is a basis for V show that the operator $T(\boldsymbol{x}_1) = \boldsymbol{x}_2$, $T(\boldsymbol{x}_2) = \boldsymbol{x}_3, \ldots, T(\boldsymbol{x}_{n-1}) = \boldsymbol{x}_n$, $T(\boldsymbol{x}_n) = \alpha\boldsymbol{x}_1$, satisfies the relation $T^n = \alpha I_V$. If $\alpha \neq 0$, what is T^{-1}?

8. Let V_1, V_2, and V_3 be vector spaces, and let $S : V_1 \to V_2$ and $T : V_2 \to V_3$ be linear transformations.
   (a) If S is one-one and T is one-one, show that $T \circ S$ is one-one.
   (b) If S is onto and T is onto, show that $T \circ S$ is onto.
   (c) If S and T are isomorphisms, show that $T \circ S$ is an isomorphism and $(T \circ S)^{-1} = S^{-1} \circ T^{-1}$.

9. Let $T : V \to V$ and $S : V \to V$ be linear operators on a vector space V of finite dimension.
   (a) If $T \circ S$ is one-one, show that T is an isomorphism.
   (b) If $T \circ S$ is onto, show that T is an isomorphism.
   (c) If T^k is an isomorphism for some positive integer k, show that T is an isomorphism.
   (d) If T^k is invertible for some positive integer k, and $(T^k)^{-1} = S$, show that $T^{-1} = T^{k-1} \circ S$.

10. If T and S are linear operators on a vector space V and $(T + S)^2 = T^2 + 2S \circ T + S^2$, show that S and T commute, i.e., $S \circ T = T \circ S$.

11. If T is a linear transformation from a finite-dimensional vector space V onto a vector space W, show that
    (a) $\dim W$ is finite, and $\dim W \leq \dim V$.
    (b) There is a linear transformation $S : W \to V$, such that $T \circ S = I_W$.

12. If T is a one-one linear transformation from a vector space V into a finite-dimensional vector space W, show that
    (a) V is finite-dimensional.
    (b) There is a linear transformation $S : W \to V$, such that $S \circ T = I_V$.

13. If T is a linear operator on a vector space V and if $T^2 - T + I_V = 0$, show that T is invertible.

14. Let $T : \boldsymbol{R}^n \to \boldsymbol{R}^n$ be a linear operator in $\boldsymbol{R}^n$. If $m < n$, and there are linear transformations $S : \boldsymbol{R}^n \to \boldsymbol{R}^m$ and $U : \boldsymbol{R}^m \to \boldsymbol{R}^n$, such that $T = U \circ S$, show $r(T) \leq m$ and T is not invertible.

15. If T is a linear operator on a finite-dimensional vector space V, show that $r(T) = r(T^2)$ if and only if $N_T \cap R_T = \mathbf{0}$, i.e., the nullspace and range of T have only the zero vector in common.

16. Let $T_{(ab)}$ be the linear operator on P_n, defined by $(T_{(ab)}(f))(x) = f(ax + b)$, where a, b are real numbers.
    (a) Show that $T_{(ab)} \circ T_{(cd)} = T_{(ac,bc+d)}$.
    (b) If $a \neq 0$, show that $T_{(ab)}$ is invertible and find its inverse.

17. Let D be the differentiation operator on P_n and let T_α be the operator defined by $T_\alpha(f)(x) = f(\alpha x)$. Show that $D \circ T_\alpha = \alpha T_\alpha \circ D$.

18. Let T be a linear operator on a vector space V.
    (a) If for some positive integer k, $R(T^k) = R(T^{k+1})$, show that $R(T^{k+1}) = R(T^{k+2})$.
    (b) If for some positive integer k, $N(T^k) = N(T^{k+1})$, show that $N(T^{k+1}) = N(T^{k+2})$.

19. Let V be a vector space and $\boldsymbol{x}_1, \boldsymbol{x}_2, \ldots, \boldsymbol{x}_n$ be a basis for V. Suppose $T : V \to V$ is the linear operator $T(\boldsymbol{x}_1) = \lambda_1\boldsymbol{x}_1, \ldots, T(\boldsymbol{x}_n) = \lambda_n\boldsymbol{x}_n$, for scalars $\lambda_1, \lambda_2, \ldots, \lambda_n$ where $\lambda_i \neq \lambda_j$ if $i \neq j$. Suppose $T \circ S = S \circ T$, where S is a linear operator on V. Show that there are scalars $\alpha_1, \alpha_2, \ldots, \alpha_n$ such that $S(\boldsymbol{x}_1) = \alpha_1\boldsymbol{x}_1$, $S(\boldsymbol{x}_2) = \alpha_2\boldsymbol{x}_2, \ldots, S(\boldsymbol{x}_n) = \alpha_n\boldsymbol{x}_n$.

20. Let $T : V \to V$ be a linear operator on a finite-dimensional space V. Show that T commutes with all other linear operators on V if and only if $T = \alpha I_V$, i.e., T is a scalar multiple of the identity on V.

21. Let V and W be vector spaces. Let $\boldsymbol{x}$ be a fixed nonzero vector in V. Show that the function from $L(V, W)$ into W, defined by $L(T) = T(\boldsymbol{x})$, is a linear transformation from $L(V, W)$ onto W.

22. If $T : \boldsymbol{R}^n \to \boldsymbol{R}^n$ is a linear operator on $\boldsymbol{R}^n$ of rank r, show that there are linear transformations $S : \boldsymbol{R}^n \to \boldsymbol{R}^r$, where S is onto and $U : \boldsymbol{R}^r \to \boldsymbol{R}^n$, where U is one-one such that $T = U \circ S$. Interpret this result in terms of matrices.

23. If $S : V_1 \to V_2$ and $T : W_1 \to W_2$ are linear isomorphisms of vector spaces V_1, W_1, V_2, and W_2, show that the function $L : L(V_1, W_1) \to L(V_2, W_2)$ defined by $L(U) = T \circ U \circ S^{-1}$ is a linear isomorphism of $L(V_1, W_1)$ and $L(V_2, W_2)$.

24. Suppose T is a linear operator on $\boldsymbol{R}^3$ with the property that $T \neq 0$ but $T^2 = 0$. Show that $r(T) = 1$.

25. If $T : \boldsymbol{R}^n \to \boldsymbol{R}^n$ is a linear operator such that $T^{n-1} \neq 0$ but $T^n = 0$, show that $r(T) = n - 1$. Give an example of such a T.

26. If $T : V \to V$ and $S : V \to V$ are linear operators on a vector space V, show that $T \circ S = 0$ if and only if $R_S \subset N_T$.

27. If $T : V \to V$ is a linear operator on a finite-dimensional vector space V and $T^2 = 0$, show that $r(T) \leq \frac{1}{2} \dim V$.

8 MATRIX REPRESENTATION OF A LINEAR TRANSFORMATION

Let $T : V \to W$ be a linear transformation between two finite dimensional vector spaces V and W. If $V = \boldsymbol{R}^n$ and $W = \boldsymbol{R}^m$, in §5.2 we saw that it is possible, using the standard bases for $\boldsymbol{R}^m$ and $\boldsymbol{R}^n$, to find an $m \times n$ matrix A such that $T(\boldsymbol{x}) = A\boldsymbol{x}$. In this section we perform a similar construction using arbitrary bases for V and W.

Let $\mathcal{B} = \{\boldsymbol{x}_1, \boldsymbol{x}_2, \ldots, \boldsymbol{x}_n\}$ be a basis for V and $\mathcal{C} = \{\boldsymbol{y}_1, \boldsymbol{y}_2, \ldots, \boldsymbol{y}_m\}$ be a basis for W. Since $\mathcal{C}$ is a basis for W, there are scalars a_{ij}, such that

$$T(\boldsymbol{x}_j) = \sum_{i=1}^{m} a_{ij}\boldsymbol{y}_i$$

or

$$T(\boldsymbol{x}_j) \underset{\mathcal{C}}{\leftrightarrow} \begin{bmatrix} a_{1j} \\ a_{2j} \\ \vdots \\ a_{mj} \end{bmatrix}$$

Define the $m \times n$ matrix A_T by $A_T = [a_{ij}]_{(mn)}$. Note that the jth column of A_T is just the n-tuple of coordinates of $T(\boldsymbol{x}_j)$ relative to the basis $\mathcal{C} = \{\boldsymbol{y}_1, \boldsymbol{y}_2, \ldots, \boldsymbol{y}_m\}$ for W.

If $\boldsymbol{x}$ belongs to V, we may associate with $\boldsymbol{x}$ its n-tuple of coordinates relative to the basis $\mathcal{B}$:

$$\boldsymbol{x} \underset{\mathcal{B}}{\leftrightarrow} \begin{bmatrix} \alpha_1 \\ \alpha_2 \\ \vdots \\ \alpha_n \end{bmatrix}$$

which means, of course, $\boldsymbol{x} = \alpha_1\boldsymbol{x}_1 + \alpha_2\boldsymbol{x}_2 + \cdots + \alpha_n\boldsymbol{x}_n$.

Defining the m-vector by

$$\begin{bmatrix} \beta_1 \\ \beta_2 \\ \vdots \\ \beta_m \end{bmatrix} = \begin{bmatrix} a_{11} & a_{12} & \cdots & a_{1n} \\ a_{21} & a_{22} & \cdots & a_{2n} \\ & & \vdots & \\ a_{m1} & a_{m2} & \cdots & a_{mn} \end{bmatrix} \begin{bmatrix} \alpha_1 \\ \alpha_2 \\ \vdots \\ \alpha_n \end{bmatrix}$$

we claim

$$T(\boldsymbol{x}) \underset{\mathcal{C}}{\leftrightarrow} \begin{bmatrix} \beta_1 \\ \beta_2 \\ \vdots \\ \beta_n \end{bmatrix}$$

or equivalently

$$T(\boldsymbol{x}) = \beta_1 \boldsymbol{y}_1 + \beta_2 \boldsymbol{y}_2 + \cdots + \beta_m \boldsymbol{y}_m$$

To prove this assertion, note that

$$\boldsymbol{x} = \sum_{j=1}^{n} \alpha_j \boldsymbol{x}_j$$

$$T(\boldsymbol{x}) = \sum_{j=1}^{n} \alpha_j T(\boldsymbol{x}_j)$$

$$= \sum_{j=1}^{n} \alpha_j \left(\sum_{i=1}^{m} a_{ij} \boldsymbol{y}_i \right) = \sum_{i=1}^{m} \left(\sum_{j=1}^{n} a_{ij} \alpha_j \right) \boldsymbol{y}_i$$

Thus, the ith coordinate of $T(\boldsymbol{x})$ relative to the basis $\mathcal{C} = \{\boldsymbol{y}_1, \boldsymbol{y}_2, \ldots, \boldsymbol{y}_n\}$ is $\beta_i = \sum_{j=1}^{n} a_{ij}\alpha_j$.

In other words, as soon as a basis $\mathcal{B}$ for V and a basis $\mathcal{C}$ for W are specified, there is a matrix A_T with the property that the product of A_T with the coordinate vector of $\boldsymbol{x}$ relative to $\mathcal{B}$ is just the coordinate vector of $T(\boldsymbol{x})$ relative to $\mathcal{C}$.

If $V = W$, that is if $T : V \to V$ is a linear operator on V, it is customary to use the same basis in both the initial and final spaces (that is, assume $\mathcal{B} = \mathcal{C}$). If $I_V : V \to V$ is the identity operator on V, and if $\mathcal{B} = \{\boldsymbol{x}_1, \boldsymbol{x}_2, \ldots, \boldsymbol{x}_n\}$ is a basis for V, the fact that $I_V \boldsymbol{x}_j = \boldsymbol{x}_j$ implies that the matrix associated with I_V is just I_n, the identity matrix of order n. Likewise, the matrix associated with the zero operator is the zero matrix.

Example 1 Let M_{22} denote the space of real 2×2 matrices. Let

$$\begin{bmatrix} u \\ v \end{bmatrix}$$

be a fixed vector in $\boldsymbol{R}^2$. Consider the transformation $T : M_{22} \to \boldsymbol{R}^2$ defined by

$$T(A) = A \begin{bmatrix} u \\ v \end{bmatrix}$$

T is linear, since if A, B belong to M_{22} and α is a scalar

$$T(A + B) = (A + B)\begin{bmatrix} u \\ v \end{bmatrix}$$

$$= A\begin{bmatrix} u \\ v \end{bmatrix} + B\begin{bmatrix} u \\ v \end{bmatrix}$$

$$= T(A) + T(B)$$

and

$$T(\alpha A) = (\alpha A)\begin{bmatrix} u \\ v \end{bmatrix}$$

$$= \alpha A\begin{bmatrix} u \\ v \end{bmatrix}$$

$$= \alpha T(A)$$

Using the basis $\mathcal{B}$,

$$E_1 = \begin{bmatrix} 1 & 0 \\ 0 & 0 \end{bmatrix}, \quad E_2 = \begin{bmatrix} 0 & 1 \\ 0 & 0 \end{bmatrix}, \quad E_3 = \begin{bmatrix} 0 & 0 \\ 1 & 0 \end{bmatrix}, \quad E_4 = \begin{bmatrix} 0 & 0 \\ 0 & 1 \end{bmatrix}$$

For M_{22}, and using the basis $\mathcal{C}$,

$$\boldsymbol{y}_1 = \begin{bmatrix} 1 \\ 0 \end{bmatrix}, \quad \boldsymbol{y}_2 = \begin{bmatrix} 0 \\ 1 \end{bmatrix}$$

for $\boldsymbol{R}^2$, we calculate the matrix representation of T.

$$T(E_1) = \begin{bmatrix} 1 & 0 \\ 0 & 0 \end{bmatrix}\begin{bmatrix} u \\ v \end{bmatrix} = \begin{bmatrix} u \\ 0 \end{bmatrix} = u\boldsymbol{y}_1$$

$$T(E_2) = \begin{bmatrix} 0 & 1 \\ 0 & 0 \end{bmatrix}\begin{bmatrix} u \\ v \end{bmatrix} = \begin{bmatrix} v \\ 0 \end{bmatrix} = v\boldsymbol{y}_1$$

$$T(E_3) = \begin{bmatrix} 0 & 0 \\ 1 & 0 \end{bmatrix}\begin{bmatrix} u \\ v \end{bmatrix} = \begin{bmatrix} 0 \\ u \end{bmatrix} = u\boldsymbol{y}_2$$

$$T(E_4) = \begin{bmatrix} 0 & 0 \\ 0 & 1 \end{bmatrix}\begin{bmatrix} u \\ v \end{bmatrix} = \begin{bmatrix} 0 \\ v \end{bmatrix} = v\boldsymbol{y}_2$$

Thus, it follows that

$$A_T = \begin{bmatrix} u & v & 0 & 0 \\ 0 & 0 & u & v \end{bmatrix}$$

The matrix A_T is obtained by placing in successive columns the coordinate vectors of $T(\boldsymbol{x}_1)$, $T(\boldsymbol{x}_2)$, $T(\boldsymbol{x}_3)$, $T(\boldsymbol{x}_4)$ relative to the basis $\boldsymbol{y}_1$, $\boldsymbol{y}_2$.

If

$$M = \begin{bmatrix} a & b \\ c & d \end{bmatrix}$$

then

$$M \underset{\mathcal{B}}{\leftrightarrow} \begin{bmatrix} a \\ b \\ c \\ d \end{bmatrix}$$

and so

$$T(M) \underset{\mathcal{C}}{\leftrightarrow} \begin{bmatrix} u & v & 0 & 0 \\ 0 & 0 & u & v \end{bmatrix} \begin{bmatrix} a \\ b \\ c \\ d \end{bmatrix} = \begin{bmatrix} au + bv \\ cu + dv \end{bmatrix}$$

Example 2 Let P_n be defined in the usual manner. Let $D : P_n \to P_n$ be the differentiation operator, $D(f) = f'$. Choose the basis $1, x, x^2, \ldots, x^n$ for P_n. Then,

$$\begin{aligned} D(1) &= 0 \\ D(x) &= 1 \\ D(x^2) &= 2x \\ &\vdots \\ D(x^n) &= nx^{n-1} \end{aligned}$$

Thus, the matrix A_D associated with the operator D is

$$A_D = \begin{bmatrix} 0 & 1 & 0 & 0 & \ldots & 0 \\ 0 & 0 & 2 & 0 & \ldots & 0 \\ 0 & 0 & 0 & 3 & \ldots & 0 \\ 0 & 0 & 0 & 0 & \ldots & 0 \\ & & & \vdots & & \\ 0 & 0 & 0 & 0 & \ldots & n \\ 0 & 0 & 0 & 0 & \ldots & 0 \end{bmatrix}$$

determined by placing in successive columns the coordinates of $D(1)$, $D(x), \ldots, D(x^n)$ relative to the basis $1, x, x^2, \ldots, x^n$. If $f = \alpha_0 + \alpha_1 x + \alpha_2 x^2 + \cdots + \alpha_n x^n$, then

$$f \underset{\mathcal{B}}{\leftrightarrow} \begin{bmatrix} \alpha_0 \\ \alpha_1 \\ \alpha_2 \\ \vdots \\ \alpha_n \end{bmatrix}$$

Thus

$$D(f) \underset{\mathcal{B}}{\leftrightarrow} \begin{bmatrix} 0 & 1 & 0 & 0 & \ldots & 0 \\ 0 & 0 & 2 & 0 & \ldots & 0 \\ 0 & 0 & 0 & 3 & \ldots & 0 \\ 0 & 0 & 0 & 0 & \ldots & 0 \\ & & \vdots & & & \\ 0 & 0 & 0 & 0 & \ldots & n \\ 0 & 0 & 0 & 0 & \ldots & 0 \end{bmatrix} \begin{bmatrix} \alpha_0 \\ \alpha_1 \\ \alpha_2 \\ \vdots \\ \alpha_{n-1} \\ \alpha_n \end{bmatrix} = \begin{bmatrix} \alpha_1 \\ 2\alpha_2 \\ 3\alpha_3 \\ \vdots \\ n\alpha_n \\ 0 \end{bmatrix}$$

i.e., $D(f) = \alpha_1 + 2\alpha_2 x + 3\alpha_3 x^2 + \cdots + n\alpha_n x^{n-1}$.

The matrix representation of a linear transformation effected by choosing bases can often be used to reduce problems regarding linear transformations to problems involving matrices and linear equations, problems for which we have developed effective computational procedures.

Again, let T be a linear transformation between vector spaces V and W, $\mathcal{B} = \{\boldsymbol{x}_1, \boldsymbol{x}_2, \ldots, \boldsymbol{x}_n\}$ be a basis for V, $\mathcal{C} = \{\boldsymbol{y}_1, \boldsymbol{y}_2, \ldots, \boldsymbol{y}_m\}$ be a basis for W, and A_T be the $m \times n$ matrix associated with T relative to the above bases. If $\boldsymbol{x}$ belongs to V,

$$\boldsymbol{x} \underset{\mathcal{B}}{\leftrightarrow} \begin{bmatrix} \alpha_1 \\ \alpha_2 \\ \vdots \\ \alpha_n \end{bmatrix}$$

then it follows that

$$T(\boldsymbol{x}) \underset{\mathcal{C}}{\leftrightarrow} A \begin{bmatrix} \alpha_1 \\ \alpha_2 \\ \vdots \\ \alpha_n \end{bmatrix}$$

Thus, $T(\boldsymbol{x}) = \mathbf{0}$ if and only if

$$A\begin{bmatrix}\alpha_1\\ \alpha_2\\ \vdots\\ \alpha_n\end{bmatrix} = \mathbf{0}$$

Stated otherwise, a vector $\boldsymbol{x}$ in V belongs to N_T if and only if its associated coordinate n-tuple in $\boldsymbol{R}^n$ (or $\boldsymbol{C}^n$ if V and W are complex vector spaces) belongs to N_{A_T}. Equivalently, the isomorphism which carries $\boldsymbol{x}$ into its associated coordinate n-tuple in $\boldsymbol{R}^n$ induces an isomorphism between N_T and N_{A_T}. This provides a computational means of determining the nullspace of a given linear transformation. Indeed, it reduces the problem to one of solving a system of homogeneous linear equations. Moreover, it implies that $n(T) = n(A_T)$. Since

$$r(T) = \dim V - n(T)$$

$$r(A_T) = \dim V - n(A_T)$$

it follows that

$$r(T) = r(A_T)$$

Thus, the rank and nullity of a linear transformation are just the rank and nullity of the associated matrix.

In Example 1 above, if the fixed vector is nonzero, the rank of the matrix

$$\begin{bmatrix}u & v & 0 & 0\\ 0 & 0 & u & v\end{bmatrix}$$

is 2. Thus $r(T) = 2$. Since $n(T) = 4 - r(T) = 2$, the nullspace is of dimension 2.

In Example 2 above, the rank of the matrix A_D is n. Thus, $r(D) = n$ and $n(D) = \dim P_n - r(D) = n + 1 - n = 1$.

Associating with each linear transformation in $L(V, W)$ its matrix relative to the bases $\mathcal{B}$ and $\mathcal{C}$ determines a function $G(T) = A_T$ from $L(V, W)$ to the space M_{mn} of $m \times n$ matrices (with real entries if V and W are real vector spaces, complex entries if V and W are complex vector spaces). In view of the examples of §5.7, the following result is not surprising.

Theorem 1 The function which assigns to each T in $L(V, W)$ its matrix relative to bases $\mathcal{B}$ for V and $\mathcal{C}$ for W is an isomorphism from $L(V, W)$ to M_{mn}, the space of $m \times n$ matrices.

Proof We denote the function by G. First, we show that G is linear. To accomplish this let S, T be linear transformations from V to W. Then, if

$$S(\boldsymbol{x}_j) = \sum_{i=1}^{m} a_{ij}\boldsymbol{y}_i$$

$$T(\boldsymbol{x}_j) = \sum_{i=1}^{m} b_{ij}\boldsymbol{y}_i$$

$G(S) = A_S = [a_{ij}]_{(mn)}$ and $G(T) = A_T = [b_{ij}]_{(mn)}$. Since

$$\begin{aligned}(S + T)(\boldsymbol{x}_j) &= S(\boldsymbol{x}_j) + T(\boldsymbol{x}_j)\\ &= \sum_{i=1}^{m} (a_{ij} + b_{ij})\boldsymbol{y}_i\end{aligned}$$

$A_{S+T} = [a_{ij} + b_{ij}]_{(mn)} = A_S + A_T$.

Thus, $G(S + T) = G(S) + G(T)$. A similar proof yields $G(\alpha S) = \alpha G(S)$.

Next, we show that G is one-one. Assume T is a linear operator such that $G(T) = A_T = 0$. Then, $T(\boldsymbol{x}_j) = \Sigma_{i=1}^{m} 0 \cdot \boldsymbol{y}_i = \boldsymbol{0}$. Since T sends all basis vectors into $\boldsymbol{0}$, $T = 0$. Thus, $G(T) = 0$ implies $T = 0$, and so G is one-one.

Finally, we show that G is onto. If A is an $m \times n$ matrix, $A = [a_{ij}]_{(mn)}$, we must exhibit a linear transformation T from V to W such that

$$T(\boldsymbol{x}_j) = \sum_{i=1}^{n} a_{ij}\boldsymbol{y}_i$$

for then by definition of A_T we have $A_T = A$. By Theorem 2, §5.2, such a linear transformation exists. Thus, G is onto.

Since G is linear, one-one, and onto, G is an isomorphism. ∎

In particular, if $T : V \to V$ is a linear operator on a vector space V and $\boldsymbol{x}_1, \boldsymbol{x}_2, \ldots, \boldsymbol{x}_n$ is a basis for V, the function which associates with the operator T the $n \times n$ matrix

$$A_T = [a_{ij}]_{(nn)}, \quad \text{where } A\boldsymbol{x}_j = \sum_{i=1}^{n} a_{ij}\boldsymbol{x}_j$$

induces an isomorphism of $L(V, V)$ and the space of $n \times n$ matrices. By considering the case $V = \boldsymbol{R}^n$ it seems reasonable to expect that the correspondence $T \to A_T$ preserves products. The next theorem shows that our expectations are well-founded.

Theorem 2 Let $S : V \to V$ and $T : V \to V$ be linear operators on the vector space V and $\mathcal{B} = \{\boldsymbol{x}_1, \boldsymbol{x}_2, \ldots, \boldsymbol{x}_n\}$ be a basis for V. Suppose A_S,

A_T, and $A_{S\circ T}$ are the $n \times n$ matrices associated with the operators S, T, and S $\circ$ T, respectively. Then,

$$A_S A_T = A_{S\circ T}$$

Proof $A_S = [a_{ij}]_{(nn)}$, $A_T = [b_{jk}]_{(nn)}$ where

$$S(\boldsymbol{x}_j) = \sum_{i=1}^{n} a_{ij}\boldsymbol{x}_i \quad \text{and} \quad T(\boldsymbol{x}_k) = \sum_{j=1}^{n} b_{jk}\boldsymbol{x}_j$$

The above equalities were derived from the definition of A_S and A_T. Thus,

$$\begin{aligned}(S \circ T)(\boldsymbol{x}_k) &= S(T(\boldsymbol{x}_k))\\ &= S\left(\sum_{j=1}^{n} b_{jk}\boldsymbol{x}_j\right)\\ &= \sum_{j=1}^{n} b_{jk} S(\boldsymbol{x}_j)\\ &= \sum_{j=1}^{n} b_{jk}\left(\sum_{i=1}^{n} a_{ij}\boldsymbol{x}_i\right)\\ &= \sum_{i=1}^{n}\left(\sum_{j=1}^{n} a_{ij}b_{jk}\right)\boldsymbol{x}_i\end{aligned}$$

Therefore, $(S \circ T)(\boldsymbol{x}_k) = \Sigma_{i=1}^{n} c_{ik}\boldsymbol{x}_i$, where $c_{ik} = \Sigma_{j=1}^{n} a_{ij}b_{jk}$. Since by definition of $A_{S\circ T}$,

$$A_{S\circ T} = [c_{ik}]_{(nn)}$$

it follows that $A_{S\circ T} = A_S A_T$. ∎

One consequence of this correspondence between compositions of linear transformations and matrix products is that $T : V \to V$ is invertible if and only if the matrix A_T is invertible, and $(A_T)^{-1} = A_{T^{-1}}$. To see this note that if T is invertible, then by Theorem 2,

$$A_T A_{T^{-1}} = A_{T\circ T^{-1}} = A_{I_V} = I_n$$

$$A_{T^{-1}} A_T = A_{T^{-1}\circ T} = A_{I_V} = I_n$$

Thus, A_T is invertible and $(A_T)^{-1} = A_{T^{-1}}$.

If A_T is invertible, there is some operator S (since the correspondence $T \to A_T$ is onto) such that $A_S = (A_T)^{-1}$. Therefore,

$$A_{S\circ T} = A_S A_T = (A_T)^{-1}A_T = I_n$$

and

$$A_{T\circ S} = A_T A_S = A_T(A_T)^{-1} = I_n$$

Thus, $A_{S\circ T} = A_{T\circ S} = I_n$. Since the correspondence $T \to A_T$ is one-one and since $A_{I_V} = I_n$, it follows that $S \circ T = T \circ S = I_V$.

Example 3 Let P_2 denote the space of all polynomials in a variable x of degree less than or equal to 2, with real coefficients. Let $\mathfrak{D}$ denote the collection of linear operators on P_2 of the form

$$\begin{aligned} T(f) &= a_0 f + (b_0 + b_1 x)f' + (c_0 + c_1 x + c_2 x^2)f'' \\ &= a_0 I(f) + (b_0 + b_1 x)D(f) + (c_0 + c_1 x + c_2 x^2)D^2(f) \end{aligned}$$

where a_0, b_0, b_1, c_0, c_1, and c_2 are real numbers.

Clearly any T of this form is a linear operator on P_2. Since the sum of two operators in $\mathfrak{D}$ is again an operator in $\mathfrak{D}$ and a scalar multiple of an operator in $\mathfrak{D}$ again belongs to $\mathfrak{D}$, $\mathfrak{D}$ is a subspace of $L(P_2, P_2)$.

We calculate the 3×3 matrix associated with

$$T = a_0 I + (b_0 + b_1 x)D + (c_0 + c_1 x + c_2 x^2)D$$

relative to the basis $1, x, x^2$.

Since

$$\begin{aligned} T(1) &= a_0 \\ T(x) &= b_0 + (a_0 + b_1)x \\ T(x^2) &= 2c_0 + (2c_1 + 2b_0)x + (2c_2 + 2b_1 + a_0)x^2 \end{aligned}$$

$$L(T) = A_T = \begin{bmatrix} a_0 & b_0 & 2c_0 \\ 0 & a_0 + b_1 & 2b_0 + 2c_1 \\ 0 & 0 & a_0 + 2b_1 + 2c_2 \end{bmatrix}$$

where L denotes the linear transformation from $L(P_2, P_2)$ to M_{33}, $L(T) = A_T$. If T belongs to $\mathfrak{D}$, note that A_T is an upper triangular matrix, i.e., all entries below the diagonal are 0. Denote the subspace of M_{33} consisting of the upper triangular matrices by $\mathfrak{T}$. Thus, L induces a linear transformation from $\mathfrak{D}$ to $\mathfrak{T}$. We claim $L : \mathfrak{D} \to \mathfrak{T}$ is an isomorphism. Indeed, if we define $M : \mathfrak{T} \to \mathfrak{D}$ by

$$M\left(\begin{bmatrix} \alpha_0 & \beta_0 & \gamma_0 \\ 0 & \beta_1 & \gamma_1 \\ 0 & 0 & \gamma_2 \end{bmatrix}\right) = \alpha_0 I + (\beta_0 + (\beta_1 - \alpha_0)x)D$$

$$+ (\tfrac{1}{2}\gamma_0 + (\tfrac{1}{2}\gamma_1 - \beta_0)x + (\tfrac{1}{2}\gamma_2 + \tfrac{1}{2}\alpha_0 - \beta_1)x^2)D^2$$

an easy calculation shows that

$$M \circ L = I_{\mathfrak{D}}, \qquad L \circ M = I_{\mathfrak{T}}, \quad \text{or } M = L^{-1}$$

It is clear that the subspace $\mathcal{T}$ is closed under products, i.e., if A and B belong to $\mathcal{T}$, AB belongs to $\mathcal{T}$. Moreover, if $\alpha_0\beta_1\gamma_2 \neq 0$,

$$\begin{bmatrix} \alpha_0 & \beta_0 & \gamma_0 \\ 0 & \beta_1 & \gamma_1 \\ 0 & 0 & \gamma_2 \end{bmatrix}^{-1} = \frac{1}{\alpha_0\beta_1\gamma_2}\begin{bmatrix} \beta_1\gamma_2 & -\beta_0\gamma_2 & \beta_0\gamma_1 - \beta_1\gamma_0 \\ 0 & \alpha_0\gamma_2 & -\alpha_0\gamma_1 \\ 0 & 0 & \alpha_0\beta_1 \end{bmatrix}$$

From this it follows that if A in $\mathcal{T}$ is invertible, A^{-1} also belongs to $\mathcal{T}$. Using these two facts along with $A_{S\circ T} = A_S A_T$, it follows that

(1) $\mathcal{D}$ is closed under composition, i.e., if S and T belong to $\mathcal{D}$, $S \circ T$ belongs to $\mathcal{D}$.
(2) If S belongs to $\mathcal{D}$ and S is invertible, then S^{-1} belongs to $\mathcal{D}$.

Using the above formulas, we may explicitly calculate the inverses of operators in $\mathcal{D}$. Note that T is invertible if and only if $\det A_T = a_0(a_0 + b_1)(a_0 + 2b_1 + 2c_2) \neq 0$.

For example, consider $T(f) = f + f' + (1 - x^2)f''$.

$$A_T = \begin{bmatrix} 1 & 1 & 2 \\ 0 & 1 & 2 \\ 0 & 0 & -1 \end{bmatrix} \quad \text{and} \quad (A_T)^{-1} = \begin{bmatrix} 1 & -1 & 0 \\ 0 & 1 & 2 \\ 0 & 0 & -1 \end{bmatrix}$$

Therefore,

$$\begin{aligned} T^{-1} &= M(A_{T^{-1}}) = M((A_T)^{-1}) \\ &= I - D + (2x - x^2)D^2 \end{aligned}$$

or

$$T^{-1}(f) = f - f' + (2x - x^2)f''$$

One natural question arises concerning the matrix representation of a linear operator on a vector space V. What is the relationship between the matrix representation of a linear operator with respect to one basis and its matrix representation with respect to another basis? The answer is contained in the next theorem.

Theorem 3 Suppose $T : V \to V$ is a linear operator on a vector space V and $\mathcal{B} = \{\boldsymbol{x}_1, \boldsymbol{x}_2, \ldots, \boldsymbol{x}_n\}$ and $\mathcal{B}' = \{\boldsymbol{x}_1', \boldsymbol{x}_2', \ldots, \boldsymbol{x}_n'\}$ are two bases for V. Let Q be the matrix associated with the change of coordinates from the basis $\mathcal{B}$ to the basis $\mathcal{B}'$, i.e., $Q = [q_{ij}]_{(nn)}$, where

$$\boldsymbol{x}_j' = \sum_{i=1}^{n} q_{ij}\boldsymbol{x}_i$$

Let A_T be the matrix associated with T relative to the basis $\mathcal{B}$, and A'_T be the matrix associated with T relative to the basis $\mathcal{B}'$, then

$$Q^{-1}A_TQ = A'_T$$

Proof Let x be a vector in V. Suppose

$$\begin{bmatrix} \alpha_1 \\ \alpha_2 \\ \vdots \\ \alpha_n \end{bmatrix} \quad \text{and} \quad \begin{bmatrix} \alpha'_1 \\ \alpha'_2 \\ \vdots \\ \alpha'_n \end{bmatrix}$$

are the coordinate n-tuples of the vector x in V with respect to the basis $\mathcal{B}$ and $\mathcal{B}'$, respectively. Then

$$\begin{bmatrix} \alpha_1 \\ \alpha_2 \\ \vdots \\ \alpha_n \end{bmatrix} \underset{\mathcal{B}}{\leftrightarrow} x \underset{\mathcal{B}'}{\leftrightarrow} \begin{bmatrix} \alpha'_1 \\ \alpha'_2 \\ \vdots \\ \alpha'_n \end{bmatrix}$$

We know, by definition of the matrices A_T and A'_T,

$$A_T\begin{bmatrix} \alpha_1 \\ \alpha_2 \\ \vdots \\ \alpha_n \end{bmatrix} \underset{\mathcal{B}}{\leftrightarrow} T(x) \underset{\mathcal{B}'}{\leftrightarrow} A'_T\begin{bmatrix} \alpha'_1 \\ \alpha'_2 \\ \vdots \\ \alpha'_n \end{bmatrix}$$

By §4.9, we also have

$$\begin{bmatrix} \alpha_1 \\ \alpha_2 \\ \vdots \\ \alpha_n \end{bmatrix} = Q\begin{bmatrix} \alpha'_1 \\ \alpha'_2 \\ \vdots \\ \alpha'_n \end{bmatrix}$$

$$A_T\begin{bmatrix} \alpha_1 \\ \alpha_2 \\ \vdots \\ \alpha_n \end{bmatrix} = QA'_T\begin{bmatrix} \alpha'_1 \\ \alpha'_2 \\ \vdots \\ \alpha'_n \end{bmatrix}$$

Thus,

$$A_TQ\begin{bmatrix} \alpha'_1 \\ \alpha'_2 \\ \vdots \\ \alpha'_n \end{bmatrix} = QA'_T\begin{bmatrix} \alpha'_1 \\ \alpha'_2 \\ \vdots \\ \alpha'_n \end{bmatrix}$$

Since this holds for all n-tuples of coordinates with respect to $\mathcal{B}'$ we have by the lemma of §3.7

$$A_T Q = Q A'_T$$

or

$$Q^{-1} A_T Q = A'_T$$

Two $n \times n$ matrices A and B are said to be similar if there is some invertible matrix Q, such that $A = QBQ^{-1}$. The foregoing theorem may be restated in the following manner: Two matrices which represent the same linear operator with respect to two (possibly different) bases are similar. Alternatively, two matrices are similar if and only if they represent the same linear operator with respect to two (possibly different) bases.

If A and B are similar matrices, it follows that $\det A = \det B$. Thus, we may define the determinant of a linear operator to be the determinant of its matrix representation with respect to some basis. By the previous remarks we see that the definition does not depend on the basis chosen.

Many problems involving matrices and linear operators become quite manageable if one makes a judicious choice of basis.

Example 4 Prove

$$\begin{vmatrix} 1 + a_1b_1 & a_1b_2 & \dots & a_1b_n \\ a_2b_1 & 1 + a_2b_2 & \dots & a_2b_n \\ a_3b_1 & a_3b_2 & \dots & a_3b_n \\ & \vdots & & \\ a_nb_1 & a_nb_2 & \dots & 1 + a_nb_n \end{vmatrix} = 1 + a_1b_1 + a_2b_2 + \cdots + a_nb_n$$

Let P be the linear operator on $\boldsymbol{R}^n$ whose matrix representation with respect to the standard basis $\boldsymbol{e}_1, \boldsymbol{e}_2, \dots, \boldsymbol{e}_n$ is

$$A_P = \begin{bmatrix} a_1b_1 & a_1b_2 & \dots & a_1b_n \\ a_2b_1 & a_2b_2 & \dots & a_2b_n \\ a_3b_1 & a_3b_2 & \dots & a_3b_n \\ & & \vdots & \\ a_nb_1 & a_nb_2 & \dots & a_nb_n \end{bmatrix}$$

We wish to calculate $\det [I + P]$. Let

$$\boldsymbol{a} = \begin{bmatrix} a_1 \\ a_2 \\ \vdots \\ a_n \end{bmatrix}$$

If $a_1 = 0$, the result follows by induction. Assume that $a_1 \neq 0$. Then,

$$
\begin{aligned}
P(\boldsymbol{e}_1) &= b_1\boldsymbol{a}\\
P(\boldsymbol{e}_2) &= b_2\boldsymbol{a}\\
&\vdots\\
P(\boldsymbol{e}_n) &= b_n\boldsymbol{a}
\end{aligned}
$$

Since $a_1 \neq 0$, the vectors $\boldsymbol{a}$, $\boldsymbol{e}_2$, $\boldsymbol{e}_3, \ldots, \boldsymbol{e}_n$ form a basis for $\boldsymbol{R}^n$. Consider the linear operator P with respect to the new basis

$$\boldsymbol{e}_1' = \boldsymbol{a}, \quad \boldsymbol{e}_2' = \boldsymbol{e}_2, \quad \boldsymbol{e}_3' = \boldsymbol{e}_3, \quad \ldots, \quad \boldsymbol{e}_n' = \boldsymbol{e}_n$$

Then,

$$P(\boldsymbol{e}_1') = P\left(\sum_{i=1}^{n} a_i\boldsymbol{e}_i\right) = \sum_{i=1}^{n} a_iP(\boldsymbol{e}_i) = \left(\sum_{i=1}^{n} a_ib_i\right)\boldsymbol{a}$$

or

$$
\begin{aligned}
P(\boldsymbol{e}_1') &= (a_1b_1 + a_2b_2 + \cdots + a_nb_n)\boldsymbol{e}_1'\\
P(\boldsymbol{e}_2') &= b_2\boldsymbol{e}_1'\\
&\vdots\\
P(\boldsymbol{e}_n') &= b_n\boldsymbol{e}_1'
\end{aligned}
$$

Thus,

$$
\begin{aligned}
(I + P)(\boldsymbol{e}_1') &= (1 + a_1b_1 + a_2b_2 + \cdots + a_nb_n)\boldsymbol{e}_1'\\
(I + P)(\boldsymbol{e}_2') &= \boldsymbol{e}_2' + b_2\boldsymbol{e}_1'\\
&\vdots\\
(I + P)(\boldsymbol{e}_n') &= \boldsymbol{e}_n' + b_n\boldsymbol{e}_1'
\end{aligned}
$$

In this basis the matrix associated with $I + P$ is

$$
\begin{bmatrix}
1 + \sum_{i=1}^{n} a_ib_i & b_2 & b_3 & \ldots & b_n\\
0 & 1 & 0 & \ldots & 0\\
0 & 0 & 1 & \ldots & 0\\
 & \vdots & & & \\
0 & 0 & 0 & \ldots & 1
\end{bmatrix}
$$

Its determinant is clearly $1 + a_1b_1 + a_2b_2 + \cdots + a_nb_n$. Since the determinant of the operator $I + P$ is independent of basis, we obtain the desired result.

Example 5 Determine the function R_θ from the plane into itself which reflects each point about the line l which makes an angle of θ degrees with the x axis.

Interpreting vector addition as given by the parallelogram law and scalar multiplication as signed magnification of length, R_θ is easily seen to be linear.

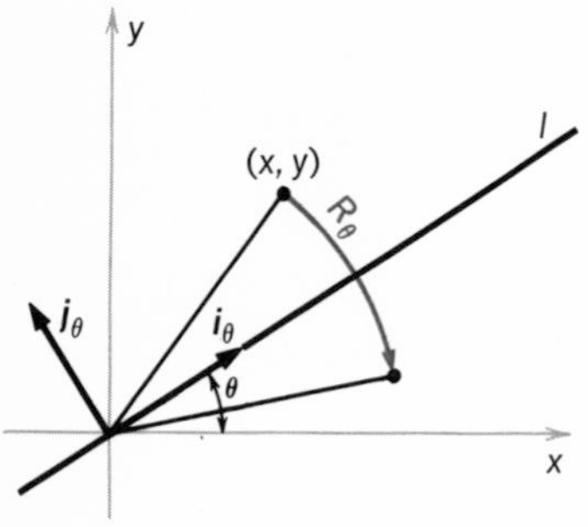

FIGURE 8-1

Let $\boldsymbol{i}_\theta = (\cos\theta)\boldsymbol{i} + (\sin\theta)\boldsymbol{j}$ be the unit vector in the direction of the line l. Let $\boldsymbol{j}_\theta = -(\sin\theta)\boldsymbol{i} + (\cos\theta)\boldsymbol{j}$. (See Figure 8-1.) Of course, the vectors $\boldsymbol{i}_\theta$ and $\boldsymbol{j}_\theta$ are just the vectors obtained by rotating the standard basis vectors $\boldsymbol{i}$ and $\boldsymbol{j}$ by θ degrees. We calculate the matrix representation of the linear transformation R_θ in the $\boldsymbol{i}_\theta\boldsymbol{j}_\theta$ coordinate system.

Clearly, we must have

$$R_\theta(\boldsymbol{i}_\theta) = \boldsymbol{i}_\theta$$

and

$$R_\theta(\boldsymbol{j}_\theta) = -\boldsymbol{j}_\theta$$

Thus, relative to the $\boldsymbol{i}_\theta\boldsymbol{j}_\theta$ system, the matrix representation of R_θ is

$$\begin{bmatrix} 1 & 0 \\ 0 & -1 \end{bmatrix}$$

Let B_θ be the matrix representation of R_θ relative to the basis $\{\boldsymbol{i}, \boldsymbol{j}\}$. Let

$$Q = \begin{bmatrix} \cos\theta & -\sin\theta \\ \sin\theta & \cos\theta \end{bmatrix}$$

be the matrix associated with the change of basis from the basis $\{\boldsymbol{i}, \boldsymbol{j}\}$ to the basis $\{\boldsymbol{i}_\theta, \boldsymbol{j}_\theta\}$.

By the results of this section

$$Q^{-1}B_\theta Q = \begin{bmatrix} 1 & 0 \\ 0 & -1 \end{bmatrix}$$

or

$$\begin{aligned} B_\theta &= Q\begin{bmatrix} 1 & 0 \\ 0 & -1 \end{bmatrix}Q^{-1} \\ &= \begin{bmatrix} \cos\theta & -\sin\theta \\ \sin\theta & \cos\theta \end{bmatrix}\begin{bmatrix} 1 & 0 \\ 0 & -1 \end{bmatrix}\begin{bmatrix} \cos\theta & \sin\theta \\ -\sin\theta & \cos\theta \end{bmatrix} \end{aligned}$$

$$= \begin{bmatrix} \cos\theta & \sin\theta \\ \sin\theta & -\cos\theta \end{bmatrix} \begin{bmatrix} \cos\theta & \sin\theta \\ -\sin\theta & \cos\theta \end{bmatrix}$$

$$= \begin{bmatrix} \cos^2\theta - \sin^2\theta & 2\sin\theta\cos\theta \\ 2\sin\theta\cos\theta & \sin^2\theta - \cos^2\theta \end{bmatrix}$$

$$= \begin{bmatrix} \cos 2\theta & \sin 2\theta \\ \sin 2\theta & -\cos 2\theta \end{bmatrix}$$

Thus,

$$R_\theta\left(\begin{bmatrix} x \\ y \end{bmatrix}\right) = \begin{bmatrix} \cos 2\theta & \sin 2\theta \\ \sin 2\theta & -\cos 2\theta \end{bmatrix} \begin{bmatrix} x \\ y \end{bmatrix}$$

$$= \begin{bmatrix} x\cos 2\theta + y\sin 2\theta \\ x\sin 2\theta - y\cos 2\theta \end{bmatrix}$$

A similar procedure may be used to derive the formula for the projection operator of §5.1.

Example 6 Let A and B be $n \times n$ matrices. A and B are said to skew-commute if $AB = -BA$.

For example,

$$\begin{bmatrix} a & 0 \\ 0 & -a \end{bmatrix} \begin{bmatrix} 0 & 1 \\ 1 & 0 \end{bmatrix} = \begin{bmatrix} 0 & a \\ -a & 0 \end{bmatrix}$$

$$\begin{bmatrix} 0 & 1 \\ 1 & 0 \end{bmatrix} \begin{bmatrix} a & 0 \\ 0 & -a \end{bmatrix} = \begin{bmatrix} 0 & -a \\ a & 0 \end{bmatrix}$$

so

$$\begin{bmatrix} a & 0 \\ 0 & -a \end{bmatrix} \quad \text{and} \quad \begin{bmatrix} 0 & 1 \\ 1 & 0 \end{bmatrix}$$

skew-commute.

We ask: If A is a given 2×2 matrix under what circumstances can we find a nonzero 2×2 matrix B, such that A and B skew-commute?

It should be clear that given A we cannot always find such a B. For example, if $A = I_2$ the identity matrix, since

$$AB = I_2B = B$$

$$BA = BI_2 = B$$

if $AB = -BA$, then $B = -B$ or $B = 0$.

Let us reformulate the problem using the language of linear operators. Consider the linear operator T_A, defined on the space of 2×2 matrices by $T_A(B) = AB + BA$. A matrix B belongs to N_{T_A} if and only if $AB + BA = 0$, or $AB = -BA$. Therefore, N_{T_A} consists precisely of those 2×2 matrices which skew-commute with A.

With this in mind, it follows that the problem we have posed is equivalent to the problem: Under what circumstances is T_A singular, i.e., $N_{T_A} \neq \mathbf{0}$, or when is $\det T_A = 0$? We calculate $\det T_A$ using the basis

$$E_1 = \begin{bmatrix} 1 & 0 \\ 0 & 0 \end{bmatrix}, \quad E_2 = \begin{bmatrix} 0 & 0 \\ 1 & 0 \end{bmatrix}, \quad E_3 = \begin{bmatrix} 0 & 1 \\ 0 & 0 \end{bmatrix}, \quad E_4 = \begin{bmatrix} 0 & 0 \\ 0 & 1 \end{bmatrix}$$

Let $A = \begin{bmatrix} a & b \\ c & d \end{bmatrix}$, then

$$T_A(E_1) = \begin{bmatrix} 2a & b \\ c & 0 \end{bmatrix} \qquad T_A(E_3) = \begin{bmatrix} c & a+d \\ 0 & c \end{bmatrix}$$

$$T_A(E_2) = \begin{bmatrix} b & 0 \\ a+d & b \end{bmatrix} \qquad T_A(E_4) = \begin{bmatrix} 0 & b \\ c & 2d \end{bmatrix}$$

We find that the matrix associated with T_A for this basis is

$$\begin{bmatrix} 2a & b & c & 0 \\ c & a+d & 0 & c \\ b & 0 & a+d & b \\ 0 & b & c & 2d \end{bmatrix}$$

Hence

$$\det T_A = \det T_{A^T} = \begin{vmatrix} 2a & c & b & 0 \\ b & a+d & 0 & b \\ c & 0 & a+d & c \\ 0 & c & b & 2d \end{vmatrix}$$

$$= \begin{vmatrix} 2a & c & b & -2a \\ b & a+d & 0 & 0 \\ c & 0 & a+d & 0 \\ 0 & c & b & 2d \end{vmatrix}$$

$$= \begin{vmatrix} 2a & 0 & 0 & -2(a+d) \\ b & a+d & 0 & 0 \\ c & 0 & a+d & 0 \\ 0 & c & b & 2d \end{vmatrix}$$

$$= (2a)\begin{vmatrix} a+d & 0 & 0 \\ 0 & a+d & 0 \\ c & b & 2d \end{vmatrix} + 2(a+d)\begin{vmatrix} b & a+d & 0 \\ c & 0 & a+d \\ 0 & c & b \end{vmatrix}$$

$$= 4ad(a+d)^2 + 2(a+d)(-bc(a+d) - (a+d)(bc))$$

$$= 4(ad - bc)(a+d)^2$$

Thus, $\det T_A = 0$ if and only if $ad - bc = 0$ or $a + d = 0$. The quantity $a + d$ is important in many connections in matrix theory and is called the **trace** of A. Thus, there is a nonzero matrix which skew-commutes with A if and only if $\det A = 0$ or $\operatorname{tr} A = 0$.

EXERCISES

1. Let T be the linear operator on $\mathbf{R}^3$ defined by

$$T\left(\begin{bmatrix} x \\ y \\ z \end{bmatrix}\right) = \begin{bmatrix} 0 & 1 & 0 \\ 1 & 0 & 0 \\ 0 & 0 & 1 \end{bmatrix}\begin{bmatrix} x \\ y \\ z \end{bmatrix}$$

What is the matrix associated with T relative to the following bases of R^3?

(a) $\begin{bmatrix} 1 \\ 1 \\ 0 \end{bmatrix}, \begin{bmatrix} 1 \\ -1 \\ 0 \end{bmatrix}, \begin{bmatrix} 0 \\ 0 \\ 1 \end{bmatrix}$ (b) $\begin{bmatrix} 1 \\ 0 \\ 1 \end{bmatrix}, \begin{bmatrix} 0 \\ 1 \\ 1 \end{bmatrix}, \begin{bmatrix} 0 \\ 0 \\ 1 \end{bmatrix}$

(c) $\begin{bmatrix} 0 \\ 1 \\ 0 \end{bmatrix}, \begin{bmatrix} 1 \\ 0 \\ 0 \end{bmatrix}, \begin{bmatrix} 1 \\ 1 \\ 1 \end{bmatrix}$

2. Consider the linear operator on the space of 2×2 matrices $T(A) = A^T$. Calculate the matrix representation of T relative to the following bases.

(a) $\begin{bmatrix} 1 & 0 \\ 0 & 0 \end{bmatrix}, \begin{bmatrix} 0 & 1 \\ 0 & 0 \end{bmatrix}, \begin{bmatrix} 0 & 0 \\ 1 & 0 \end{bmatrix}, \begin{bmatrix} 0 & 0 \\ 0 & 1 \end{bmatrix}$

(b) $\begin{bmatrix} 1 & 0 \\ 0 & 0 \end{bmatrix}, \begin{bmatrix} 0 & 0 \\ 0 & 1 \end{bmatrix}, \begin{bmatrix} 0 & 1 \\ 1 & 0 \end{bmatrix}, \begin{bmatrix} 0 & 1 \\ -1 & 0 \end{bmatrix}$

3. Let $\boldsymbol{x}_1$, $\boldsymbol{x}_2$, $\boldsymbol{x}_3$ be three linearly independent vectors in $\boldsymbol{R}^3$. Let $A = [\boldsymbol{x}_1, \boldsymbol{x}_2, \boldsymbol{x}_3]$ be the matrix whose columns are $\boldsymbol{x}_1$, $\boldsymbol{x}_2$, and $\boldsymbol{x}_3$ in that order. Let T be the linear operator on $\boldsymbol{R}^3$ defined by

$$T(\boldsymbol{x}_1) = \boldsymbol{x}_1$$
$$T(\boldsymbol{x}_2) = \boldsymbol{x}_1 + \boldsymbol{x}_2$$
$$T(\boldsymbol{x}_3) = \boldsymbol{x}_1 + \boldsymbol{x}_2 + \boldsymbol{x}_3$$

Show that

$$T\left(\begin{bmatrix} x \\ y \\ z \end{bmatrix}\right) = \left(A \begin{bmatrix} 1 & 1 & 1 \\ 0 & 1 & 1 \\ 0 & 0 & 1 \end{bmatrix} A^{-1}\right) \begin{bmatrix} x \\ y \\ z \end{bmatrix}$$

4. Find the matrix representation relative to basis $\{\boldsymbol{i}, \boldsymbol{j}\}$ for $\boldsymbol{R}^2$ of the operator T_θ, where

$$T_\theta(\boldsymbol{i}_\theta) = \boldsymbol{i}_\theta$$
$$T_\theta(\boldsymbol{j}_\theta) = 2\boldsymbol{j}_\theta$$

with

$$\boldsymbol{i}_\theta = (\cos\theta)\boldsymbol{i} + (\sin\theta)\boldsymbol{j}, \qquad \boldsymbol{j}_\theta = -(\sin\theta)\boldsymbol{i} + (\cos\theta)\boldsymbol{j}$$

5. Determine the matrix representation of the operator $T(f) = f - xf'$ on P_n, relative to the basis $1, x, x^2, \ldots, x^n$.

6. Let

$$\begin{bmatrix} a \\ b \end{bmatrix}, \quad \begin{bmatrix} c \\ d \end{bmatrix}$$

be linearly independent vectors in $\boldsymbol{R}^2$. Let T be the linear transformation of $\boldsymbol{R}^2$ such that

$$T\left(\begin{bmatrix} a \\ b \end{bmatrix}\right) = \begin{bmatrix} c \\ d \end{bmatrix}, \quad T\left(\begin{bmatrix} c \\ d \end{bmatrix}\right) = \begin{bmatrix} a \\ b \end{bmatrix}$$

Show that

$$T\left(\begin{bmatrix} x \\ y \end{bmatrix}\right) = \frac{1}{ad - bc} \begin{bmatrix} cd - ab & a^2 - c^2 \\ d^2 - b^2 & ab - cd \end{bmatrix} \begin{bmatrix} x \\ y \end{bmatrix}$$

7. Show that all matrices of the form

$$\begin{bmatrix} \cos\theta & \sin\theta \\ \sin\theta & -\cos\theta \end{bmatrix}$$

are similar.

8. Let T_α be the linear operator on P_n defined by $T_\alpha(f(x)) = f(x + \alpha)$. Determine the matrix representation of T_α relative to the basis 1, x, $x^2, \ldots, x^n$ for P_n. By showing that $(T_\alpha)^{-1} = T_{-\alpha}$, calculate the inverse of this matrix.
9. Let V be a vector space and $\boldsymbol{x}_1, \boldsymbol{x}_2, \boldsymbol{x}_3, \ldots, \boldsymbol{x}_n$ be a basis for V. Calculate the matrix representation of the following linear operators on V relative to the $\boldsymbol{x}_1, \boldsymbol{x}_2, \ldots, \boldsymbol{x}_n$ basis.
   (a) $T(\boldsymbol{x}_1) = \boldsymbol{x}_2$, $T(\boldsymbol{x}_2) = \boldsymbol{x}_3, \ldots, T(\boldsymbol{x}_{n-1}) = \boldsymbol{x}_n$, $T(\boldsymbol{x}_n) = \boldsymbol{0}$
   (b) $T(\boldsymbol{x}_1) = \boldsymbol{x}_2$, $T(\boldsymbol{x}_2) = \boldsymbol{x}_3, \ldots, T(\boldsymbol{x}_{n-1}) = \boldsymbol{x}_n$, $T(\boldsymbol{x}_n) = \alpha\boldsymbol{x}_1$
   (c) $T(\boldsymbol{x}_1) = \boldsymbol{x}_1$, $T(\boldsymbol{x}_2) = \boldsymbol{x}_1 + \boldsymbol{x}_2, \ldots, T(\boldsymbol{x}_{n-1}) = \boldsymbol{x}_1 + \boldsymbol{x}_2 + \cdots + \boldsymbol{x}_{n-1}$, $T(\boldsymbol{x}_n) = \boldsymbol{x}_1 + \boldsymbol{x}_2 + \cdots + \boldsymbol{x}_n$
   (d) $T(\boldsymbol{x}_1) = \lambda_1\boldsymbol{x}_1$, $T(\boldsymbol{x}_2) = \lambda_2\boldsymbol{x}_2, \ldots, T(\boldsymbol{x}_n) = \lambda_n\boldsymbol{x}_n$
10. In $\boldsymbol{R}^2$, let

$$\boldsymbol{i}_\theta = (\cos\theta)\boldsymbol{i} + (\sin\theta)\boldsymbol{j}$$
$$\boldsymbol{j}_\theta = -(\sin\theta)\boldsymbol{i} + (\cos\theta)\boldsymbol{j}$$

   Let T_θ be the operator which carries $\boldsymbol{i}_\theta$ into $\boldsymbol{j}_\theta$ and $\boldsymbol{j}_\theta$ into $\boldsymbol{0}$.
   (a) Calculate the matrix representation of T relative to the basis $\{\boldsymbol{i}, \boldsymbol{j}\}$ for $\boldsymbol{R}^2$.
   (b) Show that $T^2 = 0$.
11. If A is a 2×2 nilpotent matrix with real entries, show that

$$A = \pm\begin{bmatrix} xy & -x^2 \\ y^2 & -xy \end{bmatrix}$$

   for some real numbers x and y. Show that any matrix of this sort is nilpotent.
12. Let V and W be real vector spaces, $\boldsymbol{x}_1, \boldsymbol{x}_2, \ldots, \boldsymbol{x}_n$ be a basis for V, and $\boldsymbol{y}_1, \boldsymbol{y}_2, \ldots, \boldsymbol{y}_n$ be a basis for W. Consider the function f from $L(V, W)$ into $\boldsymbol{R}$ which assigns to each T in $L(V, W)$ the (i, j)th entry of the matrix A_T which represents the operator T relative to the above bases for V and W. Show that f is linear.
13. Let

$$A = \begin{bmatrix} a & c \\ b & d \end{bmatrix}$$

   be a 2×2 matrix. Consider the linear operator M on the space of 2×2 matrices defined by $M(B) = AB$.
   (a) Show that the matrix representation of M relative to the basis

$$E_1 = \begin{bmatrix} 1 & 0 \\ 0 & 0 \end{bmatrix}, \quad E_2 = \begin{bmatrix} 0 & 0 \\ 1 & 0 \end{bmatrix}$$

$$E_3 = \begin{bmatrix} 0 & 1 \\ 0 & 0 \end{bmatrix}, \quad E_4 = \begin{bmatrix} 0 & 0 \\ 0 & 1 \end{bmatrix}$$

is

$$\begin{bmatrix} a & c & 0 & 0 \\ b & d & 0 & 0 \\ 0 & 0 & a & c \\ 0 & 0 & b & d \end{bmatrix}$$

(b) Show that $\det M = (\det A)^2$.

14. Let A be an $n \times n$ matrix. Consider that linear transformation M from the space of $n \times n$ matrices into itself defined by $M(B) = AB$. Show that $\det M = (\det A)^n$.

15. If the $n \times n$ matrix A is similar to the matrix

$$\begin{bmatrix} d_1 & 0 & 0 & \dots & 0 \\ 0 & d_2 & 0 & \dots & 0 \\ 0 & 0 & d_3 & \dots & 0 \\ & & \vdots & & \\ 0 & 0 & 0 & \dots & d_n \end{bmatrix}$$

show that $\det [xI_n + yA] = (x + yd_1)(x + yd_2) \dots (x + yd_n)$.

16. Let V be a finite-dimensional vector space with basis B. Let S and T be linear operators on V and let A_S and A_T be their matrix representations relative to the basis B.
(a) Show that S and T commute if and only if A_S and A_T commute.
(b) Show that T is nilpotent if and only if A_T is nilpotent.
(c) Show that T is idempotent if and only if A_T is idempotent.

17. Let T be a linear operator on $\boldsymbol{R}^2$ such that $T^2 = I$.
(a) If $\boldsymbol{x}$ belongs to the range space of the operator $\frac{1}{2}(I + T)$, show that $T(\boldsymbol{x}) = \boldsymbol{x}$.
(b) If $\boldsymbol{x}$ belongs to the range space of the operator $\frac{1}{2}(I - T)$, show that $T(\boldsymbol{x}) = -\boldsymbol{x}$.
(c) If $T \neq \pm I$, show that there are nonzero vectors $\boldsymbol{x}_1$ and $\boldsymbol{x}_2$ such that $T(\boldsymbol{x}_1) = \boldsymbol{x}_1$ and $T(\boldsymbol{x}_2) = -\boldsymbol{x}_2$. Show that these vectors are linearly independent.
(d) If $A^2 = I_2$ and $A \neq \pm I_2$, where A is a 2×2 matrix, show that A is similar to the matrix

$$\begin{bmatrix} 1 & 0 \\ 0 & -1 \end{bmatrix}$$

18. Let V be a vector space and $\boldsymbol{x}_1, \boldsymbol{x}_2, \dots, \boldsymbol{x}_n$ be a basis for V. If T is a linear operator on V such that

$$
\begin{aligned}
T(\boldsymbol{x}_1) &= \lambda_2\boldsymbol{x}_2 \\
T(\boldsymbol{x}_2) &= \lambda_3\boldsymbol{x}_3 \\
&\vdots \\
T(\boldsymbol{x}_{n-1}) &= \lambda_n\boldsymbol{x}_n \\
T(\boldsymbol{x}_n) &= \lambda_1\boldsymbol{x}_1
\end{aligned}
$$

show that $T^n = (\lambda_1\lambda_2 \ldots \lambda_n)I$.

19. Let $T: V \to V$ be a linear operator on an n-dimensional space V. If $r(T) = 1$, show that the matrix representation of T relative to any basis of V is of the form

$$
\begin{bmatrix}
a_1b_1 & a_2b_1 & \ldots & a_nb_1 \\
a_1b_2 & a_2b_2 & \ldots & a_nb_2 \\
 & & \vdots & \\
a_1b_n & a_2b_n & \ldots & a_nb_n
\end{bmatrix}
$$

20. Consider the linear transformation $T: \boldsymbol{R}^n \to \boldsymbol{R}^n$ with $T(\boldsymbol{y}) = A\boldsymbol{y}$ for $\boldsymbol{y}$ in $\boldsymbol{R}^n$, where

$$
A = \begin{bmatrix}
1 & 1 & 1 & \ldots & 1 \\
1 & x & 0 & \ldots & 0 \\
1 & 0 & x & \ldots & 0 \\
 & & \vdots & & \\
1 & 0 & 0 & \ldots & x
\end{bmatrix}
$$

By calculating the matrix representation of T relative to the basis $\boldsymbol{e}_1$, $\boldsymbol{e}_2$, $\boldsymbol{e}_3 - \boldsymbol{e}_2, \ldots, \boldsymbol{e}_n - \boldsymbol{e}_2$, show that $\det A = x^{n-1} - (n-1)x^{n-2}$.

21. Let A be an $n \times n$ matrix of rank 1.

(a) If $A^2 = \alpha A$, where $\alpha \neq 0$, show that A is similar to the matrix

$$
\begin{bmatrix}
\alpha & 0 & \ldots & 0 \\
0 & 0 & \ldots & 0 \\
 & & \vdots & \\
0 & 0 & \ldots & 0
\end{bmatrix}
$$

(b) If $A^2 = 0$, $A \neq 0$, show that A is similar to the matrix

$$
\begin{bmatrix}
0 & 0 & \ldots & 0 \\
1 & 0 & \ldots & 0 \\
0 & 0 & \ldots & 0 \\
 & & \vdots & \\
0 & 0 & \ldots & 0
\end{bmatrix}
$$

(c) Show that

$$\begin{vmatrix} \lambda + a_1b_1 & a_2b_1 & \cdots & a_nb_1 \\ a_1b_2 & \lambda + a_2b_2 & \cdots & a_nb_2 \\ & \vdots & & \\ a_1b_n & a_2b_n & \cdots & \lambda + a_nb_n \end{vmatrix} = \lambda^{n-1}(\lambda + a_1b_1 + a_2b_2 + \cdots + a_nb_n)$$

22. If V and W are finite-dimensional vector spaces, show that $L(V, W)$ is finite-dimensional and $\dim L(V, W) = \dim V \dim W$.
23. Let A be a 2×2 matrix. Let C_A be the linear transformation of the space of 2×2 matrices into itself defined by

$$C_A(B) = AB - BA$$

(a) Show that $C_A = 0$ if and only if A is a scalar multiple of the identity.
(b) Relative to the basis

$$E_1 = \begin{bmatrix} 1 & 0 \\ 0 & 0 \end{bmatrix}, \quad E_2 = \begin{bmatrix} 0 & 0 \\ 1 & 0 \end{bmatrix}, \quad E_3 = \begin{bmatrix} 0 & 1 \\ 0 & 0 \end{bmatrix}, \quad E_4 = \begin{bmatrix} 0 & 0 \\ 0 & 1 \end{bmatrix}$$

show that the matrix associated to C_A is

$$\begin{bmatrix} 0 & c & -b & 0 \\ b & d-a & 0 & -b \\ -c & 0 & a-d & c \\ 0 & -c & b & 0 \end{bmatrix} \quad \text{if } A = \begin{bmatrix} a & c \\ b & d \end{bmatrix}$$

(c) If A is not a scalar multiple of the identity, show that the rank of the above matrix is 2.
(d) If A is not a scalar multiple of the identity, show that all matrices which commute with A can be written in the form $\alpha I_2 + \beta A$.
24. Show that any $n \times n$ matrix of rank r is similar to a matrix of the form

$$\begin{bmatrix} a_{11} & \cdots & a_{1r} & 0 & \cdots & 0 \\ a_{21} & \cdots & a_{2r} & 0 & \cdots & 0 \\ & & \vdots & & & \\ a_{n1} & \cdots & a_{nr} & 0 & \cdots & 0 \end{bmatrix}$$

25. The trace function is defined on $n \times n$ matrices by

$$\operatorname{tr} A = \sum_{i=1}^{n} a_{ii} \quad \text{if } A = [a_{ij}]_{(nn)}$$

If $T : V \to V$ is a linear operator on a finite-dimensional vector space V, define tr T = tr A_T, where A_T is the matrix representation of T relative to some basis for V.

(a) Show that the trace is independent of the basis chosen for V.
(b) Show that tr $(S + T)$ = tr S + tr T, and tr $\alpha S = \alpha$tr S.
(c) Show that tr $S \circ T$ = tr $T \circ S$.

26. Let P_n be the space of polynomials of degree less than or equal to n. Let $\mathfrak{D}$ denote the collection of linear operators on P_n of the form

$$T(f) = p_0 f + p_1 f' + p_2 f'' + \cdots + p_n f^{(n)}$$

where p_0 is a polynomial of degree 0, and p_i is a polynomial of degree $\leq i$, $i = 1, 2, \ldots, n$. Then $T = p_0 I + p_1 D + \cdots + p_n D^n$.

(a) Show that $\mathfrak{D}$ is a subspace of $L(P_n, P_n)$.
(b) Show that the operators $I, D, xD, D^2, xD^2, x^2D^2, \ldots, D^n, xD^n, \ldots, x^nD^n$ form a basis for $\mathfrak{D}$ and that dim $\mathfrak{D} = \frac{1}{2}n(n + 1)$.
(c) Show that the matrix representation of any operator in $\mathfrak{D}$ relative to the basis $1, x, \ldots, x^n$ is upper triangular.
(d) Show that every upper triangular matrix is the representation of some operator in $\mathfrak{D}$. [Hint: Show that dim $\mathfrak{D}$ = dim T, where T denotes the space of upper triangular matrices, and use the fact that $T \to A_T$ is one-one.]
(e) Show that if A and B belong to $\mathfrak{D}$, then $A \circ B$ belongs to $\mathfrak{D}$; if A belongs to $\mathfrak{D}$ and A is invertible, then A^{-1} belongs to $\mathfrak{D}$.
(f) Which operators in $\mathfrak{D}$ correspond to diagonal matrices?
(g) Show that the differential equation

$$(a_0 + a_1 x + a_2 x^2)y'' + (b_0 + b_1 x)y' + c_0 y = 0$$

has a polynomial solution if and only if $n(n - 1)a_2 + nb_1 + c_0 = 0$ for some non-negative integer n.

6

PRODUCTS

1 INNER AND CROSS PRODUCTS IN $\boldsymbol{R}^3$

So far in our study of vectors we have neglected to ask if there are any circumstances in which it makes sense to speak of the product of two vectors. In this section we correct this omission by discussing two sorts of products which are often very useful in physical applications. The first product we consider is called the inner product. The names dot product and scalar product are often used instead.

Let $\boldsymbol{a} = a_1\boldsymbol{i} + a_2\boldsymbol{j} + a_3\boldsymbol{k}$ and $\boldsymbol{b} = b_1\boldsymbol{i} + b_2\boldsymbol{j} + b_3\boldsymbol{k}$ be two vectors in $\boldsymbol{R}^3$. We define the inner product of $\boldsymbol{a}$ and $\boldsymbol{b}$, written $(\boldsymbol{a}, \boldsymbol{b})$, to be the real number $a_1b_1 + a_2b_2 + a_3b_3$. $(\boldsymbol{a}, \boldsymbol{b})$ is frequently denoted by $\boldsymbol{a} \cdot \boldsymbol{b}$, hence, the name dot product. Note that the inner product of two vectors is a scalar quantity.

If we agree to identify 1×1 matrices and scalars, the inner product may also be written in terms of matrix multiplication. Indeed, we have

$$\boldsymbol{a}^T\boldsymbol{b} = [a_1 \quad a_2 \quad a_3]\begin{bmatrix} b_1 \\ b_2 \\ b_3 \end{bmatrix} = a_1b_1 + a_2b_2 + a_3b_3$$

Certain properties of the inner product follow immediately from the definition. We suppose $\boldsymbol{a}$, $\boldsymbol{b}$, and $\boldsymbol{c}$ are vectors in $\boldsymbol{R}^3$, α and β are real scalars.

(1) $(\boldsymbol{a}, \boldsymbol{a}) \geq 0$
$(\boldsymbol{a}, \boldsymbol{a}) = 0$ if and only if $\boldsymbol{a} = \boldsymbol{0}$

(2) $(\alpha\boldsymbol{a}, \boldsymbol{b}) = \alpha(\boldsymbol{a}, \boldsymbol{b})$
$(\boldsymbol{a}, \beta\boldsymbol{b}) = \beta(\boldsymbol{a}, \boldsymbol{b})$

(3) $(\boldsymbol{a}, \boldsymbol{b} + \boldsymbol{c}) = (\boldsymbol{a}, \boldsymbol{b}) + (\boldsymbol{a}, \boldsymbol{c})$
$(\boldsymbol{a} + \boldsymbol{b}, \boldsymbol{c}) = (\boldsymbol{a}, \boldsymbol{c}) + (\boldsymbol{b}, \boldsymbol{c})$

(4) $(\boldsymbol{a}, \boldsymbol{b}) = (\boldsymbol{b}, \boldsymbol{a})$

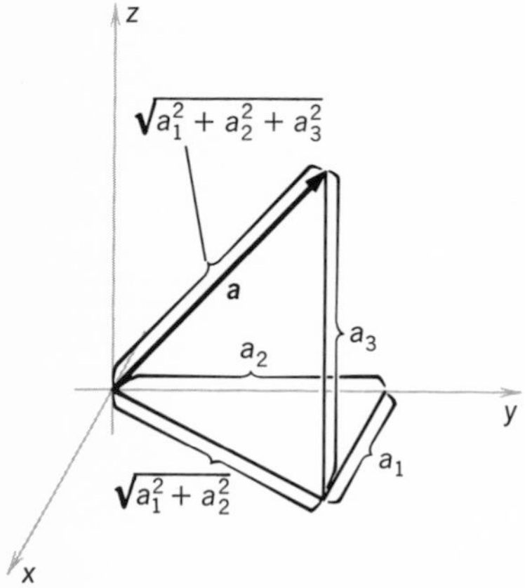

FIGURE 1-1

To prove (1), observe that if $\boldsymbol{a} = a_1\boldsymbol{i} + a_2\boldsymbol{j} + a_3\boldsymbol{k}$, $(\boldsymbol{a}, \boldsymbol{a}) = a_1^2 + a_2^2 + a_3^2$. Since a_1, a_2, and a_3 are real numbers $a_1^2 \geq 0$, $a_2^2 \geq 0$, $a_3^2 \geq 0$. Thus, $(\boldsymbol{a}, \boldsymbol{a}) \geq 0$. Moreover, if $a_1^2 + a_2^2 + a_3^2 = 0$, then $a_1 = a_2 = a_3 = 0$. Therefore, $\boldsymbol{a} = \boldsymbol{0}$.

The proofs of the other properties of the inner product are easily obtained.

It follows from the Pythagorean theorem that the length of the vector $\boldsymbol{a} = a_1\boldsymbol{i} + a_2\boldsymbol{j} + a_3\boldsymbol{k}$ is $\sqrt{a_1^2 + a_2^2 + a_3^2}$. (See Figure 1-1.) The length of the vector $\boldsymbol{a}$ is denoted by $|\boldsymbol{a}|$. This quantity is often called the **norm** of $\boldsymbol{a}$. Since $(\boldsymbol{a}, \boldsymbol{a}) = a_1^2 + a_2^2 + a_3^2$, it follows that $|\boldsymbol{a}| = (\boldsymbol{a}, \boldsymbol{a})^{1/2}$. Vectors of norm 1 are called unit vectors.

For example, in the plane the vector $\boldsymbol{i}_\theta = (\cos\theta)\boldsymbol{i} + (\sin\theta)\boldsymbol{j}$ is the unit vector making an angle of θ degrees with the x axis. (See Figure 1-2.)

If $\boldsymbol{a}$ and $\boldsymbol{b}$ are vectors, we have seen that the vector $\boldsymbol{b} - \boldsymbol{a}$ is parallel to the directed line segment from the endpoint of $\boldsymbol{a}$ to the endpoint of $\boldsymbol{b}$. It follows that the distance from the endpoint of $\boldsymbol{a}$ to the endpoint of $\boldsymbol{b}$ is $|\boldsymbol{b} - \boldsymbol{a}|$. (See Figure 1-3.) This distance is often written $d(\boldsymbol{a}, \boldsymbol{b})$. For example, the distance from the endpoint of the vector $\boldsymbol{i}$, that is, the point $(1, 0, 0)$ to the endpoint of the vector $\boldsymbol{j}$, $(0, 1, 0)$, is

$$\sqrt{(1-0)^2 + (0-1)^2 + (0-0)^2} = \sqrt{2}$$

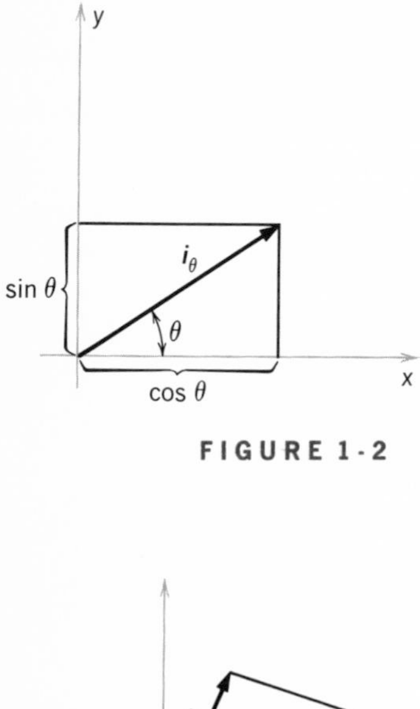

FIGURE 1-2

Based on our previous work with vectors, it is reasonable to expect that the inner product of two vectors has a geometric interpretation. Let $\boldsymbol{a}$ and $\boldsymbol{b}$ be two vectors in $\boldsymbol{R}^3$ and θ be the angle between them. (See Figure 1-4.) We will show that $(\boldsymbol{a}, \boldsymbol{b}) = |\boldsymbol{a}||\boldsymbol{b}| \cos\theta$.

By the law of cosines, (§3.6), applied to the triangle with one vertex at the origin and adjacent sides determined by the vectors $\boldsymbol{a}$ and $\boldsymbol{b}$, it follows that

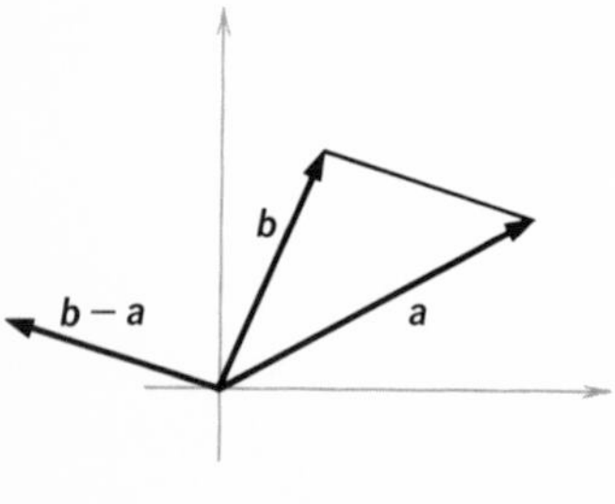

FIGURE 1-3

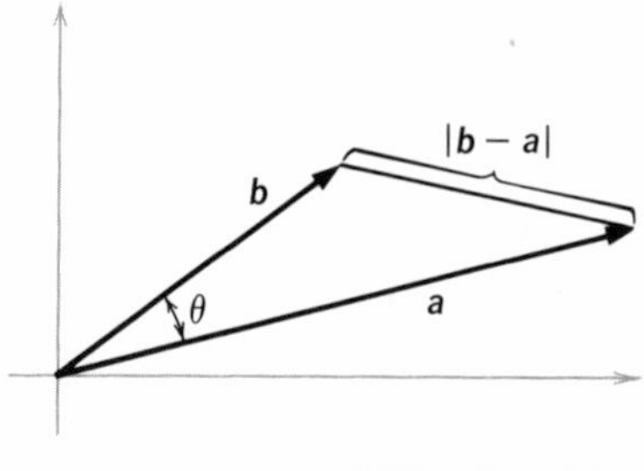

FIGURE 1-4

$$|\boldsymbol{b} - \boldsymbol{a}|^2 = |\boldsymbol{a}|^2 + |\boldsymbol{b}|^2 - 2|\boldsymbol{a}||\boldsymbol{b}| \cos \theta$$

Since $|\boldsymbol{b} - \boldsymbol{a}|^2 = (\boldsymbol{b} - \boldsymbol{a}, \boldsymbol{b} - \boldsymbol{a})$, $|\boldsymbol{a}|^2 = (\boldsymbol{a}, \boldsymbol{a})$, $|\boldsymbol{b}|^2 = (\boldsymbol{b}, \boldsymbol{b})$, we can rewrite the last equation as

$$(\boldsymbol{b} - \boldsymbol{a}, \boldsymbol{b} - \boldsymbol{a}) = (\boldsymbol{a}, \boldsymbol{a}) + (\boldsymbol{b}, \boldsymbol{b}) - 2|\boldsymbol{a}||\boldsymbol{b}| \cos \theta$$

Now,

$$\begin{aligned}(\boldsymbol{b} - \boldsymbol{a}, \boldsymbol{b} - \boldsymbol{a}) &= (\boldsymbol{b}, \boldsymbol{b} - \boldsymbol{a}) - (\boldsymbol{a}, \boldsymbol{b} - \boldsymbol{a}) \\ &= (\boldsymbol{b}, \boldsymbol{b}) - (\boldsymbol{b}, \boldsymbol{a}) - (\boldsymbol{a}, \boldsymbol{b}) + (\boldsymbol{a}, \boldsymbol{a}) \\ &= (\boldsymbol{a}, \boldsymbol{a}) + (\boldsymbol{b}, \boldsymbol{b}) - 2(\boldsymbol{a}, \boldsymbol{b})\end{aligned}$$

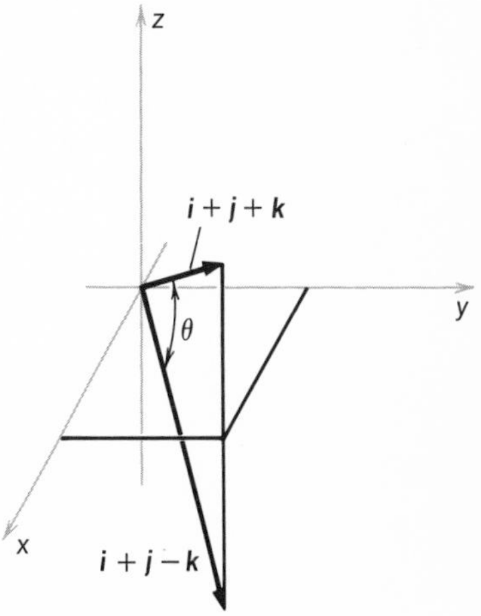

FIGURE 1-5

Thus, $(\boldsymbol{a}, \boldsymbol{a}) + (\boldsymbol{b}, \boldsymbol{b}) - 2(\boldsymbol{a}, \boldsymbol{b}) = (\boldsymbol{a}, \boldsymbol{a}) + (\boldsymbol{b}, \boldsymbol{b}) - 2|\boldsymbol{a}||\boldsymbol{b}| \cos \theta$, or $(\boldsymbol{a}, \boldsymbol{b}) = |\boldsymbol{a}||\boldsymbol{b}| \cos \theta$.

This formula means that the inner product of two vectors is the product of their lengths times the cosine of the angle between them. This relationship is often of value in problems of a geometric nature.

Example 1 Find the angle between the vectors $\boldsymbol{i} + \boldsymbol{j} + \boldsymbol{k}$ and $\boldsymbol{i} + \boldsymbol{j} - \boldsymbol{k}$, illustrated in Figure 1-5. By the above, $(\boldsymbol{i} + \boldsymbol{j} + \boldsymbol{k}, \boldsymbol{i} + \boldsymbol{j} - \boldsymbol{k}) = |\boldsymbol{i} + \boldsymbol{j} + \boldsymbol{k}||\boldsymbol{i} + \boldsymbol{j} - \boldsymbol{k}| \cos \theta$. So $1 + 1 - 1 = (\sqrt{3})(\sqrt{3}) \cos \theta$, or $\cos \theta = \frac{1}{3}$. So $\theta = \cos^{-1}(\frac{1}{3})$.

If $\boldsymbol{a}$ and $\boldsymbol{b}$ are nonzero vectors in $\boldsymbol{R}^3$ and θ is the angle between them, we see that $(\boldsymbol{a}, \boldsymbol{b}) = 0$ if and only if $\cos \theta = 0$. From this it follows that the inner product of two nonzero vectors is zero if and only if the vectors are perpendicular. Often we say that perpendicular vectors are **orthogonal**. We adopt the convention that the zero vector is orthogonal to all vectors.

For example the vectors $\boldsymbol{i}_\theta = (\cos \theta)\boldsymbol{i} + (\sin \theta)\boldsymbol{j}$ and $\boldsymbol{j}_\theta = -(\sin \theta)\boldsymbol{i} + (\cos \theta)\boldsymbol{j}$ are orthogonal since $(\boldsymbol{i}_\theta, \boldsymbol{j}_\theta) = -\cos \theta \sin \theta + \sin \theta \cos \theta = 0$. (See Figure 1-6.)

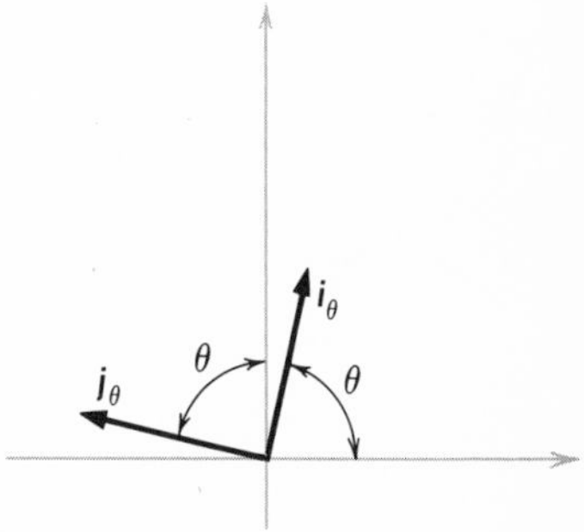

FIGURE 1-6

Example 2 Let $\boldsymbol{a}$ and $\boldsymbol{b}$ be two nonzero orthogonal vectors. Let $\boldsymbol{c}$ be another vector in the plane spanned by $\boldsymbol{a}$ and $\boldsymbol{b}$. As we have seen, there are scalars α and β such that $\boldsymbol{c} = \alpha\boldsymbol{a} + \beta\boldsymbol{b}$. We use the inner product to determine α and β. (See Figure 1-7.)

$$(\boldsymbol{a}, \boldsymbol{c}) = (\boldsymbol{a}, \alpha\boldsymbol{a} + \beta\boldsymbol{b}) = \alpha(\boldsymbol{a}, \boldsymbol{a}) + \beta(\boldsymbol{a}, \boldsymbol{b})$$

Since $\boldsymbol{a}$ and $\boldsymbol{b}$ are orthogonal, $(\boldsymbol{a}, \boldsymbol{b}) = 0$, and so $\alpha = (\boldsymbol{a}, \boldsymbol{c})/(\boldsymbol{a}, \boldsymbol{a}) = (\boldsymbol{a}, \boldsymbol{c})/|\boldsymbol{a}^2|$. Similarly, $\beta = (\boldsymbol{b}, \boldsymbol{c})/(\boldsymbol{b}, \boldsymbol{b}) = (\boldsymbol{b}, \boldsymbol{c})/|\boldsymbol{b}^2|$.

The same result may be obtained using the geometric interpretation of the scalar product. Let l be the distance measured from the origin to

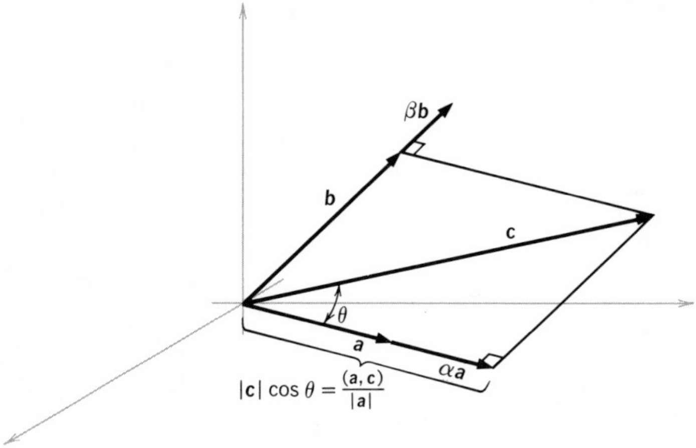

FIGURE 1-7

the point where the perpendicular from the endpoint of $\boldsymbol{c}$ intersects the line determined by extending $\boldsymbol{a}$. It follows that

$$l = |\boldsymbol{c}| \cos \theta$$

where θ is the angle between $\boldsymbol{a}$ and $\boldsymbol{c}$.

Moreover, $l = \alpha|\boldsymbol{a}|$. Taken together,

$$\alpha|\boldsymbol{a}| = |\boldsymbol{c}| \cos \theta$$

$$\alpha = \frac{|\boldsymbol{c}| \cos \theta}{|\boldsymbol{a}|} = \frac{(\boldsymbol{a}, \boldsymbol{c})}{(\boldsymbol{a}, \boldsymbol{a})}$$

Example 3 Recall the projection operator P_θ of §5.2. If l is a line making an angle of θ degrees with the x axis, $P_\theta(\boldsymbol{v})$ is the vector along l, ending at the point where the perpendicular from the endpoint of $\boldsymbol{v}$ to the line l intersects l. (See Figure 1-8.) We now derive the formula for $P_\theta(\boldsymbol{v})$ using the inner product.

We know that $\boldsymbol{i}_\theta = (\cos \theta)\boldsymbol{i} + (\sin \theta)\boldsymbol{j}$ is the unit vector in the direction of l. If ψ is the angle between the vector $\boldsymbol{v}$ and the line l,

$$P_\theta(\boldsymbol{v}) = (|\boldsymbol{v}| \cos \psi)\boldsymbol{i}_\theta = (\boldsymbol{v}, \boldsymbol{i}_\theta)\boldsymbol{i}_\theta$$

If $\boldsymbol{v} = x\boldsymbol{i} + y\boldsymbol{j}$,

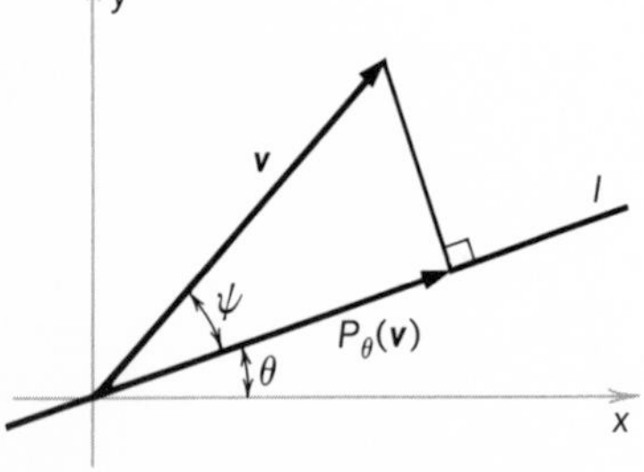

FIGURE 1-8

$$\begin{aligned} P_\theta(\boldsymbol{v}) &= P_\theta(x\boldsymbol{i} + y\boldsymbol{j}) = (x\boldsymbol{i} + y\boldsymbol{j}, \boldsymbol{i}_\theta)\boldsymbol{i}_\theta \\ &= (x \cos \theta + y \sin \theta)((\cos \theta)\boldsymbol{i} + (\sin \theta)\boldsymbol{j}) \end{aligned}$$

or

$$P_\theta\left(\begin{vmatrix} x \\ y \end{vmatrix}\right) = \begin{vmatrix} \cos^2\theta & \sin\theta\cos\theta \\ \sin\theta\cos\theta & \sin^2\theta \end{vmatrix}\begin{vmatrix} x \\ y \end{vmatrix}$$

Besides the inner product, there is another product of vectors which is sometimes useful. Let $\boldsymbol{a} = a_1\boldsymbol{i} + a_2\boldsymbol{j} + a_3\boldsymbol{k}$ and $\boldsymbol{b} = b_1\boldsymbol{i} + b_2\boldsymbol{j} + b_3\boldsymbol{k}$ be vectors in $\boldsymbol{R}^3$. The **cross product** of $\boldsymbol{a}$ and $\boldsymbol{b}$, denoted by $\boldsymbol{a} \times \boldsymbol{b}$, is defined to be the vector

$$\boldsymbol{a} \times \boldsymbol{b} = \begin{vmatrix} a_2 & a_3 \\ b_2 & b_3 \end{vmatrix}\boldsymbol{i} - \begin{vmatrix} a_1 & a_3 \\ b_1 & b_3 \end{vmatrix}\boldsymbol{j} + \begin{vmatrix} a_1 & a_2 \\ b_1 & b_2 \end{vmatrix}\boldsymbol{k}$$

or symbolically,

$$\boldsymbol{a} \times \boldsymbol{b} = \begin{vmatrix} \boldsymbol{i} & \boldsymbol{j} & \boldsymbol{k} \\ a_1 & a_2 & a_3 \\ b_1 & b_2 & b_3 \end{vmatrix}$$

Note that the cross product of two vectors is another vector. Just as the inner product is sometimes called the scalar product, the cross product is sometimes called the **vector product**.

Again, certain algebraic properties of the cross product follow immediately from the definition. Suppose $\boldsymbol{a}$, $\boldsymbol{b}$, and $\boldsymbol{c}$ are vectors and α, β, and γ are scalars.

(1) $\boldsymbol{a} \times \boldsymbol{b} = -(\boldsymbol{b} \times \boldsymbol{a})$
(2) $\boldsymbol{a} \times (\beta\boldsymbol{b} + \gamma\boldsymbol{c}) = \beta(\boldsymbol{a} \times \boldsymbol{b}) + \gamma(\boldsymbol{a} \times \boldsymbol{c})$
$(\alpha\boldsymbol{a} + \beta\boldsymbol{b}) \times \boldsymbol{c} = \alpha(\boldsymbol{a} \times \boldsymbol{c}) + \beta(\boldsymbol{b} \times \boldsymbol{c})$

Note that $\boldsymbol{a} \times \boldsymbol{a} = -(\boldsymbol{a} \times \boldsymbol{a})$ by (1). Thus, $\boldsymbol{a} \times \boldsymbol{a} = \boldsymbol{0}$. Also

$$\boldsymbol{i} \times \boldsymbol{j} = \boldsymbol{k}$$
$$\boldsymbol{j} \times \boldsymbol{k} = \boldsymbol{i}$$
$$\boldsymbol{k} \times \boldsymbol{i} = \boldsymbol{j}$$

For example,

$$(3\boldsymbol{i} - \boldsymbol{j} + \boldsymbol{k}) \times (\boldsymbol{i} + 2\boldsymbol{j} - \boldsymbol{k}) = \begin{vmatrix} \boldsymbol{i} & \boldsymbol{j} & \boldsymbol{k} \\ 3 & -1 & 1 \\ 1 & 2 & -1 \end{vmatrix}$$
$$= -\boldsymbol{i} + 4\boldsymbol{j} + 7\boldsymbol{k}$$

Our next goal is to present a geometric interpretation of the cross product analogous to that of the inner product.

First, we provide a formula for the product $\boldsymbol{a} \cdot (\boldsymbol{b} \times \boldsymbol{c})$. If $\boldsymbol{a} = a_1\boldsymbol{i} + a_2\boldsymbol{j} + a_3\boldsymbol{k}$, $\boldsymbol{b} = b_1\boldsymbol{i} + b_2\boldsymbol{j} + b_3\boldsymbol{k}$, and $\boldsymbol{c} = c_1\boldsymbol{i} + c_2\boldsymbol{j} + c_3\boldsymbol{k}$,

$$\begin{aligned}\boldsymbol{a} \cdot (\boldsymbol{b} \times \boldsymbol{c}) &= (a_1\boldsymbol{i} + a_2\boldsymbol{j} + a_3\boldsymbol{k}) \cdot \left(\begin{vmatrix} b_2 & b_3 \\ c_2 & c_3 \end{vmatrix}\boldsymbol{i} - \begin{vmatrix} b_1 & b_3 \\ c_1 & c_3 \end{vmatrix}\boldsymbol{j} + \begin{vmatrix} b_1 & b_2 \\ c_1 & c_2 \end{vmatrix}\boldsymbol{k}\right) \\ &= a_1\begin{vmatrix} b_2 & b_3 \\ c_2 & c_3 \end{vmatrix} - a_2\begin{vmatrix} b_1 & b_3 \\ c_1 & c_3 \end{vmatrix} + a_3\begin{vmatrix} b_1 & b_2 \\ c_1 & c_2 \end{vmatrix}\end{aligned}$$

This can be written more concisely as

$$\boldsymbol{a} \cdot (\boldsymbol{b} \times \boldsymbol{c}) = \begin{vmatrix} a_1 & a_2 & a_3 \\ b_1 & b_2 & b_3 \\ c_1 & c_2 & c_3 \end{vmatrix}$$

Now suppose that $\boldsymbol{a}$ is a vector in the subspace of $\boldsymbol{R}^3$ spanned by the vectors $\boldsymbol{b}$ and $\boldsymbol{c}$. This means that the first row in the determinant expression for $\boldsymbol{a} \cdot (\boldsymbol{b} \times \boldsymbol{c})$ is a linear combination of the second and third rows, and therefore $\boldsymbol{a} \cdot (\boldsymbol{b} \times \boldsymbol{c}) = 0$. In other words, the vector $\boldsymbol{b} \times \boldsymbol{c}$ is orthogonal to any vector in the subspace spanned by $\boldsymbol{b}$ and $\boldsymbol{c}$, in particular to both $\boldsymbol{b}$ and $\boldsymbol{c}$.

Next, we calculate the magnitude of $\boldsymbol{b} \times \boldsymbol{c}$. Note that

$$\begin{aligned}|\boldsymbol{b} \times \boldsymbol{c}|^2 &= \begin{vmatrix} b_2 & b_3 \\ c_2 & c_3 \end{vmatrix}^2 + \begin{vmatrix} b_1 & b_3 \\ c_1 & c_3 \end{vmatrix}^2 + \begin{vmatrix} b_1 & b_2 \\ c_1 & c_2 \end{vmatrix}^2 \\ &= (b_1^2 + b_2^2 + b_3^2)(c_1^2 + c_2^2 + c_3^2) - (b_1c_1 + b_2c_2 + b_3c_3)^2 \\ &= |\boldsymbol{b}|^2|\boldsymbol{c}|^2 - |\boldsymbol{b}|^2|\boldsymbol{c}|^2 \cos^2\theta = |\boldsymbol{b}|^2|\boldsymbol{c}|^2 \sin^2\theta\end{aligned}$$

where θ is the angle between $\boldsymbol{b}$ and $\boldsymbol{c}$.

Combining our results, we see that $\boldsymbol{b} \times \boldsymbol{c}$ is a vector perpendicular to $\boldsymbol{b}$ and $\boldsymbol{c}$ with length $|\boldsymbol{b}||\boldsymbol{c}||\sin\theta|$.

If $\boldsymbol{b}$ and $\boldsymbol{c}$ are linearly dependent, $\theta = 0$, and so $\boldsymbol{b} \times \boldsymbol{c} = \boldsymbol{0}$. If $\boldsymbol{b}$ and $\boldsymbol{c}$ are linearly independent, they span a plane and $\boldsymbol{b} \times \boldsymbol{c}$ is a vector perpendicular to this plane. The length of $\boldsymbol{b} \times \boldsymbol{c}$, $|\boldsymbol{b}||\boldsymbol{c}||\sin\theta|$, is just the area of the parallelogram with adjacent sides represented by the vectors and $\boldsymbol{b}$ and $\boldsymbol{c}$. (See Figure 1-9.)

Example 4 Find a unit vector orthogonal to the vectors $\boldsymbol{i} + \boldsymbol{j}$ and $\boldsymbol{j} + \boldsymbol{k}$.

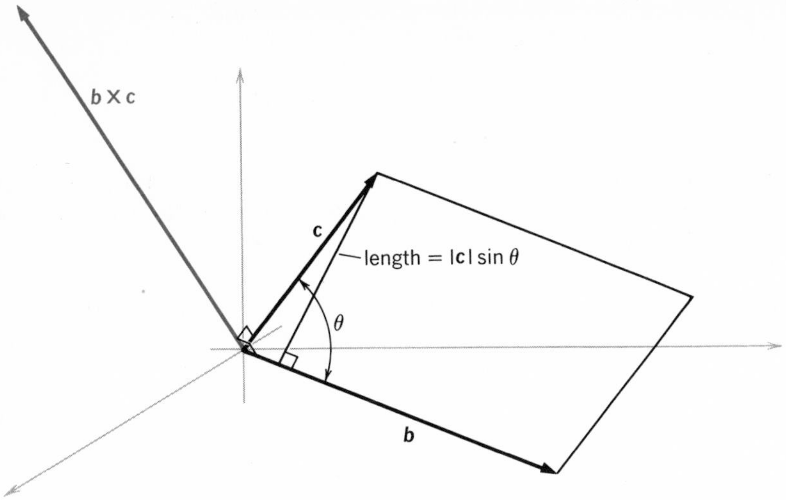

FIGURE 1-9

A vector perpendicular to both $\boldsymbol{i} + \boldsymbol{j}$ and $\boldsymbol{j} + \boldsymbol{k}$ is the vector

$$(\boldsymbol{i} + \boldsymbol{j}) \times (\boldsymbol{j} + \boldsymbol{k}) = \begin{vmatrix} \boldsymbol{i} & \boldsymbol{j} & \boldsymbol{k} \\ 1 & 1 & 0 \\ 0 & 1 & 1 \end{vmatrix} = \boldsymbol{i} - \boldsymbol{j} + \boldsymbol{k}$$

Since $|\boldsymbol{i} - \boldsymbol{j} + \boldsymbol{k}| = \sqrt{3}$, the vector

$$\frac{1}{\sqrt{3}}(\boldsymbol{i} - \boldsymbol{j} + \boldsymbol{k})$$

is a unit vector perpendicular to $\boldsymbol{i} + \boldsymbol{j}$ and $\boldsymbol{j} + \boldsymbol{k}$.

Using the cross product, we may obtain an interesting geometric interpretation of determinants. Let $\boldsymbol{a} = a_1\boldsymbol{i} + a_2\boldsymbol{j}$ and $\boldsymbol{b} = b_1\boldsymbol{i} + b_2\boldsymbol{j}$ be two vectors in the plane. If θ denotes the angle between $\boldsymbol{a}$ and $\boldsymbol{b}$, we have seen that $|\boldsymbol{a} \times \boldsymbol{b}| = |\boldsymbol{a}||\boldsymbol{b}||\sin\theta|$. As noted above, $|\boldsymbol{a}||\boldsymbol{b}|\sin\theta$ is the area of the parallelogram with adjacent sides $\boldsymbol{a}$ and $\boldsymbol{b}$. (See Figure 1-10.) Using the definition of the cross product,

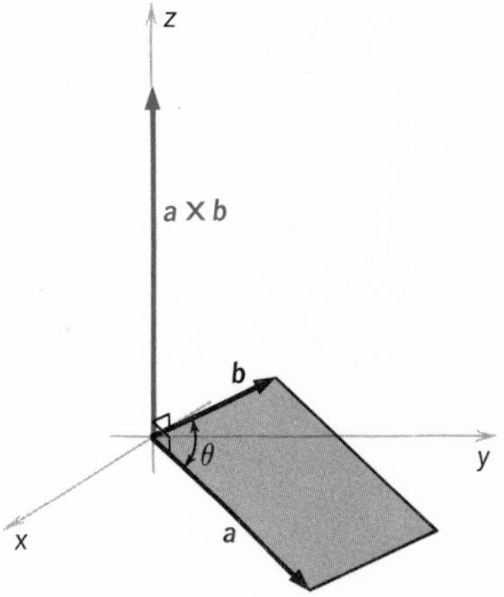

FIGURE 1-10

$$\boldsymbol{a} \times \boldsymbol{b} = \begin{vmatrix} \boldsymbol{i} & \boldsymbol{j} & \boldsymbol{k} \\ a_1 & a_2 & 0 \\ b_1 & b_2 & 0 \end{vmatrix} = \begin{vmatrix} a_1 & a_2 \\ b_1 & b_2 \end{vmatrix} \boldsymbol{k}$$

Thus, $|\boldsymbol{a} \times \boldsymbol{b}|$ is the absolute value of the determinant

$$\begin{vmatrix} a_1 & a_2 \\ b_1 & b_2 \end{vmatrix}$$

From this it follows that the absolute value of the above determinant is the area of the parallelogram with adjacent sides $a_1\boldsymbol{i} + a_2\boldsymbol{j}$ and $b_1\boldsymbol{i} + b_2\boldsymbol{j}$.

Example 5 Find the area of the triangle with vertices at the points (1, 1), (0, 2), and (3, 2). (See Figure 1-11.)

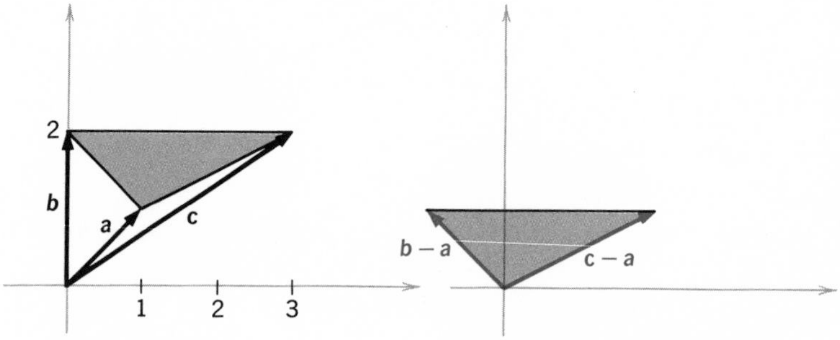

FIGURE 1-11

Let $\boldsymbol{a} = \boldsymbol{i} + \boldsymbol{j}$, $\boldsymbol{b} = 2\boldsymbol{j}$, and $\boldsymbol{c} = 3\boldsymbol{i} + 2\boldsymbol{j}$. It is clear that the triangle with vertices at the endpoints of the vectors $\boldsymbol{a}, \boldsymbol{b}$, and $\boldsymbol{c}$ has the same area as the triangle with vertices at $\boldsymbol{0}$, $\boldsymbol{b} - \boldsymbol{a}$, and $\boldsymbol{c} - \boldsymbol{a}$. Indeed, the latter is merely a translation of the former triangle. Since the area of this triangle is half the area of the parallelogram with adjacent sides $\boldsymbol{b} - \boldsymbol{a}$ and $\boldsymbol{c} - \boldsymbol{a}$, we find that the area of the triangle with vertices (1, 1), (0, 2), and (3, 2) is the absolute value of

$$\frac{1}{2}\begin{vmatrix} -1 & 1 \\ 2 & 1 \end{vmatrix} = -\frac{3}{2}$$

that is, $\frac{3}{2}$.

Analogous to the interpretation of 2×2 determinants as areas, there is an interpretation of 3×3 determinants as volumes. Let $\boldsymbol{a} = a_1\boldsymbol{i} + a_2\boldsymbol{j} + a_3\boldsymbol{k}$, $\boldsymbol{b} = b_1\boldsymbol{i} + b_2\boldsymbol{j} + b_3\boldsymbol{k}$, and $\boldsymbol{c} = c_1\boldsymbol{i} + c_2\boldsymbol{j} + c_3\boldsymbol{k}$ be vectors in $\boldsymbol{R}^3$. (See Figure 1-12.) We will show that the volume of the parallelepiped with adjacent sides $\boldsymbol{a}$, $\boldsymbol{b}$, and $\boldsymbol{c}$ is the absolute value of the determinant

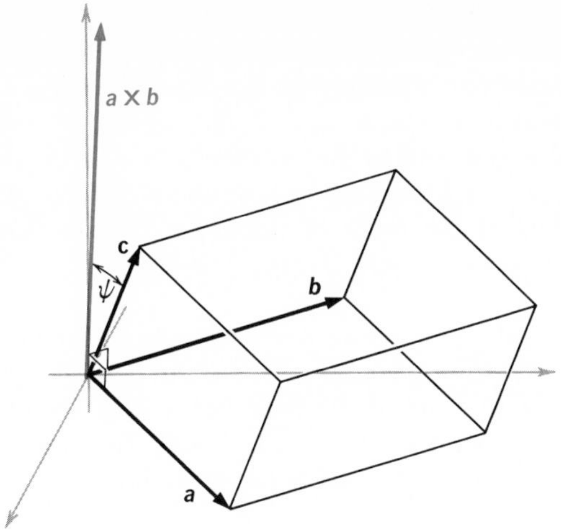

FIGURE 1-12

$$\begin{vmatrix} a_1 & a_2 & a_3 \\ b_1 & b_2 & b_3 \\ c_1 & c_2 & c_3 \end{vmatrix}$$

As we now have seen, $|\boldsymbol{a} \times \boldsymbol{b}|$ is the area of the parallelogram with adjacent sides $\boldsymbol{a}$ and $\boldsymbol{b}$. Moreover, $(\boldsymbol{a} \times \boldsymbol{b}) \cdot \boldsymbol{c}$ is $|\boldsymbol{c}||\boldsymbol{a} \times \boldsymbol{b}| \cos \psi$, where ψ is the angle which $\boldsymbol{c}$ makes with the normal to the plane spanned by $\boldsymbol{a}$ and $\boldsymbol{b}$. Since the volume of the parallelepiped with adjacent sides $\boldsymbol{a}$, $\boldsymbol{b}$, and $\boldsymbol{c}$ is the product of the area of the base $|\boldsymbol{a} \times \boldsymbol{b}|$ times the altitude $|\boldsymbol{c}| \cos \psi$, it follows that the volume is merely $|(\boldsymbol{a} \times \boldsymbol{b}) \cdot \boldsymbol{c}|$.

We saw earlier that $|(\boldsymbol{a} \times \boldsymbol{b}) \cdot \boldsymbol{c}|$ is the absolute value of the determinant

$$\begin{vmatrix} a_1 & a_2 & a_3 \\ b_1 & b_2 & b_3 \\ c_1 & c_2 & c_3 \end{vmatrix}$$

Therefore, the absolute value of the foregoing determinant is the volume of the parallelepiped with adjacent sides $\boldsymbol{a}$, $\boldsymbol{b}$, and $\boldsymbol{c}$.

It is important to note that the cross product is neither commutative nor associative. Indeed, we saw earlier that $\boldsymbol{a} \times \boldsymbol{b} = -(\boldsymbol{b} \times \boldsymbol{a})$. To see that the cross product is not associative, consider $(\boldsymbol{i} \times \boldsymbol{j}) \times \boldsymbol{j}$ and $\boldsymbol{i} \times (\boldsymbol{j} \times \boldsymbol{j})$; $(\boldsymbol{i} \times \boldsymbol{j}) \times \boldsymbol{j} = \boldsymbol{k} \times \boldsymbol{j} = -\boldsymbol{i}$ and $\boldsymbol{i} \times (\boldsymbol{j} \times \boldsymbol{j}) = \boldsymbol{i} \times \boldsymbol{0} = \boldsymbol{0}$. Thus, $(\boldsymbol{i} \times \boldsymbol{j}) \times \boldsymbol{j} \neq \boldsymbol{i} \times (\boldsymbol{j} \times \boldsymbol{j})$, and the cross product is not associative.

Using vector methods, we can derive the equation of a line in the

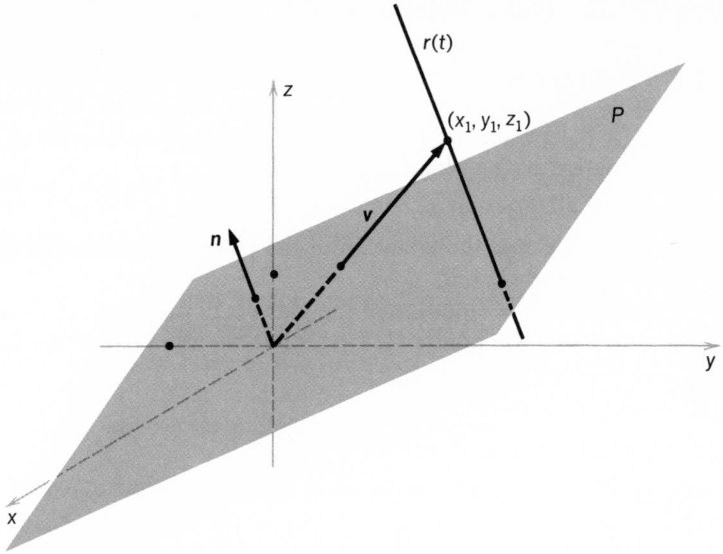

FIGURE 1-15

The vector

$$n = \frac{Ai + Bj + Ck}{\sqrt{A^2 + B^2 + C^2}}$$

is a unit vector normal to the plane. If

$$a = \frac{-Dn}{\sqrt{A^2 + B^2 + C^2}}$$

it is clear that the vector $\boldsymbol{a}$ ends on the given plane. If the endpoint of the vector $\boldsymbol{r} = x\boldsymbol{i} + y\boldsymbol{j} + z\boldsymbol{k}$ lies on the plane, we have $(\boldsymbol{r} - \boldsymbol{a}, \boldsymbol{n}) = 0$.

Let $\boldsymbol{v} = x_1\boldsymbol{i} + y_1\boldsymbol{j} + z_1\boldsymbol{k}$ be the vector ending at (x_1, y_1, z_1). Then, the line through (x_1, y_1, z_1) perpendicular to the plane is, in parametric form, $\boldsymbol{r}(t) = \boldsymbol{v} + t\boldsymbol{n}$. This line intersects the plane when

$$(\boldsymbol{r}(t) - \boldsymbol{a}, \boldsymbol{n}) = 0$$

or

$$(\boldsymbol{v} + t\boldsymbol{n} - \boldsymbol{a}, \boldsymbol{n}) = 0$$

$$t = (\boldsymbol{a} - \boldsymbol{v}, \boldsymbol{n})$$

The distance from the point (x_1, y_1, z_1) to the point at which $\boldsymbol{r}(t)$ intersects the plane, i.e., the distance from (x_1, y_1, z_1) to the plane is

$$|\boldsymbol{v} - (\boldsymbol{v} + (\boldsymbol{a} - \boldsymbol{v}, \boldsymbol{n})\boldsymbol{n})| = |(\boldsymbol{a} - \boldsymbol{v}, \boldsymbol{n})|$$

$$= \frac{|Ax_1 + By_1 + Cz_1 + D|}{\sqrt{A^2 + B^2 + C^2}}$$

EXERCISES

1. Find the angle between the vectors $\boldsymbol{j} + \boldsymbol{k}$ and $\boldsymbol{i} + \boldsymbol{j}$.
2. Resolve the vector $\boldsymbol{i} + \boldsymbol{j} + \boldsymbol{k}$ into components that are either parallel or perpendicular to the vector $\boldsymbol{i} + \boldsymbol{j} - \boldsymbol{k}$.
3. What point on the line $x + y = 1$ is closest to the point $(0, 2)$? What is the distance from the line to this point?
4. If $\boldsymbol{a}$ is a unit vector in the direction of a line l which passes through the origin, and $\boldsymbol{x}$ is another vector, show that the vector $(\boldsymbol{x}, \boldsymbol{a})\boldsymbol{a}$ is the closest vector on the line to the vector $\boldsymbol{x}$.
5. What is the distance of the point $(1, 1, 1)$ from the plane $x - y + z = 2$?
6. Show that the lines

$$a_1x + a_2y = \alpha$$
$$b_1x + b_2y = \beta$$

are perpendicular if and only if

$$a_1b_1 + a_2b_2 = 0$$

7. Show that the planes

$$a_1x + a_2y + a_3z = \alpha$$
$$b_1x + b_2y + b_3z = \beta$$
$$c_1x + c_2y + c_2z = \gamma$$

are perpendicular if and only if

$$a_1b_1 + a_2b_2 + a_3b_3 = b_1c_1 + b_2c_2 + b_3c_3 = a_1c_1 + a_2c_2 + a_3c_3 = 0$$

8. Determine the angle between the planes

$$x + y - z = 1$$
$$2x + y + z = 0$$

9. Find the interior angles of the triangle with vertices $(1, 0)$, $(0, 1)$, $(3, 4)$.
10. Find the equation of a line perpendicular to the vector $4\boldsymbol{i} + 3\boldsymbol{j}$ and containing the point (β, β).
11. Find the length of the diagonal of a cube whose edge is of length l.
12. Find a unit vector having the same direction as $\boldsymbol{i} - \boldsymbol{j} + 2\boldsymbol{k}$.
13. If α is a real number, $0 < \alpha < 1$, and $a^2 + b^2 + c^2 = 1$, show that the locus of points

$$ax + by + cz = \alpha\sqrt{x^2 + y^2 + z^2}$$

is a cone.

14. If (x_1, y_1), (x_2, y_2), and (x_3, y_3) are the vertices of a triangle, show that its area is the absolute value of

$$\frac{1}{2}\begin{vmatrix} 1 & x_1 & y_1 \\ 1 & x_2 & y_2 \\ 1 & x_3 & y_3 \end{vmatrix}$$

15. Show that the locus of points satisfying the equation

$$(x - x_1)^2 + (y - y_1)^2 + (z - z_1)^2 = r^2$$

is a sphere of radius r and center (x_1, y_1, z_1).

16. Find the equation of a plane containing the points (1, 0, 0), (0, 1, 1), and (1, 1, 1).

17. If $\boldsymbol{x}$ and $\boldsymbol{y}$ are vectors, show that $\frac{1}{2}(|\boldsymbol{x} + \boldsymbol{y}|^2 + |\boldsymbol{x} - \boldsymbol{y}|^2) = |\boldsymbol{x}|^2 + |\boldsymbol{y}^2|$. Interpret this geometrically.

18. Show that the points (x_1, y_1, z_1), (x_2, y_2, z_2), (x_3, y_3, z_3), and (x_4, y_4, z_4) are coplanar if and only if

$$\begin{vmatrix} x_1 & y_1 & z_1 & 1 \\ x_2 & y_2 & z_2 & 1 \\ x_3 & y_3 & z_3 & 1 \\ x_4 & y_4 & z_4 & 1 \end{vmatrix} = 0$$

19. If $\boldsymbol{a}$ and $\boldsymbol{b}$ are vectors in $\boldsymbol{R}^3$, show that the locus of points equidistant from the endpoints of $\boldsymbol{a}$ and $\boldsymbol{b}$ consists of the points $\boldsymbol{r}$ on the plane $(\boldsymbol{r}, \boldsymbol{b} - \boldsymbol{a}) = \frac{1}{2}((\boldsymbol{b}, \boldsymbol{b}) - (\boldsymbol{a}, \boldsymbol{a}))$.

20. If $\boldsymbol{x}$ is a unit vector and α is a scalar, under what circumstances is $\alpha\boldsymbol{x}$ a unit vector?

21. Determine the angle between the diagonal of a cube and one of its edges. Determine the angle between the diagonal of a cube and the diagonal of one of its faces.

22. Show that the endpoints of the vectors $\boldsymbol{x}$, $\boldsymbol{y}$, and $\boldsymbol{z}$ are collinear if and only if $(\boldsymbol{x} - \boldsymbol{y}) \times (\boldsymbol{z} - \boldsymbol{y}) = \boldsymbol{0}$.

23. If $\boldsymbol{w}$, $\boldsymbol{x}$, $\boldsymbol{y}$, $\boldsymbol{z}$ are vectors, prove

$$(\boldsymbol{w} \times \boldsymbol{x}) \cdot (\boldsymbol{y} \times \boldsymbol{z}) = \begin{vmatrix} \boldsymbol{w} \cdot \boldsymbol{y} & \boldsymbol{w} \cdot \boldsymbol{z} \\ \boldsymbol{x} \cdot \boldsymbol{y} & \boldsymbol{x} \cdot \boldsymbol{z} \end{vmatrix}$$

24. Find a unit vector orthogonal to the vectors $\boldsymbol{i}$ and $\boldsymbol{j} + t\boldsymbol{k}$, where t is a real number.

25. Let $\boldsymbol{n}$ be a fixed vector. Describe the locus of points $\boldsymbol{r}$ such that $\boldsymbol{r} \cdot \boldsymbol{n} > \boldsymbol{0}$.

26. Prove that if $\boldsymbol{x}$, $\boldsymbol{y}$, and $\boldsymbol{z}$ are vectors, $\boldsymbol{x} \times (\boldsymbol{y} \times \boldsymbol{z}) = (\boldsymbol{x}, \boldsymbol{z})\boldsymbol{y} - (\boldsymbol{x}, \boldsymbol{y})\boldsymbol{z}$.

27. Show that the cosine of the angle between the vector $a\boldsymbol{i} + b\boldsymbol{j} + c\boldsymbol{k}$ and the x axis is $a/\sqrt{a^2 + b^2 + c^2}$. Find the cosine of the angle between the given vector and the y axis; the z axis.

28. Find a vector from the origin to the intersection of the medians of a triangle with vertices at (1, 0, 0), (0, 1, 0), and (1, 1, 1).
29. If $\boldsymbol{x}$ and $\boldsymbol{y}$ are vectors, show that $(|\boldsymbol{x}|\boldsymbol{y} + |\boldsymbol{y}|\boldsymbol{x})/|\boldsymbol{x}||\boldsymbol{y}|$ bisects the angle between $\boldsymbol{x}$ and $\boldsymbol{y}$.
30. Let $\boldsymbol{x}$ be a vector in $\boldsymbol{R}^3$. Let P be the function from $\boldsymbol{R}^3$ to $\boldsymbol{R}^3$, defined by $P_x(\boldsymbol{z}) = \boldsymbol{x} \times \boldsymbol{z}$.
    (a) Show that P_x is a linear operator on $\boldsymbol{R}^3$.
    (b) Show that N_{P_x} is spanned by $\boldsymbol{x}$, if $\boldsymbol{x} \neq \boldsymbol{0}$.
    (c) Show that the range of P_x consists precisely of the vectors orthogonal to $\boldsymbol{x}$.
    (d) Show that relative to the standard basis $\boldsymbol{i}$, $\boldsymbol{j}$, and $\boldsymbol{k}$ the matrix representation of P_x is skew-symmetric.
    (e) Show that the mapping which carries $\boldsymbol{x}$ in $\boldsymbol{R}^3$ into P_x determines an isomorphism from $\boldsymbol{R}^3$ to the space of skew-symmetric matrices.
31. Let $\boldsymbol{a}$ and $\boldsymbol{b}$ be orthogonal unit vectors spanning a plane in $\boldsymbol{R}^3$. Let $\boldsymbol{z}$ be another vector in $\boldsymbol{R}^3$. Show that the closest vector to $\boldsymbol{z}$ in this plane is the vector $(\boldsymbol{z}, \boldsymbol{a})\boldsymbol{a} + (\boldsymbol{z}, \boldsymbol{b})\boldsymbol{b}$. Describe geometrically the function P from $\boldsymbol{R}^3$ to $\boldsymbol{R}^3$, where $P(\boldsymbol{z}) = (\boldsymbol{z}, \boldsymbol{a})\boldsymbol{a} + (\boldsymbol{z}, \boldsymbol{b})\boldsymbol{b}$.
32. Let $(x_1, y_1, z_1), (x_2, y_2, z_2), \ldots, (x_k, y_k, z_k)$ be points in $\boldsymbol{R}^3$.
    (a) Show that the points lie in some plane if and only if the rank of

$$\begin{bmatrix} x_1 & y_1 & z_1 & 1 \\ x_2 & y_2 & z_2 & 1 \\ & \vdots & & \\ x_k & y_k & z_k & 1 \end{bmatrix}$$

    is less than or equal to 3.
    (b) Show that the points lie on some line if and only if the rank of the above matrix is less than or equal to 2.
33. Show that the area of a parallelogram in space is the square root of the sum of the squares of the areas of its projections onto the xy, yz, and xz planes.
34. Find the equation of the plane perpendicular to the nonzero vector $a\boldsymbol{i} + b\boldsymbol{j} + c\boldsymbol{k}$ on which the vector $a\boldsymbol{i} + b\boldsymbol{j} + c\boldsymbol{k}$ ends.

2 INNER PRODUCT IN $\boldsymbol{R}^n$ AND $\boldsymbol{C}^n$

By analogy with the three-dimensional case, we define the inner product of two vectors

$$\boldsymbol{a} = \begin{bmatrix} \alpha_1 \\ \alpha_2 \\ \vdots \\ \alpha_n \end{bmatrix} \quad \text{and} \quad \boldsymbol{b} = \begin{bmatrix} \beta_1 \\ \beta_2 \\ \vdots \\ \beta_n \end{bmatrix}$$

in $\boldsymbol{R}^n$, denoted by $(\boldsymbol{a}, \boldsymbol{b})$, to be the real number $(\boldsymbol{a}, \boldsymbol{b}) = \alpha_1\beta_1 + \alpha_2\beta_2 +$

$\cdots + \alpha_n\beta_n$. The properties that were given in §6.1 for the inner product are again valid for inner products on an n-dimensional space.

In a similar manner, we define the norm of the vector $\boldsymbol{a}$, denoted by $|\boldsymbol{a}|$, by

$$|\boldsymbol{a}| = (\boldsymbol{a}, \boldsymbol{a})^{1/2} = \sqrt{\alpha_1^2 + \alpha_2^2 + \cdots + \alpha_n^2}$$

We note that

(1) $|\boldsymbol{a}| \geq 0$
$|\boldsymbol{a}| = 0$ if and only if $\boldsymbol{a} = \boldsymbol{0}$

(2) $|\alpha\boldsymbol{a}| = |\alpha||\boldsymbol{a}|$

For example, to prove (2),

$$\begin{aligned}|\alpha\boldsymbol{a}| = \sqrt{(\alpha\alpha_1)^2 + \cdots + (\alpha\alpha_n)^2} &= \sqrt{\alpha^2(\alpha_1^2 + \cdots + \alpha_n^2)} \\ &= |\alpha|\sqrt{\alpha_1^2 + \cdots + \alpha_n^2} \\ &= |\alpha||\boldsymbol{a}|\end{aligned}$$

The inner product $(\boldsymbol{a}, \boldsymbol{b})$ may also be written $\boldsymbol{a}^T\boldsymbol{b}$. For we have

$$\boldsymbol{a}^T\boldsymbol{b} = [\alpha_1 \quad \alpha_2 \quad \ldots \quad \alpha_n]\begin{bmatrix}\beta_1 \\ \beta_2 \\ \vdots \\ \beta_n\end{bmatrix} = \alpha_1\beta_1 + \cdots + \alpha_n\beta_n$$

There is an important inequality which relates the inner product of two vectors to the product of their norms.

Theorem 1 Cauchy–Schwarz Inequality If $\boldsymbol{x}$ and $\boldsymbol{y}$ belong to $\boldsymbol{R}^n$, $|(\boldsymbol{x}, \boldsymbol{y})| \leq |\boldsymbol{x}||\boldsymbol{y}|$.

Proof If $\boldsymbol{x} = \boldsymbol{0}$ or $\boldsymbol{y} = \boldsymbol{0}$, $(\boldsymbol{x}, \boldsymbol{y}) = 0$ and thus $|(\boldsymbol{x}, \boldsymbol{y})| = |\boldsymbol{x}||\boldsymbol{y}|$.

If $\boldsymbol{x} \neq \boldsymbol{0}$ and $\boldsymbol{y} \neq \boldsymbol{0}$

$$\left(\frac{\boldsymbol{x}}{|\boldsymbol{x}|} - \frac{\boldsymbol{y}}{|\boldsymbol{y}|}, \frac{\boldsymbol{x}}{|\boldsymbol{x}|} - \frac{\boldsymbol{y}}{|\boldsymbol{y}|}\right) \geq 0$$

and

$$\left(\frac{\boldsymbol{x}}{|\boldsymbol{x}|} + \frac{\boldsymbol{y}}{|\boldsymbol{y}|}, \frac{\boldsymbol{x}}{|\boldsymbol{x}|} + \frac{\boldsymbol{y}}{|\boldsymbol{y}|}\right) \geq 0$$

Thus,

$$\frac{(\boldsymbol{x}, \boldsymbol{x})}{|\boldsymbol{x}|^2} - \frac{2(\boldsymbol{x}, \boldsymbol{y})}{|\boldsymbol{x}||\boldsymbol{y}|} + \frac{(\boldsymbol{y}, \boldsymbol{y})}{|\boldsymbol{y}|^2} \geq 0$$

$$\frac{(\boldsymbol{x}, \boldsymbol{x})}{|\boldsymbol{x}|^2} + \frac{2(\boldsymbol{x}, \boldsymbol{y})}{|\boldsymbol{x}||\boldsymbol{y}|} + \frac{(\boldsymbol{y}, \boldsymbol{y})}{|\boldsymbol{y}|^2} \geq 0$$

Since $(\boldsymbol{x}, \boldsymbol{x}) = |\boldsymbol{x}|^2$ and $(\boldsymbol{y}, \boldsymbol{y}) = |\boldsymbol{y}|^2$, we have

$$2 - \frac{2(\boldsymbol{x}, \boldsymbol{y})}{|\boldsymbol{x}||\boldsymbol{y}|} \geq 0$$

$$2 + \frac{2(\boldsymbol{x}, \boldsymbol{y})}{|\boldsymbol{x}||\boldsymbol{y}|} \geq 0$$

or

$$-|\boldsymbol{x}||\boldsymbol{y}| \leq (\boldsymbol{x}, \boldsymbol{y}) \leq |\boldsymbol{x}||\boldsymbol{y}|$$

It follows that $|(\boldsymbol{x}, \boldsymbol{y})| \leq |\boldsymbol{x}||\boldsymbol{y}|$. ∎

In $\boldsymbol{R}^3$ we saw that $(\boldsymbol{x}, \boldsymbol{y}) = |\boldsymbol{x}||\boldsymbol{y}| \cos \theta$, where θ is the angle between $\boldsymbol{x}$ and $\boldsymbol{y}$. Since $|\cos \theta| \leq 1$, in $\boldsymbol{R}^3$ the Cauchy–Schwarz inequality is a consequence of our geometric interpretation of the inner product.

Immediately from the Cauchy–Schwarz inequality we obtain the following.

Theorem 2 If $\boldsymbol{x}$ and $\boldsymbol{y}$ belong to $\boldsymbol{R}^n$, $|\boldsymbol{x} + \boldsymbol{y}| \leq |\boldsymbol{x}| + |\boldsymbol{y}|$.

Proof By definition,

$$\begin{aligned} |\boldsymbol{x} + \boldsymbol{y}|^2 &= (\boldsymbol{x} + \boldsymbol{y}, \boldsymbol{x} + \boldsymbol{y}) \\ &= (\boldsymbol{x}, \boldsymbol{x}) + 2(\boldsymbol{x}, \boldsymbol{y}) + (\boldsymbol{y}, \boldsymbol{y}) \\ &\leq |\boldsymbol{x}|^2 + 2|\boldsymbol{x}||\boldsymbol{y}| + |\boldsymbol{y}|^2 \\ &= (|\boldsymbol{x}| + |\boldsymbol{y}|)^2 \end{aligned}$$

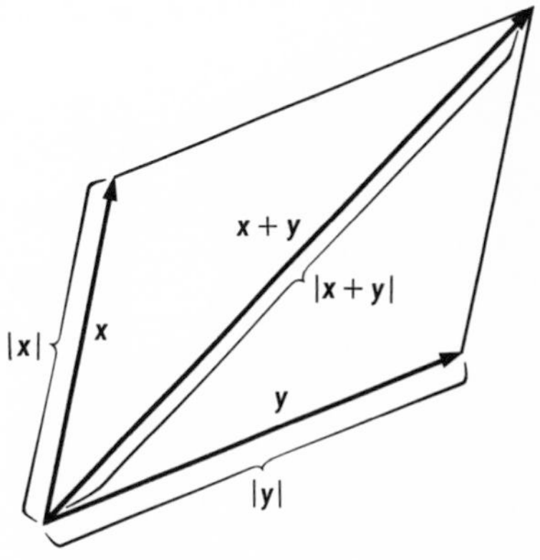

FIGURE 2-1

Since $|\boldsymbol{x} + \boldsymbol{y}|^2 \leq (|\boldsymbol{x}| + |\boldsymbol{y}|)^2$, it follows that $|\boldsymbol{x} + \boldsymbol{y}| \leq |\boldsymbol{x}| + |\boldsymbol{y}|$. ∎

Theorem 2 can be interpreted geometrically. If the vectors $\boldsymbol{x}$ and $\boldsymbol{y}$ represent the sides of a parallelogram, then $\boldsymbol{x} + \boldsymbol{y}$ represents the diagonal of the parallelogram. Thus, Theorem 2 may be understood to mean that the sum of the lengths of two adjacent sides of a parallelogram is less than or equal to the length of the diagonal. (See Figure 2-1.)

Letting $d(\boldsymbol{x}, \boldsymbol{y}) = |\boldsymbol{x} - \boldsymbol{y}|$, the distance between the endpoints of the vectors $\boldsymbol{x}$ and $\boldsymbol{y}$, by Theorem 2, $|\boldsymbol{x} - \boldsymbol{z}| \leq |\boldsymbol{x} - \boldsymbol{y}| + |\boldsymbol{y} - \boldsymbol{z}|$, or $d(\boldsymbol{x}, \boldsymbol{z}) \leq d(\boldsymbol{x}, \boldsymbol{y}) + d(\boldsymbol{y}, \boldsymbol{z})$. If we now consider the triangle with vertices the endpoints of $\boldsymbol{x}$, $\boldsymbol{y}$, and $\boldsymbol{z}$, this last inequality may be reinterpreted as follows: The length of one side of a triangle is less than or equal to the sum of the lengths of the remaining two sides. (See Figure 2-2.)

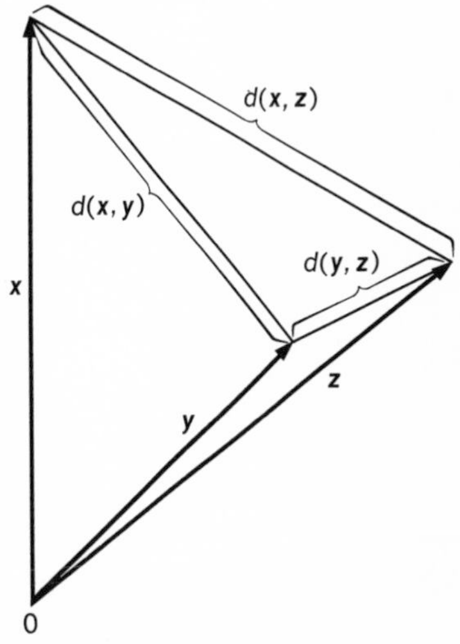

FIGURE 2-2

As in the three-dimensional case, vectors of norm 1 are said to be unit vectors. Two vectors whose inner product is zero are said to be orthogonal.

Under what circumstances does equality hold in the Cauchy–Schwarz inequality? If $\boldsymbol{x}$ and $\boldsymbol{y}$ are vectors in $\boldsymbol{R}^3$ and $|(\boldsymbol{x}, \boldsymbol{y})| = |\boldsymbol{x}||\boldsymbol{y}|$, then the cosine of the angle between $\boldsymbol{x}$ and $\boldsymbol{y}$ is ± 1. In either case $\boldsymbol{x}$ and

$\boldsymbol{y}$ are linearly dependent. (Why?) The following theorem indicates that this is always the case.

Theorem 3 Let $\boldsymbol{x}$ and $\boldsymbol{y}$ be vectors in $\boldsymbol{R}^n$. Then, $|(\boldsymbol{x}, \boldsymbol{y})| = |\boldsymbol{x}||\boldsymbol{y}|$ if and only if $\boldsymbol{x}$ and $\boldsymbol{y}$ are linearly dependent.

Proof First, suppose $|(\boldsymbol{x}, \boldsymbol{y})| = |\boldsymbol{x}||\boldsymbol{y}|$. If either $\boldsymbol{x} = \boldsymbol{0}$ or $\boldsymbol{y} = \boldsymbol{0}$, $\boldsymbol{x}$ and $\boldsymbol{y}$ are clearly linearly dependent. If not, we have either $(\boldsymbol{x}, \boldsymbol{y}) = |\boldsymbol{x}||\boldsymbol{y}|$ or $(\boldsymbol{x}, \boldsymbol{y}) = -|\boldsymbol{x}||\boldsymbol{y}|$. In the first case, it follows that

$$\left(\frac{\boldsymbol{x}}{|\boldsymbol{x}|} - \frac{\boldsymbol{y}}{|\boldsymbol{y}|}, \frac{\boldsymbol{x}}{|\boldsymbol{x}|} - \frac{\boldsymbol{y}}{|\boldsymbol{y}|}\right) = 0$$

In the second case,

$$\left(\frac{\boldsymbol{x}}{|\boldsymbol{x}|} + \frac{\boldsymbol{y}}{|\boldsymbol{y}|}, \frac{\boldsymbol{x}}{|\boldsymbol{x}|} + \frac{\boldsymbol{y}}{|\boldsymbol{y}|}\right) = 0$$

In either case,

$$\frac{\boldsymbol{x}}{|\boldsymbol{x}|} = \pm\frac{\boldsymbol{y}}{|\boldsymbol{y}|}$$

Thus, $\boldsymbol{x}$ and $\boldsymbol{y}$ are linearly dependent.

If, on the other hand, $\boldsymbol{x}$ and $\boldsymbol{y}$ are linearly dependent, one vector, say $\boldsymbol{x}$, is a scalar multiple of the other. Thus, $\boldsymbol{x} = \beta\boldsymbol{y}$.

Then, $(\boldsymbol{x}, \boldsymbol{y}) = (\beta\boldsymbol{y}, \boldsymbol{y}) = \beta(\boldsymbol{y}, \boldsymbol{y})$, while $|\boldsymbol{x}| = |\beta||\boldsymbol{y}|$. Thus, $|(\boldsymbol{x}, \boldsymbol{y})| = |\beta||\boldsymbol{y}|^2 = |\boldsymbol{x}||\boldsymbol{y}|$. ∎

Corollary If $\boldsymbol{a}$ and $\boldsymbol{b}$ are unit vectors and $(\boldsymbol{a}, \boldsymbol{b}) = 1$, then $\boldsymbol{a} = \boldsymbol{b}$.

A collection of vectors $\boldsymbol{x}_1, \boldsymbol{x}_2, \ldots, \boldsymbol{x}_k$ in $\boldsymbol{R}^n$ is said to be an **orthonormal** system if any pair of distinct vectors is orthogonal and the norm of each vector is 1. In terms of the Kronecker delta,

$$(\boldsymbol{x}_i, \boldsymbol{x}_j) = \delta_{ij}$$

For example in $\boldsymbol{R}^4$, the vectors

$$\begin{bmatrix} a \\ b \\ c \\ d \end{bmatrix}, \quad \begin{bmatrix} -b \\ a \\ d \\ -c \end{bmatrix}, \quad \begin{bmatrix} -c \\ -d \\ a \\ b \end{bmatrix}$$

constitute an orthonormal system if $a^2 + b^2 + c^2 + d^2 = 1$.

From the definition it is clear that any subset of an orthonormal set of vectors is again orthonormal.

If an orthonormal collection of vectors forms a basis for $\boldsymbol{R}^n$, it is said to be an **orthonormal basis.** For example, in $\boldsymbol{R}^n$ the standard basis $\boldsymbol{e}_1, \boldsymbol{e}_2, \ldots, \boldsymbol{e}_n$ is an orthonormal basis. In $\boldsymbol{R}^2$, the vectors

$$\boldsymbol{i}_\theta = (\cos\theta)\boldsymbol{i} + (\sin\theta)\boldsymbol{j}$$
$$\boldsymbol{j}_\theta = -(\sin\theta)\boldsymbol{i} + (\cos\theta)\boldsymbol{j}$$

form an orthonormal basis.

Theorem 4 Let $\boldsymbol{x}_1, \boldsymbol{x}_2, \ldots, \boldsymbol{x}_k$ be a collection of nonzero orthogonal vectors in $\boldsymbol{R}^n$. Then, $\boldsymbol{x}_1, \boldsymbol{x}_2, \ldots, \boldsymbol{x}_k$ are linearly independent.

Proof Suppose $\alpha_1\boldsymbol{x}_1 + \alpha_2\boldsymbol{x}_2 + \alpha_3\boldsymbol{x}_3 + \cdots + \alpha_k\boldsymbol{x}_k = \boldsymbol{0}$. Forming the inner product of $\alpha_1\boldsymbol{x}_1 + \alpha_2\boldsymbol{x}_2 + \cdots + \alpha_k\boldsymbol{x}_k$ and $\boldsymbol{x}_i$, we obtain

$$0 = (\alpha_1\boldsymbol{x}_1 + \alpha_2\boldsymbol{x}_2 + \cdots + \alpha_k\boldsymbol{x}_k, \boldsymbol{x}_i)$$

or

$$0 = \alpha_1(\boldsymbol{x}_1, \boldsymbol{x}_i) + \alpha_2(\boldsymbol{x}_2, \boldsymbol{x}_i) + \cdots + \alpha_k(\boldsymbol{x}_k, \boldsymbol{x}_i)$$

Since $(\boldsymbol{x}_j, \boldsymbol{x}_i) = 0$, if $\boldsymbol{j} \neq \boldsymbol{i}$, $0 = \alpha_i(\boldsymbol{x}_i, \boldsymbol{x}_i)$. Since $\boldsymbol{x}_i \neq 0$, $(\boldsymbol{x}_i, \boldsymbol{x}_i) \neq 0$, and $\alpha_i = 0$. Thus, $\alpha_1\boldsymbol{x}_1 + \alpha_2\boldsymbol{x}_2 + \cdots + \alpha_k\boldsymbol{x}_k = \boldsymbol{0}$ implies that $\alpha_1 = \alpha_2 = \cdots = \alpha_k = 0$. Therefore, the vectors $\boldsymbol{x}_1, \boldsymbol{x}_2, \ldots, \boldsymbol{x}_k$ are linearly independent. ∎

For example, in $\boldsymbol{R}^4$, the vectors

$$\begin{bmatrix}1\\1\\1\\0\end{bmatrix}, \quad \begin{bmatrix}-1\\1\\0\\1\end{bmatrix}, \quad \begin{bmatrix}2\\1\\-3\\1\end{bmatrix}$$

are nonzero and orthogonal. Thus, they are linearly independent.

If $\boldsymbol{x}_1, \boldsymbol{x}_2, \ldots, \boldsymbol{x}_n$ are vectors in $\boldsymbol{R}^n$, it is often of interest to consider the matrix

$$\begin{bmatrix}(\boldsymbol{x}_1, \boldsymbol{x}_1) & (\boldsymbol{x}_1, \boldsymbol{x}_2) & \cdots & (\boldsymbol{x}_1, \boldsymbol{x}_n)\\ (\boldsymbol{x}_2, \boldsymbol{x}_1) & (\boldsymbol{x}_2, \boldsymbol{x}_2) & \cdots & (\boldsymbol{x}_2, \boldsymbol{x}_n)\\ & & \vdots & \\ (\boldsymbol{x}_n, \boldsymbol{x}_1) & (\boldsymbol{x}_n, \boldsymbol{x}_2) & \cdots & (\boldsymbol{x}_n, \boldsymbol{x}_n)\end{bmatrix}$$

Its determinant is called the **Gramian** of $\boldsymbol{x}_1, \boldsymbol{x}_2, \ldots, \boldsymbol{x}_n$.

If $A = [\boldsymbol{x}_1, \boldsymbol{x}_2, \ldots, \boldsymbol{x}_n]$ is the matrix whose ith column is the vector $\boldsymbol{x}_i$, we see that

$$A^T A = \begin{bmatrix} (\boldsymbol{x}_1, \boldsymbol{x}_1) & (\boldsymbol{x}_1, \boldsymbol{x}_2) & \cdots & (\boldsymbol{x}_1, \boldsymbol{x}_n) \\ (\boldsymbol{x}_2, \boldsymbol{x}_1) & (\boldsymbol{x}_2, \boldsymbol{x}_2) & \cdots & (\boldsymbol{x}_2, \boldsymbol{x}_n) \\ & & \vdots & \\ (\boldsymbol{x}_n, \boldsymbol{x}_1) & (\boldsymbol{x}_n, \boldsymbol{x}_2) & \cdots & (\boldsymbol{x}_n, \boldsymbol{x}_n) \end{bmatrix}$$

It follows that the Gramian determinant of $\boldsymbol{x}_1, \boldsymbol{x}_2, \ldots, \boldsymbol{x}_n$ is just $\det A^T A = (\det A)^2$. As a consequence, the Gramian of $\boldsymbol{x}_1, \boldsymbol{x}_2, \ldots, \boldsymbol{x}_n$ is always nonnegative and is strictly positive if and only if $\boldsymbol{x}_1, \boldsymbol{x}_2, \ldots, \boldsymbol{x}_n$ are linearly independent.

There is an inner product on $\boldsymbol{C}^n$ related to the inner product on $\boldsymbol{R}^n$. If $\boldsymbol{a}$ and $\boldsymbol{b}$ are two vectors in $\boldsymbol{C}^n$,

$$\boldsymbol{a} = \begin{bmatrix} \alpha_1 \\ \alpha_2 \\ \vdots \\ \alpha_n \end{bmatrix}, \qquad \boldsymbol{b} = \begin{bmatrix} \beta_1 \\ \beta_2 \\ \vdots \\ \beta_n \end{bmatrix}$$

we define $(\boldsymbol{a}, \boldsymbol{b})$, the inner product of $\boldsymbol{a}$ and $\boldsymbol{b}$, by

$$(\boldsymbol{a}, \boldsymbol{b}) = \alpha_1\bar{\beta}_1 + \alpha_2\bar{\beta}_2 + \cdots + \alpha_n\bar{\beta}_n$$

where $\bar{\beta}$ denotes the complex conjugate of the complex number β. If $\boldsymbol{a}$ and $\boldsymbol{b}$ are vectors in $\boldsymbol{R}^n$, i.e., if all the components of $\boldsymbol{a}$ and $\boldsymbol{b}$ are real, the above definition of the inner product coincides with our earlier definition.

The properties of the complex inner product are similar to those of the real inner product. We assume $\boldsymbol{a}, \boldsymbol{b}, \boldsymbol{c}$ belong to $\boldsymbol{C}^n$ and α and β are complex scalars.

(1) $(\boldsymbol{a}, \boldsymbol{a}) \geq 0$
$(\boldsymbol{a}, \boldsymbol{a}) = 0$ if and only if $\boldsymbol{a} = \boldsymbol{0}$

(2) $(\boldsymbol{a}, \boldsymbol{b} + \boldsymbol{c}) = (\boldsymbol{a}, \boldsymbol{b}) + (\boldsymbol{a}, \boldsymbol{c})$
$(\boldsymbol{a} + \boldsymbol{b}, \boldsymbol{c}) = (\boldsymbol{a}, \boldsymbol{c}) + (\boldsymbol{b}, \boldsymbol{c})$

(3) $(\alpha\boldsymbol{a}, \boldsymbol{b}) = \alpha(\boldsymbol{a}, \boldsymbol{b})$
$(\boldsymbol{a}, \beta\boldsymbol{b}) = \bar{\beta}(\boldsymbol{a}, \boldsymbol{b})$

(4) $(\boldsymbol{a}, \boldsymbol{b}) = \overline{(\boldsymbol{b}, \boldsymbol{a})}$

For example, to prove (1), observe that $(\boldsymbol{a}, \boldsymbol{a}) = \alpha_1\bar{\alpha}_1 + \alpha_2\bar{\alpha}_2 + \cdots + \alpha_n\bar{\alpha}_n = |\alpha_1|^2 + \cdots + |\alpha_n|^2$. Since $|\alpha_1|^2 \geq 0, \ldots, |\alpha_n|^2 \geq 0$, we have $(\boldsymbol{a}, \boldsymbol{a}) \geq 0$. If $(\boldsymbol{a}, \boldsymbol{a}) = 0$, it follows that $|\alpha_1|^2 = |\alpha_2|^2 = \cdots = |\alpha_n|^2 = 0$ which implies that $\boldsymbol{a} = \boldsymbol{0}$.

To prove (2), let

$$a = \begin{bmatrix} \alpha_1 \\ \alpha_2 \\ \vdots \\ \alpha_n \end{bmatrix}, \qquad b = \begin{bmatrix} \beta_1 \\ \beta_2 \\ \vdots \\ \beta_n \end{bmatrix}, \qquad c = \begin{bmatrix} \gamma_1 \\ \gamma_2 \\ \vdots \\ \gamma_n \end{bmatrix}$$

Then,

$$\begin{aligned}(a, b + c) &= \alpha_1(\overline{\beta_1 + \gamma_1}) + \alpha_2(\overline{\beta_2 + \gamma_2}) + \cdots + \alpha_n(\overline{\beta_n + \gamma_n}) \\ &= \alpha_1(\bar{\beta}_1 + \bar{\gamma}_1) + \alpha_2(\bar{\beta}_2 + \bar{\gamma}_2) + \cdots + \alpha_n(\bar{\beta}_n + \bar{\gamma}_n) \\ &= \alpha_1\bar{\beta}_1 + \alpha_2\bar{\beta}_2 + \cdots + \alpha_n\bar{\beta}_n + \alpha_1\bar{\gamma}_1 + \alpha_2\bar{\gamma}_2 + \cdots + \alpha_n\bar{\gamma}_n \\ &= (a, b) + (a, c)\end{aligned}$$

To prove (4)

$$(a, b) = \alpha_1\bar{\beta}_1 + \cdots + \alpha_n\bar{\beta}_n$$

$$(b, a) = \beta_1\bar{\alpha}_1 + \cdots + \beta_n\bar{\alpha}_n$$

Thus,

$$(\overline{b, a}) = \bar{\beta}_1\alpha_1 + \cdots + \bar{\beta}_n\alpha_n = (a, b)$$

As before we define the norm of the vector a, denoted by $|a|$, to be $|a| = (a, a)^{1/2}$. From the properties of the inner product, we again obtain

(1) $|a| \geq 0$
$|a| = 0$ if and only if $a = 0$

(2) $|\alpha a| = |\alpha||a|$

In the complex case, the Cauchy–Schwarz inequality is also valid.

Theorem 5 If a and b belong to C^n, $|(a, b)| \leq |a||b|$.

Proof The result is obvious if $a = 0$, $b = 0$, or $(a, b) = 0$. So assume $a \neq 0$, $b \neq 0$, and $(a, b) \neq 0$.

Since

$$\left(\frac{b}{|b|} - \frac{|(a, b)|}{(a, b)}\frac{a}{|a|}, \frac{b}{|b|} - \frac{|(a, b)|}{(a, b)}\frac{a}{|a|}\right) \geq 0$$

it follows that

$$\frac{(b, b)}{|b|^2} - \frac{|(a, b)|}{|a||b|(\overline{a, b})}(b, a) - \frac{|(a, b)|}{|a||b|(a, b)}(a, b) + \frac{|(a, b)|^2}{(a, b)(\overline{a, b})}\frac{(a, a)}{|a|^2} \geq 0$$

Since $(\overline{\boldsymbol{a}, \boldsymbol{b}}) = (\boldsymbol{b}, \boldsymbol{a})$, $(\boldsymbol{a}, \boldsymbol{b})(\overline{\boldsymbol{a}, \boldsymbol{b}}) = |(\boldsymbol{a}, \boldsymbol{b})|^2$, $(\boldsymbol{b}, \boldsymbol{b}) = |\boldsymbol{b}|^2$, and $(\boldsymbol{a}, \boldsymbol{a}) = |\boldsymbol{a}|^2$, it follows that

$$2 - 2\,\frac{(\boldsymbol{a}, \boldsymbol{b})}{|\boldsymbol{a}||\boldsymbol{b}|} \geq 0$$

or

$$|(\boldsymbol{a}, \boldsymbol{b})| \leq |\boldsymbol{a}||\boldsymbol{b}|$$

As before the Cauchy–Schwarz inequality implies the norm inequality $|\boldsymbol{a} + \boldsymbol{b}| \leq |\boldsymbol{a}| + |\boldsymbol{b}|$.

The definitions of orthogonality, orthonormal system, and basis for complex vectors are formally the same as for real vectors. Moreover, theorems analogous to those holding for real vectors with the real inner product hold for complex vectors with the complex inner product.

Thus, for example, in $\boldsymbol{C}^2$, the vectors

$$\begin{bmatrix} a \\ b \end{bmatrix}, \qquad \begin{bmatrix} -\bar{b} \\ \bar{a} \end{bmatrix}$$

are orthogonal. Indeed, their inner product is

$$a(\overline{-\bar{b}}) + b(\overline{\bar{a}}) = -ab + ba = 0$$

As a consequence, if

$$\begin{bmatrix} a \\ b \end{bmatrix} \neq 0$$

we see that

$$\begin{bmatrix} a \\ b \end{bmatrix} \quad \text{and} \quad \begin{bmatrix} -\bar{b} \\ \bar{a} \end{bmatrix}$$

are linearly independent in $\boldsymbol{C}^2$. If $|a|^2 + |b|^2 = 1$, these vectors form an orthonormal basis for $\boldsymbol{C}^2$.

EXERCISES

1. Show that the vectors

$$\begin{bmatrix} a \\ b \end{bmatrix}, \qquad \begin{bmatrix} -b \\ a \end{bmatrix}$$

are orthogonal.

2. Find an orthonormal basis for $\boldsymbol{R}^3$ which contains the vector

$$\frac{1}{\sqrt{3}}\begin{bmatrix}1\\1\\1\end{bmatrix}$$

3. If $\boldsymbol{a}$ is a unit vector in $\boldsymbol{R}^3$, show there are numbers ϕ and θ, such that $\boldsymbol{a} = (\sin\theta\cos\phi)\boldsymbol{i} + (\sin\theta\sin\phi)\boldsymbol{j} + (\cos\theta)\boldsymbol{k}$. Interpret this geometrically.
4. If $\boldsymbol{x}$ and $\boldsymbol{y}$ are vectors, show that the vectors $|\boldsymbol{x}|\boldsymbol{y} + |\boldsymbol{y}|\boldsymbol{x}$ and $|\boldsymbol{y}|\boldsymbol{x} - |\boldsymbol{x}|\boldsymbol{y}$ are orthogonal.
5. Resolve the vector

$$\begin{bmatrix}x\\y\\z\end{bmatrix}$$

into components, one perpendicular to and one parallel to the vector

$$\frac{1}{\sqrt{3}}\begin{bmatrix}1\\1\\1\end{bmatrix}$$

6. If $\boldsymbol{y}_1, \boldsymbol{y}_2, \ldots, \boldsymbol{y}_n$ are vectors orthogonal to the vector $\boldsymbol{x}$, show that any vector in $\mathrm{sp}(\boldsymbol{y}_1, \boldsymbol{y}_2, \ldots, \boldsymbol{y}_n)$ is orthogonal to $\boldsymbol{x}$.
7. Let $\boldsymbol{x}_1, \boldsymbol{x}_2, \ldots, \boldsymbol{x}_n$ be vectors in $\boldsymbol{R}^n$. Let A be the matrix whose columns are $\boldsymbol{x}_1, \boldsymbol{x}_2, \ldots, \boldsymbol{x}_n$. Show that A^TA is diagonal if and only if every pair of distinct vectors in $\boldsymbol{x}_1, \boldsymbol{x}_2, \ldots, \boldsymbol{x}_n$ is orthogonal.
8. If $\boldsymbol{x}$ and $\boldsymbol{y}$ are vectors, show that $\frac{1}{2}(|\boldsymbol{x}+\boldsymbol{y}|^2 + |\boldsymbol{x}-\boldsymbol{y}|^2) = |\boldsymbol{x}|^2 + |\boldsymbol{y}|^2$.
9. If

$$\boldsymbol{x} = \begin{bmatrix}x_1\\x_2\\\vdots\\x_n\end{bmatrix}, \qquad \boldsymbol{y} = \begin{bmatrix}y_1\\y_2\\\vdots\\y_n\end{bmatrix}$$

show that

$$(\boldsymbol{x},\boldsymbol{x})(\boldsymbol{y},\boldsymbol{y}) - (\boldsymbol{x},\boldsymbol{y})^2 = \sum_{1\le i<j\le n}\begin{vmatrix}x_i & x_j\\ y_i & y_j\end{vmatrix}^2$$

reduce the Cauchy–Schwarz inequality.
10. Let f be a linear transformation from $\boldsymbol{R}^n$ into $\boldsymbol{R}$. Show that there is a vector $\boldsymbol{y}$ in $\boldsymbol{R}^n$, such that $f(\boldsymbol{x}) = (\boldsymbol{x}, \boldsymbol{y})$.
11. Determine the length of the diagonal of an n-dimensional cube with edge of length 1.
12. Find an orthonormal basis for the set of solutions to the equations

$$\begin{aligned} x_1 - 2x_2 + 3x_3 - x_4 &= 0\\ 2x_1 + x_2 - x_3 + 2x_4 &= 0 \end{aligned}$$

13. Find a unit vector orthogonal to the vectors

$$\begin{bmatrix} -1 \\ 1 \\ 0 \\ 0 \end{bmatrix}, \quad \begin{bmatrix} 0 \\ 2 \\ -1 \\ -1 \end{bmatrix}, \quad \begin{bmatrix} 3 \\ -2 \\ 0 \\ -1 \end{bmatrix}$$

14. Show that the vectors

$$\begin{bmatrix} x_1 \\ x_2 \\ \vdots \\ x_{2n-1} \\ x_{2n} \end{bmatrix} \quad \text{and} \quad \begin{bmatrix} x_2 \\ -x_1 \\ \vdots \\ x_{2n} \\ -x_{2n-1} \end{bmatrix}$$

in $\boldsymbol{R}^{2n}$ are orthogonal.

15. Let $T : \boldsymbol{R}^n \to \boldsymbol{R}^n$ be a linear transformation of rank 1.
    (a) Show that there are vectors $\boldsymbol{y}$ and $\boldsymbol{z}$ in $\boldsymbol{R}^n$ such that $T(\boldsymbol{x}) = (\boldsymbol{x}, \boldsymbol{y})\boldsymbol{z}$.
    (b) Show that $T^2 = (\boldsymbol{z}, \boldsymbol{y})T$.
    (c) Show that R_T consists of all scalar multiples of $\boldsymbol{z}$.
16. If $\boldsymbol{x}_1, \boldsymbol{x}_2, \ldots, \boldsymbol{x}_k$ is an orthogonal set of vectors, show that

$$|\boldsymbol{x}_1 + \boldsymbol{x}_2 + \cdots + \boldsymbol{x}_k|^2 = |\boldsymbol{x}_1|^2 + |\boldsymbol{x}_2|^2 + \cdots + |\boldsymbol{x}_k|^2$$

17. If $a^2 + b^2 + c^2 + d^2 = 1$, show that the vectors

$$\begin{bmatrix} a \\ b \\ c \\ d \end{bmatrix}, \quad \begin{bmatrix} -b \\ a \\ d \\ -c \end{bmatrix}, \quad \begin{bmatrix} -c \\ -d \\ a \\ b \end{bmatrix}, \quad \begin{bmatrix} -d \\ c \\ -b \\ a \end{bmatrix}$$

form an orthogonal basis for $\boldsymbol{R}^4$.

18. Find a nonzero linear transformation T from $\boldsymbol{R}^2$ into $\boldsymbol{R}^2$ such that for all $\boldsymbol{x}$, $(T(\boldsymbol{x}), \boldsymbol{x}) = 0$.
19. If $\boldsymbol{x}$ and $\boldsymbol{y}$ are orthogonal unit vectors, show that $|\boldsymbol{x} - \boldsymbol{y}| = \sqrt{2}$.
20. Let $\boldsymbol{x}$ be a unit vector. Consider the function f defined on the vectors of norm 1 by $f(\boldsymbol{y}) = |\boldsymbol{x} - \boldsymbol{y}|$.
    (a) Show that the maximum value of f is 2.
    (b) Show that the minimum value of f is 0.
    (c) Show that $f(\boldsymbol{y}) = \sqrt{2}$ if and only if $\boldsymbol{x}$ and $\boldsymbol{y}$ are orthogonal.
21. If $\boldsymbol{x}$ and $\boldsymbol{y}$ are vectors, show that $|\boldsymbol{x} + \boldsymbol{y}| = |\boldsymbol{x}| + |\boldsymbol{y}|$ only if $\boldsymbol{x}$ and $\boldsymbol{y}$ are linearly dependent.
22. If $\boldsymbol{x}$ and $\boldsymbol{y}$ are real vectors, determine the minimum value of the function $f(\lambda) = |\boldsymbol{x} + \lambda\boldsymbol{y}|^2$.
23. Show that the diagonals of a cube in $\boldsymbol{R}^3$ are not perpendicular.
24. If $\boldsymbol{a}$ and $\boldsymbol{b}$ are orthogonal unit vectors, show that the vectors $\alpha\boldsymbol{a} + \beta\boldsymbol{b}$ and $\gamma\boldsymbol{a} + \delta\boldsymbol{b}$ are orthogonal if and only if $\alpha\gamma + \beta\delta = 0$.

25. If a and b are real numbers, show that

$$|a| + |b| \leq (1 + a^2)^{1/2}(1 + b^2)^{1/2}$$

26. If $x_1, x_2, \ldots, x_n$ are real numbers, show that

$$(x_1^2 + x_2^2 + \cdots + x_n^2)^{1/2} \leq |x_1| + |x_2| + \cdots + |x_n|$$

If $(x_1^2 + x_2^2 + \cdots + x_n^2)^{1/2} = |x_1| + |x_2| + \cdots + |x_n|$, show that at least $n - 1$ of the numbers $x_1, x_2, \ldots, x_n$ are 0.

27. If $\boldsymbol{x}$ and $\boldsymbol{y}$ are vectors in $\boldsymbol{R}^n$ show that $||\boldsymbol{x}| - |\boldsymbol{y}|| \leq |\boldsymbol{x} - \boldsymbol{y}|$.

3 ORTHOGONAL COMPLEMENTS AND RELATED INEQUALITIES

If $\boldsymbol{x}$ is a unit vector in $\boldsymbol{R}^n$ (or $\boldsymbol{C}^n$) and $\boldsymbol{y}$ is another vector in the same space, the quantity $(\boldsymbol{y}, \boldsymbol{x})$ is called the **component** of $\boldsymbol{y}$ in the direction of $\boldsymbol{x}$. The vector $(\boldsymbol{y}, \boldsymbol{x})\boldsymbol{x}$ is called the **projection** of $\boldsymbol{y}$ onto $\boldsymbol{x}$. If $\boldsymbol{x}$ and $\boldsymbol{y}$ are vectors in $\boldsymbol{R}^3$, then $(\boldsymbol{y}, \boldsymbol{x})\boldsymbol{x}$ is the vector ending at the point where the perpendicular from the endpoint of $\boldsymbol{y}$ to the line determined by extending $\boldsymbol{x}$ intersects the line determined by extending $\boldsymbol{x}$. (See Figure 3-1.)

If $\boldsymbol{x}_1, \boldsymbol{x}_2, \ldots, \boldsymbol{x}_n$ is an orthonormal basis for $\boldsymbol{R}^n$ (or $\boldsymbol{C}^n$), there are scalars $\alpha_1, \alpha_2, \ldots, \alpha_n$ such that $\boldsymbol{x} = \alpha_1\boldsymbol{x}_1 + \alpha_2\boldsymbol{x}_2 + \cdots + \alpha_n\boldsymbol{x}_n$. If follows that $(\boldsymbol{x}, \boldsymbol{x}_i) = (\alpha_1\boldsymbol{x}_1 + \alpha_2\boldsymbol{x}_2 + \cdots + \alpha_n\boldsymbol{x}_n, \boldsymbol{x}_i) = \alpha_i(\boldsymbol{x}_i, \boldsymbol{x}_i) = \alpha_i$. Therefore, $\boldsymbol{x} = (\boldsymbol{x}, \boldsymbol{x}_1)\boldsymbol{x}_1 + (\boldsymbol{x}, \boldsymbol{x}_2)\boldsymbol{x}_2 + \cdots + (\boldsymbol{x}, \boldsymbol{x}_n)\boldsymbol{x}_n$.

We see in this case that the component of $\boldsymbol{x}$ in the direction of $\boldsymbol{x}_i$ is just the ith component of $\boldsymbol{x}$ relative to the basis $\boldsymbol{x}_1, \boldsymbol{x}_2, \ldots, \boldsymbol{x}_n$. In view of this fact, our choice of terminology is reasonable.

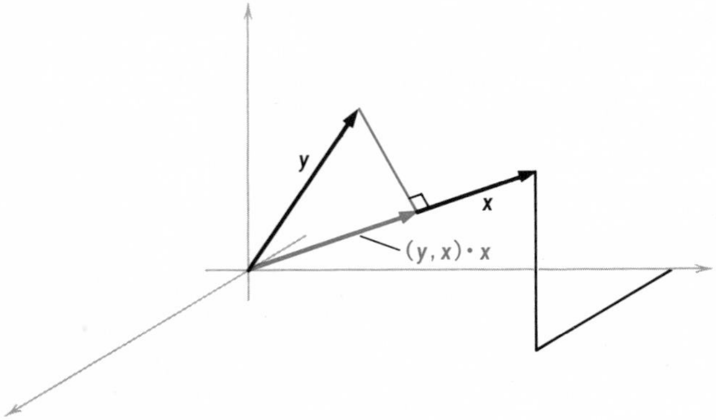

FIGURE 3-1

Theorem 1 Given a unit vector $\boldsymbol{x}$ in $\boldsymbol{R}^n$ (or $\boldsymbol{C}^n$) and another vector $\boldsymbol{y}$ in the same space, the vector $\boldsymbol{y}$ may be expressed as the sum of two vectors $\boldsymbol{u}$ and $\boldsymbol{v}$, where $\boldsymbol{u}$ is parallel to $\boldsymbol{x}$ and $\boldsymbol{v}$ is perpendicular to $\boldsymbol{x}$. This decomposition is unique.

Proof Let $\boldsymbol{u} = (\boldsymbol{y}, \boldsymbol{x})\boldsymbol{x}$ and $\boldsymbol{v} = \boldsymbol{y} - (\boldsymbol{y}, \boldsymbol{x})\boldsymbol{x}$. Clearly, $\boldsymbol{u}$ is parallel to $\boldsymbol{x}$ and $\boldsymbol{u} + \boldsymbol{v} = \boldsymbol{y}$. Since $(\boldsymbol{v}, \boldsymbol{x}) = (\boldsymbol{y} - (\boldsymbol{y}, \boldsymbol{x})\boldsymbol{x}, \boldsymbol{x}) = (\boldsymbol{y}, \boldsymbol{x}) - (\boldsymbol{y}, \boldsymbol{x})(\boldsymbol{x}, \boldsymbol{x}) = 0$, $\boldsymbol{v}$ is orthogonal to $\boldsymbol{x}$. This proves the existence of the decomposition asserted in the theorem.

We prove the uniqueness of the decomposition by showing that any decomposition is precisely the one exhibited above. To this end, suppose $\boldsymbol{y} = \boldsymbol{u} + \boldsymbol{v}$, where $\boldsymbol{u} = \alpha\boldsymbol{x}$ for some scalar α and $(\boldsymbol{v}, \boldsymbol{x}) = 0$. Then, $(\boldsymbol{y}, \boldsymbol{x}) = (\boldsymbol{u} + \boldsymbol{v}, \boldsymbol{x}) = (\boldsymbol{u}, \boldsymbol{x}) + (\boldsymbol{v}, \boldsymbol{x})$. Since $(\boldsymbol{v}, \boldsymbol{x}) = 0$ and $\boldsymbol{u} = \alpha\boldsymbol{x}$, $(\boldsymbol{y}, \boldsymbol{x}) = \alpha(\boldsymbol{x}, \boldsymbol{x}) = \alpha$. Thus, $\boldsymbol{u} = (\boldsymbol{y}, \boldsymbol{x})\boldsymbol{x}$ and $\boldsymbol{v} = \boldsymbol{y} - (\boldsymbol{y}, \boldsymbol{x})\boldsymbol{x}$. ∎

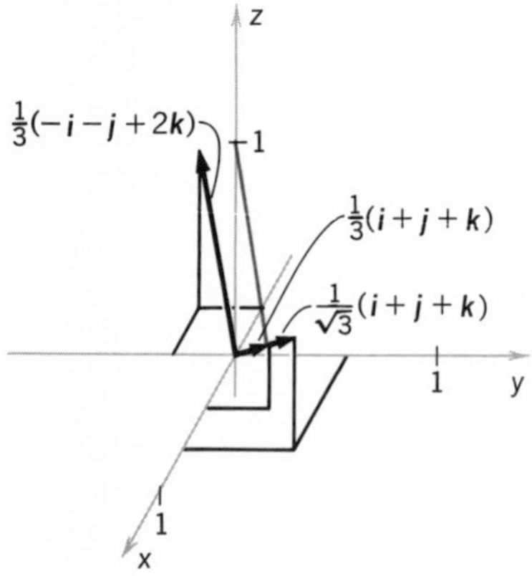

FIGURE 3-2

Example 1 Decompose the vector $\boldsymbol{k}$ into two parts, one parallel and one perpendicular to $(1/\sqrt{3})(\boldsymbol{i} + \boldsymbol{j} + \boldsymbol{k})$.

The component of $\boldsymbol{k}$ in the direction of $(1/\sqrt{3})(\boldsymbol{i} + \boldsymbol{j} + \boldsymbol{k})$ is $(\boldsymbol{k}, (1/\sqrt{3})(\boldsymbol{i} + \boldsymbol{j} + \boldsymbol{k}))(1/\sqrt{3})(\boldsymbol{i} + \boldsymbol{j} + \boldsymbol{k}) = \frac{1}{3}(\boldsymbol{i} + \boldsymbol{j} + \boldsymbol{k})$. By Theorem 1,

$$\boldsymbol{k} = \tfrac{1}{3}(\boldsymbol{i} + \boldsymbol{j} + \boldsymbol{k}) + \tfrac{1}{3}(-\boldsymbol{i} - \boldsymbol{j} + 2\boldsymbol{k})$$

where $\frac{1}{3}(\boldsymbol{i} + \boldsymbol{j} + \boldsymbol{k})$ is parallel to $(1/\sqrt{3})(\boldsymbol{i} + \boldsymbol{j} + \boldsymbol{k})$ and $\frac{1}{3}(-\boldsymbol{i} - \boldsymbol{j} + 2\boldsymbol{k})$ is perpendicular to $(1/\sqrt{3})(\boldsymbol{i} + \boldsymbol{j} + \boldsymbol{k})$. (See Figure 3-2.)

Next we describe a procedure for obtaining an orthonormal set of vectors from an arbitrary collection of linearly independent vectors.

Let $\boldsymbol{x}_1, \boldsymbol{x}_2, \ldots, \boldsymbol{x}_k$ be a linearly independent collection of vectors in $\boldsymbol{R}^n$ (or $\boldsymbol{C}^n$). We define a new sequence of vectors $\boldsymbol{x}_1', \boldsymbol{x}_2', \ldots, \boldsymbol{x}_k'$ by

$$\begin{aligned}
\boldsymbol{x}_1' &= \boldsymbol{x}_1/|\boldsymbol{x}_1| \\
\boldsymbol{x}_2' &= (\boldsymbol{x}_2 - (\boldsymbol{x}_2, \boldsymbol{x}_1')\boldsymbol{x}_1')/|\boldsymbol{x}_2 - (\boldsymbol{x}_2, \boldsymbol{x}_1')\boldsymbol{x}_1'| \\
&\ \vdots \\
\boldsymbol{x}_i' &= (\boldsymbol{x}_i - (\boldsymbol{x}_i, \boldsymbol{x}_1')\boldsymbol{x}_1' - \cdots - (\boldsymbol{x}_i, \boldsymbol{x}_{i-1}')\boldsymbol{x}_{i-1}')/ \\
&\qquad |\boldsymbol{x}_i - (\boldsymbol{x}_i, \boldsymbol{x}_1')\boldsymbol{x}_1' - \cdots - (\boldsymbol{x}_i, \boldsymbol{x}_{i-1}')| \\
\boldsymbol{x}_k' &= (\boldsymbol{x}_k - (\boldsymbol{x}_k, \boldsymbol{x}_1')\boldsymbol{x}_1' - \cdots - (\boldsymbol{x}_k, \boldsymbol{x}_{k-1}')\boldsymbol{x}_{k-1}')/ \\
&\qquad |\boldsymbol{x}_k - (\boldsymbol{x}_k, \boldsymbol{x}_1')\boldsymbol{x}_1' - \cdots - (\boldsymbol{x}_k, \boldsymbol{x}_{k-1}')|
\end{aligned}$$

Theorem 2 Gram–Schmidt Orthogonalization Process Let $\boldsymbol{x}_1, \boldsymbol{x}_2, \ldots, \boldsymbol{x}_k$ be a linearly independent collection of vectors in $\boldsymbol{R}^n$ (or $\boldsymbol{C}^n$). Then, the vectors $\boldsymbol{x}_1', \boldsymbol{x}_2', \ldots, \boldsymbol{x}_k'$, defined above, have the properties

(1) $\boldsymbol{x}_1', \boldsymbol{x}_2', \ldots, \boldsymbol{x}_k'$ is an orthonormal set of vectors.

(2) $\mathrm{sp}(\boldsymbol{x}_1', \ldots, \boldsymbol{x}_k') = \mathrm{sp}(\boldsymbol{x}_1, \ldots, \boldsymbol{x}_i')$, $i = 1, 2, \ldots, k$.

Proof It is clear by the linear independence of $x_1, x_2, \ldots, x_k$, that $x_1 \neq 0$, and thus the definition $x_1' = x_1/|x_1|$ makes sense. It is also clear that $|x_1'| = 1$ and $\text{sp}(x_1') = \text{sp}(x_1)$.

Assuming the truth of the theorem for $i - 1$, we prove that it is valid for i. First, we show that $x_i' - (x_i, x_1')x_1' - \cdots - (x_i, x_{i-1}')x_{i-1}' \neq 0$, so that the definition of x_i' is meaningful. Indeed, if $x_i - (x_i, x_1')x_1' - \cdots - (x_i, x_{i-1}')x_{i-1}' = 0$, then x_i belongs to $\text{sp}(x_1', \ldots, x_{i-1}') = \text{sp}(x_1, \ldots, x_{i-1})$. This implies that x_i is a linear combination of $x_1, x_2, \ldots, x_{i-1}$, in contradiction to the linear independence of $x_1, x_2, \ldots, x_i$. Hence, $x_i - (x_i, x_1')x_1' - \cdots - (x_i, x_{i-1}')x_{i-1}' \neq 0$.

It is immediate from the definition of x_i' that $|x_i'| = 1$. Now, if $j < i$,

$$(x_i', x_j') = \left(\frac{x_i - (x_i, x_1')x_1' - \cdots - (x_i, x_{i-1}')x_{i-1}'}{|x_i - (x_i, x_1')x_1' - \cdots - (x_i, x_{i-1}')x_{i-1}'|}, x_j' \right)$$

$$= \frac{1}{|x_i - (x, x_1')x_1' \cdots (x_i, x_{i-1}')x_{i-1}'|}((x_i, x_j') - (x_i, x_j')(x_j', x_j'))$$

In the last step, we used the assumption that $x_1', \ldots, x_{i-1}'$ is an orthonormal system. Since $(x_i, x_j') - (x_i, x_j')(x_j', x_j') = 0$, $(x_i', x_j') = 0$. Thus, $x_1', \ldots, x_i'$ is an orthonormal system.

Since $x_1', x_2', \ldots, x_i'$ belongs to $\text{sp}(x_1, x_2, \ldots, x_i')$, $\text{sp}(x_1', \ldots, x_i') \subset \text{sp}(x_1, x_2, \ldots, x_i)$. By Theorem 4, §6.2, $x_1', \ldots, x_i'$ are linearly independent. Since $\text{sp}(x_1, x_2, \ldots, x_i)$ is a subspace of dimension i, by Theorem 2, §4.8, $x_1', \ldots, x_i'$ is a basis for $\text{sp}(x_1, x_2, \ldots, x_i)$. Thus,

$$\text{sp}(x_1', \ldots, x_i') = \text{sp}(x_1, \ldots, x_i).$$ ∎

Example 2 Find an orthonormal basis for the subspace of R^4 spanned by x_1, x_2, and x_3, if

$$x_1 = \begin{bmatrix} -1 \\ 2 \\ 0 \\ 2 \end{bmatrix}, \quad x_2 = \begin{bmatrix} 2 \\ -4 \\ 1 \\ -4 \end{bmatrix}, \quad x_3 = \begin{bmatrix} -1 \\ 3 \\ 1 \\ 1 \end{bmatrix}$$

$|x_1| = 3$, thus

$$x_1' = \begin{bmatrix} -1/3 \\ 2/3 \\ 0 \\ 2/3 \end{bmatrix}$$

$$x_2' = \frac{x_2 - (x_2, x_1')x_1'}{|x_2 - (x_2, x_1')x_1'|} = \frac{x_2 + 6x_1'}{|x_2 + 6x_1'|} = \begin{bmatrix} 0 \\ 0 \\ 1 \\ 0 \end{bmatrix}$$

Finally,

$$x_3 - (x_3, x_1')x_1' - (x_3, x_2')x_2' = x_3 - 3x_1' - x_2' = \begin{bmatrix} 0 \\ 1 \\ 0 \\ -1 \end{bmatrix}$$

Thus,

$$x_3' = \begin{bmatrix} 0 \\ 1/\sqrt{2} \\ 0 \\ -1/\sqrt{2} \end{bmatrix}$$

Therefore, the vectors

$$\begin{bmatrix} -1/3 \\ 2/3 \\ 0 \\ 2/3 \end{bmatrix}, \quad \begin{bmatrix} 0 \\ 0 \\ 1 \\ 0 \end{bmatrix}, \quad \frac{1}{\sqrt{2}} \begin{bmatrix} 0 \\ 1 \\ 0 \\ -1 \end{bmatrix}$$

form an orthonormal basis for the same subspace of R^4.

Using the Gram–Schmidt process Theorem 3 follows.

Theorem 3 Let H be a subspace of R^n (or C^n); then H has an orthonormal basis.

Proof Take any basis for H and apply the Gram–Schmidt orthogonalization process to obtain an orthonormal basis for H. ∎

Let H be a subspace of R^n (or C^n). The **orthogonal complement** of H, denoted by $H^\perp$, is the set of vectors $H^\perp = \{y \mid (h, y) = 0 \text{ for all } h \text{ in } H\}$. In words, $H^\perp$ consists of those vectors which are orthogonal to all vectors of H.

If $x_1, x_2, \ldots, x_k$ is a basis for H, a vector y belongs to $H^\perp$ if and only if y is orthogonal to $x_1, x_2, \ldots, x_k$. Indeed, if y belongs to $H^\perp$, y is orthogonal to all vectors in H, in particular to $x_1, x_2, \ldots, x_k$. If, on the other hand, y is orthogonal to $x_1, x_2, \ldots, x_k$, and if x belongs to H, there are scalars $\alpha_1, \alpha_2, \ldots, \alpha_k$ such that $x = \alpha_1 x_1 + \alpha_2 x_2 + \cdots + \alpha_k x_k$. Thus,

$$\begin{aligned}(x, y) &= (\alpha_1 x_1 + \alpha_2 x_2 + \cdots + \alpha_k x_k, y) \\ &= \alpha_1(x_1, y) + \alpha_2(x_2, y) + \cdots + \alpha_k(x_k, y) \\ &= 0\end{aligned}$$

Thus, y is orthogonal to all vectors in H, and so y belongs to $H^\perp$.

Example 3 Find the orthogonal complement to the subspace of $\boldsymbol{R}^3$ spanned by

$$\boldsymbol{x}_1 = \begin{bmatrix} 1 \\ 0 \\ 0 \end{bmatrix} \quad \text{and} \quad \boldsymbol{x}_2 = \begin{bmatrix} 0 \\ 1 \\ 1 \end{bmatrix}$$

We need only determine those vectors in $\boldsymbol{R}^3$ orthogonal to $\boldsymbol{x}_1$ and $\boldsymbol{x}_2$. Now,

$$\begin{bmatrix} x \\ y \\ z \end{bmatrix}$$

is such a vector if and only if

$$x = 0$$
$$y + z = 0$$

Thus, the orthogonal complement consists of the vectors

$$\begin{bmatrix} 0 \\ \alpha \\ -\alpha \end{bmatrix}$$

where α ranges through all real values. In other words, the orthogonal complement is the subspace spanned by the vector

$$\begin{bmatrix} 0 \\ 1 \\ -1 \end{bmatrix}$$

(See Figure 3-3.)

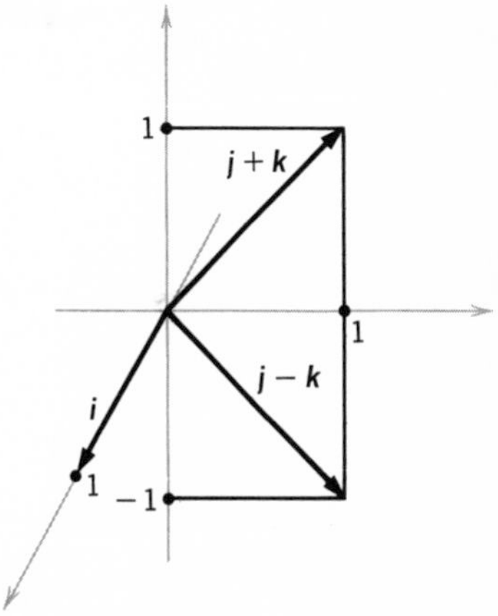

FIGURE 3-3

In this case the orthogonal complement is just the space of vectors normal to the plane spanned by $\boldsymbol{i}$ and $\boldsymbol{j} + \boldsymbol{k}$.

The following theorem tells us that the orthogonal complement of a given subspace is itself always a subspace.

Theorem 4 Let H be a subspace of $\boldsymbol{R}^n$ (or $\boldsymbol{C}^n$) and $H^\perp$ be its orthogonal complement. Then, $H^\perp$ is a subspace.

Proof Suppose $\boldsymbol{y}_1$ and $\boldsymbol{y}_2$ belong to $H^\perp$. Then, for all $\boldsymbol{x}$ in H, we have $(\boldsymbol{y}_1, \boldsymbol{x}) = 0$, $(\boldsymbol{y}_2, \boldsymbol{x}) = 0$. Then, $(\boldsymbol{y}_1 + \boldsymbol{y}_2, \boldsymbol{x}) = (\boldsymbol{y}_1, \boldsymbol{x}) + (\boldsymbol{y}_2, \boldsymbol{x}) = 0$. Thus, $\boldsymbol{y}_1 + \boldsymbol{y}_2$, being orthogonal to all vectors in H, belongs to $H^\perp$.

Similarly, if α is a scalar and $\boldsymbol{y}$ belongs to $H^\perp$, $(\alpha\boldsymbol{y}, \boldsymbol{x}) = \alpha(\boldsymbol{y}, \boldsymbol{x}) = 0$, for all $\boldsymbol{x}$ in H. Thus, $\alpha\boldsymbol{y}$ belongs to $H^\perp$. ∎

Two subspaces H and K of a vector space V are said to be **complementary** if $H \cap K = \mathbf{0}$ and $H + K = V$. In other words, H and K have only the zero vector in common, and every vector in V can be expressed as the sum of a vector in H and a vector in K.

For example, if $\boldsymbol{a}$ and $\boldsymbol{b}$ are linearly independent vectors in $\boldsymbol{R}^2$, let H be the subspace spanned by $\boldsymbol{a}$ and let K be the subspace spanned by $\boldsymbol{b}$. Then H and K are complementary subspaces. Indeed, if $\boldsymbol{z}$ belongs to both H and K, then for suitable scalars α and β, $\boldsymbol{z} = \alpha\boldsymbol{a} = \beta\boldsymbol{b}$. Thus, $\alpha\boldsymbol{a} - \beta\boldsymbol{b} = \mathbf{0}$. However, since $\boldsymbol{a}$ and $\boldsymbol{b}$ are linearly independent, $\alpha = 0$ and $\beta = 0$. Thus, $\boldsymbol{z} = \mathbf{0}$ and $H \cap K = \mathbf{0}$. Likewise, since $\boldsymbol{a}$ and $\boldsymbol{b}$ are linearly independent, they span $\boldsymbol{R}^2$. Thus, every vector in $\boldsymbol{R}^2$ is a linear combination of $\boldsymbol{a}$ and $\boldsymbol{b}$ and $H + K = \boldsymbol{R}^2$. (See Figure 3-4.)

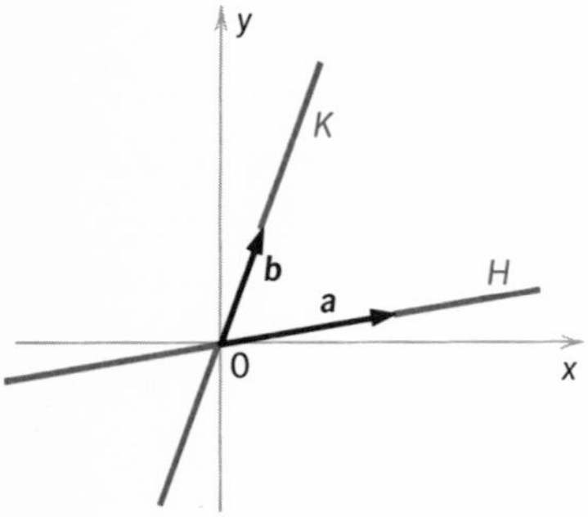

FIGURE 3-4

Theorem 5 If H is a subspace of $\boldsymbol{R}^n$ (or $\boldsymbol{C}^n$), then H and $H^\perp$ are complementary subspaces, i.e., $H \cap H^\perp = \mathbf{0}$ and $H + H^\perp = \boldsymbol{R}^n$.

Proof First, we show that H and $H^\perp$ have only the zero vector in common. Suppose $\boldsymbol{y}$ is a vector in both H and $H^\perp$. Then, since $\boldsymbol{y}$ belongs to $H^\perp$ it is orthogonal to all vectors in H. Since $\boldsymbol{y}$ is also a vector in H, it follows that $(\boldsymbol{y}, \boldsymbol{y}) = 0$. Therefore, $\boldsymbol{y} = \mathbf{0}$. Hence, H and $H^\perp$ have only the zero vector in common.

Next, we show that every vector may be expressed as the sum of a vector in H and $H^\perp$. Let $\boldsymbol{y}$ be an arbitrary vector in $\boldsymbol{R}^n$. Take $\boldsymbol{x}_1, \boldsymbol{x}_2, \ldots, \boldsymbol{x}_k$ to be an orthonormal basis for H. Let $\boldsymbol{z} = (\boldsymbol{y}, \boldsymbol{x}_1)\boldsymbol{x}_1 + (\boldsymbol{y}, \boldsymbol{x}_2)\boldsymbol{x}_2 + \cdots + (\boldsymbol{y}, \boldsymbol{x}_k)\boldsymbol{x}_k$. Clearly, $\boldsymbol{z}$ belongs to H. Then, $\boldsymbol{y} = \boldsymbol{z} + (\boldsymbol{y} - \boldsymbol{z})$. To complete the proof, we must show that $\boldsymbol{y} - \boldsymbol{z}$ belongs to $H^\perp$. Note that

$$\begin{aligned}
(\boldsymbol{y} - \boldsymbol{z}, \boldsymbol{x}_i) &= (\boldsymbol{y}, \boldsymbol{x}_i) - (\boldsymbol{z}, \boldsymbol{x}_i) \\
&= (\boldsymbol{y}, \boldsymbol{x}_i) - ((\boldsymbol{y}, \boldsymbol{x}_1)\boldsymbol{x}_1 + \cdots + (\boldsymbol{y}, \boldsymbol{x}_k)\boldsymbol{x}_k, \boldsymbol{x}_i) \\
&= (\boldsymbol{y}, \boldsymbol{x}_i) - (\boldsymbol{y}, \boldsymbol{x}_i)(\boldsymbol{x}_i, \boldsymbol{x}_i) \\
&= 0
\end{aligned}$$

Since $\boldsymbol{y} - \boldsymbol{z}$ is orthogonal to all vectors in a basis for H, $\boldsymbol{y} - \boldsymbol{z}$ is orthogonal to all vectors in H and thus belongs to $H^\perp$. ∎

For example, if H is the subspace of $\boldsymbol{R}^3$ consisting of the vectors on some plane through the origin, $H^\perp$ consists of all scalar multiples of some vector normal to the plane. Every vector in space may be expressed as the sum of a vector lying on the plane and some vector normal to the plane. Clearly, the plane and the line normal to the plane have only $\mathbf{0}$ in common.

If H and K are subspaces of $\boldsymbol{R}^n$ (or $\boldsymbol{C}^n$), other properties of the orthogonal complement are

(1) If $H \subset K$, then $K^\perp \subset H^\perp$.

(2) $(H^\perp)^\perp = H$.

To prove (1), observe that if $\boldsymbol{y}$ belongs to $K^{\perp}$, $\boldsymbol{y}$ is orthogonal to all vectors in K and thus to all vectors in H. Therefore, $\boldsymbol{y}$ belongs to $H^{\perp}$.

To prove (2), note first that if $\boldsymbol{x}$ belongs to H, $\boldsymbol{x}$ is orthogonal to all vectors in $H^{\perp}$, by definition of $H^{\perp}$, so $\boldsymbol{x}$ belongs to $(H^{\perp})^{\perp}$. Hence, $H \subset (H^{\perp})^{\perp}$. Next, suppose $\boldsymbol{y}$ belongs to $(H^{\perp})^{\perp}$. Then, $\boldsymbol{y} = \boldsymbol{h} + \boldsymbol{k}$, where $\boldsymbol{h}$ belongs to H and $\boldsymbol{k}$ belongs to $H^{\perp}$. Then,

$$\begin{aligned}(\boldsymbol{y}, \boldsymbol{y}) &= (\boldsymbol{h} + \boldsymbol{k}, \boldsymbol{h} + \boldsymbol{k}) \\ &= (\boldsymbol{h}, \boldsymbol{h}) + (\boldsymbol{h}, \boldsymbol{k}) + (\boldsymbol{k}, \boldsymbol{h}) + (\boldsymbol{k}, \boldsymbol{h}) \\ &= (\boldsymbol{h}, \boldsymbol{h}) + (\boldsymbol{k}, \boldsymbol{k})\end{aligned}$$

Thus, $|\boldsymbol{y}|^2 = |\boldsymbol{h}|^2 + |\boldsymbol{k}|^2$. Also, $(\boldsymbol{y}, \boldsymbol{y}) = (\boldsymbol{y}, \boldsymbol{h} + \boldsymbol{k}) = (\boldsymbol{y}, \boldsymbol{h}) + (\boldsymbol{y}, \boldsymbol{k})$.

By hypothesis $\boldsymbol{y}$ belongs to $(H^{\perp})^{\perp}$. Thus, $(\boldsymbol{y}, \boldsymbol{k}) = 0$ and $(\boldsymbol{y}, \boldsymbol{y}) = (\boldsymbol{y}, \boldsymbol{h}) = (\boldsymbol{h} + \boldsymbol{k}, \boldsymbol{h}) = (\boldsymbol{h}, \boldsymbol{h})$. Thus, $|\boldsymbol{y}|^2 = |\boldsymbol{h}|^2$. Therefore, $|\boldsymbol{k}|^2 = 0$, and so $\boldsymbol{k} = \boldsymbol{0}$, and $\boldsymbol{y} = \boldsymbol{h}$ belongs to H. Combining this with the result above. we see that $(H^{\perp})^{\perp} = H$.

There is a useful inequality relating the norm of a vector to its components relative to some orthonormal system. It is known as **Bessel's inequality.**

Theorem 6 Let $\boldsymbol{x}_1, \boldsymbol{x}_2, \ldots, \boldsymbol{x}_k$ be an orthonormal set of vectors in $\boldsymbol{R}^n$ (or $\boldsymbol{C}^n$). Let $\boldsymbol{x}$ be another vector in the same space. Then,

$$|\boldsymbol{x}|^2 \geq \sum_{i=1}^{k} |(\boldsymbol{x}, \boldsymbol{x}_i)|^2$$

Moreover, if $\boldsymbol{x}_1, \ldots, \boldsymbol{x}_k$ is an orthonormal basis, equality holds.

Proof Let H be the subspace spanned by $\boldsymbol{x}_1, \ldots, \boldsymbol{x}_k$. By Theorem 5, $\boldsymbol{x} = \boldsymbol{h} + \boldsymbol{l}$, where $\boldsymbol{h}$ belongs to H and $\boldsymbol{l}$ belongs to $H^{\perp}$. Thus, $\boldsymbol{h} = \alpha_1\boldsymbol{x}_1 + \alpha_2\boldsymbol{x}_2 + \cdots + \alpha_k\boldsymbol{x}_k$, since $\boldsymbol{x}_1, \ldots, \boldsymbol{x}_k$ span H, and $\boldsymbol{x} = \alpha_1\boldsymbol{x}_1 + \alpha_2\boldsymbol{x}_2 + \cdots + \alpha_k\boldsymbol{x}_k + \boldsymbol{l}$. Since $\boldsymbol{l}$ belongs to $H^{\perp}$,

$$\begin{aligned}(\boldsymbol{x}, \boldsymbol{x}_i) &= (\alpha_1\boldsymbol{x}_1 + \cdots + \alpha_k\boldsymbol{x}_k + \boldsymbol{l}, \boldsymbol{x}_i) \\ &= \alpha_i + (\boldsymbol{l}, \boldsymbol{x}_i) \\ &= \alpha_i\end{aligned}$$

Hence we can rewrite $\boldsymbol{x}$ in the form $\boldsymbol{x} = (\boldsymbol{x}, \boldsymbol{x}_1)\boldsymbol{x}_1 + \cdots + (\boldsymbol{x}, \boldsymbol{x}_k)\boldsymbol{x}_k + \boldsymbol{l}$. Therefore,

$$\begin{aligned}(\boldsymbol{x}, \boldsymbol{x}) &= ((\boldsymbol{x}, \boldsymbol{x}_1)\boldsymbol{x}_1 + \cdots + (\boldsymbol{x}, \boldsymbol{x}_k)\boldsymbol{x}_k + \boldsymbol{l}, (\boldsymbol{x}, \boldsymbol{x}_1)\boldsymbol{x}_1 + \cdots + (\boldsymbol{x}, \boldsymbol{x}_k)\boldsymbol{x}_k + \boldsymbol{l}) \\ &= |(\boldsymbol{x}, \boldsymbol{x}_1)|^2 + \cdots + |(\boldsymbol{x}, \boldsymbol{x}_k)|^2 + (\boldsymbol{l}, \boldsymbol{l})\end{aligned}$$

Thus, $|\boldsymbol{x}|^2 \geq |(\boldsymbol{x}, \boldsymbol{x}_1)|^2 + \cdots + |(\boldsymbol{x}, \boldsymbol{x}_k)|^2$.

If $\boldsymbol{x}_1, \ldots, \boldsymbol{x}_k$ is an orthonormal basis, $H^\perp = \mathbf{0}$. Thus, $\boldsymbol{l} = \mathbf{0}$, and $|\boldsymbol{x}|^2 = |(\boldsymbol{x}, \boldsymbol{x}_1)|^2 + \cdots + |(\boldsymbol{x}, \boldsymbol{x}_k)|^2$. ∎

EXERCISES

1. Show that the orthogonal complement of the subspace of $\boldsymbol{R}^2$ spanned by $\boldsymbol{i}_\theta = (\cos\theta)\boldsymbol{i} + (\sin\theta)\boldsymbol{j}$ is the subspace spanned by $\boldsymbol{j}_\theta = -(\sin\theta)\boldsymbol{i} + \cos\theta$.
2. Show that the n-vector

$$\boldsymbol{a} = \frac{1}{\sqrt{n}}\begin{bmatrix}1\\1\\\vdots\\1\end{bmatrix}$$

is a unit vector.

(a) Find the component of the vector

$$\begin{bmatrix}x_1\\x_2\\\vdots\\x_n\end{bmatrix}$$

in the direction of $\boldsymbol{a}$.

(b) Show that

$$\begin{bmatrix}x_1\\x_2\\\vdots\\x_n\end{bmatrix}$$

is orthogonal to $\boldsymbol{a}$ if and only if $x_1 + x_2 + \cdots + x_n = 0$.

3. If $\boldsymbol{x}_1, \ldots, \boldsymbol{x}_k$ is an orthogonal subset of unit vectors in $\boldsymbol{R}^n$, show that $\boldsymbol{x}_1, \boldsymbol{x}_2, \ldots, \boldsymbol{x}_k$ is a subset of some orthonormal basis for $\boldsymbol{R}^n$.
4. Determine an orthonormal basis for the subspace of $\boldsymbol{R}^n$ spanned by each of the following sets of vectors.

(a) $\begin{bmatrix}1\\-1\\0\end{bmatrix}, \begin{bmatrix}1\\0\\-1\end{bmatrix}$ (b) $\begin{bmatrix}-1\\0\\2\\1\end{bmatrix}, \begin{bmatrix}0\\-1\\0\\1\end{bmatrix}, \begin{bmatrix}3\\1\\2\\0\end{bmatrix}$

(c) $\begin{bmatrix}-1\\0\\1\\1\end{bmatrix}, \begin{bmatrix}1\\2\\0\\0\end{bmatrix}$

5. Find an orthonormal basis for $\boldsymbol{R}^3$ containing the vector

$$\frac{1}{\sqrt{3}}\begin{bmatrix}1\\-1\\1\end{bmatrix}$$

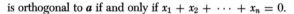

6. If $\boldsymbol{a}$ and $\boldsymbol{b}$ form an orthonormal basis for $\boldsymbol{R}^2$, show that for some θ,

$$\boldsymbol{a} = \boldsymbol{i}_\theta = (\cos\theta)\boldsymbol{i} + (\sin\theta)\boldsymbol{j}$$
$$\boldsymbol{b} = \pm\boldsymbol{j}_\theta = \pm(-(\sin\theta)\boldsymbol{i} + (\cos\theta)\boldsymbol{j})$$

7. Find an orthonormal basis for the orthogonal complement of the subspace of $\boldsymbol{R}^n$ spanned by

$$\begin{bmatrix} 1 \\ -1 \\ 0 \\ 0 \\ \vdots \\ 0 \end{bmatrix}$$

8. If H is a subspace of $\boldsymbol{R}^n$ and $H^\perp$ is its orthogonal complement, we saw that any vector in $\boldsymbol{R}^n$ may be expressed as the sum of a vector in H and a vector in $H^\perp$. Show that this expression is unique.

9. If $\boldsymbol{x}_1, \boldsymbol{x}_2, \ldots, \boldsymbol{x}_k$ is an orthonormal set of vectors in $\boldsymbol{R}^n$, show that $\boldsymbol{x}_1, \boldsymbol{x}_2, \ldots, \boldsymbol{x}_k$ is an orthonormal basis if and only if for all $\boldsymbol{x}$ in $\boldsymbol{R}^n$, $|\boldsymbol{x}|^2 = |(\boldsymbol{x}, \boldsymbol{x}_1)|^2 + |(\boldsymbol{x}, \boldsymbol{x}_2)|^2 + \cdots + |(\boldsymbol{x}, \boldsymbol{x}_n)|^2$.

10. If

$$\boldsymbol{x} = \begin{bmatrix} a \\ b \\ c \\ d \end{bmatrix}$$

is a unit vector in $\boldsymbol{R}^4$, show that the orthogonal complement of the subspace spanned by $\boldsymbol{x}$ has as an orthonormal basis the vectors

$$\begin{bmatrix} -b \\ a \\ d \\ -c \end{bmatrix}, \quad \begin{bmatrix} -c \\ -d \\ a \\ b \end{bmatrix}, \quad \begin{bmatrix} -d \\ c \\ -b \\ a \end{bmatrix}$$

11. Let H be a subspace of $\boldsymbol{R}^n$ and $\boldsymbol{x}_1, \boldsymbol{x}_2, \ldots, \boldsymbol{x}_k$ be an orthonormal basis for H. If $\boldsymbol{x}$ belongs to $\boldsymbol{R}^n$, let $P(\boldsymbol{x}) = (\boldsymbol{x}, \boldsymbol{x}_1)\boldsymbol{x}_1 + (\boldsymbol{x}, \boldsymbol{x}_2)\boldsymbol{x}_2 + \cdots + (\boldsymbol{x}, \boldsymbol{x}_k)\boldsymbol{x}_k$.
    (a) Show that P is a linear transformation.
    (b) Show that $P^2 = P$.
    (c) Show that the range space of P is H and the nullspace of P is $H^\perp$.
    (d) Show that the range space of $(I - P)$ is $H^\perp$ and the nullspace of $I - P$ is H.
    (e) If $\boldsymbol{x}$ is a vector in $\boldsymbol{R}^n$ and $\boldsymbol{y}$ is a vector in H, show that $|\boldsymbol{x} - \boldsymbol{y}| \geq |\boldsymbol{x} - P(\boldsymbol{x})|$, i.e., $P(\boldsymbol{x})$ is the closest vector in H to $\boldsymbol{x}$.

12. If $\boldsymbol{x}_1, \boldsymbol{x}_2, \ldots, \boldsymbol{x}_n$ is an orthonormal basis for $\boldsymbol{R}^n$ and $\boldsymbol{y}$ and $\boldsymbol{z}$ are vectors in $\boldsymbol{R}^n$, show that

$$(\boldsymbol{y}, \boldsymbol{z}) = (\boldsymbol{y}, \boldsymbol{x}_1)(\boldsymbol{z}, \boldsymbol{x}_1) + (\boldsymbol{y}, \boldsymbol{x}_2)(\boldsymbol{z}, \boldsymbol{x}_2) + \cdots + (\boldsymbol{y}, \boldsymbol{x}_n)(\boldsymbol{z}, \boldsymbol{x}_n)$$

13. If H is an h-dimensional subspace of $\boldsymbol{R}^n$, and if K is a complementary subspace of H, show that K is of dimension $n - h$.
14. If H and K are subspaces of $\boldsymbol{R}^n$ with the same orthogonal complement, show that $H = K$.
15. If $\boldsymbol{x}_1, \boldsymbol{x}_2, \ldots, \boldsymbol{x}_n$ is an orthonormal basis for $\boldsymbol{R}^n$, show that

$$(\mathrm{sp}(\boldsymbol{x}_1, \boldsymbol{x}_2, \ldots, \boldsymbol{x}_i))^\perp = \mathrm{sp}(\boldsymbol{x}_{i+1}, \ldots, \boldsymbol{x}_n)$$

16. If $\boldsymbol{x}_1, \boldsymbol{x}_2, \ldots, \boldsymbol{x}_k$ is an orthogonal set of vectors in H, a subspace of $\boldsymbol{R}^n$, and $\boldsymbol{y}_1, \boldsymbol{y}_2, \ldots, \boldsymbol{y}_k$ is an orthogonal set of vectors in $H^\perp$, show that $\boldsymbol{x}_1 + \boldsymbol{y}_1, \boldsymbol{x}_2 + \boldsymbol{y}_2, \ldots, \boldsymbol{x}_k + \boldsymbol{y}_k$ is an orthogonal set of vectors.
17. Show that the vectors

$$\frac{1}{\sqrt{2}}\begin{bmatrix} 1 \\ -1 \\ 0 \\ 0 \\ \vdots \\ 0 \end{bmatrix}, \quad \frac{1}{\sqrt{4}}\begin{bmatrix} 1 \\ 1 \\ -2 \\ 0 \\ \vdots \\ 0 \end{bmatrix}, \quad \frac{1}{\sqrt{12}}\begin{bmatrix} 1 \\ 1 \\ 1 \\ -3 \\ \vdots \\ 0 \end{bmatrix}, \quad \ldots, \quad \frac{1}{\sqrt{n^2 - n}}\begin{bmatrix} 1 \\ 1 \\ 1 \\ \vdots \\ 1 \\ -(n-1) \end{bmatrix}$$

form an orthonormal basis for the orthogonal complement of the subspace spanned by

$$\begin{bmatrix} 1 \\ 1 \\ \vdots \\ 1 \end{bmatrix}$$

18. If $\omega = e^{2\pi i/n}$ is an nth root of unity, show that the vectors

$$\frac{1}{\sqrt{n}}\begin{bmatrix} 1 \\ 1 \\ 1 \\ \vdots \\ 1 \end{bmatrix}, \quad \frac{1}{\sqrt{n}}\begin{bmatrix} 1 \\ \omega \\ \omega^2 \\ \vdots \\ \omega^{n-1} \end{bmatrix}, \quad \ldots, \quad \frac{1}{\sqrt{n}}\begin{bmatrix} 1 \\ \omega^{n-1} \\ \omega^{2(n-1)} \\ \vdots \\ \omega^{(n-1)^2} \end{bmatrix}$$

form an orthonormal basis for $\boldsymbol{C}^n$.
19. If H is a subspace of $\boldsymbol{R}^n$ and $\boldsymbol{x}$ is a vector in $\boldsymbol{R}^n$, consider the function $f(\boldsymbol{y}) = |\boldsymbol{x} + \boldsymbol{y}|$, defined for $\boldsymbol{y}$ in H.
    (a) Show $|f(\boldsymbol{y}_1) - f(\boldsymbol{y}_2)| \leq |\boldsymbol{y}_1 - \boldsymbol{y}_2|$.
    (b) If $\boldsymbol{x} = \boldsymbol{h} + \boldsymbol{k}$, where $\boldsymbol{h}$ belongs to H and $\boldsymbol{k}$ belongs to $H^\perp$, show that the minumum value of f is $|\boldsymbol{k}|$.
    (c) Show that the function f takes the value $|\boldsymbol{k}|$ at exactly one point.
20. Let H_1 and H_2 be subspaces of $\boldsymbol{R}^n$. Show that

$$(H_1 + H_2)^\perp = H_1^\perp \cap H_2^\perp$$

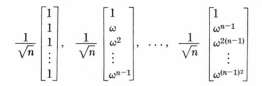

21. If H is a subspace of $\boldsymbol{R}^n$, $\boldsymbol{x}_1, \ldots, \boldsymbol{x}_k$ is an orthonormal basis for H, and $\boldsymbol{y}_1, \boldsymbol{y}_2, \ldots, \boldsymbol{y}_l$ is an orthonormal basis for $H^\perp$, show that $\boldsymbol{x}_1, \ldots, \boldsymbol{x}_k, \boldsymbol{y}_1, \ldots, \boldsymbol{y}_l$ is an orthonormal basis for $\boldsymbol{R}^n$.
22. Given vectors $\boldsymbol{x}_1, \boldsymbol{x}_2, \ldots, \boldsymbol{x}_k$ in $\boldsymbol{R}^n$, where $k < n$, show that there is a vector $\boldsymbol{y} \neq \boldsymbol{0}$, such that $(\boldsymbol{y}, \boldsymbol{x}_1) = (\boldsymbol{y}, \boldsymbol{x}_2) = \cdots = (\boldsymbol{y}, \boldsymbol{x}_k) = 0$.
23. Let H be a subspace of $\boldsymbol{R}^n$ and K be a complementary subspace. If $\boldsymbol{x}$ is a vector in $\boldsymbol{R}^n$, we know there are vectors $\boldsymbol{h}$ in H and $\boldsymbol{k}$ in K such that $\boldsymbol{x} = \boldsymbol{h} + \boldsymbol{k}$. If for all $\boldsymbol{x}$ in $\boldsymbol{R}^n$, $|\boldsymbol{x}|^2 = |\boldsymbol{h}|^2 + |\boldsymbol{k}|^2$, show that $K = H^\perp$.
24. If H and K are complementary subspaces of $\boldsymbol{R}^n$, show that there is a unique linear operator P such that
(1) $P^2 = P$ (2) $R_P = H$ (3) $N_P = K$
25. Given a linear operator P on $\boldsymbol{R}^n$ such that $P^2 = P$, show that R_P and N_P are complementary subspaces.

4 ORTHOGONAL AND UNITARY MATRICES AND OPERATORS

If A is a real $n \times n$ matrix and $A^TA = I_n$, A is said to be an **orthogonal matrix.**

For example, if

$$A = \begin{bmatrix} 4/5 & 0 & -3/5 \\ -9/25 & 4/5 & -12/25 \\ 12/25 & 3/5 & 16/25 \end{bmatrix}$$

$$A^TA = \begin{bmatrix} 4/5 & -9/25 & 12/25 \\ 0 & 4/5 & 3/5 \\ -3/5 & -12/25 & 16/25 \end{bmatrix} \begin{bmatrix} 4/5 & 0 & -3/5 \\ -9/25 & 4/5 & -12/25 \\ 12/25 & 3/5 & 16/25 \end{bmatrix} = \begin{bmatrix} 1 & 0 & 0 \\ 0 & 1 & 0 \\ 0 & 0 & 1 \end{bmatrix}$$

Thus, A is an orthogonal matrix. Also,

$$\begin{bmatrix} \cos\theta & \sin\theta \\ -\sin\theta & \cos\theta \end{bmatrix} \begin{bmatrix} \cos\theta & -\sin\theta \\ \sin\theta & \cos\theta \end{bmatrix} = \begin{bmatrix} 1 & 0 \\ 0 & 1 \end{bmatrix}$$

Thus,

$$\begin{bmatrix} \cos\theta & -\sin\theta \\ \sin\theta & \cos\theta \end{bmatrix}$$

is an orthogonal matrix. Clearly, any orthogonal matrix is invertible and $A^{-1} = A^T$.

If A is an $n \times n$ matrix and $A_1, A_2, \ldots, A_n$ are its columns, then

$$A^TA = \begin{bmatrix} (A_1, A_2) & (A_1, A_2) & \ldots & (A_1, A_n) \\ (A_2, A_1) & (A_2, A_2) & \ldots & (A_2, A_n) \\ & & \vdots & \\ (A_n, A_1) & (A_n, A_2) & \ldots & (A_n, A_n) \end{bmatrix}$$

Thus, the matrix A is orthogonal if and only if its columns form an orthonormal basis for $\boldsymbol{R}^n$.

A complex $n \times n$ matrix is said to be **unitary** if $U^*U = I_n$. Just as in the real case, a matrix U is unitary if and only if its columns form an orthonormal basis for $\boldsymbol{C}^n$. For example, if $|a|^2 + |b|^2 = 1$, the matrix

$$\begin{bmatrix} a & b \\ -\bar{b} & \bar{a} \end{bmatrix}$$

is unitary. Indeed,

$$\begin{bmatrix} \bar{a} & -b \\ \bar{b} & a \end{bmatrix} \begin{bmatrix} a & b \\ -\bar{b} & \bar{a} \end{bmatrix} = \begin{bmatrix} |a|^2 + |b|^2 & 0 \\ 0 & |a|^2 + |b|^2 \end{bmatrix} = \begin{bmatrix} 1 & 0 \\ 0 & 1 \end{bmatrix}$$

If A is an orthogonal matrix, from $A^TA = I_n$, it follows that $\det A^T \det A = 1$, or $\det A = \pm 1$.

If A is an $n \times n$ orthogonal matrix, consider the linear operator on $\boldsymbol{R}^n$, $T(\boldsymbol{x}) = A\boldsymbol{x}$. Then,

$$\begin{aligned} (T(\boldsymbol{x}), T(\boldsymbol{y})) &= (T(\boldsymbol{x}))^T(T(\boldsymbol{y})) \\ &= (A\boldsymbol{x})^T(A\boldsymbol{y}) \\ &= (\boldsymbol{x}^TA^T)(A\boldsymbol{y}) \\ &= \boldsymbol{x}^T(A^TA)\boldsymbol{y} = \boldsymbol{x}^T\boldsymbol{y} = (\boldsymbol{x}, \boldsymbol{y}) \end{aligned}$$

In other words, the linear operator induced by an orthogonal matrix leaves the inner product invariant. Likewise, the linear operator on $\boldsymbol{C}^n$ induced by multiplication by a unitary operator leaves the complex inner product invariant.

If $T : \boldsymbol{R}^n \to \boldsymbol{R}^n$ (or $\boldsymbol{C}^n \to \boldsymbol{C}^n$) is a linear operator with the property that for all vectors $\boldsymbol{x}$ and $\boldsymbol{y}$, $(T(\boldsymbol{x}), T(\boldsymbol{y})) = (\boldsymbol{x}, \boldsymbol{y})$, T is said to be an **isometry**. As we have seen above, any linear operator induced by an orthogonal or unitary matrix is an isometry.

The following theorem summarizes some important properties of isometries.

Theorem 1 Let $T : \boldsymbol{R}^n \to \boldsymbol{R}^n$ (or $\boldsymbol{C}^n \to \boldsymbol{C}^n$) be an isometry. Then,

(i) $|T(\boldsymbol{x})| = |\boldsymbol{x}|$.
(ii) T is an isomorphism.
(iii) If $\boldsymbol{x}$ and $\boldsymbol{y}$ are orthogonal vectors, $T(\boldsymbol{x})$ and $T(\boldsymbol{y})$ are also orthogonal.
(iv) If $\boldsymbol{x}_1, \boldsymbol{x}_2, \ldots, \boldsymbol{x}_n$ is an orthonormal basis, $T(\boldsymbol{x}_1), T(\boldsymbol{x}_2), \ldots, T(\boldsymbol{x}_n)$ is an orthonormal basis.

Proof

(i) $|T(\boldsymbol{x})| = (T(\boldsymbol{x}), (T(\boldsymbol{x}))^{1/2} = (\boldsymbol{x}, \boldsymbol{x})^{1/2}$, since T is an isometry. Thus, $|T(\boldsymbol{x})| = |\boldsymbol{x}|$.
(ii) Since T is a linear operator on a finite-dimensional space, if T is one-one, T is an isomorphism. Since $|T(\boldsymbol{x})| = |\boldsymbol{x}|$, if $T(\boldsymbol{x}) = \boldsymbol{0}$, then $\boldsymbol{x} = \boldsymbol{0}$. Thus, T is one-one.
(iii) If $(\boldsymbol{x}, \boldsymbol{y}) = 0$, then $(T(\boldsymbol{x}), T(\boldsymbol{y})) = (\boldsymbol{x}, \boldsymbol{y}) = 0$.
(iv) If $\boldsymbol{x}_1, \boldsymbol{x}_2, \ldots, \boldsymbol{x}_n$ is an orthonormal basis, by (i), $T(\boldsymbol{x}_1), T(\boldsymbol{x}_2), \ldots, T(\boldsymbol{x}_n)$ are each unit vectors. By (iii), if $i \neq j$, $T(\boldsymbol{x}_i)$ and $T(\boldsymbol{x}_j)$ are orthogonal, since $\boldsymbol{x}_i$ and $\boldsymbol{x}_j$ are orthogonal. By (ii), if $\boldsymbol{x}_1, \boldsymbol{x}_2, \ldots, \boldsymbol{x}_n$ form a basis, so do $T(\boldsymbol{x}_1), T(\boldsymbol{x}_2), \ldots, T(\boldsymbol{x}_n)$. Thus, $T(\boldsymbol{x}_1), T(\boldsymbol{x}_2), \ldots, T(\boldsymbol{x}_n)$ form an orthogonal basis. ∎

As we may have expected from earlier results, the matrix representation of an isometry with respect to an orthonormal basis has a special form.

Theorem 2 Let $T : \boldsymbol{R}^n \to \boldsymbol{R}^n$ be a linear operator. Let $\boldsymbol{x}_1, \boldsymbol{x}_2, \ldots, \boldsymbol{x}_n$ be an orthonormal basis for $\boldsymbol{R}^n$. Suppose A_T is the matrix representation of T relative to the basis $\boldsymbol{x}_1, \boldsymbol{x}_2, \ldots, \boldsymbol{x}_n$. Then, T is an isometry if and only if A_T is an orthogonal matrix.

Proof If

$$T(\boldsymbol{x}_j) = \Sigma a_{ij}\boldsymbol{x}_i$$

we have seen that $A_T = [a_{ij}]_{(nn)}$.

Since $\boldsymbol{x}_1, \boldsymbol{x}_2, \ldots, \boldsymbol{x}_n$ is an orthonormal basis,

$$\begin{aligned}(T(\boldsymbol{x}_j), T(\boldsymbol{x}_k)) &= \left(\sum_{i=1}^{n} a_{ij}\boldsymbol{x}_i, \sum_{i=1}^{n} a_{ik}\boldsymbol{x}_i\right)\\ &= \sum_{i=1}^{n} a_{ij}a_{ik}\end{aligned}$$

If T is an isometry, $T(\boldsymbol{x}_1), T(\boldsymbol{x}_2), \ldots, T(\boldsymbol{x}_n)$ is an orthonormal basis, and thus

$$\sum_{i=1}^{n} a_{ij}a_{ij} = 1 \quad \text{and} \quad \sum_{i=1}^{n} a_{ij}a_{ik} = 0 \text{ if } j \neq k$$

Therefore, the columns of A_T form an orthonormal basis and A_T is orthogonal.

If, on the other hand, A_T is orthogonal,

$$\sum_{i=1}^{n} a_{ij}a_{ij} = 1 \quad \text{and} \quad \sum_{i=1}^{n} a_{ij}a_{ik} = 0 \text{ if } j \neq k$$

Thus $(T(\boldsymbol{x}_j), T(\boldsymbol{x}_j)) = 1$ and $(T(\boldsymbol{x}_j), T(\boldsymbol{x}_k)) = 0$ if $j \neq k$. Thus, the vectors $T(\boldsymbol{x}_1), T(\boldsymbol{x}_2), \ldots, T(\boldsymbol{x}_n)$ form an orthonormal basis for $\boldsymbol{R}^n$. If $\boldsymbol{y} = \Sigma_{i=1}^{n}\beta_i\boldsymbol{x}_i$ and $\boldsymbol{z} = \Sigma_{i=1}^{n}\gamma_i\boldsymbol{x}_i$ are two vectors in $\boldsymbol{R}^n$, it follows that

$$\begin{aligned}(T(\boldsymbol{y}), T(\boldsymbol{z})) &= (\beta_1 T(\boldsymbol{x}_1) + \cdots + \beta_n T(\boldsymbol{x}_n), \gamma_1 T(\boldsymbol{x}_1) + \cdots + \gamma_n T(\boldsymbol{x}_n)) \\ &= \beta_1\gamma_1 + \beta_2\gamma_2 + \cdots + \beta_n\gamma_n \\ &= (\boldsymbol{y}, \boldsymbol{z})\end{aligned}$$

Therefore, T is an isometry of $\boldsymbol{R}^n$. ∎

As an example, consider the orthonormal basis for $\boldsymbol{R}^3$,

$$\begin{aligned}\boldsymbol{i}_{\theta\phi} &= (\sin\theta\cos\phi)\boldsymbol{i} + (\sin\theta\sin\phi)\boldsymbol{j} + (\cos\theta)\boldsymbol{k} \\ \boldsymbol{j}_{\theta\phi} &= -(\sin\phi)\boldsymbol{i} + (\cos\phi)\boldsymbol{j} \\ \boldsymbol{k}_{\theta\phi} &= (\cos\theta\cos\phi)\boldsymbol{i} + (\cos\theta\sin\phi)\boldsymbol{j} - (\sin\theta)\boldsymbol{k}\end{aligned}$$

(See Figure 4-1.)

If

$$A_{\theta\phi} = \begin{bmatrix} \sin\theta\cos\phi & -\sin\phi & \cos\theta\cos\phi \\ \sin\theta\sin\phi & \cos\phi & \cos\theta\sin\phi \\ \cos\theta & 0 & -\sin\theta \end{bmatrix}$$

since the columns of $A_{\theta\phi}$ form an orthonormal system, $A_{\theta\phi}$ is an orthogonal matrix. Thus, the linear operator $T(\boldsymbol{x}) = A_{\theta\phi}\boldsymbol{x}$ is an isometry of $\boldsymbol{R}^3$.

As another result we determine all 2×2 orthogonal matrices.

Example 1 Let A be a 2×2 orthogonal matrix. Then, for some real number θ, we have

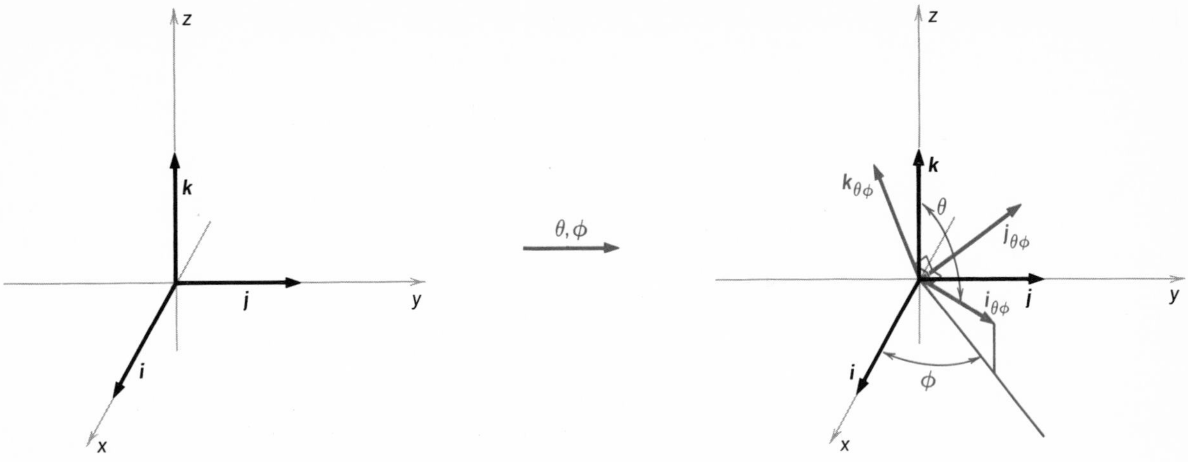

FIGURE 4-1

$$A = \begin{bmatrix} \cos\theta & -\sin\theta \\ \sin\theta & \cos\theta \end{bmatrix} \quad \text{or} \quad A = \begin{bmatrix} \cos\theta & \sin\theta \\ \sin\theta & -\cos\theta \end{bmatrix}$$

To prove this, first note that if A is a 2×2 orthogonal matrix, its columns, say $\boldsymbol{a}$ and $\boldsymbol{b}$, form an orthogonal basis for $\boldsymbol{R}^2$. Let θ be the angle which $\boldsymbol{a}$ makes with the x axis. Since $\boldsymbol{a}$ is a unit vector $\boldsymbol{a} = \boldsymbol{i}_\theta = (\cos\theta)\boldsymbol{i} + (\sin\theta)\boldsymbol{j}$. Since $\boldsymbol{b}$ is a unit vector orthogonal to $\boldsymbol{a}$, $\boldsymbol{b} = \pm\boldsymbol{j}_\theta = \pm(-(\sin\theta)\boldsymbol{i} + (\cos\theta)\boldsymbol{j})$. Thus,

$$A = \begin{bmatrix} \cos\theta & -\sin\theta \\ \sin\theta & \cos\theta \end{bmatrix} \quad \text{or} \quad \begin{bmatrix} \cos\theta & \sin\theta \\ \sin\theta & -\cos\theta \end{bmatrix}$$

In the previous chapter, we saw that a matrix of the first type induces a rotation of the plane by θ degrees, while matrices of the second type induce a reflection of the plane about the line making an angle of $\frac{1}{2}\theta$ degrees with the x axis. Thus, any 2×2 orthogonal matrix induces a rotation or reflection of the plane.

EXERCISES

1. Which of the following matrices are orthogonal?

(a) $\begin{bmatrix} \frac{1}{\sqrt{2}} & \frac{1}{\sqrt{2}} \\ \frac{1}{\sqrt{2}} & \frac{1}{\sqrt{2}} \end{bmatrix}$ (b) $\begin{bmatrix} 1 & 1 \\ -1 & 1 \end{bmatrix}$

(c) $\begin{bmatrix} 1/\sqrt{2} & 1/\sqrt{2} & 0 \\ 1/\sqrt{2} & -1/\sqrt{2} & 1/\sqrt{2} \\ 0 & 0 & 1/\sqrt{2} \end{bmatrix}$ (d) $\begin{bmatrix} 0 & 1/\sqrt{2} & 1/\sqrt{2} \\ 1 & 0 & 0 \\ 0 & -1/\sqrt{2} & 1/\sqrt{2} \end{bmatrix}$

2. Show that an isometry of $\boldsymbol{R}^3$ preserves the angle between any two vectors.
3. Show that in $\boldsymbol{R}^n$, $(\boldsymbol{x}, \boldsymbol{y}) = \frac{1}{2}(|\boldsymbol{x} + \boldsymbol{y}|^2 - |\boldsymbol{x}|^2 - |\boldsymbol{y}|^2)$.
4. If $A : \boldsymbol{R}^n \to \boldsymbol{R}^n$ is a linear operator such that $|A\boldsymbol{x}| = |\boldsymbol{x}|$ for all $\boldsymbol{x}$ in $\boldsymbol{R}^n$, show that A is an isometry. [Hint: Use exercise 3.]
5. If S and T are isometries of $\boldsymbol{R}^n$ (or $\boldsymbol{C}^n$), show that $S \circ T$ and T^{-1} are also isometries.
6. If U and V are orthogonal matrices, show that UV and U^{-1} are also orthogonal matrices. Prove the same if U and V are unitary.
7. Let $\boldsymbol{x}$ be a given unit vector in $\boldsymbol{R}^n$. Show that there is an orthogonal matrix whose first column is $\boldsymbol{x}$.
8. Let T be a linear operator on $\boldsymbol{R}^3$ which preserves the angle between any two vectors. Show that T is a nonzero scalar multiple of an isometry and conversely.
9. If A is an orthogonal matrix, $\det A = 1$, and a_{ij} and A_{ij} are the entry and cofactor in position (i, j), respectively. Show that $a_{ij} = A_{ij}$. What happens if $\det A = -1$?
10. If U is a unitary matrix, show that $|\det U| = 1$.
11. If $\boldsymbol{x}_1, \boldsymbol{x}_2, \ldots, \boldsymbol{x}_n$ are vectors in $\boldsymbol{R}^n$, and $[\boldsymbol{x}_1, \boldsymbol{x}_2, \ldots, \boldsymbol{x}_n]$ is the $n \times n$ matrix whose ith column is the vector $\boldsymbol{x}_i$, show that $|\det [\boldsymbol{x}_1, \boldsymbol{x}_2, \ldots, \boldsymbol{x}_n]| \leq |\boldsymbol{x}_1||\boldsymbol{x}_2| \ldots |\boldsymbol{x}_n|$. Interpret this geometrically in $\boldsymbol{R}^2$ and $\boldsymbol{R}^3$.
12. If A is an orthogonal matrix and $I_n + A$ is nonsingular, show that $(I_n - A)(I_n + A)^{-1}$ is skew-symmetric.
13. Calculate the orthogonal matrix associated with a rotation of $\boldsymbol{R}^3$ of θ degrees about the z axis.
14. Calculate the orthogonal matrix associated with a rotation of $\boldsymbol{R}^3$ by θ degrees about the z axis, followed by a rotation of θ degrees about the y axis.
15. Show that any 2×2 unitary matrix of determinant 1 is of the form

$$\begin{bmatrix} a & b \\ -\bar{b} & \bar{a} \end{bmatrix}$$

where $|a|^2 + |b|^2 = 1$.

16. Let T be a linear operator on $\boldsymbol{R}^n$. Let $\boldsymbol{x}_1, \boldsymbol{x}_2, \ldots, \boldsymbol{x}_n$ be an orthonormal basis for $\boldsymbol{R}^n$. If $T(\boldsymbol{x}_1), T(\boldsymbol{x}_2), \ldots, T(\boldsymbol{x}_n)$ is an orthonormal basis for $\boldsymbol{R}^n$, show that T is an isometry.
17. Let T be a linear transformation of $\boldsymbol{R}^2$ which carries the unit circle into itself. Show that T is an isometry.
18. Let $\boldsymbol{x}_1, \boldsymbol{x}_2$, and $\boldsymbol{x}_3$ be three vectors in $\boldsymbol{R}^2$ whose endpoints are noncollinear. Let $\boldsymbol{y}$ and $\boldsymbol{z}$ be two more vectors in $\boldsymbol{R}^2$. If

$$d(\boldsymbol{y}, \boldsymbol{x}_1) = d(\boldsymbol{z}, \boldsymbol{x}_1)$$
$$d(\boldsymbol{y}, \boldsymbol{x}_2) = d(\boldsymbol{z}, \boldsymbol{x}_2)$$
$$d(\boldsymbol{y}, \boldsymbol{x}_3) = d(\boldsymbol{z}, \boldsymbol{x}_3)$$

show that $\boldsymbol{y} = \boldsymbol{z}$.

19. Let A be a 2×2 orthogonal matrix and $\boldsymbol{y}$ be a 2-vector. Show that the function from $\boldsymbol{R}^2$ into $\boldsymbol{R}^2$ defined by $f(\boldsymbol{x}) = A\boldsymbol{x} + \boldsymbol{y}$ has the property that $d(f(\boldsymbol{x}_1), f(\boldsymbol{x}_2)) = d(\boldsymbol{x}_1, \boldsymbol{x}_2)$, for all $\boldsymbol{x}_1$ and $\boldsymbol{x}_2$ in $\boldsymbol{R}^2$.

20. Let f be a function from $\boldsymbol{R}^2$ to $\boldsymbol{R}^2$. Suppose $d(f(\boldsymbol{x}_1), f(\boldsymbol{x}_2)) = d(\boldsymbol{x}_1, \boldsymbol{x}_2)$ for all $\boldsymbol{x}_1$ and $\boldsymbol{x}_2$ in $\boldsymbol{R}^2$. If f fixes three noncollinear points, show that $f(\boldsymbol{x}) = \boldsymbol{x}$ for all $\boldsymbol{x}$. [Hint: Use exercise 18.]

21. Let f be a function from $\boldsymbol{R}^2$ to $\boldsymbol{R}^2$ such that $d(f(\boldsymbol{x}_1), f(\boldsymbol{x}_2)) = d(\boldsymbol{x}_1, \boldsymbol{x}_2)$ for all $\boldsymbol{x}_1$ and $\boldsymbol{x}_2$ in $\boldsymbol{R}^2$. Show that there is an orthogonal matrix A and a vector $\boldsymbol{y}$, such that $f(\boldsymbol{x}) = A\boldsymbol{x} + \boldsymbol{y}$. [Hint: Use exercises 19 and 20.]

22. If Q is a real $n \times n$ matrix and Q^TQ is a scalar multiple of the identity, show that Q is a scalar multiple of an orthogonal matrix.

23. Show that the matrix

$$\begin{bmatrix} \frac{1}{2}(1+i) & i/\sqrt{3} & (3+i)/2\sqrt{15} \\ -\frac{1}{2} & 1/\sqrt{3} & (4+3i)/2\sqrt{15} \\ \frac{1}{2} & -i/\sqrt{3} & 5i/2\sqrt{15} \end{bmatrix}$$

is unitary.

24. Let $\lambda_1, \lambda_2, \ldots, \lambda_n$ be complex numbers of absolute value 1. If the columns of a unitary matrix are multiplied by $\lambda_1, \lambda_2, \ldots, \lambda_n$, respectively, show that the resulting matrix is unitary.

25. Prove that if an orthogonal matrix is triangular, it is diagonal and all its diagonal entries are ± 1.

26. Let A be a real symmetric matrix and S be a real skew-symmetric matrix. Suppose that $AS = SA$ and $\det[A - S] \neq 0$. Show that $(A + S)(A - S)^{-1}$ is orthogonal.

27. Let $\boldsymbol{a}$ be the vector $\boldsymbol{a} = \boldsymbol{i} + \boldsymbol{j} + \boldsymbol{k}$ in $\boldsymbol{R}^3$. Find the orthogonal matrix representing a rotating of θ degrees about the line determined by the vector $\boldsymbol{a}$.

28. Let A be a real $n \times n$ matrix. Let $S(A)$ be the function which assigns to A the sum of the squares of the entries of A. If Q is an orthogonal matrix, show that $S(QAQ^{-1}) = S(A)$.

29. Let $\boldsymbol{x}_1, \boldsymbol{x}_2, \ldots, \boldsymbol{x}_n$ and $\boldsymbol{y}_1, \boldsymbol{y}_2, \ldots, \boldsymbol{y}_n$ be orthonormal bases for $\boldsymbol{R}^n$. Show that there is an isometry such that $T(\boldsymbol{x}_1) = \boldsymbol{y}_1, T(\boldsymbol{x}_2) = \boldsymbol{y}_2, \ldots, T(\boldsymbol{x}_n) = \boldsymbol{y}_n$.

5 GENERAL INNER PRODUCTS

Just as the concept of an abstract vector space was formulated by studying the principal properties of vector addition and scalar multiplication on $\boldsymbol{R}^n$, so too it is possible to consider abstract inner product spaces by singling out the most important properties of the dot product on $\boldsymbol{R}^n$. If V is a real vector space, a function which assigns to every pair of vectors $\boldsymbol{x}$ and $\boldsymbol{y}$ in V a real number $(\boldsymbol{x}, \boldsymbol{y})$ is said to be an inner product on V, if it has the following properties.

(1) $(\boldsymbol{x}, \boldsymbol{x}) \geq 0$
$(\boldsymbol{x}, \boldsymbol{x}) = 0$ if and only if $\boldsymbol{x} = \boldsymbol{0}$

(2) $(\boldsymbol{x}, \boldsymbol{y} + \boldsymbol{z}) = (\boldsymbol{x}, \boldsymbol{y}) + (\boldsymbol{x}, \boldsymbol{z})$
$(\boldsymbol{x} + \boldsymbol{y}, \boldsymbol{z}) = (\boldsymbol{x}, \boldsymbol{z}) + (\boldsymbol{y}, \boldsymbol{z})$

(3) $(\alpha\boldsymbol{x}, \boldsymbol{y}) = \alpha(\boldsymbol{x}, \boldsymbol{y})$
$(\boldsymbol{x}, \beta\boldsymbol{y}) = \beta(\boldsymbol{x}, \boldsymbol{y})$

(4) $(\boldsymbol{x}, \boldsymbol{y}) = (\boldsymbol{y}, \boldsymbol{x})$

The vector space V together with its inner product is said to constitute an **inner product space**.

As we noted earlier if $V = \boldsymbol{R}^n$ and if whenever

$$\boldsymbol{x} = \begin{bmatrix} \alpha_1 \\ \alpha_2 \\ \vdots \\ \alpha_n \end{bmatrix} \quad \text{and} \quad \boldsymbol{y} = \begin{bmatrix} \beta_1 \\ \beta_2 \\ \vdots \\ \beta_n \end{bmatrix}$$

we define

$$(\boldsymbol{x}, \boldsymbol{y}) = \alpha_1\beta_1 + \alpha_2\beta_2 + \cdots + \alpha_n\beta_n$$

we obtain an inner product on $\boldsymbol{R}^n$. In addition there are many other natural inner products.

Example 1 If $V = \boldsymbol{R}^n$ and $c_1, c_2, \ldots, c_n$ are n strictly positive real numbers, if we define

$$(\boldsymbol{x}, \boldsymbol{y}) = c_1\alpha_1\beta_1 + c_2\alpha_2\beta_2 + \cdots + c_n\alpha_n\beta_n$$

when

$$\boldsymbol{x} = \begin{bmatrix} \alpha_1 \\ \alpha_2 \\ \vdots \\ \alpha_n \end{bmatrix} \quad \text{and} \quad \boldsymbol{y} = \begin{bmatrix} \beta_1 \\ \beta_2 \\ \vdots \\ \beta_n \end{bmatrix}$$

then $(\ ,\ )$ is an inner product on $\boldsymbol{R}^n$.

Example 2 If $V = P_n$, the space of polynomials with real coefficients of degree less than or equal to n in a variable x, and if we define

$$(f, g) = \int_a^b f(x)g(x)\, dx$$

where a and b are real numbers, $a < b$, we obtain an inner product on P_n.

We verify the properties explicitly in this case.

(1) If f is a polynomial, $f^2(x) \geq 0$, thus

$$(f, f) = \int_a^b f^2(x)\, dx \geq 0$$

If $(f, f) = 0$, then

$$\int_a^b f^2(x)\, dx = 0$$

Since $f^2(x) \geq 0$, the function $f(x)$ must be zero everywhere on the interval $[a, b]$. But any polynomial of degree less than or equal to n with more than n roots must be identically zero. Thus, $f = 0$.

$$\text{(2)}\quad (f, g + h) = \int_a^b f(x)(g(x) + h(x))\, dx$$

$$= \int_a^b f(x)g(x)\, dx + \int_a^b f(x)h(x)\, dx$$

$$= (f, g) + (f, h)$$

$$\text{(3)}\qquad (\alpha f, g) = \int_a^b \alpha f(x)g(x)\, dx$$

$$= \alpha \int_a^b f(x)g(x)\, dx$$

$$= \alpha(f, g)$$

$$\text{(4)}\qquad (f, g) = \int_a^b f(x)g(x)\, dx = \int_a^b g(x)f(x)\, dx = (g, f)$$

If A is an $n \times n$ matrix, $A = [a_{ij}]_{(nn)}$, the trace of A, denoted by tr A, has already been defined as

$$\operatorname{tr} A = a_{11} + a_{22} + \cdots + a_{nn}$$

Notice that the trace of A is the sum of the diagonal entries of A.

Immediate consequences of the definition are

$$\operatorname{tr} \alpha A = \alpha \operatorname{tr} A$$

$$\operatorname{tr}(A + B) = \operatorname{tr} A + \operatorname{tr} B$$

$$\operatorname{tr} A = \operatorname{tr} A^T$$

The first two properties imply that the trace function is a linear transformation from the space of matrices into the reals (or complexes if we are dealing with complex matrices).

Less obvious is the relationship

$$\text{tr } AB = \text{tr } BA$$

To prove this, let

$$A = [a_{ij}]_{(nn)}, \qquad B = [b_{ij}]_{(nn)}$$

Then,

$$\text{tr } AB = \sum_{i=1}^{n} \sum_{j=1}^{n} a_{ij}b_{ji}$$

$$\text{tr } BA = \sum_{i=1}^{n} \sum_{j=1}^{n} b_{ij}a_{ji}$$

So

$$\text{tr } AB = \text{tr } BA$$

Example 3 Using the trace function, we define an inner product on the space of $n \times n$ matrices as follows: If A and B are $n \times n$ matrices

$$(A, B) = \text{tr } AB^T$$

To prove (1), we note that

$$\text{tr } AA^T = \sum_{i,j=1}^{n} a_{ij}^2, \qquad \text{if } A = [a_{ij}]_{(nn)}$$

Thus, the trace of AA^T is the sum of the squares of the entries of A. Therefore, $\text{tr } AA^T \geq 0$. If $\text{tr } AA^T = 0$, all entries of A must be zero, and thus $A = 0$.

That the other properties of the inner product are satisfied follows easily from the corresponding properties of the trace.

The definitions and theorems concerning abstract inner products are formally the same as those of the special inner product considered in the first five sections of this chapter. For example, the norm of the vector $\boldsymbol{x}$, denoted as before by $|\boldsymbol{x}|$, is defined to be $(\boldsymbol{x}, \boldsymbol{x})^{1/2}$. Two vectors $\boldsymbol{x}$ and $\boldsymbol{y}$ are orthogonal if $(\boldsymbol{x}, \boldsymbol{y}) = 0$. A vector of norm 1 is said to be a unit vector. Orthonormal sets of vectors and bases are defined as before.

All theorems proved for the special inner product on $\boldsymbol{R}^n$ are still valid for general inner products. Indeed a cursory perusal of the proofs of these theorems offered in earlier sections shows that only properties

(1)–(4) of the inner product were used during the course of the demonstration. Thus the theorems hold for all functions, defined on pairs of vectors, which satisfy properties (1)–(4), that is, general inner products.

Thus, we have the Cauchy–Schwarz inequality, $|(\boldsymbol{x}, \boldsymbol{y})| \leq |\boldsymbol{x}||\boldsymbol{y}|$, for vectors $\boldsymbol{x}$ and $\boldsymbol{y}$ in V, and the norm inequality, $|\boldsymbol{x} + \boldsymbol{y}| \leq |\boldsymbol{x}| + |\boldsymbol{y}|$. In the case of Example 2 above, the Cauchy–Schwarz inequality becomes

$$\left|\int_a^b f(x)g(x)\,dx\right| \leq \left(\int_a^b f^2(x)\,dx\right)^{1/2}\left(\int_a^b g^2(x)\,dx\right)^{1/2}$$

In the case of Example 3, it becomes

$$|\text{tr}\, AB^T| \leq (\text{tr}\, AA^T)^{1/2}(\text{tr}\, BB^T)^{1/2}$$

In a similar vein, we see that any set of nonzero orthogonal vectors is linearly independent, that the Gram–Schmidt orthogonalization procedure is valid, and that any finite-dimensional real vector space with an inner product admits an orthonormal basis with respect to this inner product. We illustrate with examples.

Example 4 In P_3, with inner product

$$(f, g) = \int_{-1}^{+1} f(x)g(x)\,dx$$

the polynomials $1, x, x^2 - \frac{1}{3}, x^3 - \frac{3}{5}x$ form an orthogonal collection of vectors. Thus, they are linearly independent and form a basis. By multiplying each polynomial by a suitable constant we may obtain an orthonormal basis.

Example 5 Consider the space of 2×2 matrices with real entries, with the inner product of Example 3 above.

An easy verification shows that the matrices

$$\begin{bmatrix} 1 & 0 \\ 0 & 0 \end{bmatrix}, \quad \begin{bmatrix} 0 & 1 \\ 0 & 0 \end{bmatrix}, \quad \begin{bmatrix} 0 & 0 \\ 1 & 0 \end{bmatrix}, \quad \begin{bmatrix} 0 & 0 \\ 0 & 1 \end{bmatrix}$$

form an orthonormal basis.

If V_1 and V_2 are real vector spaces with inner products $(\ ,\ )_1$ and $(\ ,\ )_2$, respectively, an invertible linear transformation $T : V_1 \to V_2$ is said to be an isometry if

$$(T(\boldsymbol{x}), T(\boldsymbol{y}))_2 = (\boldsymbol{x}, \boldsymbol{y})_1,$$

for all vectors $\boldsymbol{x}$ and $\boldsymbol{y}$ in V_1.

Just as $\boldsymbol{R}^n$ is the model real vector space, $\boldsymbol{R}^n$ with the standard inner product is the model real inner product space. We have the following theorem.

Theorem Let V_1 and V_2 be two real n-dimensional inner product spaces with inner product $(\ ,\ )_1$ and $(\ ,\ )_2$, respectively. Then, there is an isometry from V_1 onto V_2.

Proof Take $\boldsymbol{x}_1, \boldsymbol{x}_2, \ldots, \boldsymbol{x}_n$ as an orthonormal basis for V_1 and $\boldsymbol{y}_1, \boldsymbol{y}_2, \ldots, \boldsymbol{y}_n$ as an orthonormal basis for V_2. By §5.2, there is a linear transformation T from V_1 to V_2 such that $T(\boldsymbol{x}_1) = \boldsymbol{y}_1$, $T(\boldsymbol{x}_2) = \boldsymbol{y}_2, \ldots,$ $T(\boldsymbol{x}_n) = \boldsymbol{y}_n$. T is clearly an isomorphism. To see that T is an isometry let $\boldsymbol{x} = \alpha_1\boldsymbol{x}_1 + \alpha_2\boldsymbol{x}_2 + \cdots + \alpha_n\boldsymbol{x}_n$ and $\boldsymbol{y} = \beta_1\boldsymbol{x}_1 + \beta_2\boldsymbol{x}_2 + \cdots + \beta_n\boldsymbol{x}_n$ be vectors in V_1. Then,

$$
\begin{aligned}
(\boldsymbol{x}, \boldsymbol{y})_1 &= (\alpha_1\boldsymbol{x}_1 + \alpha_2\boldsymbol{x}_2 + \cdots + \alpha_n\boldsymbol{x}_n, \beta_1\boldsymbol{x}_1 + \beta_2\boldsymbol{x}_2 + \cdots + \beta_n\boldsymbol{x}_n) \\
&= \alpha_1\beta_1 + \alpha_2\beta_2 + \cdots + \alpha_n\beta_n
\end{aligned}
$$

since $\boldsymbol{x}_1, \boldsymbol{x}_2, \ldots, \boldsymbol{x}_n$ is an orthonormal basis, and

$$
\begin{aligned}
(T(\boldsymbol{x}), T(\boldsymbol{y}))_2 &= (\alpha_1\boldsymbol{y}_1 + \alpha_2\boldsymbol{y}_2 + \cdots + \alpha_n\boldsymbol{y}_n, \beta_1\boldsymbol{y}_1 + \beta_2\boldsymbol{y}_2 + \cdots + \beta_n\boldsymbol{y}_n) \\
&= \alpha_1\beta_1 + \alpha_2\beta_2 + \cdots + \alpha_n\beta_n
\end{aligned}
$$

since $\boldsymbol{y}_1, \boldsymbol{y}_2, \ldots, \boldsymbol{y}_n$ is an orthonormal basis. Thus, $(T(\boldsymbol{x}), T(\boldsymbol{y})) = (\boldsymbol{x}, \boldsymbol{y})$. Therefore, T is an isometry from V_1 to V_2. ∎

This theorem enables us to reduce many results about arbitrary inner product spaces to results about $\boldsymbol{R}^n$ with the familiar dot product as inner product.

It is also possible to define general complex inner products by abstracting the properties of the standard complex inner product on $\boldsymbol{C}^n$. All properties, definitions, theorems, and proofs are the same as before.

EXERCISES

1. Let $\boldsymbol{x} = \alpha_1\boldsymbol{i} + \alpha_2\boldsymbol{j}$ and $\boldsymbol{y} = \beta_1\boldsymbol{i} + \beta_2\boldsymbol{j}$. Show that the following are inner products on $\boldsymbol{R}^2$.
   (a) $(\boldsymbol{x}, \boldsymbol{y}) = 2\alpha_1\beta_1 + 3\alpha_2\beta_2$.
   (b) $(\boldsymbol{x}, \boldsymbol{y}) = 2\alpha_1\beta_1 + \alpha_1\beta_2 + \alpha_2\beta_1 + \alpha_1\beta_2$.
2. Let m be a continuous strictly positive function on the interval $[a, b]$. If f and g are polynomials in P_n, show that

$$
(f, g) = \int_a^b f(x)g(x)m(x)\,dx
$$

is an inner product on P_n.

3. On P_3, let the inner product of f and g be

$$(f, g) = \int_{-1}^{+1} f(x)g(x)\,dx$$

(a) Calculate

$$(1 - x, 1 + x + x^2), \qquad ((1 + x)^2, (1 - x)^2)$$

(b) Find the norm of 1, x, x^2, and x^3.
(c) Determine a basis for the orthogonal complement of the subspace of P_3 spanned by 1, x, and x^2.

4. Consider the inner product on the space of $n \times n$ matrices of Example 3 in the text. Show that the norm of the matrix A is the square root of the sum of the squares of the entries of A.

5. In P_n, let the inner product of f and g be given by

$$(f, g) = \int_{-1}^{+1} f(x)g(x)\,dx$$

(a) Show that the orthogonal complement of the subspace of even polynomials in P_n is the subspace of odd polynomials in P_n.
(b) Show that the linear transformation which carries $f(x)$ into $f(-x)$ is an isometry.

6. Let V be a real vector space. Show that the sum of two inner products on V is again an inner product on V.

7. Let V be a real finite-dimensional vector space and $(\ ,\ )$ be an inner product on V. Let $f : V \to R$ be a linear transformation from V into R. Show that there is a unique vector $\boldsymbol{y}$ in V such that for all $\boldsymbol{x}$ in V

$$f(\boldsymbol{x}) = (\boldsymbol{x}, \boldsymbol{y})$$

8. On P_2 let the inner product of two polynomials f and g be

$$(f, g) = \int_{-1}^{+1} f(x)g(x)\,dx$$

(a) Let L_a be the function from P_2 to R defined by $L_a(f) = f(a)$. Show that L is a linear transformation from P_2 into R.
(b) Let g_a be the polynomial

$$g_a(x) = \tfrac{1}{8}(9 - 15a^2) + \tfrac{3}{2}ax + \tfrac{1}{8}(45a^2 - 15)x^2$$

Show that $L_a(f) = (f, g_a)$.

9. Let $\boldsymbol{x}_1, \boldsymbol{x}_2, \ldots, \boldsymbol{x}_n$ be a basis for $\boldsymbol{R}^n$. Show that there is an inner product on $\boldsymbol{R}^n$ with respect to which $\boldsymbol{x}_1, \boldsymbol{x}_2, \ldots, \boldsymbol{x}_n$ is an orthonormal basis.

10. Regarding the complex numbers as a vector space over the reals, define $(\boldsymbol{z}_1, \boldsymbol{z}_2) = \frac{1}{2}(\boldsymbol{z}_1\bar{\boldsymbol{z}}_2 + \boldsymbol{z}_2\bar{\boldsymbol{z}}_1)$.
(a) Show that $(\ ,\ )$ is an inner product.

(b) If $z_1 = \alpha_1 + \alpha_2 i$ and $z_2 = \beta_1 + \beta_2 i$, show that

$$(z_1, z_2) = \alpha_1\beta_1 + \alpha_2\beta_2$$

Show that (z, z) is the square of the absolute value of the complex number z in the usual sense.

(c) Let M_a be the linear transformation of C^n into itself, defined by $M_a(z) = az$. Show that $(M_a z_1, M_a z_2) = a\bar{a}(z_1, z_2)$.

(d) With M_a defined as in (c) show that M_a is an isometry if and only if $|a| = 1$.

(e) Letting $T(z) = \bar{z}$, show that T is an isometry.

11. In the space of $n \times n$ matrices with the inner product of Example 3 in the text, let E_{ij} be the $n \times n$ matrix with 0 in all entries except (i, j) and 1 in entry (i, j). Show that the matrices E_{ij}, $1 \leq i \leq n$ and $1 \leq j \leq n$, form an orthonormal basis for M_{nn}.

12. In the space of $n \times n$ matrices with the inner product of Example 3 in the text, let T be the linear operator defined by $T(B) = AB$, where A is an $n \times n$ orthogonal matrix. Show that T is an isometry.

13. Let $x_1, x_2, \ldots, x_n$ be an orthonormal basis for R^n with the standard inner product. Show that the matrices $A_{ij} = x_i x_j^T$ form an orthonormal basis for the space of $n \times n$ matrices with the inner product of Example 3 in the text.

14. In the space P_n of polynomials of degree not exceeding n with real coefficients in a variable x, consider the inner product

$$(f, g)_1 = \int_a^b f(x)g(x)\,dx$$

$$(f, g)_2 = \int_c^d f(x)g(x)\,dx$$

where $a < b$ and $c < d$. Let T be the function from P_n (regarded as an inner product space with the first inner product) into P_n (regarded as an inner product space with the second inner product), defined by

$$(T(f))(x) = \sqrt{\frac{b-a}{d-c}}\, f\left(\frac{b-a}{d-c}x + \frac{ad-bc}{d-c}\right)$$

Show that T is an isometry.

15. Consider the space of $n \times n$ matrices with the inner product of Example 3 in the text.

(a) Show that the orthogonal complement of the subspace of diagonal matrices is the subspaces of matrices all of whose diagonal entries are 0.

(b) Show that the orthogonal complement of the subspace of symmetric matrices is the subspace of skew-symmetric matrices.

16. Show that all inner products on R^2 are of the form

$$\left(\begin{bmatrix} x_1 \\ x_2 \end{bmatrix}, \begin{bmatrix} y_1 \\ y_2 \end{bmatrix}\right) = \begin{bmatrix} x_1 \\ x_2 \end{bmatrix}^T \begin{bmatrix} a & b \\ b & c \end{bmatrix} \begin{bmatrix} y_1 \\ y_2 \end{bmatrix}$$

where $ac - b^2 > 0$ and $a > 0$, and conversely that each of these defines an inner product on $\boldsymbol{R}^2$.

17. Let $C[a, b]$ denote the collection of all real-valued continuous functions on the interval $[a, b]$.
    (a) Show that $C[a, b]$ is a vector space with ordinary addition and scalar multiplication of functions.
    (b) Letting

$$(f, g) = \int_a^b f(x)g(x)\,dx$$

    show that $(\ ,\ )$ is an inner product on $C[a, b]$.
    (c) If f and g are continuous functions on $[a, b]$, show that

$$\int_a^b |f(x)g(x)|\,dx \leq \left(\int_a^b f^2(x)\,dx\right)^{1/2}\left(\int_a^b g^2(x)\,dx\right)^{1/2}$$

    (d) Show that the functions

$$\sqrt{\frac{2}{\pi}}\sin x, \ldots, \sqrt{\frac{2}{\pi}}\sin nx$$

    form an orthonormal system in $C[0, \pi]$.
    (e) If f is a continuous function, show that

$$\left|\int_0^\pi f(x)\sin x\,dx\right|^2 + \cdots + \left|\int_0^\pi f(x)\sin nx\,dx\right|^2 \leq \frac{\pi}{2}\int_0^\pi (f(x))^2\,dx$$

7

EIGENVALUES AND CANONICAL FORMS

1 EIGENVALUES AND EIGENVECTORS

In many of the applications of the theory of linear transformations the following problem arises. Given a linear operator T on a vector space V, determine those scalars λ and those vectors $\boldsymbol{x}$ in V which satisfy the equation $T(\boldsymbol{x}) = \lambda\boldsymbol{x}$. This is known as the eigenvalue problem.

If T is a linear operator on a vector space V and $T(\boldsymbol{x}) = \lambda\boldsymbol{x}$ for some nonzero $\boldsymbol{x}$ in V and some scalar λ, λ is said to be an **eigenvalue** of T. The vector $\boldsymbol{x}$ is said to be an **eigenvector** of T corresponding to the eigenvalue λ. Often eigenvalues are called **characteristic values** and eigenvectors are called **characteristic vectors.** If A is an $n \times n$ matrix with real entries, a real number λ for which there is some nonzero real column n-vector $\boldsymbol{x}$ such that $A\boldsymbol{x} = \lambda\boldsymbol{x}$ is said to be an eigenvalue of A, and $\boldsymbol{x}$ is said to be an eigenvector of A. Complex eigenvalues and eigenvectors are defined similarly.

Intuitively, eigenvectors correspond to the fixed directions of the linear operator T.

Example 1 Let T be the linear operator on $\boldsymbol{R}^2$ defined by $T(\boldsymbol{x}) = A\boldsymbol{x}$, where A is the 2×2 matrix

$$A = \begin{bmatrix} 4 & -1 \\ 2 & 1 \end{bmatrix}$$

Since

$$T\left(\begin{bmatrix} 1 \\ 1 \end{bmatrix}\right) = \begin{bmatrix} 4 & -1 \\ 2 & 1 \end{bmatrix}\begin{bmatrix} 1 \\ 1 \end{bmatrix} = 3\begin{bmatrix} 1 \\ 1 \end{bmatrix}$$

and

$$T\left(\begin{bmatrix}1\\2\end{bmatrix}\right) = \begin{bmatrix}4 & -1\\2 & 1\end{bmatrix}\begin{bmatrix}1\\2\end{bmatrix} = 2\begin{bmatrix}1\\2\end{bmatrix}$$

it follows that 3 and 2 are eigenvalues of T with corresponding eigenvectors

$$\begin{bmatrix}1\\1\end{bmatrix} \quad \text{and} \quad \begin{bmatrix}1\\2\end{bmatrix}$$

respectively.

Thus, the linear operator T fixes the lines through the origin determined by the scalar multiples of the above vectors. (See Figure 1-1.)

Since $T(\boldsymbol{x}) = \lambda\boldsymbol{x}$ is equivalent to $(T - \lambda I)(\boldsymbol{x}) = \boldsymbol{0}$, it follows that $\boldsymbol{x}$ is an eigenvector of T corresponding to the eigenvalue λ if and only if $\boldsymbol{x}$ belongs to the nullspace of $\lambda I - T$. We call the collection of all eigenvectors of T corresponding to the eigenvalue λ the **eigenspace** of T corresponding to the eigenvalue λ. It follows that the eigenspace of T corresponding to the eigenvalue λ is just the nullspace of $\lambda I - T$. In particular, it is clear that an eigenspace is a subspace of the vector space upon which T operates.

Example 2 Let T be the linear operator on P_n, defined by

$$(T(f))(x) = f(-x)$$

Note that 1 is an eigenvalue of T. Since $(T(f))(x) = f(x)$ if and only if $f(-x) = f(x)$, we see that the eigenspace of T corresponding to the eigenvalue 1 is precisely the subspace of even polynomials in P_n.

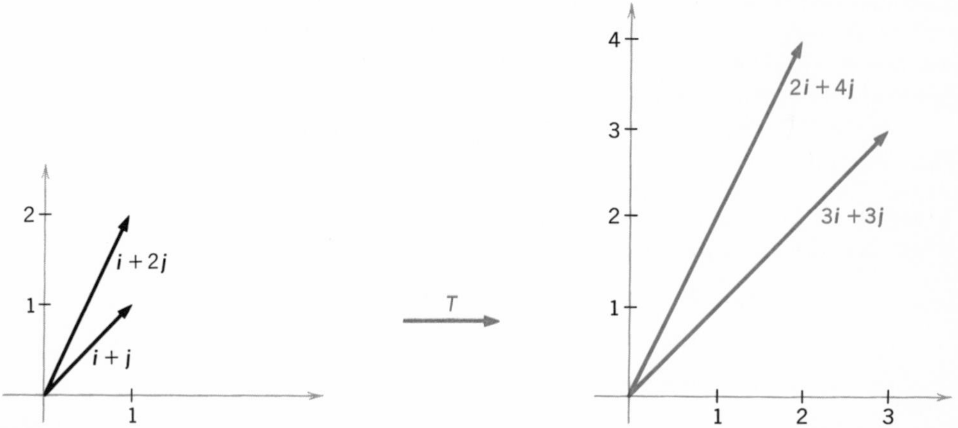

FIGURE 1-1

Similar reasoning shows that if $n \geq 1$, -1 is an eigenvalue of T with eigenspace the subspace of odd polynomials in P_n.

A means of determining which scalars are eigenvalues of a given linear operator is furnished by the following theorem.

Theorem 1 Let T be a linear operator on a real vector space V. Let λ be a real number. There is a nonzero vector $\boldsymbol{x}$ in V such that $T(\boldsymbol{x}) = \lambda\boldsymbol{x}$ if and only if $\det[T - \lambda I] = 0$.

Proof Note that the nullspace of $T - \lambda I$ is nonzero if and only if $\det[T - \lambda I] = 0$. ∎

An analogous result holds if the word "real" in the above theorem is replaced throughout by "complex."

The polynomial $f(x) = \det[xI - T]$ is called the **characteristic polynomial** of T and is denoted by C_T. It is calculated by choosing a basis for the vector space V and determining the matrix representation of the operator T relative to this basis. If A represents T relative to the chosen basis, $C_T(x) = \det[xI_n - A]$. If A' represents T with respect to another basis, in §5.8 we saw that there is an invertible matrix B such that $A' = BAB^{-1}$.

Then,

$$\begin{aligned}\det[xI_n - A'] &= \det[xI_n - BAB^{-1}] \\ &= \det[B(xI_n - A)B^{-1}] \\ &= \det B \det[xI_n - A] \det B^{-1} \\ &= \det[xI_n - A]\end{aligned}$$

Consequently, the characteristic polynomial of T is independent of the basis used to compute it.

In a similar vein we denote the characteristic polynomial of the matrix A by C_A. The previous paragraph implies that similar matrices have the same characteristic polynomial.

From our definition of the characteristic polynomial and from Theorem 1, it follows that the roots of the characteristic polynomial are precisely the eigenvalues of the linear operator.

If $A = [a_{ij}]_{(nn)}$ is an $n \times n$ matrix,

$$C_A(x) = \begin{vmatrix} x - a_{11} & -a_{12} & -a_{13} & \cdots & -a_{1n} \\ -a_{21} & x - a_{22} & -a_{23} & \cdots & -a_{2n} \\ -a_{31} & -a_{32} & x - a_{33} & \cdots & -a_{3n} \\ & & \vdots & & \\ -a_{n1} & -a_{n2} & -a_{n3} & \cdots & x - a_{nn} \end{vmatrix}$$

It is clear that $C_A(x)$ is a polynomial of degree n with leading coefficient 1:

$$C_A(x) = x^n + a_{n-1}x^{n-1} + \cdots + a_1x + a_0$$

Expanding the above determinant for $C_A(x)$ yields that the coefficient of x^{n-1} is

$$-(a_{11} + a_{22} + \cdots + a_{nn}) = -\text{tr } A$$

The constant term a_0 in the above polynomial may be calculated by setting $x = 0$. Then, $a_0 = C_A(0) = \det -A$, or $a_0 = (-1)^n \det A$.

Example 3 Let A be a 2×2 matrix, say

$$A = \begin{bmatrix} a & b \\ c & d \end{bmatrix}$$

Then,

$$\begin{aligned} C_A(x) &= \begin{vmatrix} x - a & -b \\ -c & x - d \end{vmatrix} \\ &= x^2 - (a + d)x + ad - bc \\ &= x^2 - (\text{tr } A)x + \det A \end{aligned}$$

Thus in the two-dimensional case the characteristic polynomial of A is completely determined by the determinant and trace of A. The eigenvalues of A may be obtained by finding the solutions of the quadratic equation $x^2 - (\text{tr } A)x + \det A = 0$.

Example 4 If

$$A = \begin{bmatrix} 0 & -4 & 0 \\ 1 & 4 & 0 \\ 0 & 0 & 2 \end{bmatrix}$$

$$C_A(x) = \det [xI_3 - A] = \begin{bmatrix} x & 4 & 0 \\ -1 & x - 4 & 0 \\ 0 & 0 & x - 2 \end{bmatrix} = (x - 2)(x^2 - 4x + 4)$$

$$= (x - 2)^3$$

It follows that the only eigenvalue of A is 2. The eigenvectors of A

corresponding to the eigenvalue 2 may be found by determining the nullspace of $2I_3 - A$. This leads to the system of linear equations

$$\begin{aligned} 2x + 4y \quad &= 0 \\ -x - 2y \quad &= 0 \\ 0 \cdot z &= 0 \end{aligned}$$

A basis for the eigenspace associated with 2 is found to be

$$\begin{bmatrix} -2 \\ 1 \\ 0 \end{bmatrix} \quad \text{and} \quad \begin{bmatrix} 0 \\ 0 \\ 1 \end{bmatrix}$$

If I_n is the identity matrix of order n, the characteristic polynomial of I_n is $(x - 1)^n$. More generally, if D is an $n \times n$ diagonal matrix with diagonal entries $d_1, d_2, \ldots, d_n$, the characteristic polynomial of D is $(x - d_1)(x - d_2) \ldots (x - d_n)$.

Other properties of the characteristic polynomial are exhibited in the following theorem.

Theorem 2 Let A be an $n \times n$ matrix. Then,

(i) $C_{A^T}(x) = C_A(x)$

and if A^{-1} exists,

(ii) $C_{A^{-1}}(x) = (\det A)^{-1}(-x)^n C_A(1/x)$

Proof

$$\begin{aligned} \text{(i) } C_{A^T}(x) = \det [xI_n - A^T] &= \det [(xI_n - A)^T] \\ &= \det [xI_n - A] \\ &= C_A(x) \\ \text{(ii) } C_{A^{-1}}(x) = \det [xI_n - A^{-1}] &= \det [A^{-1}(xA - I_n)] \\ &= (\det A)^{-1} \det [xA - I_n] \\ &= (\det A)^{-1}(-x)^n \det \left[\frac{1}{x} I_n - A\right] \\ &= (\det A)^{-1}(-x)^n C_A(1/x) \end{aligned}$$

Example 5 A 3×3 orthogonal matrix U is said to be a **rotation matrix** if $\det U = 1$. We will show that 1 is an eigenvalue of any 3×3 rotation matrix. Geometrically this means that every rotation of $\boldsymbol{R}^3$ fixes a vector, that is, the eigenvector corresponding to the eigenvalue 1. Thus a rotation matrix moves the points of space about some fixed axis through the origin.

We must show that $C_U(1) = 0$. Using Theorem 3, and the fact that $UU^T = I_3$,

$$C_U(1) = C_{U^T}(1) = C_{U^{-1}}(1) = 1(-1)^3 C_U(1) = -C_U(1)$$

Consequently, $C_U(1) = 0$.

A similar argument shows that a 3×3 orthogonal matrix with determinant -1 has -1 as an eigenvalue.

It is important to note that a real matrix may have no real eigenvalues. For example, consider the matrix

$$B_\theta = \begin{bmatrix} \cos\theta & -\sin\theta \\ \sin\theta & \cos\theta \end{bmatrix}$$

We have seen that B_θ induces a rotation of the plane by θ degrees. If $0 < \theta < \pi$, it is clear that no vector in $\boldsymbol{R}^2$ is carried into a scalar multiple of itself. Thus, B_θ has no real eigenvalues.

In this case the characteristic polynomial is $x^2 - (2\cos\theta)x + 1$. Its roots are $\cos\theta + i\sin\theta$ and $\cos\theta - i\sin\theta$. If $0 < \theta < \pi$, $\sin\theta \neq 0$, and neither root is real, a conclusion which was to be expected from the above geometric considerations.

It is known that any polynomial with complex coefficients has a complex root. Thus, any $n \times n$ matrix with complex entries has some complex eigenvalue. Consequently, any real matrix has some complex eigenvalue. Sometimes this fact, coupled with other information about a given matrix, can be used to show that the matrix has a real eigenvalue.

There is an important relationship between eigenvectors and linear independence which is brought forth in the following theorem.

Theorem 3 Let T be a linear operator on a vector space V. Let $\lambda_1, \lambda_2, \ldots, \lambda_n$ be distinct eigenvalues of T and let $\boldsymbol{x}_1, \boldsymbol{x}_2, \ldots, \boldsymbol{x}_n$ be eigenvectors (necessarily nonzero vectors) corresponding to the eigenvalues $\lambda_1, \lambda_2, \ldots, \lambda_n$, respectively. Then, $\boldsymbol{x}_1, \boldsymbol{x}_2, \ldots, \boldsymbol{x}_n$ are linearly independent.

Proof Suppose $\alpha_1\boldsymbol{x}_1 + \alpha_2\boldsymbol{x}_2 + \cdots + \alpha_n\boldsymbol{x}_n = \boldsymbol{0}$. Applying T to this equation and using $T(\boldsymbol{x}_i) = \lambda_i\boldsymbol{x}_i$, we get

$$\alpha_1\lambda_1\boldsymbol{x}_1 + \alpha_2\lambda_2\boldsymbol{x}_2 + \cdots + \alpha_n\lambda_n\boldsymbol{x}_n = \boldsymbol{0}$$

Multiplying the first of the above equations by λ_1 and subtracting from the second, we obtain

$$\alpha_2(\lambda_2 - \lambda_1)\boldsymbol{x}_2 + \alpha_3(\lambda_3 - \lambda_1)\boldsymbol{x}_3 + \cdots + \alpha_n(\lambda_n - \lambda_1)\boldsymbol{x}_n = \boldsymbol{0}$$

Applying T to this equation,

$$\alpha_2(\lambda_2 - \lambda_1)\lambda_2\boldsymbol{x}_2 + \alpha_3(\lambda_3 - \lambda_1)\lambda_3\boldsymbol{x}_3 + \cdots + \alpha_n(\lambda_n - \lambda_1)\lambda_n\boldsymbol{x}_n = \boldsymbol{0}$$

Multiplying the first of the preceding two equations by λ_2 and subtracting from the second, we obtain

$$\alpha_3(\lambda_3 - \lambda_1)(\lambda_3 - \lambda_2)\boldsymbol{x}_3 + \cdots + \alpha_n(\lambda_n - \lambda_1)(\lambda_n - \lambda_2)\boldsymbol{x}_n = \boldsymbol{0}$$

By repeating this process we eventually obtain,

$$\alpha_n(\lambda_n - \lambda_1)(\lambda_n - \lambda_2)\ldots(\lambda_n - \lambda_{n-1})\boldsymbol{x}_n = \boldsymbol{0}$$

Since $\lambda_1, \lambda_2, \ldots, \lambda_n$ are distinct, $(\lambda_n - \lambda_1)(\lambda_n - \lambda_2)\ldots(\lambda_n - \lambda_{n-1}) \neq 0$, and since it is also true that $\boldsymbol{x}_n \neq \boldsymbol{0}$, it follows that $\alpha_n = 0$.

Thus, $\alpha_1\boldsymbol{x}_1 + \cdots + \alpha_{n-1}\boldsymbol{x}_{n-1} = \boldsymbol{0}$, Using the same procedure on this equation repeatedly yields

$$\alpha_1 = \alpha_2 = \cdots = \alpha_{n-1} = 0$$

Thus, we have shown that whenever $\alpha_1\boldsymbol{x}_1 + \alpha_2\boldsymbol{x}_2 + \cdots + \alpha_n\boldsymbol{x}_n = \boldsymbol{0}$, $\alpha_1 = \alpha_2 = \cdots \alpha_n = 0$, and so $\boldsymbol{x}_1, \boldsymbol{x}_2, \ldots, \boldsymbol{x}_n$ are linearly independent. ∎

Two real matrices C and D are said to be **similar** over the reals if there is a real invertible matrix B such that $BCB^{-1} = D$. If C and D are complex matrices and B is an invertible complex matrix such that $BCB^{-1} = D$, C and D are said to be similar over the complexes. We say that a matrix is **diagonalizable** over the reals if it is similar over the reals to some real diagonal matrix. An analogous definition is made for diagonalizability over the complexes.

Part of the importance of diagonal matrices stems from the ease with which they may be manipulated. For example, if C is diagonalizable and $BCB^{-1} = D$ is diagonal, to calculate C^n, we need only calculate D^n, which is readily done and form the product $B^{-1}D^nB = C^n$.

More specifically, let

$$C = \begin{bmatrix} 4 & -2 \\ 1 & 1 \end{bmatrix}, \quad B = \begin{bmatrix} 1 & -1 \\ -1 & 2 \end{bmatrix}, \quad D = \begin{bmatrix} 3 & 0 \\ 0 & 2 \end{bmatrix}$$

Then,

$$B^{-1} = \begin{bmatrix} 2 & 1 \\ 1 & 1 \end{bmatrix}$$

$$BCB^{-1} = \begin{bmatrix} 1 & -1 \\ -1 & 2 \end{bmatrix}\begin{bmatrix} 4 & -2 \\ 1 & 1 \end{bmatrix}\begin{bmatrix} 2 & 1 \\ 1 & 1 \end{bmatrix}$$

$$= \begin{bmatrix} 3 & -3 \\ -2 & 4 \end{bmatrix} \begin{bmatrix} 2 & 1 \\ 1 & 1 \end{bmatrix}$$

$$= \begin{bmatrix} 3 & 0 \\ 0 & 2 \end{bmatrix}$$

Thus,

$$C^n = B^{-1} \begin{bmatrix} 3^n & 0 \\ 0 & 2^n \end{bmatrix} B$$

$$= \begin{bmatrix} 2 & 1 \\ 1 & 1 \end{bmatrix} \begin{bmatrix} 3^n & 0 \\ 0 & 2^n \end{bmatrix} \begin{bmatrix} 1 & -1 \\ -1 & 2 \end{bmatrix}$$

$$= \begin{bmatrix} 2 \cdot 3^n - 2^n & -2 \cdot 3^n + 2^{n+1} \\ 3^n - 2^n & -3^n + 2^{n+1} \end{bmatrix}$$

A similar procedure yields

$$f(C) = \begin{bmatrix} 2f(3) - f(2) & -2f(3) + 2f(2) \\ f(3) - f(2) & -f(3) + 2 \cdot f(2) \end{bmatrix}$$

if f is any polynomial.

The computational advantages of diagonalizability are therefore apparent.

We say that a linear operator T on a vector space V is diagonalizable if there is some basis for V such that the matrix representation of T is diagonal. Since the matrix representations of T with respect to the various bases of V are all similar, T is diagonalizable if and only if one (and hence all) of its matrix representations is diagonalizable.

It is clear that a linear operator T on a vector space V is diagonalizable if and only if there is a basis for V consisting of eigenvectors of T. Framing this statement in terms of matrices, we see the following:

An $n \times n$ matrix A is diagonalizable over the reals if and only if $\boldsymbol{R}^n$ has a basis consisting of eigenvectors of A.

An analogous statement holds for the diagonalization of complex matrices.

There is one circumstance in which information about the characteristic polynomial implies diagonalizability, namely, Theorem 4.

Theorem 4 Let A be an $n \times n$ matrix with real entries and suppose $C_A(x) = (x - \lambda_1)(x - \lambda_2) \dots (x - \lambda_n)$, where $\lambda_1, \dots, \lambda_n$ are distinct real numbers. Then, A is diagonalizable over the reals.

Proof Take $\boldsymbol{x}_1, \ldots, \boldsymbol{x}_n$ eigenvectors of A corresponding to the eigenvalues $\lambda_1, \lambda_2, \ldots, \lambda_n$, respectively. By Theorem 3 above and by virtue of the fact that $\lambda_1, \lambda_2, \ldots, \lambda_n$ are distinct numbers, $\boldsymbol{x}_1, \boldsymbol{x}_2, \ldots, \boldsymbol{x}_n$ are linearly independent. But by Theorem 2, §4.8, $\boldsymbol{x}_1, \boldsymbol{x}_2, \ldots, \boldsymbol{x}_n$ also span $\boldsymbol{R}^n$, and thus $\boldsymbol{R}^n$ has a basis of eigenvectors of A. Consequently, A is diagonalizable. ■

In this case we may also provide an explicit means for calculating the matrix B such that $B^{-1}AB = D$ is diagonal. Indeed, let $\boldsymbol{x}_1, \ldots, \boldsymbol{x}_n$ be the eigenvectors corresponding to $\lambda_1, \ldots, \lambda_n$. These vectors may be found by solving the system of linear equations $(A - \lambda_i)\boldsymbol{x} = \boldsymbol{0}$.

Let $B = [\boldsymbol{x}_1, \boldsymbol{x}_2, \ldots, \boldsymbol{x}_n]$, that is, B is the matrix whose ith column is $\boldsymbol{x}_i$. Then $B\boldsymbol{e}_i = \boldsymbol{x}_i$ and $B^{-1}\boldsymbol{x}_i = \boldsymbol{e}_i$. Consequently,

$$
\begin{aligned}
(B^{-1}AB)\boldsymbol{e}_i &= B^{-1}(A\boldsymbol{x}_i) \\
&= B^{-1}(\lambda_i \boldsymbol{x}_i) \\
&= \lambda_i \boldsymbol{e}_i
\end{aligned}
$$

Thus, $B^{-1}AB$ is a diagonal matrix with successive diagonal entries $\lambda_1, \lambda_2, \ldots, \lambda_n$.

Example 6 Let

$$
A = \begin{bmatrix} -2 & 3 & 1 \\ 0 & 1 & 1 \\ -3 & 4 & 1 \end{bmatrix}
$$

Then,

$$
C_A(x) = \begin{vmatrix} x+2 & -3 & -1 \\ 0 & x-1 & -1 \\ 3 & -4 & x-1 \end{vmatrix} = x^3 - 4x = x(x-2)(x+2)
$$

Thus, the eigenvalues of A are 0, 2, and -2, which are real and distinct. It follows that there is a real invertible 3×3 matrix B such that

$$
B^{-1}AB = \begin{bmatrix} 0 & 0 & 0 \\ 0 & 2 & 0 \\ 0 & 0 & -2 \end{bmatrix}
$$

B may be determined explicitly by finding the eigenvectors corresponding to 0, 2, and -2 and writing these as successive columns of the matrix B.

Example 7 Let

$$B_\theta = \begin{bmatrix} \cos\theta & -\sin\theta \\ \sin\theta & \cos\theta \end{bmatrix}$$

Earlier we saw that B_θ has the complex eigenvalues

$$\cos\theta + i\sin\theta, \qquad \cos\theta - i\sin\theta$$

Thus, by Theorem 4, there is a complex matrix C such that

$$CB_\theta C^{-1} = \begin{bmatrix} \cos\theta + i\sin\theta & 0 \\ 0 & \cos\theta - i\sin\theta \end{bmatrix}$$

Note that though B_θ has no real eigenvalues, it is diagonalizable over the complexes.

Example 8 Let

$$A = \begin{bmatrix} 0 & -4 & 0 \\ 1 & 4 & 0 \\ 0 & 0 & 2 \end{bmatrix}$$

be the matrix of Example 4, §7.1. We saw that $C_A(x) = (x - 2)^3$. We will show that A is not diagonalizable.

Indeed, suppose A were diagonalizable (over the reals or complexes). Then there would be some invertible matrix B such that

$$BAB^{-1} = \begin{bmatrix} d_1 & 0 & 0 \\ 0 & d_2 & 0 \\ 0 & 0 & d_3 \end{bmatrix}$$

Thus,

$$C_{BAB^{-1}}(x) = (x - d_1)(x - d_2)(x - d_3)$$

Since similar matrices have the same characteristic polynomial,

$$(x - d_1)(x - d_2)(x - d_3) = (x - 2)^3$$

Thus,

$$d_1 = d_2 = d_3 = 2$$

and

$$BAB^{-1} = 2I_3$$

It would follow that $A = 2I_3$.

Since A obviously is not $2I_3$, it follows that A cannot be diagonalizable.

EXERCISES

1. Determine the characteristic polynomial and eigenvalues of the following matrices.

(a) $\begin{bmatrix} 0 & 1 \\ 1 & 0 \end{bmatrix}$ (b) $\begin{bmatrix} 1 & 1 \\ 0 & 1 \end{bmatrix}$ (c) $\begin{bmatrix} 0 & -1 \\ 1 & -1 \end{bmatrix}$

(d) $\begin{bmatrix} a & b & c \\ 0 & d & e \\ 0 & 0 & f \end{bmatrix}$, a, b, c, d, e, f are real numbers (e) $\begin{bmatrix} 0 & 1 & 1 \\ 1 & 0 & 1 \\ 1 & 1 & 0 \end{bmatrix}$

2. Show that the eigenvalues of the matrix

$$\begin{bmatrix} \cos\theta & \sin\theta \\ \sin\theta & -\cos\theta \end{bmatrix}$$

are ± 1, with corresponding eigenvectors

$$\begin{bmatrix} \cos\frac{1}{2}\theta \\ \sin\frac{1}{2}\theta \end{bmatrix} \quad \text{and} \quad \begin{bmatrix} \sin\frac{1}{2}\theta \\ -\cos\frac{1}{2}\theta \end{bmatrix}$$

Is the matrix diagonalizable?

3. Determine the eigenvalues and corresponding eigenspaces of the matrix

$$\begin{bmatrix} 1 & 0 & 2 \\ 1 & 1 & -1 \\ -1 & 0 & -2 \end{bmatrix}$$

4. Find all eigenvectors of the matrix

$$\begin{bmatrix} 1 & 1 & 1 \\ 0 & 1 & 1 \\ 0 & 0 & 1 \end{bmatrix}$$

5. If A is an upper triangular matrix show that $C_A(x) = (x - a_{11})(x - a_{22}) \ldots (x - a_{nn})$.

6. If a is any number, show that the matrix

$$\begin{bmatrix} 1-a & 1 \\ a-a^2 & a \end{bmatrix}$$

is similar to the matrix

$$\begin{bmatrix} 1 & 0 \\ 0 & 0 \end{bmatrix}$$

7. If A is a matrix and the sum of the entries in each row of A is σ, show that σ is an eigenvalue of A.
8. Determine the characteristic polynomial of the matrix

$$\begin{bmatrix} 0 & a & b \\ -a & 0 & c \\ -b & -c & 0 \end{bmatrix}$$

9. Determine which lines through the origin are carried into themselves by the linear transformation of $\boldsymbol{R}^2$ induced by the matrix

$$\begin{bmatrix} 2 & -1 \\ 3 & -2 \end{bmatrix}$$

10. Show that a linear transformation of $\boldsymbol{R}^2$ is diagonalizable if and only if it carries two distinct lines through the origin into themselves.
11. Show that two diagonalizable matrices are similar if and only if they have the same characteristic polynomial.
12. (a) If A is an $n \times n$ matrix and $A^T = \lambda A$, show that $\lambda = \pm 1$, if $A \neq 0$.
    (b) Show that the only eigenvalues of the linear transformation of the space of $n \times n$ matrices into itself defined by $T(A) = A^T$ are ± 1.
    (c) Determine the eigenspaces of each eigenvalue.
13. Find the eigenvalues of the matrix

$$\begin{bmatrix} 1 & 1 & 1 \\ 1 & \omega^2 & \omega^2 \\ 1 & \omega^2 & \omega \end{bmatrix}$$

where

$$\omega = e^{2\pi i/3} = \cos(2\pi/3) + i \sin(2\pi/3)$$

14. If T is an invertible linear operator on a vector space V, show that all eigenvalues of T are nonzero. Show also that λ is an eigenvalue of T if and only if λ^{-1} is an eigenvalue of T^{-1}.
15. If T is a diagonalizable linear operator on a vector space V of finite

dimension and if the characteristic polynomial of T has only one root, show that T is a scalar multiple of the identity.

16. Let A and B be two $n \times n$ matrices (real or complex), and let x be a fixed scalar.
    (a) Show that if λ is sufficiently large,

$$\det [xI_n - (A - \lambda I_n)B] = \det [xI_n - B(A - \lambda I_n)]$$

    [Hint: If λ is big enough, $A - \lambda I_n$ is invertible.]
    (b) Using part (a) show that for all λ

$$\det [xI_n - (A - \lambda I_n)B] = \det [xI_n - B(A - \lambda I_n)]$$

    [Hint: Two polynomials which are equal at infinitely many points are equal everywhere.]
    (c) Show that $C_{AB}(x) = C_{BA}(x)$.

17. Let A be an $n \times n$ real matrix with n real distinct eigenvalues. If B is real and commutes with A, show that B is diagonalizable.

18. If

$$A = \begin{bmatrix} a_{11} & a_{12} & a_{13} \\ a_{21} & a_{22} & a_{23} \\ a_{31} & a_{32} & a_{33} \end{bmatrix}$$

show that the coefficient of x in $C_A(x)$ is

$$\begin{vmatrix} a_{11} & a_{12} \\ a_{21} & a_{22} \end{vmatrix} + \begin{vmatrix} a_{11} & a_{12} \\ a_{31} & a_{33} \end{vmatrix} + \begin{vmatrix} a_{22} & a_{23} \\ a_{32} & a_{33} \end{vmatrix}$$

19. Let A be an $n \times n$ matrix with complex entries. Let E_λ be the eigenspace of A corresponding to the eigenvalue λ. Let T_A be the linear operator on the space of $n \times n$ matrices defined by $T_A(B) = AB$, for all $n \times n$ matrices B.
    (a) Show that $C_{T_A}(x) = (C_A(x))^n$.
    (b) If F_λ denotes the eigenspace of T_A corresponding to the eigenvalue λ, show that a matrix B belongs to F_λ if and only if each of its columns belong to E_λ.
    (c) Show that $\dim F_\lambda = n \dim E_\lambda$.
    (d) Show that T_A is diagonalizable if and only if A is diagonalizable.

20. Recall that a minor of a matrix of order r is the determinant of some $r \times r$ submatrix. A principal minor of a square matrix is obtained by deleting rows and columns with the same subscripts. For example, the 2×2 principal minors of

$$\begin{bmatrix} a_1 & b_1 & c_1 \\ a_2 & b_2 & c_2 \\ a_3 & b_3 & c_3 \end{bmatrix}$$

are

$$\begin{vmatrix} a_1 & b_1 \\ a_2 & b_2 \end{vmatrix}, \qquad \begin{vmatrix} a_1 & c_1 \\ a_3 & c_3 \end{vmatrix}, \qquad \begin{vmatrix} b_2 & c_2 \\ b_3 & c_3 \end{vmatrix}$$

If A is an $n \times n$ matrix, show that the coefficient of x^{n-r} in $C_A(x)$ is $(-1)^r$ times the sum of all principal $r \times r$ minors of A.

21. On the space P_n, show that the linear operator $T(f) = xf'$, where f' is the derivative of f, is diagonalizable.
22. If A is an $n \times n$ matrix, show the following.
    (a) $C_{A+\alpha I_n}(x) = C_A(x - \alpha)$.
    (b) If $\alpha \neq 0$, $C_{\alpha A}(x) = \alpha^n C_A(\alpha^{-1}x)$.
23. If A is an $n \times n$ matrix with complex entries, show that

$$C_{A^*}(x) = \overline{C_A(\overline{x})}$$

where $\overline{a}$ denotes the complex conjugate of a.

24. Let T be the linear operator on $\boldsymbol{R}^{2n}$ defined by $T(e_i) = e_{2n-i+1}$ for all i.
    (a) Show that the matrix representation of T is

$$\begin{bmatrix} 0 & 0 & \cdots & 0 & 1 \\ 0 & 0 & \cdots & 1 & 0 \\ & & \vdots & & \\ 0 & 1 & \cdots & 0 & 0 \\ 1 & 0 & \cdots & 0 & 0 \end{bmatrix}$$

(b) Show that

$$e_1 + e_{2n}, \qquad e_2 + e_{2n-1}, \qquad \ldots, \qquad e_{n-1} + e_n$$

are independent eigenvectors corresponding to the eigenvalue 1, and that

$$e_1 - e_{2n}, \qquad e_2 - e_{2n-1}, \qquad \ldots, \qquad e_{n-1} - e_n$$

are independent eigenvectors corresponding to the eigenvalue -1.

(c) Show that T is diagonalizable.

(d) Show that $C_T(x) = (x - 1)^n(x + 1)^n$.

2 SYMMETRIC MATRICES

If A and B are two $n \times n$ matrices and there is an $n \times n$ orthogonal matrix U, such that $A = UBU^{-1}$, we say that A and B are **orthogonally similar**. Since U is orthogonal, $U^{-1} = U^T$ and so $A = UBU^T$. If U is unitary and $A = UBU^{-1}$, we say that A and B are **unitarily similar**.

In this section we show that any real symmetric matrix is orthogonally similar to a diagonal matrix.

In this section we denote by (,), the standard inner product on $\mathbf{R}^n$ and/or $\mathbf{C}^n$, defined in §6.2. If A is a symmetric matrix and $T(\mathbf{x}) = A\mathbf{x}$ is the linear operator induced by A,

$$(T(\mathbf{x}), \mathbf{y}) = (T(\mathbf{x}))^T\mathbf{y} = (A\mathbf{x})^T\mathbf{y} = \mathbf{x}^TA^T\mathbf{y} = \mathbf{x}^TA\mathbf{y} = (\mathbf{x}, T(\mathbf{y}))$$

Our first objective is to show that all eigenvalues of a symmetric operator are real.

Theorem 1 Let A be an $n \times n$ real symmetric matrix. Then any eigenvalue of A is real.

Proof Let T be the linear operator on $\mathbf{C}^n$ defined by $T(\mathbf{x}) = A\mathbf{x}$. Let (,) be the standard inner product on $\mathbf{C}^n$, i.e., $(\mathbf{x}, \mathbf{y}) = \mathbf{x}^*\mathbf{y}$, for all $\mathbf{x}$ and $\mathbf{y}$ in $\mathbf{C}^n$. As above, $(T(\mathbf{x}), \mathbf{y}) = (\mathbf{x}, T(\mathbf{y}))$.

If $T(\mathbf{x}) = \lambda\mathbf{x}$,

$$(T(\mathbf{x}), \mathbf{x}) = (\lambda\mathbf{x}, \mathbf{x}) = \lambda(\mathbf{x}, \mathbf{x})$$

and

$$(\mathbf{x}, T\mathbf{x})) = (\mathbf{x}, \lambda\mathbf{x}) = \bar{\lambda}(\mathbf{x}, \mathbf{x})$$

Since

$$(T(\mathbf{x}), \mathbf{x}) = (\mathbf{x}, T(\mathbf{x})), \text{ as } T \text{ is symmetric}$$
$$\lambda(\mathbf{x}, \mathbf{x}) = \bar{\lambda}(\mathbf{x}, \mathbf{x})$$

Since

$$\mathbf{x} \neq \mathbf{0},\ (\mathbf{x}, \mathbf{x}) \neq 0, \text{ and } \lambda = \bar{\lambda}. \text{ Thus, } \lambda \text{ is real.}$$ ■

Note that the same result holds if A is an $n \times n$ Hermitian matrix, that is, $A = A^*$. Indeed,

$$(A\mathbf{x}, \mathbf{y}) = (A\mathbf{x})^*\mathbf{y} = \mathbf{x}^*A^*\mathbf{y} = \mathbf{x}^*A\mathbf{y} = (\mathbf{x}, A\mathbf{y})$$

If $A\mathbf{x} = \lambda\mathbf{x}$, the proof above yields $\lambda = \bar{\lambda}$.

If T is a symmetric operator on an inner product space, by observing that the matrix representation of T with respect to an orthonomal basis is Hermitian, it follows that all eigenvalues of T are real.

We say that a matrix is **orthogonally diagonalizable** if it is orthogonally similar to a diagonal matrix. In §7.1 we saw that a matrix A is diagonalizable over the reals if and only if $\mathbf{R}^n$ admits a basis of eigenvectors of A. In the same way it may be shown that a matrix A is orthogonally diagonalizable if and only if $\mathbf{R}^n$ admits an orthonormal basis of eigenvectors of A. With the same idea in mind we say that a linear operator T on an inner product space V is orthogonally diagonalizable if V admits an orthonormal basis of T eigenvectors.

We now prove the main theorem of this section.

Theorem 2 Let T be a symmetric linear operator on a finite-dimensional inner product space V. Then V has an orthonormal basis of eigenvectors of T.

Proof The proof is by induction on dim V. The theorem is obvious if dim $V = 1$.

We know that T has a (possibly complex) eigenvalue. (Take any root of the characteristic polynomial.) By Theorem 1, this root is real. Thus, T has a real eigenvalue. Let $\boldsymbol{x}$ be an eigenvector corresponding to this eigenvalue. So we have $T(\boldsymbol{x}) = \lambda\boldsymbol{x}$, with λ real. We may assume $|\boldsymbol{x}| = 1$.

Let H be the orthogonal complement of $\boldsymbol{x}$ in V. We claim: If $\boldsymbol{h}$ belongs to H, so does $T(\boldsymbol{h})$.

Indeed,

$$\begin{aligned}(T(\boldsymbol{h}), \boldsymbol{x}) &= (\boldsymbol{h}, T(\boldsymbol{x})) && \text{(by symmetry of } T\text{)}\\ &= (\boldsymbol{h}, \lambda\boldsymbol{x})\\ &= \lambda(\boldsymbol{h}, \boldsymbol{x})\\ &= 0 && \text{(since } \boldsymbol{h} \text{ belongs to } H\text{)}\end{aligned}$$

Thus, $T(\boldsymbol{h})$ belongs to H, as well.

It follows that if T is restricted to H, T is a linear operator on H. Since $\dim H = \dim V - 1$, by induction hypothesis, H has an orthonormal basis of T eigenvectors, say $\boldsymbol{x}_2, \boldsymbol{x}_3, \ldots, \boldsymbol{x}_n$. Thus, $\boldsymbol{x}, \boldsymbol{x}_2, \ldots, \boldsymbol{x}_n$ is an orthonormal basis for V of eigenvectors of T. ∎

As a consequence of this theorem, we have

(1) If A is a real symmetric matrix, A is orthogonally similar to a real diagonal matrix.
(2) If A is a Hermitian matrix, A is unitarily similar to a real diagonal matrix.

It is of interest to note that if a matrix A is orthogonally similar to a diagonal matrix, then the matrix A is symmetric. Indeed, if $U^{-1}AU = D$, where $UU^T = I_n$ and D is diagonal, then

$$\begin{aligned}A &= UDU^{-1}\\ &= UDU^T\end{aligned}$$

Thus,

$$A^T = (UDU^T)^T = (U^T)^T D^T U^T$$
$$= UD^T U^T$$

Since D is diagonal, $D^T = D$, and

$$A^T = UD^T U^T = A$$

Thus, a real matrix is orthogonally diagonalizable if and only if it is symmetric.

Example 1 Let

$$A = \begin{bmatrix} a & b \\ b & a \end{bmatrix}$$

with a and b real.

Note that A is a symmetric matrix with characteristic polynomial

$$C_A(x) = x^2 - 2ax + (a^2 - b^2) = (x - (a + b))(x - (a - b))$$

It follows that the eigenvalues of A are $a + b$ and $a - b$. Let

$$\begin{bmatrix} x \\ y \end{bmatrix}$$

be the eigenvector corresponding to $a + b$.

Then,

$$ax + by = (a + b)x$$
$$bx + ay = (a + b)y$$

or,

$$-bx + by = 0$$
$$ax - ay = 0$$

Thus,

$$\begin{bmatrix} 1/\sqrt{2} \\ 1/\sqrt{2} \end{bmatrix}$$

is a unit eigenvector corresponding to the eigenvalue $a + b$.

Similarly it can be shown that

$$\begin{bmatrix} 1/\sqrt{2} \\ -1/\sqrt{2} \end{bmatrix}$$

is the unit eigenvector corresponding to $a - b$. Thus,

$$\begin{bmatrix} 1/\sqrt{2} & 1/\sqrt{2} \\ -1/\sqrt{2} & 1/\sqrt{2} \end{bmatrix}^{-1} \begin{bmatrix} a & b \\ b & a \end{bmatrix} \begin{bmatrix} 1/\sqrt{2} & 1/\sqrt{2} \\ -1/\sqrt{2} & 1/\sqrt{2} \end{bmatrix} = \begin{bmatrix} a - b & 0 \\ 0 & a + b \end{bmatrix}$$

Note that

$$\begin{bmatrix} 1/\sqrt{2} & 1/\sqrt{2} \\ -1/\sqrt{2} & 1/\sqrt{2} \end{bmatrix}$$

is orthogonal.

Example 2 Let

$$A = \begin{bmatrix} 0 & 1 & 1 & \dots & 1 \\ 1 & 0 & 1 & \dots & 1 \\ 1 & 1 & 0 & \dots & 1 \\ & & \vdots & & \\ 1 & 1 & 1 & \dots & 0 \end{bmatrix}$$

To which diagonal matrix is A similar?

Now,

$$C_A(x) = \begin{vmatrix} x & -1 & -1 & \dots & -1 \\ -1 & x & -1 & \dots & -1 \\ -1 & -1 & x & \dots & -1 \\ & & \vdots & & \\ -1 & -1 & -1 & \dots & x \end{vmatrix}$$

$$= \begin{vmatrix} x - (n-1) & x - (n-1) & x - (n-1) & \dots & x - (n-1) \\ -1 & x & -1 & \dots & -1 \\ & & \vdots & & \\ -1 & -1 & -1 & \dots\dots & x \end{vmatrix}$$

where the second determinant was obtained by adding each of the last $n-1$ rows to the first row. Therefore,

$$C_A(x) = (x-(n-1))\begin{vmatrix} 1 & 1 & 1 & \cdots & 1 \\ -1 & x & -1 & \cdots & -1 \\ & & \vdots & & \\ -1 & -1 & -1 & \cdots & x \end{vmatrix}$$

$$= (x-(n-1))\begin{vmatrix} 1 & 1 & 1 & \cdots & 1 \\ 0 & x+1 & 0 & \cdots & 0 \\ 0 & 0 & x+1 & \cdots & 0 \\ & & \vdots & & \\ 0 & 0 & 0 & \cdots & x+1 \end{vmatrix}$$

$$= (x+1)^{n-1}(x-(n-1))$$

Since A is symmetric, it follows that there is an orthogonal matrix Q such that

$$QAQ^{-1} = \begin{vmatrix} n-1 & 0 & \cdots & 0 \\ 0 & -1 & \cdots & 0 \\ & \vdots & & \\ 0 & 0 & \cdots & -1 \end{vmatrix}$$

EXERCISES

1. For the following symmetric matrices A, find an orthogonal matrix U such that $U^{-1}AU$ is diagonal.

(a) $\begin{bmatrix} 3 & 2 \\ 2 & 0 \end{bmatrix}$ (b) $\begin{bmatrix} 6 & 6 \\ 6 & 1 \end{bmatrix}$ (c) $\begin{bmatrix} 3 & 1 \\ 1 & 1 \end{bmatrix}$

(d) $\begin{bmatrix} 1 & 1 & 1 \\ 1 & 1 & 1 \\ 1 & 1 & 1 \end{bmatrix}$ (e) $\begin{bmatrix} 3 & 1 & -1 \\ 1 & 3 & -1 \\ -1 & -1 & 5 \end{bmatrix}$ (f) $\begin{bmatrix} 0 & 0 & 1 \\ 0 & 0 & 0 \\ 1 & 0 & 0 \end{bmatrix}$

(g) $\begin{bmatrix} 1 & 6 & 1 \\ 6 & 1 & 0 \\ 1 & 0 & 1 \end{bmatrix}$

2. To what diagonal matrix is the matrix

$$\begin{vmatrix} a+3 & 2 \\ 2 & a \end{vmatrix}$$

similar?

3. If A is the 3×3 real symmetric matrix of rank 1, show that there are real numbers x_1, x_2, x_3, such that

$$A = \begin{bmatrix} x_1^2 & x_1x_2 & x_1x_3 \\ x_1x_2 & x_2^2 & x_2x_3 \\ x_1x_3 & x_2x_3 & x_3^2 \end{bmatrix}$$

show that A is similar to the matrix

$$\begin{bmatrix} x_1^2 + x_2^2 + x_3^2 & 0 & 0 \\ 0 & 0 & 0 \\ 0 & 0 & 0 \end{bmatrix}$$

4. Diagonalize the Hermitian matrix

$$\begin{bmatrix} 1 & i \\ -i & 1 \end{bmatrix}$$

5. To what diagonal matrix is

$$\begin{bmatrix} a & c \\ c & b \end{bmatrix}, \qquad a, b, c \text{ real}$$

similar?

6. Let A be an $n \times n$ matrix such that $A^* = -A$. Show that A is unitarily similar to a diagonal matrix whose diagonal entries are purely imaginary. [Hint: Show iA is symmetric.]

7. Show that the only numbers which can be real eigenvalues of an orthogonal matrix are ± 1.

8. Let A be a 3×3 real symmetric matrix whose eigenvalues are 1, 1, and -1. Show that there is a unit vector $\boldsymbol{x}$ in $\boldsymbol{R}^3$ such that $A = I_3 - 2\boldsymbol{x} \cdot \boldsymbol{x}^T$.

9. Show that two real symmetric matrices are orthogonally similar if and only if they have the same characteristic polynomial.

10. If two real symmetric matrices are similar, show that they are orthogonally similar.

11. Let

$$A = \begin{bmatrix} a & b \\ b & d \end{bmatrix}$$

be a real invertible 2×2 matrix.

(a) Show that the locus of points (x, y) such that $(ax+by)^2+(bx+dy)^2=1$ is an ellipse.
(b) Show that this ellipse is the image of the unit circle under A^{-1}.
(c) Show that the distance from the origin of the furthest point from the origin of this ellipse is the largest eigenvalue of A^{-1} in absolute value.
(d) What is the distance from the origin to the point on the ellipse

$$(2x + y)^2 + (x + 2y)^2 = 1$$

most distant from the origin?

12. Let A be a real 3×3 matrix and $C_{A^TA}(x) = x^3 - \sigma_1 x^2 + \sigma_2 x - \sigma_3$ be the characteristic polynomial of A^TA. Show that σ_i is the sum of the squares of all $i \times i$ minors of A.

13. To what diagonal matrix is

$$\begin{bmatrix} 0 & 1 & 0 & 0 \\ 1 & 0 & 1 & 0 \\ 0 & 1 & 0 & 1 \\ 0 & 0 & 1 & 0 \end{bmatrix}$$

similar?

14. To what diagonal matrix is

$$\begin{bmatrix} 0 & 1 & 0 & 0 & 0 \\ 1 & 0 & 1 & 0 & 0 \\ 0 & 1 & 0 & 1 & 0 \\ 0 & 0 & 1 & 0 & 1 \\ 0 & 0 & 0 & 1 & 0 \end{bmatrix}$$

similar?

3 BILINEAR FORMS

In an earlier section of this book we discussed the inner product and the cross product. We noted that

$$(\alpha \boldsymbol{a} + \beta \boldsymbol{b}) \times \boldsymbol{c} = \alpha(\boldsymbol{a} \times \boldsymbol{b}) + \beta(\boldsymbol{b} \times \boldsymbol{c})$$
$$\boldsymbol{a} \times (\beta \boldsymbol{b} + \gamma \boldsymbol{c}) = \beta(\boldsymbol{a} \times \boldsymbol{b}) + \gamma(\boldsymbol{a} \times \boldsymbol{c})$$
$$(\alpha \boldsymbol{a} + \beta \boldsymbol{b}, \boldsymbol{c}) = \alpha(\boldsymbol{a}, \boldsymbol{c}) + \beta(\boldsymbol{b}, \boldsymbol{c})$$
$$(\boldsymbol{a}, \beta \boldsymbol{b} + \gamma \boldsymbol{c}) = \beta(\boldsymbol{a}, \boldsymbol{b}) + \gamma(\boldsymbol{a}, \boldsymbol{c})$$

We wish to study, in general, functions of this type. Accordingly we define a bilinear function b on a vector space V to a vector space W to be a function defined on the ordered pairs of elements of V such that

$$b(\alpha \boldsymbol{x} + \beta \boldsymbol{y}, \boldsymbol{z}) = \alpha b(\boldsymbol{x}, \boldsymbol{z}) + \beta b(\boldsymbol{y}, \boldsymbol{z})$$

$$b(\boldsymbol{x}, \beta \boldsymbol{y} + \gamma \boldsymbol{z}) = \beta b(\boldsymbol{x}, \boldsymbol{y}) + \gamma b(\boldsymbol{x}, \boldsymbol{z}),$$

with $\boldsymbol{x}, \boldsymbol{y}, \boldsymbol{z}$ in V and α, β, γ scalars.

In this section all vector spaces are real. If $W = \boldsymbol{R}$ in the above definition, the bilinear function is said to be a **bilinear form**. Thus, the inner product is a bilinear form. The cross product is not.

A bilinear function is symmetric if $b(\boldsymbol{x}, \boldsymbol{y}) = b(\boldsymbol{y}, \boldsymbol{x})$ for all $\boldsymbol{x}$ and $\boldsymbol{y}$ in V. It is alternate or skew-symmetric if $b(\boldsymbol{x}, \boldsymbol{y}) = -b(\boldsymbol{y}, \boldsymbol{x})$, for all $\boldsymbol{x}$ and $\boldsymbol{y}$ in V. The cross product is alternate, the inner product symmetric.

If b is a bilinear form and there is a nonzero vector $\boldsymbol{x}$ such that $b(\boldsymbol{x}, \boldsymbol{y}) = 0$ for all $\boldsymbol{y}$ in V, the bilinear form is said to be **degenerate**. If a bilinear form is not degenerate, it is called **nondegenerate**. The inner product is a nondegenerate bilinear form. Indeed, if $\boldsymbol{x}$ is a vector in $\boldsymbol{R}^n$ such that $b(\boldsymbol{x}, \boldsymbol{y}) = 0$ for all $\boldsymbol{y}$ in $\boldsymbol{R}^n$, then $b(\boldsymbol{x}, \boldsymbol{x}) = 0$. Thus, $\boldsymbol{x} = 0$.

A bilinear form is said to be **positive** if for all vectors $\boldsymbol{x}$, $b(\boldsymbol{x}, \boldsymbol{x}) \geq 0$. If for all $\boldsymbol{x}$, $b(\boldsymbol{x}, \boldsymbol{x}) \leq 0$, b is said to be **negative**. As we have seen the inner product is a positive bilinear form.

If A is an $n \times n$ matrix, the function $b(\boldsymbol{x}, \boldsymbol{y}) = \boldsymbol{x}^T A \boldsymbol{y}$, defined for all pairs of vectors $\boldsymbol{x}, \boldsymbol{y}$ in $\boldsymbol{R}^n$, is clearly bilinear. If $A = I_n$ this bilinear form is just the inner product. We shall show that all bilinear forms on $\boldsymbol{R}^n$ are of this type.

Theorem 1 If b is a bilinear form on $\boldsymbol{R}^n$ there is a unique real $n \times n$ matrix A such that $b(\boldsymbol{x}, \boldsymbol{y}) = \boldsymbol{x}^T A \boldsymbol{y}$, for all $\boldsymbol{x}$ and $\boldsymbol{y}$ in $\boldsymbol{R}^n$.

Proof Define $A = [b(\boldsymbol{e}_i, \boldsymbol{e}_j)]_{(nn)}$, with $\boldsymbol{e}_1, \ldots, \boldsymbol{e}_n$ the standard basis for $\boldsymbol{R}^n$. Note that $\boldsymbol{e}_i{}^T A \boldsymbol{e}_j = b(\boldsymbol{e}_i, \boldsymbol{e}_j)$. Let c be the bilinear form $c(\boldsymbol{x}, \boldsymbol{y}) = \boldsymbol{x}^T A \boldsymbol{y}$. Then, the bilinear forms b and c agree on all basis vectors. We show that they agree everywhere.

Now

$$b\left(\sum_i x_i \boldsymbol{e}_i, \sum_j y_j \boldsymbol{e}_j\right) = \sum_{i,j} x_i y_j b(\boldsymbol{e}_i, \boldsymbol{e}_j)$$

$$= \sum_{i,j} x_i y_j c(\boldsymbol{e}_i, \boldsymbol{e}_j)$$

$$= c\left(\sum_i x_i \boldsymbol{e}_i, \sum_j y_j \boldsymbol{e}_j\right)$$

Thus, $b(\boldsymbol{x}, \boldsymbol{y}) = \boldsymbol{x}^T A \boldsymbol{y}$.

A is unique, for if $b(\boldsymbol{x}, \boldsymbol{y}) = \boldsymbol{x}^T A' \boldsymbol{y}$, $b(\boldsymbol{e}_i, \boldsymbol{e}_j) = \boldsymbol{e}_i^T A' \boldsymbol{e}_j$. If $A' = [a'_{ij}]_{(nn)}$, $b(\boldsymbol{e}_i, \boldsymbol{e}_j) = a'_{ij}$. Since we saw earlier that $b(\boldsymbol{e}_i, \boldsymbol{e}_j) = a_{ij}$, $a_{ij} = a'_{ij}$. Thus, $A = A'$. ∎

A bilinear form $b_A(\boldsymbol{x}, \boldsymbol{y}) = \boldsymbol{x}^T A \boldsymbol{y}$ is symmetric if and only if the matrix A is symmetric. Now since $\boldsymbol{x}^T A \boldsymbol{y}$ is a real number, $(\boldsymbol{x}^T A \boldsymbol{y})^T = \boldsymbol{x}^T A \boldsymbol{y}$. Thus, $\boldsymbol{y}^T A^T \boldsymbol{x} = \boldsymbol{x}^T A \boldsymbol{y}$. Therefore, $\boldsymbol{x}^T A \boldsymbol{y} = \boldsymbol{y}^T A \boldsymbol{x}$ if and only if $\boldsymbol{y}^T A^T \boldsymbol{x} = \boldsymbol{y}^T A \boldsymbol{x}$. By the uniqueness part of the previous theorem this is equivalent to $A^T = A$.

Likewise it can be shown that b_A is skew-symmetric if and only if A is skew-symmetric.

A function Q from $\boldsymbol{R}^n$ to R is said to be a quadratic form if there is a bilinear b on $\boldsymbol{R}^n$ such that $Q(\boldsymbol{x}) = b(\boldsymbol{x}, \boldsymbol{x})$. By Theorem 1 any quadratic form can be written as $\boldsymbol{x}^T A \boldsymbol{x}$ for some $n \times n$ matrix $A = [a_{ij}]_{(nn)}$. Thus, if

$$\boldsymbol{x} = x_1\boldsymbol{e}_1 + \cdots + x_n\boldsymbol{e}_n, \qquad Q(x) = \sum_{i,j=1}^{n} a_{ij}x_ix_j$$

Thus, for example, every quadratic form on R^2 can be written in the form $a_{11}x_1^2 + a_{12}x_1x_2 + a_{21}x_2x_1 + a_{22}x_2^2$. This expression, of course, equals $a_{11}x_1^2 + (a_{12} + a_{21})x_1x_2 + a_{22}x_2^2$. Thus, the quadratic form associated with the matrix

$$A' = \begin{bmatrix} a_{11} & a_{12} \\ a_{21} & a_{22} \end{bmatrix}$$

is the same as that associated with the symmetric matrix

$$A' = \begin{bmatrix} a_{11} & \frac{1}{2}(a_{12} + a_{21}) \\ \frac{1}{2}(a_{12} + a_{21}) & a_{12} \end{bmatrix}$$

That every quadratic form is the quadratic form associated with a symmetric bilinear form is the content of the following theorem.

Theorem 2 Let Q be a quadratic form on $\boldsymbol{R}^n$. Then there is a unique symmetric bilinear form b on $\boldsymbol{R}^n$ such that $Q(\boldsymbol{x}) = b(\boldsymbol{x}, \boldsymbol{x})$.

Proof If Q is a quadratic form, there is some bilinear form c, (not necessarily symmetric) such that $Q(\boldsymbol{x}) = c(\boldsymbol{x}, \boldsymbol{x})$. Define $b(\boldsymbol{x}, \boldsymbol{y}) = \frac{1}{2}(c(\boldsymbol{x}, \boldsymbol{y}) + c(\boldsymbol{y}, \boldsymbol{x}))$.

Clearly, b is a bilinear form. It is likewise clear that $b(\boldsymbol{x}, \boldsymbol{y}) = b(\boldsymbol{y}, \boldsymbol{x})$. Thus, b is symmetric. Moreover, $b(\boldsymbol{x}, \boldsymbol{x}) = c(\boldsymbol{x}, \boldsymbol{x}) = Q(\boldsymbol{x})$.

Next we show that b is uniquely determined by Q. For suppose b_1 and b_2 are symmetric bilinear forms and $Q(\boldsymbol{x}) = b_1(\boldsymbol{x}, \boldsymbol{x}) = b_2(\boldsymbol{x}, \boldsymbol{x})$. Then,

$$\begin{aligned} b_1(\boldsymbol{x}, \boldsymbol{y}) &= \tfrac{1}{4}(b_1(\boldsymbol{x} + \boldsymbol{y}, \boldsymbol{x} + \boldsymbol{y}) - b_1(\boldsymbol{x} - \boldsymbol{y}, \boldsymbol{x} - \boldsymbol{y})) \\ &= \tfrac{1}{4}(b_2(\boldsymbol{x} + \boldsymbol{y}, \boldsymbol{x} + \boldsymbol{y}) - b_2(\boldsymbol{x} - \boldsymbol{y}, \boldsymbol{x} - \boldsymbol{y})) \\ &= b_2(\boldsymbol{x}, \boldsymbol{y}) \end{aligned}$$

Thus, b is uniquely determined.

If C is an $n \times n$ nonsingular matrix and b is a bilinear form on $\boldsymbol{R}^n$, we may define a new bilinear form by $b'(\boldsymbol{x}, \boldsymbol{y}) = b(C\boldsymbol{x}, C\boldsymbol{y})$. We say that b' is **congruent** to b.

We interpret congruence in terms of matrices. If $b(\boldsymbol{x}, \boldsymbol{y}) = \boldsymbol{x}^T A \boldsymbol{y}$ and $b'(\boldsymbol{x}, \boldsymbol{y}) = \boldsymbol{x}^T B \boldsymbol{y}$ and b' is congruent to b, there is a nonsingular $n \times n$ matrix C such that $\boldsymbol{x}^T B \boldsymbol{y} = \boldsymbol{x}^T C^T A C \boldsymbol{y}$. Since this holds for all $\boldsymbol{x}$ and $\boldsymbol{y}$, by Theorem 1, $B = C^T A C$.

If conversely, $B = C^T A C$, then $b'(\boldsymbol{x}, \boldsymbol{y}) = b(C\boldsymbol{x}, C\boldsymbol{y})$ and b' is congruent to b. If $B = C^T A C$, then $(C^{-1})^T B C^{-1} = A$. Thus, b is congruent to b', if b' is congruent to b.

If b is a bilinear form we say that the **rank** of b is the rank of the associated matrix A. If b and b' are congruent bilinear forms, and b is associated with the matrix A, then b' is associated with $C^T A C$, for some nonsingular matrix C. Thus, it follows that the two bilinear forms b and b' have the same rank.

Just as we defined equivalence and orthogonal equivalence of matrices, we can define congruence and orthogonal congruence of bilinear forms and quadratic forms. Thus, we say that two bilinear forms b and b' are **orthogonally congruent** if there is an orthogonal matrix A such that $b(\boldsymbol{x}, \boldsymbol{y}) = b(A\boldsymbol{x}, A\boldsymbol{y})$. Likewise two quadratic forms Q and Q' are orthogonally congruent if $Q'(\boldsymbol{x}) = Q(A\boldsymbol{x})$, for some orthogonal A. This is equivalent to the statement that their associated symmetric bilinear forms are orthogonally congruent.

Clearly, the simplest class of quadratic forms are those of the form

$$\lambda_1 x_1^2 + \lambda_2 x_2^2 + \cdots + \lambda_n x_n^2$$

These correspond to diagonal matrices. Accordingly the following theorem is of interest.

Theorem 3 Any quadratic form Q is orthogonally congruent to one of the type

$$\lambda_1 x_1^2 + \cdots + \lambda_n x_n^2$$

Proof Now by Theorem 2 of this section, $Q(\boldsymbol{x}) = \boldsymbol{x}^T A \boldsymbol{x}$, for some symmetric matrix A. By Theorem 2 of the previous section there is an orthogonal matrix C such that $C^T A C$ is diagonal. ∎

Example 1 Show that the curve

$$ax^2 + bxy + cy^2 = d, \qquad d \neq 0$$

is an ellipse if $4ac - b^2 > 0$, and an hyperbola if $4ac - b^2 < 0$.

Let

$$Q(\boldsymbol{x}) = \begin{bmatrix} x \\ y \end{bmatrix}^T \begin{bmatrix} a & \frac{1}{2}b \\ \frac{1}{2}b & c \end{bmatrix} \begin{bmatrix} x \\ y \end{bmatrix}$$

Clearly, $Q(\boldsymbol{x}) = d$ represents the above curve. By Theorem 3, there is an orthogonal matrix A such that $Q(A\boldsymbol{x}) = \lambda_1 x_1^2 + \lambda_2 x_2^2$ and

$$A^T \begin{bmatrix} a & \frac{1}{2}b \\ \frac{1}{2}b & c \end{bmatrix} A = \begin{bmatrix} \lambda_1 & 0 \\ 0 & \lambda_2 \end{bmatrix}$$

The curve $\lambda_1 x_1^2 + \lambda_2^2 x_2^2 = d$ is an ellipse if $\lambda_1\lambda_2 > 0$, an hyperbola if $\lambda_1\lambda_2 < 0$. Since A is orthogonal, $A^T = A^{-1}$, and

$$\det\left(A^T \begin{bmatrix} a & \frac{1}{2}b \\ \frac{1}{2}b & c \end{bmatrix} A\right) = ac - \tfrac{1}{4}b^2 = \lambda_1\lambda_2$$

Since the curve $\lambda_1 x_1^2 + \lambda_2 x_2^2 = d$ was obtained from $ax^2 + bxy + cy^2 = d$ by rotation of axes, $ax^2 + bxy + cy^2 = d$ is an ellipse if $4ac - b^2 > 0$, an hyperbola if $4ac - b^2 < 0$.

The next result follows immediately from Theorem 3.

Theorem 4 Every quadratic form is congruent to one of the type

$$x_1^2 + \cdots + x_p^2 - x_{p+1}^2 - \cdots - x_r^2$$

Proof One replaces x_i by $1/\sqrt{\lambda_i}$ if $\lambda_i > 0$ and $1/\sqrt{-\lambda_i}$ if $\lambda_i < 0$. ∎

Thus every symmetric bilinear form b is congruent to one of the form $x_1y_1 + \cdots + x_ky_k - x_{k+1}y_{k+1} - \cdots - x_{k+r}y_{k+r}$. The rank of this form is clearly $k + r$. The number $k - r$ is called the **signature**. We shall show that the signature is uniquely determined. In other words, we wish to show that if b is also congruent to $x_1y_1 + \cdots + x_{k'}y_{k'} - x_{k'+1}y_{k'+1} - \cdots - x_{k'+r'}y_{k'+r'}$ then $k' - r' = k - r$. Since congruent forms have the same rank, $k + r = k' + r'$. Thus, to show $k' - r' = k - r$, it is sufficient to show $k = k'$.

Lemma Let W be a subspace of $\boldsymbol{R}^n$ for which $b(\boldsymbol{x}, \boldsymbol{x}) \leq 0$ for all $\boldsymbol{x}$ in W with b the form $x_1y_1 + \cdots + x_ky_k - x_{k+1}y_{k+1} - \cdots - x_{k+r}y_{k+r}$. Then, $\dim W \leq n - k$.

Proof Let V be the subspace of $\boldsymbol{R}^n$ spanned by $\boldsymbol{e}_{k+1}, \ldots, \boldsymbol{e}_n$. Clearly, $\dim V = n - k$. Let U be the subspace of $\boldsymbol{R}^n$ spanned by $\boldsymbol{e}_1, \ldots, \boldsymbol{e}_k$. Let

$\boldsymbol{y}_1, \ldots, \boldsymbol{y}_m$ be a basis for W. Then, each $\boldsymbol{y}_i$ can be expressed uniquely in the form $\boldsymbol{u}_i + \boldsymbol{v}_i$ where $\boldsymbol{u}_i$ belongs to U and $\boldsymbol{v}_i$ belongs to V. If $m > n - k$, there are real scalars $\alpha_1, \ldots, \alpha_m$, not all 0, such that

$$\sum_{i=1}^{m} \alpha_i \boldsymbol{v}_i = 0$$

then,

$$\boldsymbol{z} = \sum_{i=1}^{m} \alpha_i \boldsymbol{y}_i = \sum_{i=1}^{m} \alpha_i(\boldsymbol{u}_i + \boldsymbol{v}_i) = \sum_{i=1}^{m} \alpha_i \boldsymbol{u}_i$$

belongs to W. Thus, $b(\boldsymbol{z}, \boldsymbol{z}) \leq 0$.

But $\boldsymbol{z}$ also belongs to U, and the bilinear form b on U is just the inner product of §6.2. Thus, $b(\boldsymbol{z}, \boldsymbol{z}) \geq 0$. Consequently, $b(\boldsymbol{z}, \boldsymbol{z}) = 0$. Again, since b on U is the inner product of §6.2, $\boldsymbol{z} = \boldsymbol{0}$. Thus, $\Sigma_{i=1}^{n} \alpha_i \boldsymbol{y}_i = 0$, in contradiction to the fact that $\boldsymbol{y}_1, \ldots, \boldsymbol{y}_m$ is a basis for W and not all the α_i's are 0. Thus, we must have $m \leq n - k$.

Theorem 5 Sylvester's Law of Inertia The signature of a symmetric bilinear form is uniquely determined.

Proof In the notation preceding the lemma we must show that $k = k'$. By the lemma $n - k \leq n - k'$ and $n - k \leq n - k$. Thus, $k = k'$.

EXERCISES

1. Show that the bilinear forms on $\boldsymbol{R}^n$ form a vector space using the usual addition and scalar multiplication for real functions. Show that this space is isomorphic to the space of real $n \times n$ matrices.
2. Determine the rank and signature of the following quadratic forms:
   (a) $x^2 + xy + y^2$ (b) $x^2 + xy + 3y^2$
   (c) $x^2 + 2xy + y^2$ (d) $x^2 - y^2$
3. Let $b_A(\boldsymbol{x}, \boldsymbol{y}) = \boldsymbol{x}^T A \boldsymbol{y}$ be a symmetric bilinear form. Show that b_A is positive if and only if the eigenvalues of A are non-negative. Show that b_A is nondegenerate if and only if the eigenvalues of A are all nonzero.
4. Let b be a symmetric bilinear form on $\boldsymbol{R}^n$. Show that b is positive if and only if its rank and signature are equal.
5. Let b be a bilinear form on a vector space V. Let W be the collection of all vectors $\boldsymbol{x}$ in V such that $b(\boldsymbol{x}, \boldsymbol{y}) = 0$ for all $\boldsymbol{y}$ in V.
   (a) Show that W is a subspace of V.
   (b) Let r be the rank of b. Show that $\dim W + r = \dim V$.
6. Let A be a nonzero symmetric 2×2 matrix of rank 1. Interpret geometrically the sets $b_A(\boldsymbol{x}, \boldsymbol{y}) = c$ for all values of c.

7. Show that two symmetric bilinear forms are congruent if and only if they have the same rank and signature.
8. Determine the rank and signature of the quadratic form $xy + yz + xz$.
9. Show that every bilinear form can be expressed in a unique way as the sum of a symmetric and skew-symmetric bilinear form. Interpret this in terms of matrices.

ANSWERS TO SELECTED EXERCISES

SECTION 1.1

1. (a) $x = -1, y = 0$
   (b) $x = 1, y = 1$
   (c) no solution
   (d) $x = 3, y = 2$
   (e) $x = 0, y = 0$
   (f) all points on the line $x + 2y = 1$
   (g) $x = 0, y = 0$
3. $x = -\frac{1}{4}, y = \frac{5}{4}, z = 0$
5. $x_1 = 1, x_2 = -2, x_3 = -1, x_4 = 2$

SECTION 1.2

1. (a) equivalent
   (b) not equivalent
   (c) not equivalent
   (d) equivalent
   (e) systems are equivalent; both have no solution
   (f) equivalent

SECTION 1.3

1. (a) $x = \frac{6}{7}, y = \frac{10}{7}, z = -\frac{2}{7}$
   (b) $x_1 = 0, x_2 = 0, x_3 = 0, x_4 = -1$
   (c) $x_1 = 1, x_2 = 3, x_3 = 0$
   (d) $x_1 = 5c, x_2 = \frac{2}{5} - \frac{17}{5}c, x_3 = c, x_4 = \frac{1}{5} - \frac{6}{5}c$, c arbitrary
   (e) no solution

(f) $x_1 = 2, x_2 = -1, x_3 = 4, x_4 = 0$
(h) no solution
(i) no solution
(j) $x_1 = c, x_2 = 7c, x_3 = -c, x_4 = 3c$

5. $a = \frac{1}{5}, b = -\frac{3}{5}$

SECTION 2.1

1. (a) $\begin{bmatrix} -1 \\ 3 \\ 9 \end{bmatrix}$ (b) $\begin{bmatrix} -18 \\ 27 \\ 29 \\ 64 \end{bmatrix}$ (c) $\begin{bmatrix} -12 \\ -12 \\ -5 \\ -4 \end{bmatrix}$ (d) $\begin{bmatrix} 4 \\ 7 \end{bmatrix}$

(e) $\begin{bmatrix} -6 \\ 0 \end{bmatrix}$ (f) $\begin{bmatrix} -5 \\ -5 \\ -5 \end{bmatrix}$ (g) $\begin{bmatrix} 5 \\ 19 \end{bmatrix}$ (h) $\begin{bmatrix} 7 \\ 4 \\ 2 \end{bmatrix}$

7. $\boldsymbol{x} = -\boldsymbol{a} - \boldsymbol{b} - 3\boldsymbol{c}$

15. $\boldsymbol{x}_i = \boldsymbol{y}_i - \boldsymbol{y}_{i+1}$ if $i < n$

SECTION 2.2

3. $\boldsymbol{l}(t) = t\boldsymbol{i} + (1 - t)\boldsymbol{j} - t\boldsymbol{k}$; intersects the plane $x = 0$ at $(0, 1, 0)$; intersects the plane $z = 2$ at $(-2, 3, 2)$

7. $(-\frac{1}{9}, -\frac{1}{9}, \frac{4}{9})$

13. (a) $(1, \frac{1}{2}, \frac{1}{2})$
(b) $(1, 1, 0)$
(c) $(0, 0, 1)$

SECTION 2.3

1. (a) $\begin{bmatrix} 1 & 2 & 3 & 4 \\ 2 & 4 & 6 & 8 \\ 3 & 6 & 9 & 12 \end{bmatrix}$ (b) $\begin{bmatrix} 2 & 5 & 10 \\ 5 & 8 & 13 \\ 10 & 13 & 18 \end{bmatrix}$

(c) $\begin{bmatrix} 2 & 3 \\ 3 & 4 \end{bmatrix}$ (d) $\begin{bmatrix} 2 & 3 & 4 & 5 \\ 3 & 4 & 5 & 6 \\ 4 & 5 & 6 & 7 \\ 5 & 6 & 7 & 8 \end{bmatrix}$

5. $X = \frac{1}{2}(A + B)$
$Y = \frac{1}{2}(A - B)$

8. 64

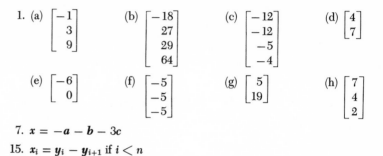

SECTION 2.4

1. (a) $\begin{bmatrix} 3 & 20 \\ 5 & 47 \end{bmatrix}$ (b) $\begin{bmatrix} 6 & 9 & 3 \\ 42 & 63 & 21 \\ 18 & 27 & 9 \end{bmatrix}$

   (c) $[78]$ (d) $\begin{bmatrix} -5 \\ 20 \\ 27 \end{bmatrix}$

   (e) $\begin{bmatrix} 35 & 14 \\ 20 & 7 \\ 11 & 8 \end{bmatrix}$ (f) $\begin{bmatrix} 2 & 1 & 20 \\ 7 & -3 & 61 \\ 1 & 2 & 13 \end{bmatrix}$

   (g) $\begin{bmatrix} 1 & a+b \\ 0 & 1 \end{bmatrix}$ (h) $\begin{bmatrix} 1 & a+b \\ 0 & 1 \end{bmatrix}$

   (i) $\begin{bmatrix} 10 & -1 \\ 2 & 4 \\ 27 & -6 \\ 3 & 0 \end{bmatrix}$ (j) $\begin{bmatrix} 1 \\ -1 \\ -1 \end{bmatrix}$

6. 3×3

10. $a = 1, b = -2, c = 0, d = 1$

13. (a) $\begin{bmatrix} 1 \\ 1 \end{bmatrix}$

    (b) $[3 \quad 1]$

SECTION 3.1

1. (a) 25
   (b) -9
   (c) 2
   (d) 1
   (e) 1
   (f) -1

SECTION 3.2

1. (a) 7
   (b) 1
   (c) 0

(d) 0
(e) 1
(f) −36
(g) 6
(h) 1
(i) −8
(j) −32
(k) 6

19. (a) 1, 2, 3
(b) $1/\sqrt{2} - 1/\sqrt{2}$
(c) 1, 2

27. x_1, x_2, x_3, x_4

SECTION 3.4

1. (a) 0
(b) −2
(c) −15
(d) 60
(e) 0
(f) −21
(g) 297
(h) 16
(i) −3
(j) 2
(k) 0

11. 1

SECTION 3.5

1. (a) $\frac{1}{10}\begin{bmatrix} 2 & 2 & 2 \\ 2 & 2 & -8 \\ -4 & 1 & 1 \end{bmatrix}$ (b) $\begin{bmatrix} 7 & -3 & -3 \\ -1 & 1 & 0 \\ -1 & 0 & 1 \end{bmatrix}$

(c) $\begin{bmatrix} 1 & 0 & -2 \\ 5 & \frac{5}{3} & \frac{10}{3} \\ -1 & -\frac{5}{3} & -\frac{4}{3} \end{bmatrix}$ (d) $\begin{bmatrix} 1 & 0 & 1 \\ -1 & 1 & 1 \\ 0 & 1 & 0 \end{bmatrix}$

(e) $\begin{bmatrix} 13 & 5 \\ 5 & 2 \end{bmatrix}$ (f) $\begin{bmatrix} 19 & 6 \\ 3 & 1 \end{bmatrix}$

(g) $\begin{bmatrix} -\frac{15}{2} & \frac{3}{4} & \frac{1}{4} & \frac{3}{2} \\ 6 & -\frac{5}{2} & -\frac{1}{2} & -1 \\ -1 & \frac{3}{2} & \frac{1}{2} & 0 \\ -\frac{5}{2} & \frac{3}{4} & \frac{1}{4} & \frac{1}{2} \end{bmatrix}$

(h) $\begin{bmatrix} 6 & 1 & 4 & -3 \\ 2 & -1 & 0 & 0 \\ 1 & 1 & 1 & 0 \\ -3 & -1 & -2 & 1 \end{bmatrix}$

(i) $\begin{bmatrix} -23 & 10 & 1 & 2 \\ 29 & -12 & -2 & -2 \\ -\frac{64}{5} & \frac{26}{5} & \frac{6}{5} & \frac{3}{5} \\ -\frac{18}{5} & \frac{7}{5} & \frac{2}{5} & \frac{1}{5} \end{bmatrix}$

(j) $\begin{bmatrix} 1 & -3 & 11 & -38 \\ 0 & 1 & -2 & 7 \\ 0 & 0 & 1 & -2 \\ 0 & 0 & 0 & 1 \end{bmatrix}$

(k) $\frac{1}{4}\begin{bmatrix} 1 & 1 & 1 & 1 \\ 1 & 1 & -1 & -1 \\ 1 & -1 & 1 & -1 \\ 1 & -1 & -1 & -1 \end{bmatrix}$

SECTION 3.6

1. (a) $x = -\frac{5}{6}, y = \frac{2}{3}, z = \frac{7}{6}$
   (b) $x = \frac{35}{18}, y = \frac{29}{18}, z = \frac{5}{18}$
   (c) $x_1 = 1, x_2 = 2, x_3 = 1$
   (d) $x_1 = 1, x_2 = \frac{2}{3}, x_3 = -\frac{1}{3}$
   (e) $x_1 = 1, x_2 = 1, x_3 = 1$
   (f) $x = 1, y = -1, z = 1$
   (g) $x_1 = 2, x_2 = 0, x_3 = 1, x_4 = 1$
   (h) $x_1 = 1, x_2 = \frac{1}{2}, x_3 = \frac{1}{2}, x_4 = 1$
   (i) $x_1 = 2, x_2 = \frac{1}{5}, x_3 = 0, x_4 = \frac{4}{5}$

SECTION 3.7

1. (a) $\frac{-1}{11}\begin{bmatrix} -3 & -4 & 5 \\ -2 & 1 & -4 \\ 2 & -1 & -7 \end{bmatrix}$

   (b) $\frac{-1}{7}\begin{bmatrix} 2 & -2 & -3 \\ -4 & -3 & 6 \\ -1 & 1 & -2 \end{bmatrix}$

(c) $\begin{bmatrix} 0 & 1 & 0 \\ 0 & 0 & 1 \\ 1 & 0 & 0 \end{bmatrix}$

(d) $\begin{bmatrix} \frac{1}{2} & 0 & -\frac{1}{2} \\ -1 & 1 & 0 \\ \frac{3}{2} & -1 & \frac{1}{2} \end{bmatrix}$

(e) $\begin{bmatrix} -2 & 1 & 0 \\ -3 & 0 & 1 \\ 3 & -\frac{1}{3} & -\frac{2}{3} \end{bmatrix}$

(f) $\begin{bmatrix} \frac{1}{2} & 1 & -\frac{1}{2} \\ 0 & -1 & 1 \\ -\frac{1}{2} & -1 & \frac{3}{2} \end{bmatrix}$

(g) $\begin{bmatrix} 1 + \mu a + \lambda b & -a & -b \\ -\mu & 1 & 0 \\ -\lambda & 0 & 1 \end{bmatrix}$

(h) $\begin{bmatrix} 0 & -1 & 0 & 1 \\ 0 & 4 & 0 & 3 \\ -\frac{1}{7} & -\frac{4}{7} & 1 & \frac{18}{7} \\ -\frac{2}{7} & -\frac{36}{7} & 1 & \frac{8}{7} \end{bmatrix}$

(i) $\begin{bmatrix} 1 & 0 & -1 & 1 \\ 1 & 0 & 0 & 0 \\ -1 & 1 & 1 & -1 \\ -1 & 0 & 1 & 0 \end{bmatrix}$

(j) $\begin{bmatrix} -\frac{1}{2} & \frac{1}{2} & 0 & 0 \\ -1 & 0 & 1 & 0 \\ -1 & 0 & 0 & 1 \\ \frac{11}{8} & -\frac{1}{8} & -\frac{1}{2} & -\frac{1}{2} \end{bmatrix}$

(k) $\begin{bmatrix} 1 & 5 & 4 & -1 \\ 0 & 1 & \frac{1}{3} & -\frac{1}{3} \\ 0 & 0 & 1 & 0 \\ 0 & 0 & 0 & 1 \end{bmatrix}$

SECTION 4.3

1. (a), (e), (g)
2. (a), (c), (f)

SECTION 4.5

3. (a), (b), (e)

SECTION 4.6

1. (a), (b), (d)
2. (a), (b), (f)

SECTION 4.7

1. (a) 2
   (b) 3
   (c) 1
   (d) 2
   (e) 2
   (f) 3
   (g) 3

SECTION 4.9

1. (a) $\begin{bmatrix} x - y \\ y \end{bmatrix}$

   (b) $\frac{1}{2}\begin{bmatrix} x + y \\ x - y \end{bmatrix}$

   (c) $\begin{bmatrix} y - x \\ -x \end{bmatrix}$

   (d) $\begin{bmatrix} 3x - 5y \\ -x + 2y \end{bmatrix}$

3. (a), (b), (c), (e)
5. (a) all λ, $\lambda \neq \pm 1/\sqrt{2}$
   (b) all λ, $\lambda \neq \pm 1$
   (c) all λ
   (d) all non-zero λ

SECTION 5.5

1. (a) 2
   (b) 2
   (c) 3
   (d) 2
   (e) 3
   (f) 2
   (g) 2

17. (a) 3
    (b) 2
    (c) 4

SECTION 5.8

1. (a) $\begin{bmatrix} 1 & 0 & 0 \\ 0 & -1 & 0 \\ 0 & 0 & 1 \end{bmatrix}$

   (b) $\begin{bmatrix} 0 & 1 & 0 \\ 1 & 0 & 0 \\ 0 & 0 & 1 \end{bmatrix}$

SECTION 6.1

1. 60°

3. $x = -\frac{1}{2}$, $y = \frac{3}{2}$

11. $\sqrt{3}\,l$

16. $y = z$

24. $\dfrac{+\boldsymbol{j} - \boldsymbol{k}}{\sqrt{1 + t^2}}$

SECTION 6.2

11. $\sqrt{n}$

SECTION 7.1

1. (a) $x^2 - 1$, ± 1
   (b) $(x - 1)^2$, 1
   (c) $x^2 + x + 1$, $\frac{1}{2}(-1 \pm \sqrt{3}i)$
   (d) $(x - a)(x - d)(x - f)$, a, d, f
   (e) $x^3 - 3x - 2$, 0, 3

INDEX

5
6
7
8
9